Biometeorology

Volume 5

For Instructions to Authors please contact the Publishing Editor: margaret.deignan@springer.com

Biometeorology has its linguistic origins in Greek; bio refers to life while meteoros makes reference to the study of phenomena near and above the earth's surface. These are the essential elements of biometeorology, such that it is concerned with the interaction between living organisms and variations in atmospheric processes at a range of time and space scales.

Since its emergence as a discipline in the mid 1950s, Biometeorology has grown to have great societal relevance. Much of its subject matter focuses on how biophysical, human, animal and plant systems are influenced by change and variability in the atmospheric environment. Understanding such relationships provides insights into how the provisioning, supporting and regulating services provided by the biosphere may be affected by atmospheric variations and ultimately the societal consequences of atmosphere related changes in biophysical and ecosystem processes.

The purpose of this series in biometeorology is to:

- complement the material published in the International Society of Biometeorology's International Journal of Biometeorology,
- to communicate the interdisciplinary philosophy and science of biometeorology to as wide an audience as possible,
- introduce scientists and policy makers to the societal relevance of and recent developments in its subfields.
- demonstrate how a biometeorological approach can provide entry points to the understanding and possible solution of cross-cutting environmental issues such as climate change, desertification, deforestation, biodiversity loss, land degradation, emerging diseases and the availability of clean water and air.

The series in biometeorology comprises theme-based volumes organised around subject matter within the subfields of biometeorology, which broadly include human, animal, plant, ecosystems, tourism, artificial or indoor climates, air quality and urban. In addition to volumes that synthesise the state of knowledge in the subfields, a primary focus will be on important, new and emerging topics in biometeorology that have a bearing on the health of the biosphere and thus human society. An editorial board, the broad role of which will be to elicit and review suggestions for volume themes, will oversee the series.

More information about this series at https://link.springer.com/bookseries/8101

Chao Ren • Glenn McGregor
Editors

Urban Climate Science for Planning Healthy Cities

Springer

Editors
Chao Ren
Faculty of Architecture
The University of Hong Kong
Pok Fu Lam, Hong Kong

Glenn McGregor
Department of Geography
Durham University
Durham, UK

Biometeorology
ISBN 978-3-030-87597-8 ISBN 978-3-030-87598-5 (eBook)
https://doi.org/10.1007/978-3-030-87598-5

This Springer imprint is published by the registered company Springer Nature Switzerland AG
The registered company address is: Gewerbestrasse 11, 6330 Cham, Switzerland

Foreword

This book addressing *'Urban Climate Science for Planning Healthy Cities'* is timely and critical when we need to address issues around outdoor air pollution, extreme heat, and flooding that so devastate the health, lives and livelihoods of populations world wide. We know that without this book and other resources that it will be very much more difficult to deliver the three landmark agreements adopted by the United Nations in 2015, which are the Paris Agreement on Climate Change, the Sustainable Development Goals (SDGs) of Agenda 2030 and the Sendai Framework for Disaster Risk Reduction 2015–2030.

This book is written to engage a range of professions, including public health professionals and design professionals spanning the climate, urban planning to public health disciplines—encouraging the linking of these and associated professions in learning with each other to deliver a more holistic solution on urban climate for planning a cohort of healthy cities. As this book demonstrates, real benefits can arise from this integrated approach.

Section One provides the context setting on urban climates and the issues around megacities. It also expands on how the concepts of urban climate, urban planning and healthy cities need to be brought together and recognises the learning from the past to address the silos of thinking in the last half century of the 1900s.

Section Two includes selected case studies from a number of worldwide cities, including Melbourne, Tokyo, Beijing, Hong Kong, New York, London, Manchester, Singapore and Stuttgart. These focus attention on the interaction of urban planning and urban climate science and the associated link to the idea of healthy cities. While the use of modelling is clearly recognised for producing information for planning healthy cities, the city case studies in a number of instances point to alternative forms of knowledge generation such as citizen science and co-production processes. Acknowledging the interaction between vulnerability, indoor and outdoor temperatures and urban form and function, the city case studies put forward a range of intervention strategies for managing urban climate emphasising the value of green and blue infrastructures and policies conducive to maintaining good air quality. That the integration of urban climate science and urban planning can help pave the way for developing policies to achieve healthy city environments is exemplified by

Stuttgart's seminal approach to managing heat and outdoor air pollution as outlined in the associated chapter.

Section Three addresses the summation of these issues by considering the value of urban climate modelling and the integration of urban climate science into urban planning to achieve healthy cities—bringing knowledge into action. This section emphasises the integration of climate and environmental services and associated monitoring systems including health as critically important to build effective knowledge. We now know that the call for a data revolution with rigorous accountability mechanisms and renewed global partnerships is essential to improving the health of cities now and into the future. Recognising the potential international impact of mainstreaming urban climate science knowledge into urban planning, the final chapter brings together the thinking on the need to build ways looking forward to achieve healthy cities in a more informed and integrated manner—ways that we hope will be clearly recognised at COP26 to be held in Glasgow in November 2021.

To me this book is an essential reflection on the diverse needs that climate change is posing to world leaders and those at national, regional and city levels. Climate change is a known driver of hazards that devastate the world and densely populated cities are especially at risk—although not climate related the 2020 COVID-19 pandemic has clearly highlighted cities as hotspots of risk. I believe we all need to learn from this book about how understanding the interactions between climate and the physical nature of cities can help deliver healthy cities and more widely deliver the implementation of the 2015 UN landmark agreements and meet the SDGs.

Public Health England
London, UK
August 2020

Virginia Murray,

Foreword

Urban Climate Science for Planning Healthy Cities is a *tour de force* that will guide the science of city planning and practicing urban planners for several decades to come. Edited books often arrive on the market as a result of esoteric specialist academic gatherings or ambitious academic editors and are rarely read or referenced thereafter. This is not one of those books. It arose from a decade of dedicated pioneering work by the authors and their collaborators, at the end of which, the world of popular science, scientific journalism, professional city designing, and building have arrived at the same sense of urgency. This is academic research at its best—prescient and relevant and doing the hard slog in data-collection, analytical modelling, theory-building, translation and history-unravelling, for practical consumption, before the issues have really captured the imagination of society at large. That point has now arrived, and this is the best book yet, to both introduce the field and take the interested reader deeper. And readers *will* be interested in what Ren and McGregor and their colleagues have to say. Their message is clear in its big picture narrative, detailed science and in its pointers to and illustration of needed action. The city case studies are fascinating and hugely informative and impactful in terms of the *what, why* and *how* of recent experiments in climate-adaptive urban planning, design and management.

I regard this book as seminal in the sense that earlier books on other cardinal performance dimensions of urban planning were. Interestingly, as the second chapter nicely demonstrates, modern urban planning arose as a public health intervention and some of the earliest writings and studies on the topic were focused on making late nineteenth century industrial cities bearable as places to live and work. We have come full circle after a century in which urban planning diversified into a multi-objective activity, largely viewed as a government responsibility. During the twentieth century, urban planning has been used as a tool to deliver industrial growth, employment growth, jobs-housing balance, environmental conservation, heritage preservation, social equity, efficient land supply, efficient housing supply, urban regeneration, preservation and stability of land and property values, public nuisance control, food security, national and regional spatial planning, and other objectives too. For the past 20 years, healthy cities have been back on the agenda and post Covid-19, health-in-the-city is amongst the uppermost issue in everyone's mind. Only, as Ren and McGregor rightly observe, the messages, concerns, science and practice are not yet well aligned. Much of the lively academic study of heath and built environment in the past two decades has focused on chronic disease—the current 'urban plagues' of sedentary urban lifestyles, including obesity and diabetes, cardio-vascular, mental health and respiratory disease. The new challenge and awareness suddenly upon us in 2020 and beyond, however, is not about chronic disease but about old-fashioned infectious disease. And many of the basic issue of city building that kicked-off the modern civic design and urban planning movement are back in public conversation: urban density, housing and neighbourhood density, open space capacity and distribution, building heights and so on. Urban ventilation too has become a performance parameter in pandemic-resilient cities.

But now we know so much more. Our data and our science are far superior to the first partnership between public health and city design. Before John Snow's prototypical geographic information system map of cholera cases in London's Soho district, infection was believed to be 'in the air', which did not imply an understanding of air-borne disease vectors. The ancient 'Miasma' theory of disease held at the time, postulating that human sicknesses come from foul-smelling 'night-air', emanating from rotting organic matter. Such matter was obviously intense in human settlements, particularly big cities, and thus urban pollution-> urban health was the dominant public health theory and narrative, via the un-explicated and unproven Miasmic pathway. Germ theory replaced Miasma theory in the late nineteenth century after the discovery of pathogens. And in the twentieth century, the science of public health (which is in large part, a science of urban health) developed further, adapting germ theory to take account of social risk-factors, genetic risk factors, environmental risk factors and epigenetic gene-environment interactions. Now, the 'Rolls Royce' of urban epidemiology studies of city health need data that can uncover associations and causation between individual disease (and thence population health), genetics and environment (social, built and natural).

Ren and McGregor's book starts a new chapter in this 150-year-old quest to improve the science of healthy cities. It shows what we know about how to observe, model and interpret environmental risks to health that are related to climate and heat. Importantly, however, the book is not primarily concerned with demonstrating climate-heat-health interactions, associations and causes. It is a good primer for the background science on these issues, but its focus is on the 'so what?' and 'what next?' questions. It references the public health, built environment epidemiology, and environmental science literature but uses that literature to make a case of action in the urban planning realm. It is a bit like the seminal books of the twentieth century that laid the case for urban planning as an economic growth tool, or as a tool for making property markets more efficient, or arguing the social science case for planning as a social and environmental redistributive tool. Every new paradigm in urban planning, or every new social objective to which it is put, needs its foundational arguments and related science, theory and evidence, laid out. Here is the book for planning for climate-change adaptation and mitigation in the world's cities.

Because urban climate externalities are, in the Samuelsonian Economic sense, a pure public good (co-consumed by all within the city), it is not surprising that a lot of the focus in the book is on regulative urban plans and government action. It is also appropriate that some attention is given to non-government action. The chapter on private gardens offers a fascinating perspective and reminds me of research published not so long ago on the contribution of private gardens in Greater London, to species and ecological biodiversity, pointing out that biodiversity in suburban London was multiple times that of Amazonia, measured by diversity/meter-squared. Urban planners are often inclined to ignore the contributions to safe, efficient, equitable, energy-efficient, beautiful, and healthy cities, made by the millions of private actions of individual citizens. Regulation is not the only way to achieve goals relating to the shape and quality of the built environment. With adequate sanctions and longevity, government regulations can be very powerful at conserving things (countryside, heritage, and other scarce resources) by *proscribing* development. It is not so successful at *prescribing* things, largely because in most countries, it is not the government that is the principal actor in developing land, and making use of it. Governments control via regulation, fiscal interventions and direct investment in city-shaping infrastructure. What will be the most effective modalities in delivering the application of ideas presented in Ren and McGregor's book? Urban climate resilience plans, for sure, and these will require a major new focus for government urban planners and their procedural and legal planning frameworks. What fiscal measures might also make a difference—perhaps as much or even more of a difference? Protecting the beauty of the urban-scape through government urban design control is arguably a failed experiment in most countries with well-developed planning systems. So too housing quality: the UK, with its famously strong post WW2 urban planning system delivered the smallest and most uniformly uninteresting and poor quality new-build housing in the whole of Europe during the second half of the twentieth century. How can individual fiscal incentives be used to deliver better urban climate and urban energy outcomes? How can the investment of government's scarce tax revenues (which will be even scarcer in the post Covid economic recovery decade) be directed to optimise the infrastructural skeletons of cities, for climate resilience, in a way that shapes and encourages the decisions of millions of individual land-users to make their own contributions? Where are the gaps, where climate resilience cannot be delivered by market signals? Who and where are the vulnerable population groups that climate-sensitive government planning needs to protect? What new development red-lines will be needed in the future, to protect the interests of long-term home investors from compromised locations? How will insurance maps be redrawn? What areas of our cities can be 're-wilded' for the benefit of all? Will this require a nationally organised new social contract achieved by a massive compensation scheme, like the compensation to UK land owners that achieved that county's post-War land development rights nationalisation experiment (which still continues)?

The readers of this book will therefore include lawyers, economists and political scientists, as well as urban climatologists, environmental scientists, planners and architects. The book lays out what we currently know about climate resilient cities, and some of what needs to be done. It is just the start of the conversation, however. To achieve all that the authors believe is needed, will require a multi-professional and multi-disciplinary approach to agenda development and policy delivery. The book's emphasis on data, especially globally comparable public domain data, is an important one in this respect. Although data are not immune from bias, special-interest distortion and imbalanced argumentation, they provide a common, open, accountable and testable multi-disciplinary language. It is high time that urban planning became more scientific. The growing call for more evidence-based policy in societies everywhere, in recent years, came at a time when urban planning in many countries had moved very far from science and engineering, but had not reverted to its other original influence: architectural urban design. The space between the two is not a convincing or efficacious one to occupy for a profession charged with such an important social function. Ren and McGregor's book helps fill that gap. I have no doubt that it will make a significant contribution to cities and to the professions that help shape them.

Faculty of Architecture
The University of Hong Kong,
Pok Fu Lam, Hong Kong
July 2020

Chris Webster

Preface

The idea for this book arose from a conversation at the 2016 Annual General Meeting of the American Meteorological Society held in Seattle. The context for the conversation was the world's burgeoning urban population, the nature of urban climates, associated climate risks, and their management. In many ways, our conversation repeated the urban climate community's long-held view that much stands to be gained by integrating knowledge on urban climate processes into the urban design and planning process with the endpoint, the achievement of agreeable lived urban climate experiences for urban populations and more broadly 'healthy cities'. Laced with a sense of frustration, much of our conversation lamented the slow progress towards the realization of the benefits of mainstreaming urban climate science knowledge into city planning. We noted the need for a book that not only articulates the case for the relevance of urban climate science but also 'sets out a stall' of exemplars of urban climate science into practice.

Little did we know in 2016, when we first hatched the idea for this book that in 2020–2021, society would be grappling with the challenges posed by the global COVID-19 pandemic. This worldwide event continues to produce a number of reminders and learnings. Amongst these are the highest attainable standard of health is one of the fundamental rights of every human being, to achieve more than a one-dimensional understanding of a pressing problem interdisciplinary collaboration is required, and an imperative for the avoidance of a crisis or at least its effective management is early action. Such reminders and learnings ring true in the case of achieving healthy cities beyond the conception of healthy as the absence of disease. Accordingly, this book brings together a number of contributors from a range of disciplines, all of whom understand the urgency of action required to manage what could be referred to as an 'urban climate crisis', and are committed to an interdisciplinary approach to problem solving, the problem being how best to obtain 'healthy cities' from a climate perspective.

Our aspiration with Urban Climate Science for Developing Healthy Cities is to illustrate the idea of 'science in time, science in place', that is knowledge that is timely and future proof and specific to a space where an increasingly large proportion of the world's population lives. The book consists of 18 chapters arranged over three sections. The first section sets the scene for what follows, laying out the books conceptual framework and outlining the idea of healthy cities. Section 2 presents a range of city-based case studies that articulate how urban climate science knowledge can be, and is integrated, into urban planning while the final section considers some of the issues at the interface of climate knowledge and climate sensitive urban planning. Our intent is for the book to be a useful reference for professionals in spatial planning, building and landscape architecture, urban design, public health, policy makers in the fields of healthy city development and climate change adaptation, academics and students and those with a general interest in environment and society.

Pok Fu Lam, Hong Kong Chao Ren

Durham, UK Glenn McGregor
April 2021

Acknowledgements

Many of the presentations made at the 2016 Annual Meeting of the American Meteorology Society held in Seattle formed the inspiration for this book. Accordingly, we acknowledge those presenters at the AMS Seattle meeting who provided an impetus for taking forward the idea for a book on the use of urban climate science for planning healthy cities.

We also acknowledge the encouragement given to us by Springer staff at the AMS Seattle meeting who made the book proposal process straight forward and subsequently provided the requisite support throughout all stages of the publication process.

This book would not be possible without the energy and dedication of the contributing authors to whom we are very grateful for bearing with us as we compiled the contents for this book through a time made difficult by the COVID-19 pandemic.

We are especially thankful to Ms. Sze Wai Wong of the Chinese University of Hong Kong. Ms. Wong has worked wonders in communicating with all authors and delivering on a range of formatting and editing issues. Her long hours of diligence and productiveness have not only been a blessing to the editors but the chapter authors as well.

The editors Chao Ren and Glenn McGregor also acknowledge the generic support provided by their respective institutions Hong Kong University, Hong Kong and Durham University, UK.

More widely, we are grateful to the international community of urban climate, social and public health scientists, planners, architects and policy makers and implementers for their continued articulation of the need for climate-sensitive planning and design, especially as we enter a period of human history in which climate presents a tangible threat to human and ecosystem health and the attainment of healthy cities. Complementing this, we acknowledge the huge efforts being made in cities worldwide, including those focused on in this book, to work towards creating physical environments that benefit the health and well-being of urban populations.

Last, but not least, we would like to dedicate this book to Professor Emeritus Baruch Givoni, a pioneer in the understanding of urban and building climatology, and human well-being, who passed in December 2019. As has the work of Professor Givoni, we hope this book adds to the growing evidence base that climate-sensitive planning of cities improves the lived urban climate experience.

Contents

Part I The Rise of Mega-cities and the Concept of Climate Resilience and Healthy Living

1 **Urban Climate Science for Planning Healthy Cities** 3
Glenn McGregor and Chao Ren

2 **Healthy City and Its Planning** 17
Lan Wang, Xiji Jiang, and Xinhu Li

Part II Urban Climate Science in Action

3 **Embedding Climate Change in Urban Planning and Urban Design in New York City** 45
Jeffrey Raven, Christian Braneon, and Cynthia Rosenzweig

4 **Cool Cities by Design: Shaping a Healthy and Equitable London in a Warming Climate** 71
Anna Mavrogianni, Jonathon Taylor, Phil Symonds, Eleni Oikonomou, Helen Pineo, Nici Zimmermann, and Mike Davies

5 **Manchester: The Role of Urban Domestic Gardens in Climate Adaptation and Resilience** 99
Gina Cavan, Fraser Baker, Konstantinos Tzoulas, and Claire L. Smith

6 **Tokyo Heat Island Adaptation Measures** 119
Yasunobu Ashie

7 **Hong Kong: Climatic Application in Urban Planning and Design at Multiple Scales for Creating a Healthy Living Environment** 151
Emily Ying Yang Chan, William B. Goggins III, Zhe Huang, and Chi Shing Wong

8 **Air Pollution Prevention and Urban Planning Response: Building a Healthy Beijing** 167
Xiaoyi Fang, Xin Li, Xiyan Ren, Jian He, Wupeng Du, Yajun Xiong, Yonghong Liu, Chen Cheng, and Shuo Zhang

9 **Singapore: An Integrated Multi-scale Urban Microclimate Model for Urban Planning in Singapore** 189
Tian Kuay Lim, Nyuk Hien Wong, Marcel Ignatius, Miguel Martin, Hann-Ming Henry Juang, Jing Lou, and Robert Lee Kong Tiong

10 **Creating Cooler, Healthier and More Liveable Australian Cities Using Irrigated Green Infrastructure** 219
Nigel Tapper

11 **What Alternatives Does Paris Have to Adapt to Future Heat Waves?** 239
Aude Lemonsu, Cécile de Munck, Raphaëlle Kounkou-Arnaud, Valery Masson, and Vincent Viguié

12 **Urban Climate in Urban Planning: The Experience from Stuttgart** 259
Ulrich Reuter and Rainer Kapp

13 **India: Heat Wave and Action Plan Implementation in Indian Cities** 285
Mahaveer Golechha, Dileep Mavalankar, and Subhash Chander Bhan

Part III Future Challenges and the Way Forward

14 **Guidance on Integrated Urban Hydro-meteorological, Climate and Environmental Services: Challenges and the Way Forward** 311
Paul Joe, Alexander Baklanov, Sue Grimmond, Veronique Bouchet, Luisa T. Molina, K. Heinke Schluenzen, Gerald Mills, Jianguo Tan, Brian Golding, Valery Masson, Chao Ren, James Voogt, Shiguang Miao, Humphrey Lean, Bert Heusinkveld, Anahit Hovespyan, Giacomo Terrugi, Patrick Parrish, Jens H. Christensen, Joachim Fallmann, Ranjeet Sokhi, Reinhard Spengler, and Junichi Yoshitani

15 **Characterising Urban Morphology for Urban Climate Modelling** 339
Gerald Mills, Jason Ching, and Benjamin Bechtel

16 **The Synergistic Impacts of Urban Air Pollution Compounding Our Climate Emergency** 355
Melissa Anne Hart, Nathan Cooper, Donna Green, and Mathew Lipson

17 International Research Collaboration and Programmes for Urban Climate Science for Planning Healthy Cities 379
Boram Lee, Hunter Jones, and Chao Ren

18 Conclusions and Moving Forward 389
Glenn McGregor and Chao Ren

Correction to: Manchester: The Role of Urban Domestic Gardens in Climate Adaptation and Resilience C1

Index 399

List of Figures

Fig. 1.1 Conceptual framework demonstrating how climate-sensitive health and well-being outcomes link to processes at a number of nested larger scales 9

Fig. 1.2 Cascade of factors linking regional climate to urban health outcomes via lived urban climate conditions as determined by a range of city-level moderating factors. VBD is vector-borne disease, WBD is water-borne disease and FBD is food-borne disease 10

Fig. 2.1 The relationship between subjects related to healthy city research. (Source: drawn by authors) 23

Fig. 2.2 Number of relevant research documents in major disciplines (cumulative value 1980–2019). (Source: drawn by authors) 24

Fig. 2.3 Cluster analysis of keywords. (Source: drawn by author) 25

Fig. 2.4 Number of relevant research literatures of main keywords (cumulative value from 1980 to 2019). (Source: drawn by authors) 26

Fig. 2.5 Timeline analysis chart of research hotspots related to healthy cities. (Source: drawn by authors) 26

Fig. 2.6 Time-zone evolution of research hotspots related to healthy cities. (Source: drawn by authors) 27

Fig. 2.7 Changes in the number of documents related to healthy city planning (1990–2019). (Source: drawn by authors) 28

Fig. 2.8 Changes in the number of climate change related literature in healthy city planning (2000–2018). (Source: according to the results of CiteSpace analysis, the author collates and draws) 29

Fig. 2.9 Cluster map of key words co-occurrence in health city planning. (Source: CiteSpace analysis results sorting) 32

Fig. 2.10 Evolution of research hotspots related to healthy city planning. (Source: CiteSpace analysis results) 35
Fig. 2.11 "Climate change" research in the related research of healthy city planning. (Source: CiteSpace analysis results) 36
Fig. 2.12 Relationship between urban planning, health and climate change. (Source: drawn by authors) 37

Fig. 3.1 The four urban climate factors: (1) reducing waste heat and greenhouse gas emissions through energy efficiency, transit access, and walkability; (2) modifying form and layout of buildings and urban districts; (3) use of heat-resistant construction materials and reflective surface coatings; and (4) increasing vegetative cover (Source: Raven 2016) 48
Fig. 3.2 Annual average daily maximum summer temperatures (June, July, August) in Central Park from 1900 to 2013, LaGuardia (LGA) Airport from 1970 to 2013, and John F. Kennedy (JFK) International Airport from 1970 to 2013. Solid lines represent linear trends for each station. Station records were obtained from the U.S. Historical Climatology Network (USHCN) Version 2.5 (Menne et al. 2013; González et al. 2019) 52
Fig. 3.3 Sunnyside Yard, Queens; Manhattan in background 59
Fig. 3.4 Sunnyside Yard, Land Surface Temperature Scenario (Data Source: NASA, Landsat 8) 59
Fig. 3.5 Baseline (current condition): Sunnyside Yard site in Queens, New York, as it is in present day, with the rail yard and surrounding areas 60
Fig. 3.6 Baseline (business-as-usual) 2050: Hypothetical scenario based on typical "market driven" decisions 60
Fig. 3.7 Best Practice (climate-driven urban design) 2050: Based on climate adaptive development, considering the 4 climate factors (efficiency of urban systems, form and layout, heat resistant construction materials and vegetative coverage); developed by the NYIT Urban Climate Lab 61
Fig. 3.8 A strategic subdistrict (See location C in Fig. 3.6) under the 2050 Baseline (top row) and 2050 Best Practice (bottom row) scenarios is illustrated by the ENVI-met grid cells in columns I, II, and III. The microclimate benefits of 2050 Best Practice urban design configuration is illustrated by these ENVI-met simulations modeling conditions at 6 pm in July:
• Column I: Illustrates the variation of vegetative cover across the sub-district.
• Column II: Illustrates the variation in air temperature distribution across the sub-district.
• Column III: Shows the variation in surface temperature across the sub-district. Simulated surface temperatures range from 27 to 43 °C with significant surface cooling 63

Fig. 3.9 "Word map" illustrating shared language and approaches between urban climate and urban design experts, from Urban Design Climate Workshop, NYC 2018 ... 65
Fig. 3.10 Urban Design Climate Workshop with urban climate and urban design experts exchanging approaches ... 66

Fig. 4.1 Urban climate models developed within the context of the LUCID project. (Source: EPSRC LUCID project) ... 79
Fig. 4.2 Screen level temperature at 1.5 m height, 26 May–19 July 2006 (LondUM simulation). (Source: EPSRC LUCID project) ... 80
Fig. 4.3 The meta-modelling framework underpinning the HPRU housing stock model. (Source: Symonds et al. 2016) ... 82
Fig. 4.4 (a) Mean Maximum Daytime living room Temperature (MMDT, °C), (b) Mean Maximum Night-time bedroom Temperature (MMNT, °C), and (c) Indoor/Outdoor ratios for PM2.5, for the pensioners' occupancy assumption aggregated at the Lower Super Output Area level. (Source: NERC AWESOME project) ... 83
Fig. 4.5 Heat vulnerability index for London. (Source: Arup and Partners 2016, © Dr. Tanja Wolf/King's College London) ... 84
Fig. 4.6 The London Triple Heat Jeopardy mapping framework combining (from top to bottom) age-related relative risk of heat mortality, dwelling temperature anomaly, and local urban temperature data. (Source: NERC AWESOME project) ... 86
Fig. 4.7 Estimated mortality per million population during a warm 55-day period in London. (Source: Taylor et al. 2015) ... 87
Fig. 4.8 Location of London care homes in relation to socio-spatial heat vulnerability indicators. (Source: Mavrogianni et al. 2018a; Oikonomou et al. 2018) ... 89
Fig. 4.9 A 3D visualisation of the area around Worton Way, Hounslow, showing the overheating risk of dwellings at 28 °C outdoor temperature (blue dwellings represent the coolest, while red are the hottest). (Source: Taylor et al. 2017) ... 90
Fig. 4.10 Selected causal effects of adding municipal and other green infrastructure. A plus (minus) next to an arrow indicates a same (inversely) directed causal relationship. A dashed arrow indicates the high ambiguity of the adverse summer and useful winter effects of the UHI. (Source: Zimmermann et al. 2018b) ... 91

Fig. 5.1 The case study area of Manchesters, UK, showing gardens as a percentage of ward area ... 103

Fig. 5.2 Proportional garden land surface cover for the average garden in each ward ... 106
Fig. 5.3 (a) Garden GI as a percentage of ward area, (b) total GI as a percentage of ward area ... 107
Fig. 5.4 Modelled maximum surface temperatures for (a) Baseline I scenario, (b) Baseline II scenario, (c) Difference between Baseline I and Baseline II scenarios ... 109
Fig. 5.5 Modelled surface runoff coefficient (where a coefficient of 1 is complete runoff) for (a) Baseline I scenario, (b) Baseline II scenario, (c) Difference between Baseline I and Baseline II scenarios ... 109
Fig. 5.6 Ward level modelled maximum surface temperature for baseline and future scenarios ... 110
Fig. 5.7 Ward level surface runoff for baseline and future scenarios ... 111
Fig. 5.8 The MBY Action Plan, illustrating Lancashire Wildlife Trust's actions and delivery mechanisms, available to download at www.mmu.ac.uk/mybackyard ... 115

Fig. 6.1 Change of annual average air temperature in Tokyo, Japan, and Worldwide (Ministry of the Environment 2013) from 1900 to 2010. Based on JMA observation data ... 120
Fig. 6.2 Change in number of extremely hot days and tropical nights (Ministry of the Environment 2018) from 1900 to 2015. Based on JMA observation data. In December 2014, the observation point moved from Otemachi, Chiyoda-ku to Kitanomaru Park. (a) Number of extremely hot days. (b) Number of tropical nights ... 121
Fig. 6.3 Number of deaths caused by heat-related illnesses (National Institute for Environmental Studies, Japan 2009) from 1970 to 2010. Based on the demographic statistics by MHLW ... 122
Fig. 6.4 Evaluation of the sensible temperature by combining the CFD grid with surface elements ... 126
Fig. 6.5 Sample analysis of radiant temperature (Tsukuba Science City) ... 127
Fig. 6.6 Climographs ... 128
Fig. 6.7 Marathon course heat countermeasures ... 129
Fig. 6.8 Example of air temperature analysis (13:00 on August 10, 2013) ... 130
Fig. 6.9 Effects of heat countermeasures (13:00 on August 10, 2013) ... 130
Fig. 6.10 Measures against heat in Bus station waiting room (Ministry of the Environment 2017) ... 131
Fig. 6.11 View of the interior of the bus stop waiting room (Ministry of the Environment 2017) ... 132

Fig. 6.12 Results of survey on a summer day (14:00 to 15:00 on August 5, 2016). Figures are created using the survey data performed by Ministry of the Environment (Ministry of the Environment 2017). (a) Thermal environment. (b) Physiology/psychology. Relative increase of heart rate: percentage increase of heart rate with heart rate when quiet as the standard (%). Heat sensation: subjective declaration in 9 levels (1: cold, 9: hot). Comfortable sensation: subjective declaration in 7 levels (1: uncomfortable, 7: comfortable)............ 133
Fig. 6.13 Wind conditions in urban spaces (July 31, 2005 at 12:00)............ 134
Fig. 6.14 Wind paths and evaluation level 135
Fig. 6.15 Heat island countermeasure map (Ministry of Land, Infrastructure, Transport and Tourism 2013)............ 136
Fig. 6.16 Tokyo Wind path map: countermeasure policy map "Wind Path" (Ministry of Land, Infrastructure, Transport and Tourism 2013)............ 138
Fig. 6.17 Redevelopment around Tokyo Station. (a) Before redevelopment. (b) After redevelopment............ 139
Fig. 6.18 Redevelopment around Nihonbashi River. The redevelopment area is indicated by the dotted line. Elevated roads will be removed and the height and volume along the river reduced. (**a**) Before redevelopment. (**b**) After redevelopment 140
Fig. 6.19 Redevelopment around Meguro River. The arrows were added to the building layout created by Nihon Sekkei Inc. to show ideal summertime wind flow 141
Fig. 6.20 Location of wind paths (Tokyo Metropolitan Government 2006b, revised 2014). Created based on 2001 land use survey 142
Fig. 6.21 Development area and surrounding area 143
Fig. 6.22 Trial calculations case............ 147
Fig. 6.23 Heat island phenomenon explained in a Japanese textbook (Dainippon-Tosho 2012)............ 148

Fig. 7.1 Health effect pyramid 153

Fig. 8.1 Schematic diagram of the Beijing terrain and the location of the observatory. (Source: drawn by author) 174
Fig. 8.2 Number of haze days as per the Beijing observatory from 2014 to 2018. (Source: drawn by author) 174
Fig. 8.3 Planning map of green spatial structure within the city territory of Beijing. (Source: drawn by author)............ 179
Fig. 8.4 Schematic diagram of the ventilation corridor in the central city area of Beijing. (Source: drawn by author) 181
Fig. 8.5 Distribution of the wind direction during the pollution process on January 27, 2013 (left 2 panels), and June 30 (right 2 panels). (Source: drawn by author)............ 183

Fig. 9.1 Schematic of urban heat island phenomena (Wong and Ignatius 2016) 192
Fig. 9.2 Methodology for different components of QUEST (IHPC 2019) 196
Fig. 9.3 Overall methodology workflow of IMEUM, in which the final output is the microclimate estimation at city-urban scale. This has been calibrated using on-site observation data (Lim et al. 2017) 197
Fig. 9.4 Independent variables used in IMEUM City-urban model (CUM) (Lim et al. 2017; Ignatius et al. 2016) 200
Fig. 9.5 Outdoor TCI-PD chart (IHPC 2019) 200
Fig. 9.6 General flowchart for the QUEST version 1.0 user PC package (IHPC 2019) 201
Fig. 9.7 Case study illustration of simulation result (Thermal Comfort Index) (IHPC 2019) 201
Fig. 9.8 QUEST workflow baseline climate selection interface (IHPC 2019) 202
Fig. 9.9 Workflow on using QUEST to generate a local weather data for cooling load simulation (Lim et al. 2017) 202
Fig. 9.10 (Left) On-site weather measurement points, where point A, B, and C were selected to be used for case study evaluation. (Right) 20-storey building used to test outputs provided by IMEUM for building energy assessment (Lim et al. 2017) 202
Fig. 9.11 Ambient temperature comparison between sensing data and calculated method from IMEUM across three different locations (Lim et al. 2017) 203
Fig. 9.12 IMEUM final output comprises weather parameters at local scale (Lim et al. 2017) 204
Fig. 9.13 Ideal cooling load as estimated by the EnergyPlus model using weather conditions assessed from IMEUM (Lim et al. 2017) 205
Fig. 9.14 Annual counts of hospital admissions for heat-related illness in Singapore, 1991–2010 (Xu et al. 2017) 205
Fig. 9.15 Case study illustration of the microclimate simulation of an urban region (*T*max map) (IHPC 2019) 206

Fig. 10.1 Spatial distribution of the population heat vulnerability index for Melbourne based on statistical relationships between morbidity and a range of health, socioeconomic and environmental variables. From Loughnan et al. (2013) 223
Fig. 10.2 WSUD cooling technologies employed at a range of scales, from the microscale (household to street) to the local scale (neighbourhood to city). Modified from Coutts et al. (2013a) 225

Fig. 10.3 Mast and temperature sensor placement on the isolated tree (left) and the thermal pattern (10-min average) within the tree canopy and surrounding environment (right) at 16:00 on a warm, sunny afternoon (9 March 2014) with moderate northerly winds. From Coutts et al. (2021) 227
Fig. 10.4 Air temperature profiles (10-min average) from 04:00 9 March 2014 to 04:00 10 March 2014 profiles for Pole 5 (see Fig. 10.3) located on the sunlit side of the tree and Pole 3 located in the centre of the tree. From Coutts et al. (2021) 228
Fig. 10.5 Average diurnal cooling (with standard deviations) with different timings and quantities of water applications across the suburb of Mawson Lakes, Adelaide. From Broadbent et al. (2018a) 228
Fig. 10.6 Cooling effectiveness of increased rooftop albedo (0.2 → 0.7), increased tree canopy (20 → 40%) and nocturnal irrigation for the city of Melbourne during EHE. The graph shows combined simulations for the current climate city (current urban structure), mid-twenty-first century climate city (current urban structure), and mid-twenty-first century climate city (future urban structure). From Jacobs et al. (2021) 229
Fig. 10.7 Factors required for identification of areas of high (C), medium (B) and moderate (A) priority for implementation of heat mitigation within a municipality. The key factors are high daytime surface temperatures (heat exposure) intersecting with areas of greater population vulnerability and areas of high activity (behavioural exposure) such as train/bus stations, car parks, etc. From Norton et al. (2015) 234
Fig. 11.1 Comparison of the 2003 heat wave (HW2003) with heat waves extracted from climate projections in A1B scenario over the time periods 1960–1989, 2020–2049, and 2070–2099. Each disk represents a heat wave characterized by a duration (indicated on the abscissa) and a maximum intensity (on the ordinate). The size of the disk is proportional to the severity of the heat wave 242
Fig. 11.2 Land use map of the Ile-de-France administrative region (left) based on the simplified MOS 2017 database produced by the Institute of Planning and Urban Development of the Parisian region (IAU). The thick black lines delimit the departments. A zoom on the area of Paris "intra-muros" and its first suburbs is presented to the right. Source: http://data.iau-idf.fr/datasets/75ab6d1bb58043729a8b 05172937cfb0_40 246

Fig. 11.3 For each research project, comparison of temperature anomaly profiles at night (0000–0200 LT) simulated for the reference case (black line) and the adaptation scenarios (colour lines) along a west-to-east cross section passing through Paris intra-muros. The temperature anomaly is calculated for each project as the difference between the temperature profile of each simulation and an average rural temperature estimated from the reference case ... 248
Fig. 11.4 Left panel: Fraction of the Paris urban unit affected by an UHI of 2 °C or more during the night (blue) and during the day (red), according to the different urban expansion scenarios. Middle panel: Same than previously but for the fraction of urban population. Right panel: Number of hours per day spent by population in strong heat stress (SHS) conditions ... 251

Fig. 12.1 Orographic situation of Stuttgart city centre. (Source: Department of Urban Climatology, Municipality of Stuttgart) ... 260
Fig. 12.2 Climate Analysis Map of Stuttgart (Verband Region Stuttgart 2008) ... 264
Fig. 12.3 Map with recommendation for Planning in Stuttgart (Verband Region Stuttgart 2008) ... 266
Fig. 12.4 Green rails in Stuttgart. (Source: Department of Urban Climatology, Municipality of Stuttgart) ... 268
Fig. 12.5 Roadside trees in Stuttgart. (Source: Department of Urban Climatology, Municipality of Stuttgart) ... 269
Fig. 12.6 Green roof in Stuttgart. (Source: Department of Urban Climatology, Municipality of Stuttgart) ... 270
Fig. 12.7 Urban land-use plan "Schelmenäcker" with green corridor (Landeshauptstadt Stuttgart, Amt für Stadtplanung und Stadterneuerung 2010) ... 271
Fig. 12.8 Climate analysis map Schelmenäcker district (Verband Region Stuttgart 2008) ... 272
Fig. 12.9 Map with recommendations for Planning "Schelmenäcker" district (Verband Region Stuttgart 2008) ... 273
Fig. 12.10 Green corridor "Schelmenäcker" district. (Source: Department of Urban Climatology, Municipality of Stuttgart) ... 273
Fig. 12.11 Hillside Development Plan Stuttgart (Landeshauptstadt Stuttgart 2008) ... 274
Fig. 12.12 Example of climate-optimised inner development (Area "Rossbollengässle" in Stuttgart-West). (Source: Department of Urban Climatology, Municipality of Stuttgart) ... 275
Fig. 12.13 Greenery network map, Valley Bottom Development Outline Plan (Landeshauptstadt Stuttgart 2018) ... 277
Fig. 12.14 Detailed Part I, "Berliner Platz/Diakonissenplatz" (Landeshauptstadt Stuttgart 2018) ... 278

Fig. 12.15 Valley Bottom Development Outline Plan, Model of optimised building structure (Landeshauptstadt Stuttgart 2018) 279
Fig. 12.16 Part of the urban development plan "Neckarpark" with legally binding "Green" and Central green space 280
Fig. 12.17 "Neckarpark": Implemented Central green space, Opening 2017. (Source: Municipality of Stuttgart)........................ 281

Fig. 13.1 Average number of (a) HW days and (b) SHW days over India during the hot weather season (March–July) for the period 1961–2010 (Pai et al. 2013). (Courtesy: Editor MAUSAM—India Meteorological Department) 287
Fig. 13.2 Duration of the longest HW spells (a) and longest SHW spells (b) during 1961–2010. The areas bounded by red color lines represent HW spells of ≥10 days duration and SHW spells of ≥5 days duration (Pai et al. 2017). (Courtesy: Editor MAUSAM—India Meteorological Department)... 288
Fig. 13.3 Decadal variation of the all India 6+. HW/SHW days (days/year) during the hot weather season (March–July) for the period 1961–2010. (Pai et al. 2013). (Courtesy: Editor MAUSAM—India Meteorological Department) 289
Fig. 13.4 Average number of HW days during the hot weather season (March–July) over India during (a) 1961–1970 and (b) 2001–2010. The climatology was computed by averaging the number of HW days during the respective decades. (Courtesy: Editor MAUSAM—India Meteorological Department)... 289
Fig. 13.5 Daily Temperature and all-cause mortality correlation graph........ 292
Fig. 13.6 Heat action plan components... 293

Fig. 14.1 Schematic showing the domino effect using a typhoon (or hurricane) event as an example. The event (red) creates several hydro-meteorological hazards (blue) that generate immediate impacts (green) with related effects (purple) requiring different services. (Source: WMO (2019a) Guidance for Urban Integrated Hydrometeorological, Climate and Environmental Services. Volume I: Concept and Methodology. WMO-No: 1234, ISBN 978-92-63-11234-7. https://library.wmo.int/doc_num.php?explnum_id=9903)... 316
Fig. 14.2 Schematic of the components of an Integrated Urban Service (IUS) System. The numbers (#) are referred to in the text (Source: WMO 2019a Guidance for Urban Integrated Hydrometeorological, Climate and Environmental Services. Volume I: Concept and Methodology. WMO-No: 1234, ISBN 978-92-63-11234-7. https://library.wmo.int/doc_num.php?explnum_id=9903)... 320

Fig. 15.1 The vertical structure of the urban atmosphere above a l ow-density, well-vegetated urban landscape. The surface layer (SL) occupies the lower 10% of the boundary layer, which can extend from 1 to 2 km depth during the daytime. The SL over urban landscapes consists of a deep roughness sub-layer that extends from two to four times the heights of the roughness elements (that is, the buildings and trees). The inertial sub-layer is where the Monin-Obukov Similarity Theory (MOST) can be applied 342

Fig. 15.2 Urban Canopy Parameters (UCPs) describing urban cover, length scales and urban structure (Oke et al. 2017). Cambridge University Press, with permission........ 344

Fig. 15.3 Vector building file for the Republica neighbourhood in Sao Paulo available at https://geosampa.prefeitura.sp.gov.br/. The information consists of building footprints and heights. Here, this information is used to generate a 3D image that is superimposed on a satellite image 348

Fig. 15.4 Shadow patterns for noon on 21 December generated by the Solweig model (Lindberg et al. 2008) using Lidar data for London at 1 m spatial resolution. The data are available from https://data.gov.uk/dataset/80c522cc-e0bf-4466-8409-57a04c456197/lidar-composite-dsm-1m 349

Fig. 16.1 The interaction of urban air pollution with local urban climate and global climate change 358

Fig. 16.2 A view of the city of Sydney on a typical day (left) and a day where the air quality is significantly impacted by smoke from hazard reduction burns (right) 366

Fig. 17.1 Pathways for climate adaptation and mitigation in cities (WCRP 2019) 384

List of Tables

Table 1.1 Non-physical Determinants of Urban Health (Adapted from Freudenberg et al. 2006) ... 11

Table 2.1 Co-occurrence analysis information of research key words related to healthy city planning (number of documents issued ≥40) (Source: CiteSpace analysis results) ... 31

Table 3.1 Projections of extreme heat days and heat waves in New York City in the 2020s, 2050s, and 2080s compared to the 1971–2000 baseline (Source: NPCC 2015) ... 57

Table 5.1 Average proportional land surface cover across Manchester's gardens ... 106

Table 5.2 Description of the four scenarios for assessing cooling potential and runoff attenuation ... 108

Table 5.3 An overview of the Action Plan, outlining the six key themed actions, delivery mechanisms, and specific examples ... 114

Table 6.1 Environmental impact of heat island phenomenon (Ministry of the Environment 2004) ... 123

Table 6.2 Relationship of mitigation countermeasures with adaptive countermeasures in global warming countermeasures and heat island countermeasures ... 124

Table 8.1 Changes in the number of days with different haze levels as per Beijing observatory from 2014 to 2018 (Source: author) ... 175

Table 9.1 Four-cell contingency table, where a is the number of hits, b is the number of misses, c are the false alarms and d is the number of hits for forecasting no rain correctly (Lim et al. 2017) ... 199

Table 9.2 Parameters (based on the contingency Table 9.1) applied to assess the quality of coupled Global-Regional Spectral Model-Mesoscale Spectral Model (GRSM-MSM) in simulating fair weather (no rain) and weather conditions (with rain) (Lim et al. 2017) ... 200
Table 9.3 PC and POD skill scores over the study area (Lim et al. 2017) ... 200

Table 10.1 Nature of micro- and local-scale urban cooling treatment examined by CRCWSC urban climate researchers and the level of summer cooling achieved in each case ... 226
Table 10.2 Australian government policies directly or indirectly addressing urban heat mitigation. For brevity State/Territory and local government examples are from Victoria only (modified from Bush et al. 2015) ... 231

Table 11.1 Description of research projects related to Paris area's adaptation to climate change ... 245

Table 13.1 Year wise mortality in India (NDMA 2017) ... 290

Table 15.1 Example of sources of urban morphology data and associated characteristics ... 347

Editors and Contributors

About the Editors

Chao Ren is an associate professor in the Faculty of Architecture at The University of Hong Kong. Her research interest is sustainable urban and environmental design and urban climatic application in urban planning. She has published widely in highly ranked journals and peer-reviewed conferences. Her studies have led to a number of practical research and government consultancy projects on urban climate application and climate change adaptation in Hong Kong, Macau, Mainland China, Taiwan, the Netherlands and France. She serves as an invited expert of the urban working group of WMO. She is an elected board member of the International Association for Urban Climate (2017–2021). She serves as an editorial advisor for Cities & Health and an associate editor for Urban Climate.

Glenn McGregor is the professor of climatology in the Department of Geography, Durham University, UK. He holds a PhD in geography from the University of Canterbury in Christchurch, New Zealand. His research interests cover climate and health (biometeorology), synoptic climatology, hydroclimatology and climate and society. He is the former World Meteorological Organisation's lead expert for Climate and Health, previous chief editor of the International Journal of Climatology and president of the International Society of Biometeorology and a past and current (6AR) IPCC Working Group II lead author. He was also lead editor for the widely consulted WMO/WHO publication 'Heatwaves and Health: Guidance on Warning-System Development' and is currently a review editor for the journals Weather and Climate Extremes and Anthropocene and a member of Public Health England's Health Protection Unit's Research Advisory Board. Previously he held posts at the University of Auckland in New Zealand where he was director of the School of Environment, King's College London, University of Birmingham, Hong Kong Baptist University, the University of Papua New Guinea and the Institute of Low Temperature Science, Hokkaido University, Japan.

About the Contributors

Yasunobu Ashie is the director of the Department of Environmental Engineering of the Building Research Institute of Japan (BRI). He also acts as head of the Evaluation Office of the CO2 Reduction Pilot Project and program director of the Sustainability Program at BRI. He received his PhD in engineering at Hokkaido University. His specialty is urban thermal environments and low-carbon cities. He has contributed to the decision-making process of the heat island measure guidelines of the Tokyo Metropolis, Yokohama City and Chiba Prefecture Governments, and the heat island measure guidelines of the Ministry of the Environment (MOE) as a committee member for more than 20 years. He was a representative of the earth simulator research project from 2004 to 2012 and made the Heat Map of Tokyo using large-scaled CFD simulations. He was awarded research paper awards by the Architectural Institute of Japan (AIJ) and the Society of Heating, Air-Conditioning and Sanitary Engineers of Japan (SHASE) and research section commendations by the Ministry of Education, Culture, Sports, Science and Technology (MEXT) and the Ministry of Land, Infrastructure, Transport and Tourism (MLIT).

Fraser Baker is a researcher in the Department of Natural Sciences, Manchester Metropolitan University. Fraser previously worked as a research associate for the 'My Back Yard' project, within which he focused on developing techniques to combine object-based image classification with citizen-science data to estimate domestic garden surface composition in Manchester, UK. His current research involves using various GIS and remote sensing techniques for high-resolution spatiotemporal urban ecosystem services assessment. Fraser works with open-source data to develop accessible urban analysis workflows for further public sector geospatial and environmental research. Fraser is also actively involved in urban greenspace mapping projects within the UK and Dominican Republic.

Alexander Baklanov (PhD in physics and mathematics (geophysics)—1983, Dr. Sci. in physics and mathematics (meteorology and climatology)—1998, professor in meteorology—2008) is a science officer of the Science and Innovation Department of the World Meteorological Organization (WMO), affiliated as a professor at the University of Copenhagen and editor-in-chief of the Urban Climate journal. He has more than 30 years of experience in environmental and atmospheric research, in particular in developing a new generation of online coupled meteorology-chemistry models, especially for urban applications. He has published about 400 scientific publications, including 14 books and more than 200 peer-reviewed papers. He led a number of big international research projects for urban meteorology and environment (e.g. FUMAPEX, MEGAPOLI, EuMetChem) and was a senior scientist in the Research Department of the Danish Meteorological Institute (DMI) during 16 years, deputy director of the Danish strategic research Centre for Energy, Environment and Health (CEEH), and member of the WMO GURME Science Advisory Group. He has supervised 12 PhD students and is a visiting/adjoint/honourary professor in several European universities and a member of the Academy of Europe and the International Eurasian Academy of Sciences.

Benjamin Bechtel holds a professorship in urban climatology at the Department of Geography, Ruhr-University Bochum, Germany. Before he was a research associate with the Cluster of Excellence CliSAP, University of Hamburg. His research interests include crowd sourcing and urban remote sensing, in particular the characterization of urban surfaces and thermal remote sensing for applications in urban climatology. Dr. Bechtel received the dissertation award 2013 for physical geography of the Verband der Geographen an Deutschen Hochschulen (VGDH) for his Ph.D. thesis on 'Remote sensing of urban canopy parameters for enhanced modelling and climate related classification of urban structures'; his habilitation was on 'Advancements in urban- and topoclimatic observations and modelling – Remote Sensing, Crowd-Sourcing and Data Fusion'. He serves as board member of the International Association for Urban Climate, as a steering committee member of the Belgian research project REACT and the GEO Global Human Settlement Working Group, and as a reviewer and guest editor for several international journals.

Subhash Chander Bhan is the head of Health Applications, India Meteorological Department, New Delhi. He joined India Meteorological Department (IMD) as meteorologist in the year 1990 and has been working in the department in different capacities, particularly in the fields of operational weather forecasting, human resources development and applications such as agrometeorology, hydrometeorology, power and health sector applications. He established the 'Location Specific Weather Information and Forecast System' in the country during the Commonwealth Games-2010, New Delhi—a first of its kind in the country. He was instrumental in establishing 'Weather Services for Power Sector' and a countrywide 'Heat-Health Early Warning System' in India through multi-stakeholder interactions. He has written more than 80 research papers in peer-reviewed national and international journals, has authored a book, and has presented more than 20 papers in various national and international symposia in India and abroad. Three of his papers have won the 'Best Paper Awards'. He was awarded the 'Certificate of Merit' by Ministry of Earth Sciences, Government of India, during 2014 for his significant contributions in the field of science; 'Illustrious Alumnus' award by the Haryana Agricultural University during 2016 and the '8th JSW-the Times of India Earth Care Award—2018'.

Veronique Bouchet works as executive director for Environment and Climate Change Canada in charge of operations. She is chair of the Global Atmospheric Watch Urban Meteorology Science Advisory Group.

Christian Braneon is an urban climate expert and scientist at NASA's Goddard Institute for Space Studies (GISS) in New York. His primary responsibilities there are focused on integrating stakeholder engagement, satellite imagery, and climate projections into urban and regional planning. Dr. Braneon supports the Urban

Climate Change Research Network's (UCCRN) *Urban Design Climate Workshop* series focused on helping cities utilize NASA datasets to enhance climate resiliency. He also serves as a principal investigator for the GISS Climate Change Research Initiative. Dr. Braneon has served as faculty at Barnard College at Columbia University, Georgia Institute of Technology (Georgia Tech), and Huston-Tillotson University. He served as co-director of the United States Environmental Protection Agency's (USEPA) inaugural Environmental Justice Academy for community leaders. Dr. Braneon led regional community engagement efforts associated with the Clean Power Plan in four states and was recognized for his service with a White House Climate Action Plan Award during his tenure at USEPA. He serves as a reviewer for the *Journal of Hydrology* as well as *Urban Climate*. Dr. Braneon earned his BS, MS and PhD degrees in civil engineering from Georgia Tech. He also earned a BS in applied physics from Morehouse College.

Gina Cavan is a senior lecturer at Manchester Metropolitan University and deputy director of the Ecology and Environment Research Centre. Gina's research addresses global challenges persistent in urban environments such as rapid urbanization, degraded ecosystems, climate change and social justice, through three interlinked themes: applied climatology, urban ecosystems and urban climate adaptation planning. Gina's recent funded projects have focussed on the role of green and blue spaces in creating healthy liveable cities. Her research emphasizes interdisciplinary approaches, usually incorporating application of geographical information systems (GIS), and seeks to promote impactful research that influences policy and practice.

Emily Ying Yang Chan MBBS (HKU), BS (Johns Hopkins), SM PIH (Harvard), MD (CUHK), DFM (HKCFP), FFPH, FHKAM (Community Medicine), FHKCCM, serves as professor, assistant dean, and head of Division of Global Health and Humanitarian Medicine, Faculty of Medicine, The Chinese University of Hong Kong (CUHK); director, Collaborating Centre for Oxford University and CUHK for Disaster and Medical Humanitarian Response (CCOUC), Centre for Global Health (CGH) and Centre of Excellence (ICoE-CCOUC), Integrated Research on Disaster Risk (IRDR); co-chairperson, WHO Thematic Platform for Health Emergency and Disaster Risk Management Research Network; member, Asia Science Technology and Academia Advisory Group (ASTAAG); visiting professor, Oxford University Nuffield Department of Medicine; and fellow, FXB Center, Harvard University. Her research interests include disaster and humanitarian medicine, climate change and health, global and planetary health, human health security and health emergency and disaster risk management (Health-EDRM), remote rural health, implementation and translational science, ethnic minority health, injury and violence epidemiology, and primary care. Awarded the 2007 Nobuo Maeda International Research Award of the American Public Health Association, Professor Chan has published more than 300 international peer-reviewed academic, technical and conference articles and 12 academic books. She also has extensive experience as a frontline emergency relief practitioner in the mid-1990s spanning 20 countries.

Chen Cheng is an engineer. He is mainly engaged in climate assessment of land space planning, meteorological disaster prevention and mitigation, and outdoor ventilation environment optimization. Used with meteorological numerical simulation, remote sensing and GIS technology in urban ecological planning, ventilation corridor design, sponge city rainfall analysis, etc., it effectively supports provincial and urban land spatial planning and urban climate environment optimization in China.

Jason Ching has been involved with meteorological and air quality research field studies and modelling these past 50 years with NOAA, the USEPA and currently at the UNC Institute for the Environment. He participated in major field programs including NOAA: BOMEX, IFYGL and GATE; EPA's Air Pollution Studies: RAPS, TPS, NEROS, and the Eulerian Model Evaluation Field Study (EMEFS), the latter for the National Acid Precipitation and Assessment Program, NAPAP. He participated in the development of EPA's Regional (1) Oxidant (ROM), (2) Acid Deposition (RADM) (and its meteorological driver MM5), and the (3) Particulate (RPM) modelling systems. He led the initial implementation of the Community Multiscale Air Quality Modeling (CMAQ) system which is now widely accepted and utilized in a wide variety of applications, including decision support systems for air quality issues at local, state federal and internationally. Currently, he is engaged in facilitating fine grid scale meteorological and air quality modelling issues for a variety of 'fit-for-purpose' urban scale applications. For this, he initiated the prototype National Urban Database and Access Portal Tools, NUDAPT and its current extension, the community-based, rapidly evolving and growing collaborative project called World Urban Database and Access Portal Tools (WUDAPT).

Jens Hesselbjerg Christensen is a professor in climate physics at the Niels Bohr Institute, University of Copenhagen. He has been leading climate science for more than 15 years at the Danish Meteorological Institute (DMI). During this period climate science at DMI established itself as the Danish focal point for climate change information, in particular with respect to future changes. He was coordinating the European framework project on regional climate change using multiple RCMs with a European focus called PRUDENCE. He has been the PI in several large EU Framework Programme consortiums, including ENSEMBLES, WATCH, IMPRESSIONS and EUCP. He was leading a large multi-disciplinary national (Danish) centre CRES. He has been a (coordinating) lead author at the 3rd, 4th and 5th Assessment Reports of the IPCC for WGI and also contributes to AR6. Together with three colleagues, he has received a large ERC Synergy grant, covering activities for 5 years.

Nathan Cooper is a researcher at the Australian Research Council Centre of Excellence for Climate Extremes at the University of New South Wales where he completed his PhD. Cooper has a background in air pollution and geography, and his research has primarily investigated vulnerable subpopulations' exposure to air pollution using GIS and spatial statistics. His research has extended to monitoring and modelling air pollution levels to better understand the threat of exposure in indoor and outdoor environments. His current research interests include investigating the health and economic impacts of air pollution and climate change, and the co-benefits of technological and policy-based responses.

Mike Davies is professor of building physics and environment at University College London (UCL). His research interests lie in the provision of healthy and comfortable built environments in the context of a changing climate. He was the founding director of UCL IEDE and is currently a member of the UK Committee on Climate Change (Adaptation) and director of the multi-disciplinary Complex Built Environment Systems group at UCL. For over a decade he has led an extensive programme of research, founded on intimate collaboration with researchers from health and other disciplines, which has sought to understand the complex relationship between the built environment and human well-being. Via this body of work, he has helped articulate the multiple underpinning connections and opportunities for healthy sustainable development.

Cécile de Munck is a researcher in urban climatology at the National Center for Meteorological Research (CNRM, Météo-France/CNRS). Her research interests, explored through numerical modelling, cover sustainable adaptation strategies to urban climate and climate change, with a particular focus on nature-based solutions and water resources stakes. She participated in several national and interdisciplinary projects tackling adaptation of urban environments, especially CLIM2 (air-conditioning impacts), VEGDUD (role of vegetation), MUSCADE (energy demand and comfort) and EUREQUA (environmental quality evaluation). She recently contributed to the National Observatory on the Effects of Climate Change (ONERC) annual report for the French Government to discuss the interests of vegetation in the resilience of urban ecosystems as part of nature-based solutions to adapt to climate change.

Wupeng Du is a doctor, deputy director and senior researcher of Beijing Municipal Climate Center. In the past 5 years, he has published more than ten related academic papers, edited and participated in the preparation of two monographs, and presided over seven scientific research projects, including the National Natural Science Foundation and the climate change program of China Meteorological Administration. Meanwhile, he compiled three state and industry standards and won two provincial and ministerial scientific awards. He is familiar with the research fields of climate and atmospheric environment characteristics in Beijing, as one of the main finishers, the project 'research and application of urban planning climate assessment and ventilation corridor design technology' won the Beijing Science and Technology Award in 2018.

Joachim Fallmann PhD, currently works at the University of Mainz, Germany, at the Institute of Atmospheric Physics. Next to teaching meteorology and geography courses his scientific expertise lies in the development and application of mesoscale atmospheric/chemical transport models in the field of urban climate and air quality. In the focus of his studies is the impact of urban canopy parameterizations on the dynamical and chemical structure of the urban boundary layer using the model systems WRF-Chem and COSMO. Further he investigates the role of urban heat island mitigation measures such as urban greening on regional air quality. After having obtained his PhD at the Karlsruhe Institute of Technology he has been working at the UK Met Office on the development of the fully coupled ocean-wave-atmosphere Environmental Prediction System before obtaining a PostDoc position at Uni Mainz. Joachim Fallmann was part of the EU Interreg Project UHI Central Europe, member of the UK Environmental Prediction consortium and joins activities in MED-Cordex model intercomparisons.

Xiaoyi Fang is a doctor and senior scientist. She is a former director of Beijing Climate Center and now director of Research Center of Practical Meteorology, China Academy of Meteorological Sciences. She is the leader of the urban climate assessment team and technical director of national meteorological industry land space planning and urban planning climate assessment. She is aiming at the solution of the ecological environment problems of the new urbanization in China from the meteorological point of view. The technology of urban environmental climate map, Urban Ventilation corridor planning and environmental performance evaluation of urban ecological construction are introduced into land space planning and urban planning. In the past 5 years, she has published two SCI papers and seven core journal papers as the first author/corresponding author. She has published 1 and participated in 6 monographs, participated in 2 and edited 1 national technical guidelines, and presided over the completion of 2 meteorological industry standards and 1 national standard. She has won 2 Beijing Science and Technology awards, 2 Huaxia Construction Science and Technology awards, 2 Beijing 38 Red Flag awards, advanced energy saving individuals, and 2 Meteorological Science and Technology Progress awards. She is also the member of 8 professional committees.

William B. Goggins III BA (Brandeis), SM, ScD (Harv), joined The Chinese University of Hong Kong as an instructor in biostatistics for the Centre for Epidemiology and Biostatistics of the JC School of Public Health and Primary Care in October 2003. He was promoted to assistant professor in September 2004 and is now associate professor. Professor Goggins obtained his doctorate in biostatistics from Harvard University in 1997. His current research focuses on the health impact of meteorological conditions in subtropical climates and the impact of urban design on health. In addition, Professor Goggins serves as a collaborator and statistical consultant for numerous projects within the JC School of Public Health and Primary Care and other departments of the CUHK Faculty of Medicine. Professor Goggins is a statistical advisor for the *Hong Kong Medical Journal* (from 2005 onwards), associate editor of the *International Journal of Biometeorology* (from 2014 onwards) and a member of the Grant Review Board of the Hong Kong SAR Government Food and Health Bureau's Health and Medical Research Fund (from 2013 onwards).

Brian Golding During a long career with the Met Office, Prof. Golding has made major contributions to high-resolution Numerical Weather Prediction and its use in a wide variety of applications, particularly flood prediction. He was instrumental in the early adoption of operational convection-permitting models in the Met Office during his tenure as the Met Office's director of Weather Science. He currently co-chairs the World Meteorological Organisation's High Impact Weather (HIWeather) project which has brought together physical, social, behavioural and economic scientists to enhance the effectiveness of weather warnings. He is a visiting professor at Bristol University and was awarded the OBE for services to severe weather forecasting.

Mahaveer Golechha Associate Professor, IIPH-Gandhinagar—has a master's degree in health policy planning and financing from The London School of Economics and Political science (LSE) and diploma degree from LSHTM, UK. He is recipient of Wellcome Trust-UKC Master's fellowship. He also has a master's and PhD from All India Institute of Medical Sciences, New Delhi, with first position. He is currently involved in teaching and research activities at IIPH-Gandhinagar. His areas of expertise are health system and policy research, environmental public health, climate change and health and pharmaceutical policy and economics. He led Rajasthan Climate Change project for development and implementation of India's first Climate Resilience Action Plan for rural settings. He is an expert of developing interventions for preventing and controlling climate sensitive diseases and heat sensitive diseases through capacity building, health system innovation, community awareness and system preparedness. He is also leading community vulnerability assessment for climate change and health. He has more than 25 publications in peer-reviewed national and international journals and also presented his work in conferences. From his education, professional work and experience, he has been recognized for his contributions with numerous accolades, namely the Outstanding Public Health Achievement Award by Govt. of Rajasthan, Exemplary Leadership Award, USA's Alzheimer's Young Scientist Award, and Govt. of India's INSPIRE Fellow Award, among others.

Donna Green is a founding member of the Climate Change Research Centre, UNSW Sydney, and an associate investigator of the Australian Research Council Centre of Excellence for Climate Extremes and the National Health Medical Research Council Centre for energy, air pollution and health Research. She recently co-founded the interdisciplinary Climate Health Partnership, a national network of climate and health researchers. She conducts research on air pollution, energy transitions and the climate impacts on public health. Her other research interests include human-environment interactions, specifically on social and economic vulnerability, adaptation and risk. She teaches domestic energy policy and co-authored *Screw Light Bulbs, smarter ways to save Australians time and money.*

Sue Grimmond is a New Zealand scientist and professor of urban meteorology at the University of Reading. She currently holds the post of Met Office Joint Chair.

Grimmond is a pioneer of the fields of urban meteorology and micrometeorology, which deal with the atmospheric boundary layer. Grimmond has contributed to wide range of topics within boundary layer and urban meteorology, publishing around 240 papers which have attracted approximately 19,000 citations, generating a h-index of 73. She has been particularly influential in understanding the effects of urbanization on climate and the representation of urban environments within climate and weather simulations.

Melissa Anne Hart is an associate professor in the Climate Change Researcher, UNSW Sydney, and graduate director in the Australian Research Council Centre of Excellence for Climate Extremes. Melissa has worked or studied in the fields or urban climate and air pollution meteorology on four continents. She has also spent time in industry, working in an air pollution consultancy. Melissa's research interests include: the impacts of land-use, surface characteristics and anthropogenic activities on urban climate; air pollution meteorology, including air quality impacts from fires; and more recently a citizen science project which has placed urban meteorological and air quality observations in Sydney schools and involves the school children in the analyses of data collected. In addition to research Melissa has developed a national, cross-institutional, climate science graduate program, which has reimagined the traditional Australian PhD, and has trained over 160 PhD students.

Jian He male, senior engineer, Master of Engineering from North China Electric Power University, has been working in Beijing Municipal Institute of City Planning and Design since 2008. He is engaged in municipal infrastructure planning, urban climate and environment research, etc. He has participated in the research of Urban Ventilation corridor and incorporate it into Master Plan of Beijing. He has participated in the relevant research and won the Beijing Science and Technology Progress Award twice and the National Excellent Urban and Rural Planning and Design Award more than ten times.

Bert Heusinkveld is a senior researcher of meteorology and air quality at Wageningen University. His research has a strong focus on Biometeorology. It is motivated by the interaction of weather and climate with the biosphere. Research topics include human thermal stress, air pollution, dew and leaf wetness, and urban climate. He has developed a special tricycle equipped with meteorological sensors to access human thermal stress levels.

Anahit Hovespyan is a Scientific Officer in Regional Climate Prediction Services division of Services Department at the World Meteorological Organization (WMO), since January 2014. Anahit coordinates and facilitates WMO activities on implementation of Climate Services Information System (CSIS) and enhancement of climate services at global, regional, and national scale in support of decision-making and managing climate risks. This is achieved through improved availability and access to authoritative climate information, methods, tools and systems, and user interaction through regional and national climate forums.

Anahit holds a master's degree in engineering and has more than 30 years of experience in the area of climatology. She is an author/coauthor of several peer-reviewed papers in international and national scientific journals, and WMO publications.

Zhe Huang BMan (Guangzhou Medical University), MPH (CUHK), received his Master of Public Health from The Chinese University of Hong Kong in 2013, majoring in epidemiology and biostatistics. Previously, he obtained a bachelor's degree in public utilities management in Guangzhou Medical University in China. He has participated in field-based assessments and interventions in rural areas in China and studied the association between weather and health in Hong Kong. Mr. Huang is now serving as a project officer at the Collaborating Centre for Oxford University and CUHK for Disaster and Medical Humanitarian Response (CCOUC) with particular focuses on climate change and data analysis. He is also a PhD candidate studying spatiotemporal association between weather condition and mortality in Hong Kong.

Marcel Ignatius is currently a research fellow at Department of Building, National University of Singapore. He has a doctoral and master's degree in building science, both awarded from the Department of Building, National University of Singapore. He has also obtained bachelor's degree in architecture from Tarumanagara University. His main research works deal with urban micro-climatic study and its relation with energy performance in urban areas. Under supervision of Prof. Wong Nyuk Hien, he has worked also in several research projects with Singapore government agencies. These projects mainly focused on urban heat island (UHI) mitigation, urban climatic mapping and sustainable design approach.

Xiji Jiang Ph.D. candidate, College of Architecture and Urban Planning, Tongji University. He is majoring in urban and rural planning, and his research focuses on the correlation and relevant mechanism between healthy outcomes and built environment, as well as the application of health impact assessment in urban planning.

Hunter Jones is the Climate and Health Project Manager of the Climate Program Office of the U.S. National Oceanic and Atmospheric Administration (NOAA) (UCAR affiliate). His responsibilities include development and day-to-day operations of the National Integrated Heat Health Information System—including running of local heat-health pilots to develop capacity for extreme heat planning and preparedness, as well as explorational development of new climate, weather, and health prototypes to support decision-making for heat health risk reduction. He is also leading NIHHIS in expanding a community science Urban Heat Island mapping program, which is developing new information and capacity specifically for mitigating heat in urban areas. In addition to these roles, Hunter is a co-founder and active member of the Global Heat Health Information Network (GHHIN), coordinates international activities for GHHIN and serves as a technical expert on heat and health in regional workshops. He serves on the Board on Environment and Health for the American Meteorological Society and is a member of the GHHIN Steering Committee. Hunter holds a master's in environmental management from Duke University as well as a bachelor's in computer information systems from Bentley University.

Hann-Ming Henry Juang is currently a research scientist with the National Oceanographic and Atmospheric Administration (NOAA), National Centers for Environmental Prediction, Environmental Modeling Center. He is renowned for the development of a hydrostatic sigma coordinate Regional Spectral Model, a nonhydrostatic Mesoscale Spectral Model, and a generalized vertical coordinate with full conservations for global spectral modelling. He has authored or co-authored articles in various scholarly journals, such as Monthly Weather Review, and Weather Forecasting. Dr. Juang is also an adjunct professor of National Central University and more recently he got a Marquis Who's Who Lifetime Achievement Award as a leader in the field of meteorology.

Paul Joe is currently a scientific consultant conducting research, writing/reviewing papers and proposals and advising on various projects. He retired from the Observation Based Research Group of Environment and Climate Change Canada in July 2018. His specialties were radar, nowcasting and observations. In support of operations, he helped write and evaluate the procurement of polarization radars for Canada that also resulted in three guides for the World Meteorological Organization. He also led several proposals for space-based precipitation radars, submitted to the Canadian and European Space Agencies, and supported the Global Precipitation Mission by chairing the Algorithm Review Panel for NASA. He was a member of the WMO World Weather Research Programme's Nowcasting Research Working Group (2002–2010) and chaired it from 2010 to 2016. He participated, led or advised on several Olympic nowcasting projects that include Sydney 2000, Beijing 2008, Vancouver 2010, Sochi 2014, Korea 2018 and Beijing 2022. He was also the science lead for the Pan American Games 2015 weather project which integrated weather, air quality and health warnings and contributed to the Global Atmospheric Watch program as a consultant for the Guides for Integrated Urban Services.

Raphaelle Kounkou-Arnaud is deputy head of the Datascience and Consulting division of the Meteorological Services Direction of Météo-France. Her expertise areas focus on meteorology and urban climate, the impacts of climate change, and adaptation. She participated to the EPICEA research project with the CNRM, CSTB and the City of Paris on the impacts of climate change on the Paris metropolitan area and the adaptation strategy of the city of Paris. She collaborates with the City of Paris and the Paris Climate Agency on climate change issues and is involved in particular in the evaluation component of the City of Paris OASIS Schoolyards project. She regularly conducts studies for industrial and local authority clients on various subjects related to meteorology and climate. Since March 2014, she has been secretary of the Environment-Health Commission of the Superior Council of Meteorology.

Rainer Kapp is the departmental manager of the Urban Climatology in the Office for Environmental Protection of the Municipality of Stuttgart. He is a process engineer with an emphasis in environmental engineering and he has been working more than 25 years in the field of urban climatology and air pollution control. He has developed statements and expertise regarding urban climate and air pollution in planning processes and handled with issues related to climate adaptation, also within the framework of urban development planning. He has been involved in several funded projects, e.g. UHI—Urban Heat Island 'Development and application of mitigation and adaptation strategies and measures for counteracting the global Urban Heat Islands phenomenon'; EU—Central Europe (3CE292P3). He is a member of the German Engineering Society (VDI)—committee 'Climate', also he is contributing to several working groups of the German Engineering Society (VDI) for developing guidelines concerning urban climate and urban planning. He is the (co-) author of some publications and books.

Humphrey Lean Humphrey manages the Mesoscale Modelling group at MetOffice@Reading. The overarching aim of the group is to undertake holistic development of km and sub km scale versions of the Unified Model in collaboration with colleagues in the University of Reading Department of Meteorology. Currently the three major areas of work in the group are (1) to improve the explicit representation of convection in versions of the Unified Model with grid lengths between 1.5 km and 100 m, (2) understanding predictability in convective scale models including development of high-resolution ensembles and (3) evaluation and improvement of km and sub km scale Unified Model versions in urban areas.The Mesoscale Modelling group works closely with the MetOffice@Reading DA@ Reading group which is aiming to utilize novel observations in high-resolution forecast systems. Humphrey's personal interests lie in the areas of orographic rainfall (particularly the seeder-feeder effect), mesoscale structures in cyclones, convection and urban meteorology. An important aspect of Humphrey's work is to facilitate collaboration with colleagues in academia in particular with those in the University of Reading Department of Meteorology where MetOffice@Reading is located.

Boram Lee is senior scientific officer of the World Climate Research Programme (WCRP) at the World Meteorological Organization (WMO). Her expertise comprises intergovernmental science policy, as well as the development and management of research & development in climate and disaster risk reduction. Boram leads the planning and coordination of the WCRP core projects, regional coordination, initiatives on urban climate, and monitoring and supporting the implementation of WCRP Grand Science Challenges on Weather and Climate Extremes and on Water for the Food Baskets of the World. Before joining WCRP, Boram was in charge of WMO's international coordination on marine meteorological affairs and managed the Coastal Inundation Forecasting Demonstration Project (CIFDP). Boram worked at the Intergovernmental Oceanographic Commission (IOC) of UNESCO as a programme specialist, at the National Institute for Agro-Environmental Sciences of Japan as visiting scientist, and at the Korea Meteorological Administration (KMA) as research scientist. She serves as adjunct professor in the College of Ocean Science, Jeju National University, Republic of Korea. Boram Lee holds PhD in climatology from the Université Pierre-et-Marie-Curie, France, and the master's and bachelor's degrees in atmospheric science from the Seoul National University in Korea.

Aude Lemonsu is a CNRS senior researcher. She has been working at the National Center for Meteorological Research (Météo-France/CNRS) since 2007, after having been a researcher in the meteorology research department of Environment and Climate Change Canada. Her activities focus on urban climate, interactions with climate change, and adaptation of cities. She studies the physical processes that influence environmental conditions in cities, and the representation of these processes in urban climate models. She is also interested in high-resolution coupled regional climate modelling for urban impact studies. She has participated in numerous interdisciplinary research projects on these topics (EPICEA, VURCA, MUSCADE, ACCLIMAT, MApUCE) and coordinated the French EUREQUA project on the study of environmental quality at the neighbourhood scale.

Xin Li is a doctor and deputy secretary general of the Beijing Municipal Committee of the Chinese People's Political Consultative Conference. Researcher, Institute of Atmospheric Physics, Chinese Academy of Sciences, mainly engaged in the research of atmospheric turbulence and atmospheric environment. In charge of air quality assurance and international cooperation for the 2008 Beijing Olympic Games, she is the main research and designer of air quality assurance measures for the 29th Beijing Summer Olympic Games. She has published more than 60 papers in Chinese and English in core academic journals at home and abroad and is the author of the book 'Theory and Practice of Cross Province Joint Control of Air Pollution'.

Xinhu Li is a research professor, College of Architecture and Urban Planning, Tongji University. He graduated from Institute of Geographical Sciences and Natural Resources Research, Chinese Academy of Sciences, with doctor degree majoring in Cartography and GIS. He has worked in the Institute of Urban Environment, CAS, and worked in University of Pennsylvania as a senior visiting scholar. His current research include urbanization and health, health city, planetary health, Big health, Data Science Method and techniques: spatial statistic, GIS and Remote Sensing.

Tian Kuay Lim is currently the Head (Thought Leadership and International Training), National Environment Agency (NEA). For over 25 years, he has been active in the interface between science and practice with meteorology and climate. As Deputy Director (Science & Systems), Meteorological Service Singapore (MSS), he managed R&D, including the establishment of the national meteorological and seismographic sensing networks, and as Head, Operational Meteorological Department, MSS, he managed the provision of Singapore's meteorological services to meet the needs of aviation, shipping, specialized services, military operation and the general public. As Advisor, Inter-agency Technical Group (Climate Change), comprising lead agencies of the Resilience Working Group, Inter-Ministerial Committee on Climate Change, he oversaw and managed the first and second Singapore's National Climate Change Vulnerability Studies. These multi-agency studies included the management of water resources, biodiversity, public health, coastal and urban infrastructures and planning among others. To strengthen Singapore's climate resilience, he led the development of a common geospatial platform for the management and sharing of data and information, multi-scale and multi-disciplinary modelling and simulation tools to support vulnerability and adaptation studies in Singapore. As Advisor, Urban Redevelopment Authority, he led the development of a multi-scale urban system modelling for sustainable urban planning and design.

Mathew Lipson is a researcher at the Australian Research Council Centre of Excellence for Climate System Science, UNSW Sydney. Dr. Lipson has previously worked to improve how urban areas are represented within global climate models, focussing on how human behaviour influences energy consumption and waste-heat release into the urban environment. His current work includes examining climate change impacts on energy use and human health within cities and evaluating the performance of urban climate models more generally. Mat holds a PhD in climate science and degrees in physics and architecture. He brings with him a passion for healthy urban environments.

Yonghong Liu is a senior researcher, mainly engaged in environmental remote sensing applications and urban climate research. In the past 10 years, he has mainly carried out related research on urban thermal environment and wind environment assessment, urban boundary layer parameter extraction, extraction of urban spatial morphological parameters and the relationship between urban spatial form and micro-climate, and published more than 20 papers.

Jing Lou is the principal scientist in IHPC. He leads the Fluid Dynamics Department and focuses on various fluid-related problems in research, new technology development and industry solutions. He was trained in physics and fluid dynamics. Since joining IHPC in 1999, he worked in various fields, such as multiphase flow and mass/heat transfer modelling, urban environment modelling, fluid-structure interaction, bubble/droplet dynamics, microfluidics, and nonlinear hydrodynamics. Before he joined A*STAR, he worked as a research fellow at Helmholtz-Zentrum Geesthacht in Germany and research scientist at NOAA/Great Lakes Environmental Research Laboratory, USA. He worked closely with various industry sectors leading the department's R&D activities and collaborations. He got a few awards, including Outstanding HPC Application Award, NSCC Singapore, 2017; SMI Best Research Project Award, 2016; Top Project, SMI Research Show Case 2016; HPC Quest Gold Award 2002. Lou Jing also serves as an adjunct associate professor at National University of Singapore. He also published over 150 journal and international conference papers.

Miguel Martin is a PhD candidate in the Department of Building of the National University of Singapore (Singapore). His main research focus is interaction between urban microclimate and building energy consumption. He obtained a master's degree in computer science from the University of Geneva (Switzerland). Then, he worked for 1 year as research assistant at the Energy Centre of the Ecole Polytechnique Federale de Lausanne (Switzerland). After that, he held a position of research engineer for 2 years at the Masdar Institute of Science and Technology (Abu Dhabi), where he was involved in collaborative projects with the Massachusetts Institute of Technology (USA). Before starting his PhD, he worked as research fellow for 1 year in the Department of Building at the National University of Singapore.

Valéry Masson is head of the urban climate research team at the National Center for Meteorological Research in Toulouse, France. He is an expert in modelling and developed the TEB urban canopy model that is used in many weather forecast models to represent cities. He also designed and participated in *in situ* experiments and measurement campaigns to improve the understanding of the urban heat island and energy balance of cities. For 10 years, he has studied adaptation of cities to climate change through transdisciplinary research projects with climatologists, sociologists, architects, environmental law researchers, geographers and urban planners.

Dileep Mavalankar is the director of Indian Institute of Public Health-Gandhinagar (IIPHG), which is a join initiative of Public Health Foundation of India and Gujarat State Government. Recently, IIPHG has been converted to a university of public health. He has done MBBS (Basic Medical degree) in 1981, MD (post-graduate medical degree) in Preventive and Social Medicine in 1987 from Gujarat University, Master of Public Health in 1982 and Doctor of Public Health Degree from Johns Hopkins University, USA, in 1990. He has also worked as Senior Management Advisor to the Averting Maternal Mortality and Morbidity (AMDD) program of Mailman School of Public Health, Columbia University, New York. His areas of interest include management of Maternal and Child Health programs, including immunization and health systems improvement and environment public health.He has been a member of several programme and scientific advisory committees including MotherCare project USA; NIHFW, New Delhi; IIPS, Mumbai; IIHMR, Jaipur; IMMPAC Project University of Aberdeen UK. He was appointed by the Prime Minister to the Missions Steering Group of the National Rural Health Mission constituted by the Government of India in 2005. He was also member of the steering group on Health for the 12th 5-year plan.

Anna Mavrogianni is associate professor in Sustainable Building and Urban Design and deputy programme director for the MSc in Health, Wellbeing and Sustainable Buildings at the Institute for Environmental Design and Engineering (IEDE), at the Bartlett, University College London (UCL). With a background in architecture and environmental design, she is an expert in indoor environmental quality and the climate change adaptation of buildings, with a focus on indoor overheating and air quality, urban heat vulnerability mapping and climate justice. Anna is currently the principal investigator of the EPSRC project 'Advancing School Performance: Indoor environmental quality, Resilience and Educational outcomes' (ASPIRE). She has been involved in numerous projects assessing the impacts of climate change and urban heat islands on human health, comfort and well-being and identifying pathways for the protection of vulnerable groups, in particular in homes, care homes and schools. To date, Anna has co-authored more than 100 peer-reviewed publications and has contributed to many policy reports.

Shiguang Miao is a senior scientist in the Institute of Urban Meteorology, China Meteorological Administration, Beijing, and has been engaged in urban climate research for more than 17 years and has more than 90 publications in peer-reviewed journals. His research interests are observation and numerical simulation of urban land surface, boundary-layer processes, and the atmospheric environment across the scales (e.g. neighbourhood scale, urban scale, and meso-scale), and the impact of urbanization on weather, especially on rainfall.

Gerald Mills is an associate professor in the School of Geography at UCD, Dublin. His primary area of interest and research is in the field of climatology, specifically the climates of urban areas. My early research work focused on energy exchanges within a city streets and blocks, mostly through numerical modelling and observations. Subsequently, this led to an interest in the transfer of urban climate knowledge into planning and design practice at different scales. The realization that as models became more sophisticated, improved urban data resources were needed led to the formation of the World Urban Database and Portal Tools (WUDAPT) project. WUDAPT is designed to gather data on cities worldwide that is suited to urban climate studies, including comparisons among cities, managing urban observations and running atmospheric modelling. Apart from research, he has been involved in the International Association for Urban Climates (IAUC), was president from 2009 to 2013 and organized the International Conference on Urban Climates in Dublin (ICUC8) in 2012. He is co-author of *Urban Climates*, the first modern text on the field in the English language.

Luisa T. Molina is affiliated with the Molina Center for Energy and the Environment (MCE2) and the Massachusetts Institute of Technology (MIT). Her research interests include atmospheric chemistry, urban air pollution and stratospheric ozone chemistry. More recently, she is involved in integrated assessment of air pollution in megacities and the emissions and impacts of short-lived climate pollutants on urban and regional climate and air quality.

Eleni Oikonomou is a research fellow in Energy and Building Modelling at UCL IEDE, with expertise in the modelling of indoor environmental quality, overheating and energy use in residential environments. She has been extensively involved in research on domestic stock modelling, including the development of representative archetypes for the domestic stock in England and London that have assisted large-scale modelling in the UK. Overall, she has been actively involved in more than 10 research projects funded by governmental organizations and research councils, including BEIS, ETI, GLA and NERC, focusing on climate resilience and the mitigation and adaptation potential of the domestic stock in the UK with a view to informing policy and prioritizing interventions. Through this work, she has co-authored over 25 peer-reviewed papers, book chapters and industry reports. Eleni is currently in the completion stage of her PhD examining the sociotechnical parameters affecting domestic heat pump performance in the field and the implications these may have on policy and technical standards.

Patrick Parrish is Chief of Training Activities Division in the Education and Training Office of the World Meteorological Organization, in Geneva, Switzerland. He has been an instructional designer and training manager for 25 years, with a primary focus on distance learning solutions. In 2008, he earned a PhD in Education with an emphasis on the design of engaging learning experiences and has published many journal articles and book chapters on related topics, which have received over 1000 citations. For 20 years, he worked with the COMET Program in the development of education and training resources in meteorology and other geosciences topics, primarily for professional continuing education, before coming to WMO in 2012. Patrick's work is driven by a fascination with the processes of learning and the challenges of training, particularly the challenge of engaging learners with intrinsically motivating learning experiences. His current efforts include promoting and guiding the use of eLearning solutions for education and professional development for weather, climate, and hydrologic forecasters and related professions.

Helen Pineo MRTPI is a lecturer in Sustainable and Healthy Built Environments and programme director for the MSc in Health, Wellbeing and Sustainable Buildings at UCL IEDE. Her research focuses on urban design, planning and governance in relation to urban health and sustainability. Prior to entering academe in 2018, she worked as an urban planner for over a decade integrating health and sustainability into new developments and planning policy, in the UK and internationally. She has worked at the Building Research Establishment, Local Government Association and in national and local government in the areas of sustainable urbanization, health, climate change and low carbon energy. Since 2015, she has been a Design Council Built Environment Expert. She holds numerous roles on government and industry steering groups.

Jeffrey Raven is the director and associate professor of the Graduate Program in Urban + Regional Design at New York Institute of Technology (NYC) and principal of RAVEN Architecture + Urban Design LLC. He is a specialist in sustainable and resilient urban design whose research is applied in professional practice and disseminated throughout the profession, government and allied disciplines to global urban design projects and knowledge transfer. He expands on traditional capabilities of architect-urbanists by bridging climate science, policy and design practice. He contributes to the development of sustainable-resilient guidelines-metrics, including *STAR Communities*, *Global EcoDistrict Protocol*, McKinsey's *Green*

Districts and Urban Land Institute *Technical Assistance Panel*. His publications include coordinating lead author of *Urban Planning & Urban Design - Climate Change and Cities* (Cambridge University Press 2018); *From Climate Science to Practice*, Urban Design & Climate Change (Urban Design Journal-London 2019), *Climate Resilient Urban Design*, Resilient Cities (Springer 2011); and *Shaping Resilient Cities in China, India and the United States* (P. Lang 2014). Recent funded research: National Science Foundation grants *EU-USA Sustainable Urbanisation Global Initiative-Belmont* and *City-As-Lab RCN*. He was elevated to the College of Fellows of the American Institute of Architects (FAIA); he is co-chair of the AIANY Planning & Urban Design Committee.

Xiyan Ren is a doctor, the general engineer and professional senior engineer of Eco & Infrastructure Institute in CAUPD. In the recent 10 years, he has finished almost 10 projects about the urban water system, storm urban flood, sponge city, and rainwater reuse. In the recent 5 years, he has also researched in urban rain flooding and water logging, Sponge City construct, such as the theory research of urban eco-space distribution and the storm runoff change (NSFC, 2014), the research of urban runoff and urban non-point pollution control (State Key Science Project, 2013). He has published more than 20 papers in Chinese and English in core academic journals at home and abroad.

Ulrich Reuter is a meteorologist. He studied meteorology at Cologne and Hamburg University (Germany). After some years as a scientific member of the Meteorological Institutes of the University of Cologne and University of Hamburg, a new work started at the Department for Urban Climatology of the Office for Environmental Protection, City of Stuttgart. From 2008 to 2017, he has been the head of this department. His field of interest is urban climate and urban planning. He has been involved in several international scientific projects on urban climate and urban planning. Since many years still up to now he has been lecturer at the University of Applied Sciences Stuttgart (Germany) in the topic of urban climatology and urban planning. For some time he also been a member of the German Engineering Society (VDI)-committees 'Climate' and 'Environmental Meteorology' and of several working groups of the German Engineering Society (VDI) for developing guidelines concerning urban climate and urban planning. He also is author of numerous publications and books.

Cynthia Rosenzweig is a senior research scientist at the NASA Goddard Institute for Space Studies and an adjunct senior research scientist at the Columbia University Earth Institute. She is also a professor in the Department of Environmental Science at Barnard College. At NASA GISS, she heads the Climate Impacts Group whose mission is to investigate the interactions of climate (both variability and change) on systems and sectors important to human well-being.Dr. Rosenzweig has organized and led large-scale interdisciplinary, national, and international studies of climate change impacts and adaptation in rural and urban settings. She is co-director of the Urban Climate Change Research Network (UCCRN) and co-editor of the UCCRN

Assessment Reports on Climate Change and Cities, the first-ever global, interdisciplinary, cross-regional, science-based assessment series to address climate risks, adaptation, mitigation, and policy mechanisms relevant to cities. She was co-chair of the New York City Panel on Climate Change, a body of experts convened by the Mayor advising the city on adaptation for its critical infrastructure, and co-led the Metropolitan East Coast Regional Assessment of the U.S. National Assessment of the Potential Consequences of Climate Variability and Change, sponsored by the U.S. Global Change Research Program.

K. Heinke Schluenzen is head of the mesoscale and microscale modelling group at the Meteorological Institute at Universität Hamburg. She is coordinating the urban system research at the Centre for Earth System Research and Sustainability (CEN) and member of the Scientific Steering Committee of CLICCS and head of SICSS. She is coordinating the development of air quality guidelines in Germany and member of the GAW Urban Research Meteorology and Environment (GURME) project. Her research focusses on local regional scale atmospheric processes resulting from direct human impacts, with emphasis on direct anthropogenic effects (e.g. growing urban areas, regenerative energy production). H. Schluenzen is a meteorologist by training, with a strong emphasis on model development and evaluation.

Claire L. Smith research interests are primarily associated with the interaction between atmospheric change and the local environment. The two main themes of Dr. Smith's work encompass adaptation to climate change and the use of speleothems (cave deposits such as stalagmites) for paleoclimate reconstruction. Adaptation to climate change research uses monitoring and modelling methods to evaluate climate change impacts. The synthesis of meteorological, technological, socio-economic, energy and emissions data within a spatial framework enables a GIS-based analysis to be used to interrogate adaptation strategies. This work is primarily focussed on the urban environment, sustainable buildings and the role of green and blue spaces within cityscapes.

Ranjeet Sokhi is director of the Centre for Atmospheric and Climate Physics Research (CACP). He has nearly 200 publications in major international journals, proceedings, books and reports. His research focusses on physical and chemical processes affecting air quality, interactions with climate and resulting exposure and health impacts on local to regional scales. He is an advisor for DEFRA, WHO, WMO and PHE and was a member of the UK Government's Air Quality Modelling Steering Group and the Prediction Capability subgroup of AQSELG. He has participated in or coordinated a large number of projects including NERC Clearflo, MESONET/MESOMAQ, AIRTRACK; OSCAR, CAIR4HEALTH, AIR4EU (FP6); MEGAPOLI, TRANSPHORM (FP7) and CREMO (Environment Agency). As a PI for NCAS, he coordinates the UK Air Quality Forecasting system (https://sci.ncas.ac.uk/airquality/). Currently, he is the principal investigator for MOES/NERC funded PROMOTE project (with Beig from IITM) on improving the prediction of air quality over Delhi, India. He is a partner in a new H2020 project, Evaluation, control and mitigation of the environmental impacts of shipping emissions (EMERGE). Currently, he chairs the WMO/GURME Science Advisory Group and coordinates the WMO/GAW PREFIA project on Air Quality Prediction for Africa. He was the coordinator of COST 728 on Enhancing Mesoscale Modelling for Air Pollution and Dispersion Applications. He participates in the Air Quality Modelling Evaluation International Initiative (AQMEII). He has been the chair of the series of International Conferences on Air Quality—Science and Application since 1996 and co-organizer of a number of other events (e.g. for EMS and WMO).

Reinhard Spengler is a graduate meteorologist, employed by the German Weather Service (DWD) since 1990. During his work in the climatology department, he processed and calculated meteorological time series. He was involved in the merging of various data archives of the DWD and the former meteorological service of the GDR. He has many years of experience in the field of meteorological and hydrological measurement networks as well as in the integration and use of external data sources (third-party data). One focus of his tasks at the DWD was the development and introduction of quality control procedures for meteorological data and products. He is currently the head of agro-meteorological consulting in the Climate and Environment division of the DWD.

Phil Symonds joined UCL IEDE as a research associate in 2014. He works on the Wellcome Trust funded project, Complex Urban Systems for Sustainability and Health, which aims to enable cities to identify healthy decarbonization pathways. Phil's research interests include computational modelling and the analysis of built environment data through the use of statistical techniques. Phil is the lead developer of a microsimulation model in collaboration with researchers at the London School of Hygiene and Tropical Medicine, which is used to assess the impacts of air pollution policy on cardiovascular morbidity. He has also developed a metamodelling framework using artificial neural networks based on EnergyPlus simulations, which has been used to look at how energy efficiency interventions in homes modify population exposure to overheating and air pollution. Prior to joining UCL IEDE, Phil completed a PhD in experimental particle physics at Brunel University (2010–2014). During his PhD, he was based at the Center for European Nuclear Research (CERN) for 18 months as a member of the Compact Muon Solenoid (CMS) experiment.

Jianguo Tan PhD, is the Director and Professor of Shanghai Meteorological IT Support Center, Shanghai Meteorological Service (SMS), which is also the East China Region Climate Center of Chinese Meteorological Administration (CMA). He is also co-director of CMA Key Laboratory of Cities' Mitigation and Adaptation to Climate Change in Shanghai (CMACC), jointly built by Shanghai-based Tongji University and Shanghai Meteorological Service. Previously, he was the director of Shanghai Institute of Meteorological Science, Shanghai Climate Center, and director of Shanghai Municipal Key Laboratory of Meteorology and Health. His research interests include urban climatology, urban micrometeorology, boundary layer meteorology, urban environment and urban climate change and human health.

He has once been a main research staff of the WMO/WHO showcase project on Shanghai Heatwave watch/warning system, GURME Pilot Shanghai Project and WMO showcase project of Shanghai Multi-hazard Warning System. Now he is co-chair on WMO-Shanghai Integrated Urban Weather and Climate Service Demonstration Project.

He got award of National Chief Leading Scientist of Meteorological Science from CMA. He is a visiting Professor of Tongji University, East China Normal University (Shanghai based) and Shanghai Normal University and is International Scientific Committee member of WMO GURME Scientific Advisory Group.

Nigel Tapper holds a personal chair in Environmental Science at Monash University, Australia (as a climate science specialist working on surface-atmosphere interactions), where he currently leads the Applied Climate Research Group within the School of Earth, Atmosphere and Environment. Nigel is a key researcher in the Cooperative Research Centre for Water Sensitive Cities where until recently he led Program B (Water Sensitive Urbanism), developing innovative microclimatic approaches to cool cities under extreme heat conditions. Outside the University Nigel contributes strongly to the work of the Intergovernmental Panel on Climate Change, where he is a current lead author of Working Group II, Impacts, Mitigation and Adaptation. He serves in the World Meteorological Organisation as co-chair of the Terrestrial Observation Panel on Climate and associated task forces. He is president of the International Association of Urban Climate that represents the work of over 1000 climate scientists. Nigel has published 7 books, 15 book chapters and more than 200 refereed research publications in a 35-year career. Nigel has a global reputation for his work in areas of weather and climate impacts and atmospheric boundary-layer processes and has a particularly strong track record in delivering industry-relevant research.

Jonathon Taylor is a senior research associate and lecturer in urban physics and health, at UCL IEDE. His research interests include combining spatial data, building simulation tools, and health models to understand how changes to urban environments and the building stock can reduce energy consumption, while improving population health and well-being. In particular, he is interested in using building physics models to predict indoor exposures to environmental hazards, including indoor air pollution, extreme temperatures, damp and mould growth, and indoor disease transmission in both developed and developing contexts. He has a BSc (Biological Sciences) from Simon Fraser University, Canada, and a BEng (Geomatics) and PhD (Building Science) from University College London.

Giacomo Terrugi With a background in environmental engineering and hydrology, Giacomo Teruggi has been working since 2004 for the World Meteorological Organization (WMO), where he is head of the Associated Programme on Flood Management (APFM) Technical Support Unit. The TSU manages the HelpDesk on Integrated Flood Management (IFM), providing capacity building and implementing pilot projects on IFM. Having understood the multi-disciplinarily approach of

IFM, which involves not only hydrological sciences but also land use planning, restoration of riverine ecosystems and socio-economic considerations, Giacomo has been pushing for an increased involvement of environmental and social scientist to increase outreach of IFM concepts. At the same time, he has been coordinating and supporting Expert Groups of the WMO Commission for Hydrology on assessment methodologies of national capabilities and needs in end-to-end early warning systems for floods.

Robert Lee Kong Tiong is an associate professor in School of Civil & Environmental Engineering, Nanyang Technological University, Singapore. He has been deputy director of Centre for Infrastructure Systems (2011–2018), former deputy director of Robotics Research Centre, CoE, NTU, and former deputy director, ICRM at NTU. He has been the program director of the MSc Programme in International Construction Management since its inception in 1991 till 2009, and then 2017 till now. Dr. Tiong graduated in 1981 from University of Glasgow with First Class Honours in Civil Engineering, with specialization in management. He obtained his M.Eng on construction engineering and management from University of California, Berkeley in 1987 and his PhD in civil engineering in 1994 from NTU. His research and consulting interests focus on integrated risk analysis and management, development of risk assessment and management frameworks, quantification of risks and project management for infrastructure projects. He has worked as PI (NTU) on the development of Quantitative Urban Environment Simulation Tool (QUEST) and provided advisory on geospatial and system integration work.

Konstantinos Tzoulas is a senior lecturer in environmental management at Manchester Metropolitan University. His research interests include urban ecology, sustainable development, green infrastructure planning and public health. Konstantinos' current research evaluates the social and economic benefits of peri-urban sphagnum farming. Recently, he worked on evaluating the well-being benefits of urban parks for people with early-stage dementia. He has also worked on co-developing a public action plan for enhancing the ecological and well-being benefits of private gardens. In the past, Konstantinos worked with local authorities on urban regeneration policy, and he currently leads the biodiversity policy for Manchester Metropolitan University.

Vincent Viguié is a researcher and head of the sustainable urban form team at CIRED research centre (École des Ponts ParisTech, CNRS, EHESS, AgroParisTech, CIRAD), which he joined in 2008 after working at the World Bank. His research focuses on building models of sustainable urban form and transport, on economic issues around adaptation policies and on prospective models for cities expansion. He is the author of more than 20 publications in major international scientific journals on these topics and has coordinated CIRED's participation in several research projects on sustainable urban development, notably the VURCA and MUSCADE projects with Météo-France, the DRAGON project with the Chinese Academy of

Sciences and the World Bank's TPSIG project. He also regularly takes part in expert studies on these subjects, both with public and private institutions.

James Voogt is an urban climatologist who specializes in the measurement and modelling of urban surface temperatures. He has contributed to research projects on understanding the three-dimensional surface temperature of cities, the climate performance of green roofs and spatial variations in the heat impacts on urban residents. He was a co-PI of the Environmental Prediction for Canadian Cities Network, a project to develop and evaluate an urban-atmosphere model for the Canadian weather forecasting system. Dr. Voogt served as president from 2014 to 2018 of the International Association for Urban Climate. He is a co-author of the recent text 'Urban Climates' published by Cambridge University Press and is a member of the WMO Urban Expert Team.

Lan Wang is Head of the Healthy City Lab and professor of Department of Urban Planning, College of Architecture and Urban Planning, Tongji University. She holds a bachelor's and a master's degree of urban planning from Tongji University and a Ph.D. in urban planning and policy from University of Illinois at Chicago. Her research interests centre around spatial planning and governance of healthy cities, high-speed rail new towns and global cities. Prof. Wang serves as deputy director and secretary general of China Healthy City Committee and executive director of the Asian Development Bank-Tongji Urban Knowledge Hub.

Chi Shing Wong BSSc (CUHK), MSc (LSE), has a background in Journalism and Communication. He received his Master in Comparative Politics at the London School of Economics and Political Science, focusing on democratization, ethnic politics and nationalism with regional foci in China and Southeast Asia, and researched on political identity at the University of Oxford afterwards. Having worked in various teaching, research and administrative positions in universities, he is now the assistant director of the Collaborating Centre for Oxford University and CUHK for Disaster and Medical Humanitarian Response (CCOUC).

Nyuk Hien Wong is currently the vice dean for research and professor at the School of Design and Environment, National University of Singapore. He is one of the most prominent and leading academia within the urban and built environment industry. His key areas of expertise include urban heat islands, urban greenery and thermal comfort in the tropics. He has been extensively involved in a number of research projects related to urban heat islands, urban climatic mapping and greenery as the lead and principal investigator with various government agencies such as the Urban Redevelopment Authority (URA), the National Parks Board (NParks), the Housing Development Board (HDB), the National Environmental Agency (NEA) and the Building and Construction Authority (BCA). With his vast experience, he has served in various advisory committees both locally and globally. He currently chairs the Singapore Building and Construction Authority's (BCA) Green Mark subcommittee on ventilation simulation. Prof. Wong has written 15 books/book chapters and published more than 500 refereed journal and conference papers in these related fields.

Yajun Xiong is a senior researcher, graduated from Institute of Environment and Engineering in Cold and Dry Areas of Chinese Academy of Sciences, mainly engaged in atmospheric physics and atmospheric environment research and has published 14 papers, undertaken more than 10 scientific research projects, and won the title of excellent forecaster of China Meteorological Administration in 2017.

Junichi Yoshitani is Professor of Water Environment and Civil Engineering at Shinshu University, Nagano, Japan. He received a M.S. from the University of California at Davis, USA, and a PhD from Hokkaido University, Sapporo, Japan. He worked for the Japanese Ministry of Land, Infrastructure and Transport as a research engineer for 34 years before he joined Shinshu University. One of his works was the establishment of the International Centre for Water Hazard and Risk Management under the auspices of UNESCO. He has been a member of Open Panel of Commission for Hydrology Experts (OPACHE) of the World Meteorological Organization.

Shuo Zhang is an engineer. She has been involved in research on remote sensing and land surface ecological environment monitoring, achieving important methods such as retrieval of land surface temperature, heat island dynamic change monitoring and forecasting based on multi-source remote sensing data, as well as extraction of Urban Environment Index and Analysis of Characteristics of Time and Space Changes. She has participated in many practical projects, made contributions to promoting calculation of ventilation environment and ventilation potential assessment, establishing refined urban environmental climate map, and supporting the suggestions for optimization of ventilation corridors. The results were published in academic papers such as Ecology and Environmental Sciences, Acta Ecologica Sinica, Theoretocal and Applied Climatology.

Nici Zimmermann is an associate professor in System Dynamics at UCL IEDE. She is an expert in system dynamics modelling and participatory collaboration who seeks counterintuitive and innovative solutions for important problems. Her work advances (1) a systems perspective in the built environment and in cities, (2) sustainability and well-being and (3) a combination of organizational and built environment research and transdisciplinary collaboration. It is interdisciplinary and international. She engages widely in her work, e.g. with national and regional government departments on different continents, private and public organizations and she has a track record of sustained contribution to policy. In addition, she is a recipient of the Gert von Kortzfleisch Prize and a repeat strategy thread chair at the International Conference of the System Dynamics Society.

Part I
The Rise of Mega-cities and the Concept of Climate Resilience and Healthy Living

Chapter 1
Urban Climate Science for Planning Healthy Cities

Glenn McGregor and Chao Ren

Abstract The rise of mega- and high-density compact cities are now irreversible trends of human urban development. While megacities and high-density urban living may make lives more convenient and society more economically efficient, they pose a range of environmental challenges, especially as increasingly dense, complex and interdependent urban systems leave cities vulnerable and sensitive to climate variability and change. Climate threats come mainly in the form of poor air quality, windstorms, heat waves, drought and floods. Flowing from these are a range of possible health impacts including death, physical injury, heat-related illness, vector- and water-borne disease and mental illness. The visibility of how climate affects cities and their infrastructure and inhabitants has generated wide-ranging concern from both the general public, local and national governments and international agencies. This 'climate of concern' has very much led to the emergence of the call for climate resilient cities especially as minds become focused on how climate change may have fundamental influences on how cities might have to plan for the future. Given this, a pressing imperative is to generate information about the causes of urban-based climate hazards and disasters and to acquire spatial and quantitative understanding of exposure and vulnerability of people and infrastructure in cities. To achieve this, a cross-disciplinary collaboration among scientists, planners, government, non-governmental organisations, the private sector and also the general public is required. However, the application of urban climate science in the realm of planning for healthy living in cities has had low uptake in the policy arena. This apparent intransigence in the city planning and policy community is most likely related more to the lack of data and hence information related to a number of unanswered urban climate questions, as opposed to the resistance to mainstreaming climate information into devising policy. Some of these questions include how to assess current and future climatic extremes and relevant health risks, how to make

G. McGregor (✉)
Department of Geography, Durham University, Durham, UK
e-mail: glenn.mcgregor@durham.ac.uk

C. Ren
Faculty of Architecture, The University of Hong Kong, Pok Fu Lam, Hong Kong
e-mail: renchao@hku.hk

C. Ren, G. McGregor (eds.), *Urban Climate Science for Planning Healthy Cities*, Biometeorology 5, https://doi.org/10.1007/978-3-030-87598-5_1

decisions to enhance climate resilience to climate extremes and disasters, the modes by which information should be conveyed to city leaders and stakeholders and how and what mitigation and adaptation actions are needed to cope with climate change in cities. These questions provide the context for this book entitled '*Urban Climate Science for Planning Healthy Cities*'.

Keywords Megacities · Healthy city · Urban climate science

1.1 Living in an Era of Megacities

Since 2006, more than half of the world's population have become urban dwellers (UNFPA 2007). Furthermore, by 2030 nearly 60% of humanity will live in cities and almost 9% of the world's population will be living in just 41 megacities (UN 2008). The rise of mega- and high-density compact cities are now irreversible trends of human urban development (UNFPA 2009).

While megacities and high-density urban living may make lives more convenient and society more economically efficient (Ng 2009, 2012), they pose a range of environmental challenges, especially as increasingly dense, complex and interdependent urban systems leave cities vulnerable (WMO 2012, 2016) and sensitive to climate variability and change (Grimmond et al. 2010). Climate threats come mainly in the form of poor air quality (WMO 2012), windstorms, heat waves, drought and floods. Flowing from these are a range of possible health and well-being impacts including death, physical injury, heat-related illness, vector, water- and food-borne disease and mental illness (WMO 2003, 2016; WHO, WMO, UNEP 2003). The visibility of how climate affects cities and their infrastructure and inhabitants has generated wide-ranging concern from both the general public, local and national governments and international agencies. This 'climate of concern' has very much led to the emergence of the call for climate resilient cities especially as minds become focused on how climate change may have fundamental influences on how cities might have to plan for the future.

An example of how the international community has responded to the impending challenges posed by the rise of megacities, in a world characterised by a set of changing environmental and other circumstances from the local to global, is manifest in UN Habitat III's New Urban Agenda. This outlines a set of global standards and calls for cooperation in sustainable urban development for all cities and for present and future generations (UN 2016a, b). It also highlights the vision of healthy, resilient and sustainable cities and human settlements to foster prosperity and quality of life for all (UN 2016a). Resonating with this is both the WHO and WMO's focus on building and promoting safe, healthy and resilient cities through political actions and engagement with public health endeavours (WHO 2016; WMO 2016). Although worthy sentiments, the fact remains, as noted in the most recent report from the Intergovernmental Panel on Climate Change, there is a dearth of scientific and other information on urban areas, particularly in relation to climate as a driver of urban change and response (IPCC 2014). Worryingly this knowledge deficit will constrain the formulation of effective policies for planning climate resilient cities and the achievement of sustainable urban development and management of healthy mega- and high-density cities. Given this, a pressing imperative is to generate information about the causes of urban-based climate hazards and disasters and to acquire spatial and quantitative understanding of exposure and vulnerability of people and infrastructure in cities.

To achieve this, a cross-disciplinary collaboration among scientists, planners, government, non-governmental organisations, the private sector and the general public is required (Parker and Penning-Rowsell 2005; Mills et al. 2010; WHO 2016). However, the application of urban climate science in the realm of planning for healthy living in cities has had low uptake in the policy arena (de Schiller and Evans 1991; Wolf et al. 2015). As opposed to any conscious resistance by decision makers to mainstream climate information into planning policy, a combination of factors may account for this situation. These may include:

1. The general absence of climate knowledge within urban planning communities and associated knowledge brokers with knowledge translation skills
2. The challenge of matching climate data and information with a particular planning problem
3. Structural barriers and institutional cultures related to the interface of science and planning
4. The 'usability gap', or hiatus between information production and application and the related 'loading dock approach'
5. The formulation of climate services (Cash et al. 2006; Lemos et al. 2012; Bierbaum et al. 2013; Briley et al. 2015; Guentchev et al. 2016; Bremer et al. 2019; Daniels et al. 2020; Fragomeni et al. 2020; Skelton 2020)

Such issues and the rapidly evolving field of urban climate science and its potential contribution to the planning of healthy cities provides the broad context for this book.

1.2 Climate, Health and Cities

Historically cities have been associated with blighted biophysical environments as depicted in many nineteenth-century images of a rapidly industrialising western world and a burgeoning urban population. London in the mid-1800s exemplifies this with the combination of heavily polluted air and water, the oppressive summer heat of the city, the proliferation of vermin and epidemics of cholera, pneumonia and tuberculosis all conspiring to create an urban health nightmare made worse by high population density particularly in the poorest neighbourhoods. Fast forward to the twenty-first century and the emergence of megacities, especially in low- to middle-income countries, and some of the problems of nineteenth-century Victorian London seem familiar. Indeed, the synergistic effects of poor housing, unhealthy and sedentary lifestyles, a lack of sanitation, the urban heat island, poor air quality and infectious disease may create a health-related 'urban penalty', a state some distance from the imagined 'urban advantage' that might arise from the benefits of city living and access to health care and cognate services. Setting aside possible health-related contradictions of city living, history tells us that without active policy interventions and associated actions, the biophysical environments of cities rapidly deteriorate, exacting a heavy toll on the health and general well-being of urban dwellers.

That the environment is important for health is not a new line. In 400 BCE Hippocrates, in his treatise '*On Airs, Waters, and Places*', outlined the health implications of the seasons of the year, hot and cold winds, the qualities of the waters and the unique setting in the landscape of a city. This 'ancient' nexus of air, water and place, while embodying urban conditions, in many ways laid the foundations for the modern and burgeoning discipline of environmental health and the concept of Healthy Cities. Broadly defined as activity related to investigating the impacts of the natural and built environment on health, in practice the discipline of environmental health draws on an 'environmental determinants of health framework'. In doing so it assesses the contributions to preventable disease of exposure to pollution and chemicals (e.g. air, water, soil, products), physical exposures (e.g. noise, radiation), the built environment, other anthropogenic changes (e.g. climate change, vector breeding sites) and related behaviours and the work environment (Pruss-Ustun et al. 2017). The criticality of both natural and built environments for health is recognised in the Healthy Cities concept, which among other factors, emphasises the importance of the creation and improvement of city physical environments to ensure city dwellers achieve their health potential (Asvall 1998).

One aspect of the physical environment often overlooked in relation to the drivers of urban health outcomes is climate. This statement may seem at odds with the extensive literature on climate and health much of which draws on climate and health data for cities. However, climate and health studies are often conducted at the city population level and draw on climate data, usually temperature, from nearby airport climate stations that are assumed to be 'representative' of the city. Regrettably, such studies treat cities as homogenous climate entities with little regard given to the well-known heterogeneity of city climates. That urban form and materials is understood to construct climate at the intra-urban to urban scale, manifests in the vast literature on urban climatology (Eliasson 2000; Mills et al. 2010; Lenzholzer and Brown 2016; Oke et al. 2017). Much of this literature calls for the application of urban climate science in the planning and design of cities to achieve ideal city climate environments with associated health benefits, akin to the idea of a 'Metutopia', a climate imagining of a healthy city (Landsberg 1973).

1.3 Urban Climate Science

Traditionally referred to as urban climatology, contemporary urban climate science encapsulates the fields of urban climatology and urban meteorology and enjoys close links with the discipline of atmospheric chemistry. It is concerned with understanding the climate of urban environments in terms of not only their average state but also their temporal and spatial variability in the horizontal and vertical dimensions. At the heart of urban climate science is the description and quantification of the impact of urban growth on atmospheric processes and the consequent development of distinct urban climates (Oke et al. 2017). This necessarily involves the measurement of multi-scalar climate processes and outcomes ranging over the intra-urban to urban scales (micro- to local- to meso-scale) and from the fast to slow (high to low frequency variations). Of chief interest to the urban climate scientist is how changes in the radiative, thermal, moisture and aerodynamic properties of the so-called 'urban surface' affect a range of urban climate phenomena, with the urban heat island receiving most attention in the literature. Understanding and explaining urban climates also requires acknowledgement that these are not just the outcome of atmospheric processes. Rather it is the interaction and feedbacks between the components of the urban climate system, including the urban atmosphere, biosphere, hydrosphere, land surface/subsurface, built infrastructure and chemical emissions that influence urban climate outcomes (Oke et al. 2017).

As for the advent of climate science in general (Coen 2020), the history of urban climate science, particularly urban climatology, predates the era of computerised climate models. Its origins stretch back to the early nineteenth century as manifest by the publication of Luke Howard's two volumes entitled *The Climate of London* in 1818 and 1820 (Mills 2008). Facilitated by the rapid expansion of meteorological observation networks throughout the latter half of the nineteenth century, literature considering the thermal climates of other major cities such as Paris, Munich and Berlin began to emerge with similar studies following in the early twentieth century for cities in Japan and North America (Oke et al. 2017). Following this nascent period, characterised by climate and weather descriptions of a number of cities, three periods of development of the field of urban climatology ensued. These include:

1. The period from 1930 to 1965 characterised by new field techniques enabling micro- and local climatological analyses
2. Closer collaboration between the climatological and meteorological research communities and the materialisation of physically based models of the urban atmosphere from 1965 to 2000
3. The post-2000 period and the development of sophisticated urban climate measurement campaigns and networks, the refinement of urban climate numerical modelling procedures and the maturation of urban climatology and meteorology as a predictive science (Oke et al. 2017)

Spurred on by a robust evidence base that a range of urban climate hazards associated with the urban heat island, urban-related alterations to precipitation, moisture and wind fields and a deterioration of urban air quality, precipitate significant societal impacts, it is in this most recent period the call for an 'urban climate information-urban planning imperative' has emerged and gathered strong momentum. Moreover, the urgency of mainstreaming urban climate science into urban planning has become even more pertinent given the prospect that anthropogenic climate change is likely to intensify some urban climate processes and hence exacerbate climate-related health impacts in cities and prejudice the achievement of healthy cities.

1.4 Conceptual Framework

Building on the framing of health as the outcome of multiple interacting determinants and that climate constitutes an essential element of healthy cities, two premises underlie the material presented in this book. Firstly, the process of urbanisation, whether it be planned or unplanned, inadvertently influences climate processes (the interactive exchange of energy, mass and momentum between the surface and the atmosphere) at the intra-urban to urban scale. Secondly, as one of a range of environmental determinants of health, climate plays a significant role, directly and indirectly, in influencing individual- to population-level health. A corollary of such premises is, through understanding urban climate processes and hazards and applying associated urban climate science knowledge in the planning of cities the management of urban climate-related health risk may be possible.

The conceptual framework underpinning the material presented in this volume is set out in Figs. 1.1 and 1.2. Because climate information to assist in the planning of healthy cities is the principal emphasis here, the framework necessarily focuses on climate as one aspect of the physical environment of cities and indeed one of a range of environmental determinants of health. An important postulate of the book is a cascade of processes from the global to urban level bear implications for the health and well-being of urban dwellers and thus influences the attainment of climate-related healthy cities (Fig. 1.1).

In the context of policy, actions and trends at multiple scales are important in terms of how climate, as one aspect of the physical environment, might influence health at the population/individual scale. For example, global scale changes in climate may determine the nature of global- and national-level climate-related policies with such policies interacting with a range of moderating factors at the city level. These in turn affect what we refer to as the lived climate experience and hence shape the character of population- and individual-level climate-sensitive health and well-being outcomes (Fig. 1.1). In this respect, global actions, as outlined in the Paris Agreement with a goal to restrict global temperature increases to within 1.5 °C of pre-industrial levels by 2050, through reductions in greenhouse gas emissions, the Sustainable Development Goals of Agenda 2030 and the Sendai Framework for Disaster Risk Reduction 2015–2030, are important.

The conceptual framework draws attention to the link between health/well-being and the lived urban climate experience, which we suggest is the outcome of how a range of city-level climate moderating factors mediate global and hence regional climate processes (Fig. 1.2). Of course, we acknowledge that there are likely to be feedbacks between the different scales represented in Fig. 1.2 such that city-level moderating factors via land–atmosphere interactions may feedback on regional climate and possibly global climate. Moderating factors may include land cover type and distribution, urban form (city size, density, diversity, complexity and morphology) and urban function (nature of the economy), type of physical infrastructure (e.g. transport system), the nature of building stock including housing (age, style, building materials), and policies related to climate-sensitive urban planning and

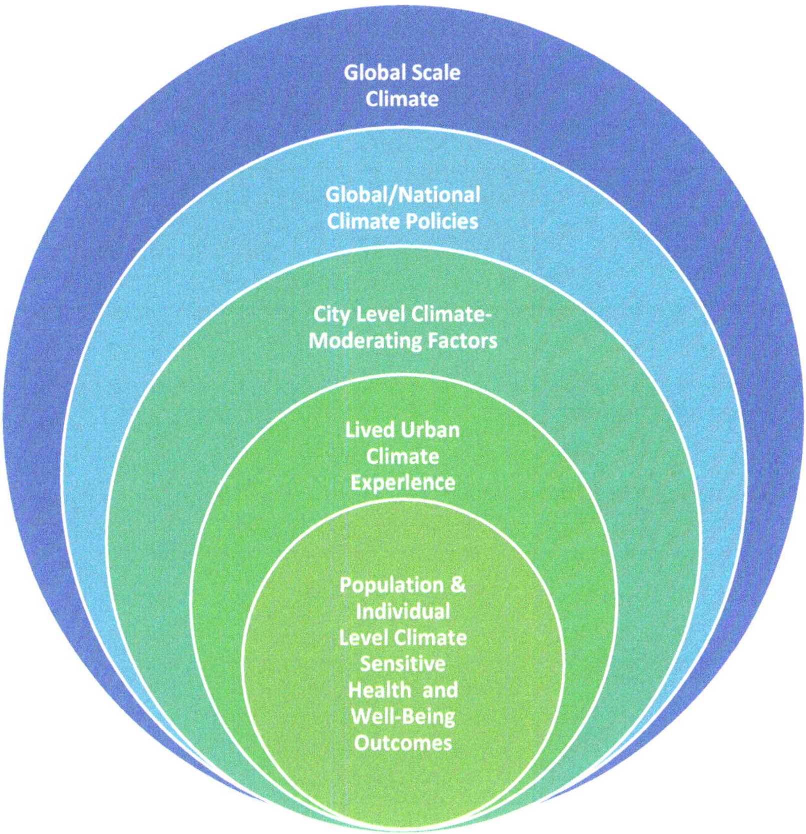

Fig. 1.1 Conceptual framework demonstrating how climate-sensitive health and well-being outcomes link to processes at a number of nested larger scales

design and climate adaptation in general. As outlined in Fig. 1.2, the lived urban climate experience manifests in a number of ways. These include indoor and outdoor climate and air quality, the incidence of flooding and noise and the prevalence of pests and mould in dwellings, with the aforementioned climate moderating factors working to either exacerbate or dampen these climate conditions, all of which bear implications for a range of climate-sensitive health outcomes.

Conceptualised in this way, our contention is by understanding intra-urban to urban scale climate processes and the impact of urbanisation on these, and knowing the drivers of climate-sensitive health outcomes, it will be possible to manage climate-related health risk as might arise from atmospheric hazards through climate-sensitive planning and design. The chapters that follow put forward this case as a basis for emphasising the role that urban climate science knowledge, integrated with urban planning and design, can make tangible contributions to achieving the planning of healthy cities. However, before proceeding, it is necessary to acknowledge that a range of non-climate factors also determine urban health outcomes

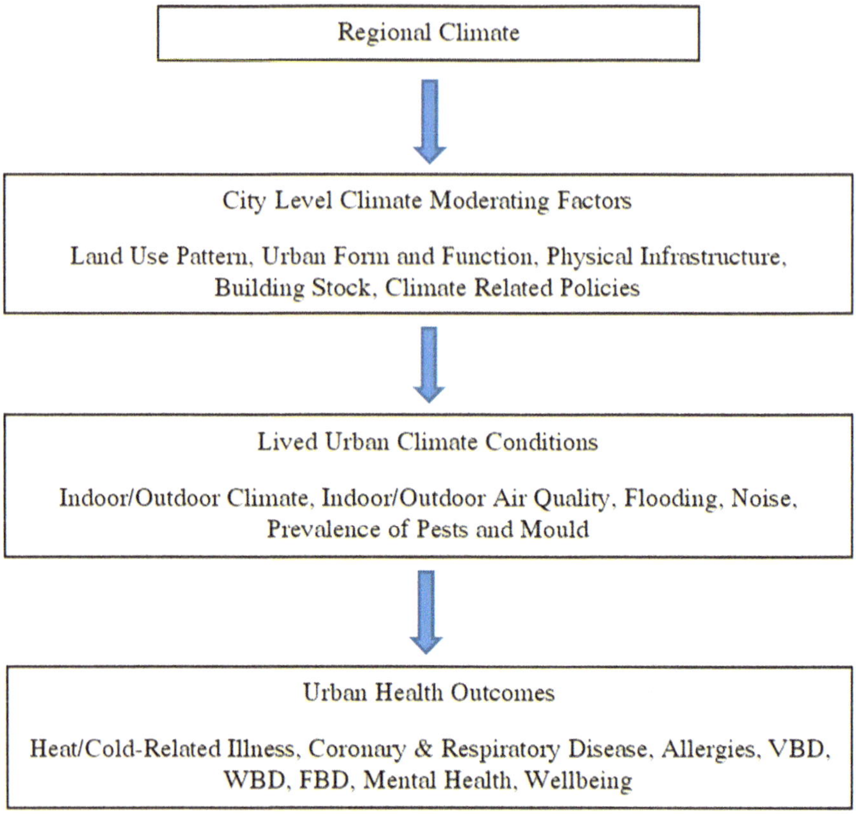

Fig. 1.2 Cascade of factors linking regional climate to urban health outcomes via lived urban climate conditions as determined by a range of city-level moderating factors. VBD is vector-borne disease, WBD is water-borne disease and FBD is food-borne disease

(Table 1.1). Given this, the inter-play between climate and the non-physical drivers of urban health and thus the achievement of healthy cities needs to be borne in mind when considering the material presented in this book.

1.5 Structure of the Book

Comprising three parts, the book proceeds from general context setting in Part I to considering examples of urban climate science in action in Part II with material related to future challenges and the way forward presented in Part III.

Chapter 2 presents an overview of the historical development of the healthy city concept and its implementation since the 1980s. The chapter also presents a categorisation of the healthy city literature and for the first time proposes a theoretical

Table 1.1 Non-physical Determinants of Urban Health (Adapted from Freudenberg et al. 2006)

Health determinant	Nature of health determinant
Population composition	Age, gender and ethnic distribution, genetic characteristics, inward and outward migration, health beliefs, and cultural attitudes
Social environment	Social networks, social capital, social support, social isolation/exclusion, violence, poverty
Health and social services	Formal and informal social agencies providing a range of health-related services including health clinics, social welfare support, food banks, etc.
Government	Municipal services such as water, sanitation and refuse collection, urban development policies (e.g. role of green infrastructure), public education, transportation and safety, criminal justice and welfare services, housing support and employment.
Markets	Role of finance and markets in allocation of housing, food, medical care, transportation, education, public safety, and acquisition of health compromising products (tobacco, illicit drugs, alcohol, obesogenic food, polluting vehicles) and promotion of unhealthy lifestyles.
Civic society	Interaction of urban residents to achieve social capital, social cohesion, social support, community capacity, community competence and political influence.

framework for future study and practice related to the urban planning–public health–climate change nexus.

Part II, 'Urban Climate Science in Action' and its associated Chaps. 3–13 presents a series of city-based case studies outlining urban climate science–urban planning experiences from New York, London, Manchester, Tokyo, Hong Kong, Beijing, Singapore, Melbourne, Paris, Stuttgart, and a number of Indian cities. The case study chapters outline climate-urban planning needs of the respective cities and articulate the nature of urban climate–urban planning solutions required, based on the available scientific evidence. These chapters reveal a vivid story of the moves afoot concerning the integration of climate information with the urban planning process and are testament to the on the ground commitment, in the case study cities, to improve the lived climate experience.

Chapter 3 on New York describes the Urban Climate Change Research Network (UCCRN). This brings together expertise and public opinion and forms a possible model for application in other cities to the issue of urban planning in a changing climate. The chapter emphasises the importance of configuring people-centred public spaces that enhance energy efficiency and improve public health and highlights the centrality of location-sensitive climate change mitigation and adaptation strategies in urban planning and urban design.

The UK cities of London and Manchester are the focus of Chaps. 4 and 5, respectively. The chapter on London examines the concept of urban heat resilience in the context of the development of healthy cities and in doing so sets out the challenges associated with achieving healthy urban environments in a warming climate. Similar to the Chapter on New York, this chapter emphasises the benefits for climate-sensitive urban planning that are likely to arise from knowledge exchange and co-design between academia, policy makers and citizens. It also explores possible solutions to over-heating in urban environments based on a number of London-based urban climate research programmes. The chapter on Manchester examines how domestic gardens, a hitherto overlooked element of the urban climate system, could play an effective role in climate adaptation and resilience. The criticality of directly engaging citizens as active stakeholders in pledging conversion of their domestic gardens to climate-friendly spaces comes through strongly in this chapter as does the value that arises from cross-disciplinary collaboration among scientists, planners, government, non-governmental organisations, the private sector and the general public.

Chapters 6–9 turn their attention to four Asian cities, namely, Tokyo, Hong Kong, Beijing and Singapore. Heat island mitigation strategies in the megacity of Tokyo are outlined in Chap. 6. These include securing wind streams through urban spaces to enhance ventilation and trialling a range of techniques for reducing outdoor temperatures. The utility of a heat and wind-path mapping tool, for informing urban design solutions to achieve thermal comfort in Tokyo, is also described in this chapter with the importance of local government support emphasised. Chap. 7 presents the physical and socio-economic factors that are likely to aggravate the impacts of extreme climate events in the highly urbanised territory of Hong Kong and reviews a number of studies that clearly demonstrate how climate variations will affect various health outcomes related to communicable and non-communicable diseases. This chapter also highlights the adverse health impacts arising from Hong Kong's urban heat island effect, a result of poor urban planning and design, and puts forward the case for more urban green space as a strategy conducive to obtaining healthy urban living environments. Beijing, a city known for its severe air pollution problems, is the focus of Chap. 8. This chapter describes how local scientists and government have worked together to achieve air pollution control not only via emission reductions but also through planning interventions. The chapter introduces the Beijing Urban Master Plan (2016–2035) as an example of the integration of strategies for air pollution abatement and climate change adaptation into urban planning and design to realise gains in environmental quality. Similar to Tokyo and Hong Kong, Singapore is a high-density compact city where high temperatures and poor wind flow in built-up areas negatively affect thermal comfort and air quality, and drive up energy demand for artificial cooling. Chapter 9 backgrounds these issues as the impetus for the development of computer-based tools, such as Singapore's Integrated Multi-scale Environmental Urban Model (IMEUM), that can be employed to support planners, architects and engineers to assess the combined impacts of the urban heat island phenomenon and rising global temperatures due to climate change.

Australian cities, which have attracted much attention recently because of an increase in the frequency of extreme heat events, is the focus of Chap. 10. Because Australian cities possess clear threshold temperatures above which adverse health outcomes escalate dramatically, this chapter asserts that even small reductions in temperature can significantly improve health outcomes. To this end, the chapter describes research emanating from the Australian Cooperative Research Centre for Water Sensitive Cites (CRCWSC) and highlights the evidence base for the effectiveness of a range of urban cooling approaches including irrigated green infrastructure and its potential to improve human thermal comfort and save lives. The chapter concludes by presenting an honest assessment of the key impediments to the implementation of effective urban cooling strategies in Australia and highlights the lack of clarity concerning the cost-benefit of cooling technologies.

The 2003 European heatwave provides the context for Chap. 11. This outlines a number of research projects, involving public stakeholders, designed to investigate implementable adaptation strategies for addressing the combined effects of climate change and the urban heat island phenomenon in Paris. Described is a range of action levers for implementation at the neighbourhood to city scales, ranging from the use of reflective materials at the building scale through to large-scale territorial planning policies related to urban expansion and landscape planning around the city.

A book on the application of urban climate science in the pursuit of healthy cities via climate-sensitive urban planning and design would not be complete without material that enunciates the urban climate-planning legacy of Stuttgart. This 'frontier' city is the subject of Chap. 12. Outlined in this chapter are the urban climate principles that underpin urban climate informed urban planning in Stuttgart. This chapter is rich with examples of urban climate science into action as evidenced by urban cooling and ventilation strategies relating to city greening, creation of wind corridors, climate optimised inner-city development and public spaces. Described are the range of climate-related urban planning and design legislations within which urban climate actions have become enshrined. Like the chapter on Australian cities, this one does not hide from the challenges of moving from science into policy. Accordingly, it reveals the factors that facilitate success and the obstacles confronted when local government officials consider climate in urban planning practice.

Complementing urban planning and design as an approach to managing climate-related health risk are heat-health early warning systems. These are the subject of Chap. 13 with a focus on Indian cities. India, with its five megacities, an unrelenting pace of urbanisation and a climate characterised by extreme heat, faces burgeoning challenges in relation to the attainment of a benign urban climate experience and the containment of heat-related health risk. These issues are touched upon in this chapter, along with a description of the development of a temperature threshold-based heat action plan, which aims to provide a framework for heatwave early warning and associated actions for heat-related health risk reduction.

Part III, comprising Chaps. 14–18, outlines 'Future Challenges and the Way Forward' from different perspectives. As described earlier, a fundamental aim of urban climate science is to understand urban climate processes at a range of scales as a prerequisite for explaining the diversity of urban climate outcomes. With this in mind, Chap. 14 considers the type of climate services required to support urban planning and decision-making. The authors draw on recently published material related to the World Meteorological Organization's (WMO) *Integrated Urban Hydrometeorological, Climate and Environmental Services (IUS)* initiative, a programme committed to providing science-based services to support safe, healthy, resilient and climate-friendly cities. This chapter outlines the concepts, methods and practices to facilitate production and provision of services relevant to the daily management of cities, respond to short-term hazards and prepare for climate change.

The chapter openly acknowledges that urban climate service provision is an emerging and diverse field with considerable research, development, capacity building and cross-service integrations yet to be done. As noted in Chap. 14, modelling has become an important tool for the generation of urban planning relevant climate information, yet representation of urban characteristics in models remains a challenge. Chapter 15 addresses this critical issue. Identified are sources for developing urban morphological data with methods to generate such data for any city on earth proposed.

Urban air pollution as an environmental health management problem, often compounded by climate extremes and the urban heat island, and likely to be exacerbated by global climate change, is the subject of Chap. 16. Described are the types of air pollutants and associated factors that conspire to produce detrimental and interconnected synergistic impacts on the health of a city and its inhabitants along with how urbanisation can modify the meteorological conditions in a city, and as a consequence, the long-term local air pollution climate. The chapter also examines how regional and global climate can impact, and be impacted by, air pollution.

In the view of the editors, recognising the need for a collaborative approach to problem solving and drawing on perspectives from seemingly disparate fields is a step in the right direction in the case of building an evidence base to inform healthy city-related climate-sensitive decision-making. In this spirit, Chap. 17 describes current good practice related to research partnerships, including needs-driven research, collaborative networks, and efforts for co-development of decision-supporting knowledge. Important findings in the urban climate and health sciences, key research priorities and challenges and issues are discussed in this chapter. Emphasised is the core challenge facing decision makers of identifying and prioritising health-related climate change interventions under countervailing influences.

Chapter 18, the final chapter, draws together the main threads running through the book. It briefly revisits the association between climate and healthy cities before summarising the 'found' relationship between urban climate science and urban planning and design. The chapter makes the case for an urban planning–urban climate imperative, draws out the main approaches enunciated throughout the book concerning the generation of urban climate-related health risk information, and points to examples of the application of urban climate science to the attainment of healthy cities. The chapter closes with a look to the future and a reflection upon some of the questions that, in the context of healthy cities, might comprise an ever-evolving urban climate–urban planning research and action agenda.

References

Asvall JE (1998) Athens declaration for healthy cities. J Public Health Med 20(4):372–374. https://doi.org/10.1093/oxfordjournals.pubmed.a024788

Bierbaum R, Smith JB, Lee A, Blair M, Carter L, Chapin FS, Fleming P, Ruffo S, Stults M, McNeeley S, Wasley E, Verduzco L (2013) A comprehensive review of climate adaptation in the United States: more than before, but less than needed. Mitig Adapt Strateg Glob Chang 18(3):361–406. https://doi.org/10.1007/s11027-012-9423-1

Bremer S, Wardekker A, Dessai S, Sobolowski S, Slaattelid R, van der Sluijs J (2019) Toward a multi-faceted conception of co-production of climate services. Clim Serv 13:42–50. https://doi.org/10.1016/j.cliser.2019.01.003

Briley L, Brown D, Kalafatis SE (2015) Overcoming barriers during the co-production of climate information for decision-making. Clim Risk Manag 9:41–49. https://doi.org/10.1016/j.crm.2015.04.004

Cash DW, Borck JC, Patt AG (2006) Countering the loading-dock approach to linking science and decision making—comparative analysis of El Nino/Southern Oscillation (ENSO) forecasting systems. Sci Technol Human Values 31(4):465–494. https://doi.org/10.1177/0162243906287547

Coen DR (2020) Advent of climate science. In: Oxford encyclopaedia of climate science. Oxford University Press, Oxford. https://oxfordre.com/climatescience/view/10.1093/acrefore/9780190228620.001.0001/acrefore-9780190228620-e-716

Daniels E, Bharwani S, Swartling AG, Vulturius G, Brandon K (2020) Refocusing the climate services lens: introducing a framework for co-designing "transdisciplinary knowledge integration processes" to build climate resilience. Clim Serv 19:100181. https://doi.org/10.1016/j.cliser.2020.100181

de Schiller S, Evans JM (1991) Bridging the gap between climate and design at the urban and building scale—research and application. Energ Buildings 15(1–2):51–55

Eliasson I (2000) The use of climate knowledge in urban planning. Landsc Urban Plan 48(1–2):31–44. https://doi.org/10.1016/s0169-2046(00)00034-7

Fragomeni MBA, Bernardes S, Shepherd JM, Rivero R (2020) A collaborative approach to heat response planning: a case study to understand the integration of urban climatology and land-use planning. Urban Clim 33:100653. https://doi.org/10.1016/j.uclim.2020.100653

Freudenberg N, Galea S, David Vlahov D (eds) (2006) Cities and the health of the public. Vanderbilt University Press, Nashville

Grimmond CSB, Roth M, Oke TR, Au YC, Best M, Betts R, Carmichael G, Cleugh H, Dabberdt W, Emmanuel R, Freitas E, Fortuniak K, Hanna S, Klein P, Kalkstein LS, Liu CH, Nickson A, Pearlmutter D, Sailor D, Voogt J (2010) Climate and more sustainable cities: climate information for improved planning and management of cities (producers/capabilities perspective). Procedia Environ Sci 1:247–274. https://doi.org/10.1016/j.proenv.2010.09.016

Guentchev GS, Rood RB, Ammann CM, Barsugli JJ, Ebi K, Berrocal V, O'Neill MS, Gronlund CJ, Vigh JL, Koziol B, Cinquini L (2016) Evaluating the appropriateness of downscaled climate information for projecting risks of salmonella. Int J Environ Res Public Health 13(3):267. https://doi.org/10.3390/ijerph13030267

IPCC (2014) Climate change 2014: impacts, adaptation, and vulnerability. part A: global and sectoral aspects. In: Field CB, Barros VR, Dokken DJ, Mach KJ, Mastrandrea MD, Bilir TE, Chatterjee M, Ebi KL, Estrada YO, Genova RC, Girma B, Kissel ES, Levy AN, MacCracken S, Mastrandrea PR, White LL (eds) Contribution of working group II to the fifth assessment report of the intergovernmental panel on climate change. Cambridge University Press, Cambridge, United Kingdom and New York, NY, 1132pp

Landsberg H (1973) Meteorologically utopian city. Bull Am Meteorol Soc 54(2):86–89. https://doi.org/10.1175/1520-0477(1973)054<0086:Tmuc>2.0.Co;2

Lemos MC, Kirchhoff CJ, Ramprasad V (2012) Narrowing the climate information usability gap. Nat Clim Chang 2(11):789–794. https://doi.org/10.1038/nclimate1614

Lenzholzer S, Brown RD (2016) Post-positivist microclimatic urban design research: a review. Landsc Urban Plan 153:111–121. https://doi.org/10.1016/j.landurbplan.2016.05.008

Mills G (2008) Luke Howard and the climate of London. Weather 63(6):153–157. https://doi.org/10.1002/wea.195

Mills G, Cleugh H, Emmanuel R, Endlicher W, Erell E, McGranahan G, Ng E, Nickson A, Rosenthal J, Steemer K (2010) Climate information for improved planning and management

of mega cities (needs perspective). Procedia Environ Sci 1(2010):228–246. https://doi.org/10.1016/j.proenv.2010.09.015
Ng E (ed) (2009) Designing high-density cities for social and environmental sustainability. Earthscan, London
Ng E (2012) Towards planning and practical understanding of the need for meteorological and climatic information in the design of high-density cities: a case-based study of Hong Kong. Int J Climatol 32(4):582–598. https://doi.org/10.1002/joc.2292
Oke TR, Mills G, Christen A, Voogt JA (2017) Urban climates. Cambridge University Press, Cambridge
Parker DJ, Penning-Rowsell EC (2005) Dealing with disasters London's environment: prospects for a sustainable world city. Imperial College Press, London
Pruss-Ustun A, Wolf J, Corvalan C, Neville T, Bos R, Neira M (2017) Diseases due to unhealthy environments: an updated estimate of the global burden of disease attributable to environmental determinants of health. J Public Health 39(3):464–475. https://doi.org/10.1093/pubmed/fdw085
Skelton M (2020) How cognitive links and decision-making capacity shape sectoral experts' recognition of climate knowledge for adaptation. Clim Chang 162(3):1535–1553. https://doi.org/10.1007/s10584-020-02859-3
UN (2008) State of the World's Cities 2008/2009—harmonious cities. United Nations Human Settlements Programme (UN-HABITAT), London
UN (2016a) UN-Habitt III: adopted draft of the New Urban Agenda. United Nations, Quito, Ecuador
UN (2016b) UN-Habitt III: sustainable cities and human settlements for all. United Nations, Quito, Ecuador
UNFPA (2007) State of world population 2007: unleashing the Potential of Urban Growth. United Nations Population Fund, New York
UNFPA (2009) The state of world population. United Nations Population Fund, New York
WHO (2016) Conference conclusions and Action Agenda. Paper presented at the the 2nd global conference on health and climate, Paris, France
WHO, WMO, UNEP (2003) Climate change and human health–risks and responses: summary (p 38). France
WMO (2003) Our future climate WMO-No. 952. WMO, Geneva
WMO (2012) WMO/IGAC impacts of megacities on air pollution and climate. WMO, Geneva
WMO (2016) Integrated weather, climate, hydrology and related environment services for sustainable cities. Paper presented at the United Nations Conference on Housing and Sustainable Urban Development (UN-Habitat III), Quito, Ecuador. https://public.wmo.int/en/our-mandate/focus-areas/urban-development-megacities. Accessed 30 Nov 2020
Wolf T, Chuang WC, McGregor G (2015) On the science-policy bridge: do spatial heat vulnerability assessment studies influence policy? Int J Environ Res Public Health 12(10):13321–13349. https://doi.org/10.3390/ijerph121013321

Chapter 2
Healthy City and Its Planning

Lan Wang, Xiji Jiang, and Xinhu Li

Abstract Today, climate change and health burden have become common challenges faced by all countries around the world. Urban development has a significant impact on climate change, while dramatic climate change has had extensive and severe impacts on human health. As primary gathering places, cities play a crucial role in health promotion and climate change mitigation. As an Initiative of public policy to improve public health back in 1900s, urban planning continues this function by enhancing the quality of urban built environment, improving accessibility of healthcare facilities, and encouraging active lifestyles. This chapter discusses global climate change and urban health issues for cities, and then elaborates the concept of healthy city proposed by WHO and its global development. With an accumulation of relevant researches on the database of the Web of Science from 1980 to 2019, this chapter maps the relationships and hotspots of key research themes in healthy city development and its planning. The cluster analysis of the literature categorises the studies into five dimensions: (1) built environment dimension focuses on the impact of built environment on travel models and physical activities of local residents and then their health outcome; (2) atmospheric environment dimension investigates the function of urban planning to adjust thermal environment and air quality, in which green space and urban form are included as major factors of built environment; (3) urban sustainability dimension is centralised with transportation system, ecosystem and open space; (4) city planning dimension explores how urban planning and design would promote physical activity, the accessibility of healthy food and healthy outcome; (5) elderly dimension studies the impact of built environment on travel behaviour and mental health of this specific population group. The literature investigating health city and its planning has rapidly increased since 1980s, with the themes becoming more comprehensive and multiple-focused. The research related to climate change within the literature pay specific attention to ozone and particulate matter under the concept of sustainability. Based on the mapping of literature, this chapter is concluded with a theoretical framework proposed for future study and practice, based on the relationship between urban planning, public health and climate change.

L. Wang (✉) · X. Jiang · X. Li
College of Architecture and Urban Planning, Tongji University, Shanghai, China
e-mail: wanglan@tongji.edu.cn; jxjcaup@tongji.edu.cn; xhli@tongji.edu.cn

C. Ren, G. McGregor (eds.), *Urban Climate Science for Planning Healthy Cities*, Biometeorology 5, https://doi.org/10.1007/978-3-030-87598-5_2

Keywords Climate change · Healthy city · Healthy city planning · Public health · Research trend · Built environment

Key Messages

- Climate change and health burden have become global challenges.
- Healthy city development has a "co-benefit" for climate change and public health.
- With time processing, the research on healthy city planning shows the characteristics of more intense and continuous, more comprehensive, detailed and decentralised focus.
- "Climate change" research is more closely related to ozone, specific matter and sustainability, and involves less material elements that can be directly controlled by planning.
- Healthy city planning needs enhance its potential impact on climate change, which also deepens the understanding of planning effects on public health.

Today, climate change and public health have become major global concerns. On the one hand, numbers of people living in urban areas exceed those of rural areas, with continuous urbanisation globally. Intensive human activities have increasingly resulted in climate change and impacted human health. On the other hand, cities have become platforms upon which climate change can be mitigated and public health improved. The concepts of the healthy city and global healthy city movement were initiated by the World Health Organisation (WHO). Development of a healthy city, therefore, requires consideration of the impacts of climate change on public health. While the WHO highlights the functions of urban planning in improving public health, it may also mitigate climate change through optimisation of the built environment, energy consumption and traffic modes (Hickman and Banister 2014; Carter et al. 2015).

This chapter systematically discusses global climate change and health issues for cities and proposes that development of a healthy city is an important pathway for realising the common benefits of climate and health. The concept and development process of a "healthy city" are combed in detail. From the perspective of public health and urban planning, the importance of healthy city planning is discussed before bibliometric analysis of relevant research on healthy city planning is undertaken, the results of which explain the evolution of hot topics in this discipline and explore the status of climate change issues. Following this, a healthy city planning framework is produced. Finally, with reference to the relationships between city planning, public health and climate change, the theoretical guidance provided by this framework is discussed to provide for future research and planning practice.

2.1 Global Picture of Climate Change and Health Issues for Cities

It is widely accepted that climate change is occurring as a result of human-related greenhouse gas (GHG) emissions. In a certain sense, human development has depended upon the destruction of the planet's natural environment, with rapid urbanisation and industrialisation creating a long legacy of emissions and environmental pollution. This bears implications for the health of the planet on which human health depends (Watts et al. 2015). With the complex interaction of climate change, urbanisation and global environment, this is likely to have a profound impact on human health and well-being.

2.1.1 Climate Change, Urbanisation and Global Environment Changes

According to the World Bank Group (WBG), urban populations account for 50% of the world's total population, and urban greenhouse gas emissions account for 80% of global emissions (Chinese Society for Urban Studies 2009). Cities have become the world's largest source of CO_2 emissions, and reduction of urban greenhouse gas emissions has become the main battlefield and key to the international community's response to climate change. However, with high-density populations, buildings, wealth and infrastructure, cities are high-risk areas affected by climate change. Extreme weather and climate events have caused huge disasters and had far-reaching impacts on cities. Moreover, urbanisation has a close relationship with climate change. On the one hand, the process of urbanisation contributes to climate change as increases in the size of urban populations lead to increases in carbon emissions. In urban regions, intensive human activities affect the local climate and environment, for example, production of "heat island effects" and "rain islands". On the other hand, climate change bears consequences for urban planning and construction, especially in low-altitude areas adjacent to rivers and coasts. In the context of global climate change, the frequency and intensity of extreme weather events and disasters are increasing, making cities vulnerable and placing them at risk (Shi and Mu 2016).

Climate change, in conjunction with other widespread adverse environmental changes, can have compound consequences. Food systems based on global ecosystems could be irreversibly damaged, beginning with biodiversity reduction. The volume of water from glacial rivers will dwindle. Weather and climate also play important roles in determining patterns of air quality. Meteorological parameters, such as humidity, temperature, wind direction and speed, are all used to estimate air quality over multiple scales in time and space. To control key pollutants, like fine particles and ozone, it is necessary to assess potential future climate conditions.

Other air contaminants related to human health may also be influenced by climate change, such as airborne pollen, mould spores and smoke from wildfires.

In some countries, such as China, environmental risks are shifting from traditional sources to modern sources associated with industrialisation and urbanisation. Traditional environmental risks include indoor air pollution from combustion of coal, wood and crop residue, and poor sanitation. Modern environmental risks include outdoor air pollution and industrial waste associated with industrialisation and urbanisation. These environmental risks co-exist with several emerging environmental risk factors, including climate change and long-range transport of pollutants.

Furthermore, other global environmental changes, including ocean acidification, biodiversity loss, land use change, pollution, and freshwater depletion, are all driven by economic development, which makes the impact of climate change on the Earth's natural system more serious. These pervasive and interconnected changes are primarily driven by a growing demand for food, energy and other resources, and even more frustrating is that these resources are inequitably distributed.

2.1.2 Climate Change and Health

It has been known for thousands of years, at least since the time of Hippocrates, that climate has wide-ranging impacts on human health. There is a high level of confidence in the direct and indirect effect of climate change on human health, and this scenario ranges from deaths due to extreme high temperatures to changes in the patterns of various diseases (IPCC 2014). Existing adverse health effects that are likely to be worsened by climate change include heat-related illnesses and death, extreme weather-related health effects, air pollution-related health effects, allergic diseases induced by pollen production, infectious diseases induced by microbial contamination and transmission, malnutrition caused by climate-related crop failure, storm surge-related drowning and injuries, etc. The challenges posed by climate change can worsen the stress on public health systems, as higher temperatures are predicted to spread deadly tropical diseases to higher latitudes, affecting larger numbers of people, especially the poor (IPCC 2014). Furthermore, heatwaves can exacerbate health problems related to high blood pressure, leading to cardiovascular diseases; most heatwave deaths occur in people with pre-existing cardiovascular disease or chronic respiratory disease (Smoyer et al. 2000).

Extremes of temperature and rainfall have long-term effects, as well as effects on mortality. For instance, people experiencing flooding, hurricanes and wildfires could suffer from mental health disorders, including elevated rates of anxiety and mood disorders, sleep disruption, acute stress reactions, suicide and suicidal ideation, etc. These outcomes could linger for many months or even many years (Palinkas and Wong 2020). Climate change could affect biodiversity and the ecosystem, on which human survival and health depend. Changes in temperature and rainfall may have effects on the spatial distribution of disease vectors and influence the spread of communicable diseases. A recent study indicated that ocean warming would facilitate the methylation of mercury and its uptake by fish, and that consumption of these contaminated fish would impair foetal-infant neurocognitive development (Bambrick and Kjellström 2004). Climate change could also alter the timing and duration of pollen and spore seasons, which could affect allergic disorders such as asthma and hay fever (Beggs 2004). Climate change will have many effects on health over the coming decades and there are many studies on the estimation of future health effects (Gosling et al. 2009; Ballester et al. 2011).

The effects of climate change on health will also increase the inequities between poor and rich. Climate change has the greatest health impact on those who have the least access to resources and who make the smallest contribution to greenhouse gas emissions, thus increasing health inequity in poor populations (Costello et al. 2009; Thomas et al. 2014). It is worth noting that luxury emissions and survival emissions differ, the latter mainly coming from poor countries and regions (Caney 2009). For these reasons, management of the health effects of climate change is essential in order to reduce poverty and inequities in health.

There are some mechanisms by which climate can have effects on health. Generally, regional populations adapt to the local climate via behavioural, physiological, cultural and technological responses. However, extreme events often stress populations beyond adaptation limits.

2.1.3 Climate and Health Co-benefits

The term "co-benefits" is used to describe multiple, equally important rationales that could be achieved by a single policy or measure (Kwan and Hashim 2016). Climate and health co-benefits create a win-win situation for both health promotion and greenhouse gas reduction. At an individual level, health co-benefits could include internal emission reduction drivers based on an individual's desire to improve their health. Health co-benefits can arise via several pathways, including reduced air pollution, increased physical activity and dietary change. Actions that reduce emissions and, at the same time, improve health include: reducing local health-damaging emissions and climate-altering air pollutants (CAPs) from energy systems by improving energy efficiency and shifting to cleaner energy sources; shifting consumption from animal products towards less CAP-intensive healthy diets; and designing transport systems to promote more physical activity and reduce motorised vehicle use, leading to lower emissions and better health. One example of a programme providing climate and health co-benefits is public bike sharing. It is has been established that green space layout (Stern and Taylor 2007; Lindberg et al. 2016; Sturiale and Scuderi 2019), infrastructure construction (Jacobson and Delucchi 2011; Kennedy and Corfee-Morlot 2013; Kennedy et al. 2014) and land use layout (Raparthi 2018) have positive impacts on climate change mitigation. Scholars have proven that changes in travel modes reduce air pollution and carbon dioxide emissions (Anable et al. 2006; Younger et al. 2008; Rissel 2009). Cervero and Kockelman (1997) and Ewing and Cervero (2010) point out that accessibility of destinations, short distances between work and traffic stations, high development density, mixed degrees of land use and the design of small blocks can significantly affect residents' choice of travel behaviours; these can be regulated and guided by urban planning with material space construction as an important intervention method.

2.2 Healthy City and Relevant Research

2.2.1 Concept of a Healthy City

WHO, in its charter, pointed out that "health refers to the integrity of physical, psychological and social well-being and is more than simply being free from disease or asthenia", and "it is the fundamental right of everyone to enjoy health of the highest possible standard" (WHO 1948). Health is based on an individual's interactions with the natural environment, physical built environment, and the social and economic conditions in which they live. When cities provide housing, for more than half of human beings' medical services, jobs and recreational facilities, they also expose residents to air pollution, noise and psychological stress.

The concept of the healthy city was proposed in 1984 by the WHO when facing an increase of non-communicative diseases resulting from urban lifestyles. According to the Health Promotion Glossary, "a healthy city is one that is continually creating and improving those physical and social environments and expanding those community resources which enable people to mutually support each other in performing all the functions of life and in developing to their maximum potential" (WHO 1998). A healthy city is one that is continuously striving to improve the health and determinants of health of its residents, not one that has achieved a particular health status. Healthy cities focus on preventive health measures to improve the physical and social environments of the city and promote healthy lifestyles.

In the Global Report on Urban Health, issued by the WHO in 2016, a vision for healthier cities was proposed with the aims of: (1) reducing health inequality for sustainable development; (2) advancing universal health coverage in cities; (3) leveraging urban advantage to tackle communicable diseases; (4) for non-communicable diseases: overcoming the new urban epidemic (mainly cardiovascular diseases, respiratory diseases, obesity, diabetes, etc.); and (5) tackling twenty-first century malnutrition. Under these concepts and understandings, the Healthy Cities project promoted by the WHO is committed to the incorporation of health into the development agenda of cities around the world and the development of institutional mechanisms at the urban level to support public health.

2.2.2 *Evolution in Research Focus*

To analyse the interdisciplinary connection, research hotspots and the evolution process of healthy city research, the core databases *SCI-EXPANDED*, *SSCI*, *A&HCI* and *ESCI* on the Web of Science (WOS) were searched using keywords and CiteSpace software for mapping. The keywords used for searching include "Healthy City", "Healthy City Development", "Healthy City Planning" and "Healthy City Planning" in the research fields of "Development Studies", "Area Studies", "Urban Studies", "Health Care Sciences & Services", "Public, Environmental & Occupational Health" and "Government & Law"; setting the time between 1980 and 2019, the language as English, and the type as article, book or chapter. In total, 1849 research papers relating to healthy cities development have been selected as basic data for analysis.

The analysis shows that about 61 themes have been integrated into healthy city research. Public health, environmental health and occupational health are the most important research fields, and are strongly associated with healthcare sciences and services, medicine, engineering and law (Figs. 2.1 and 2.2). The theme of urban

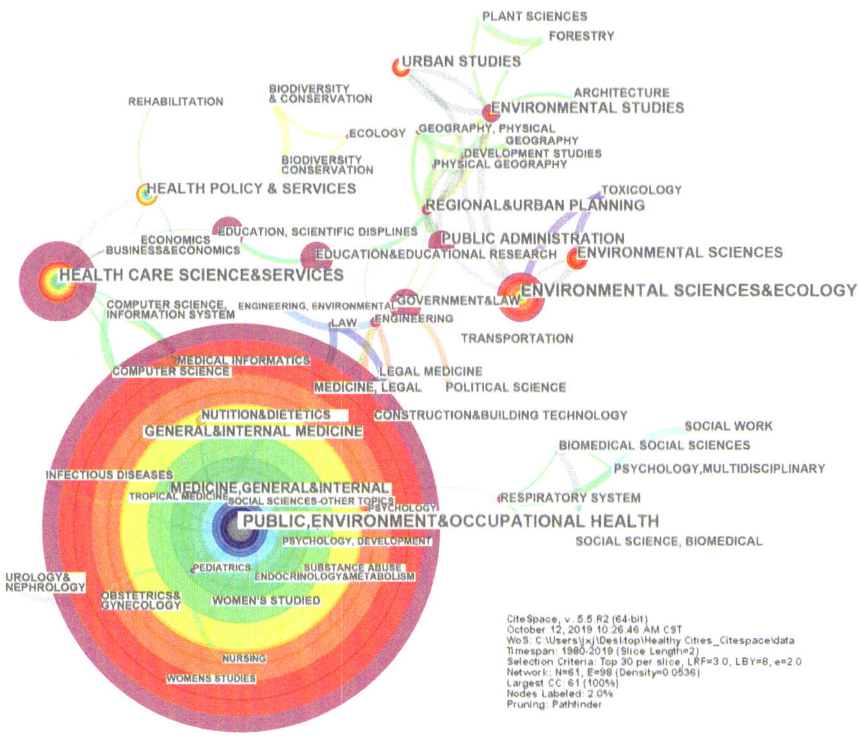

Fig. 2.1 The relationship between subjects related to healthy city research. (Source: drawn by authors)

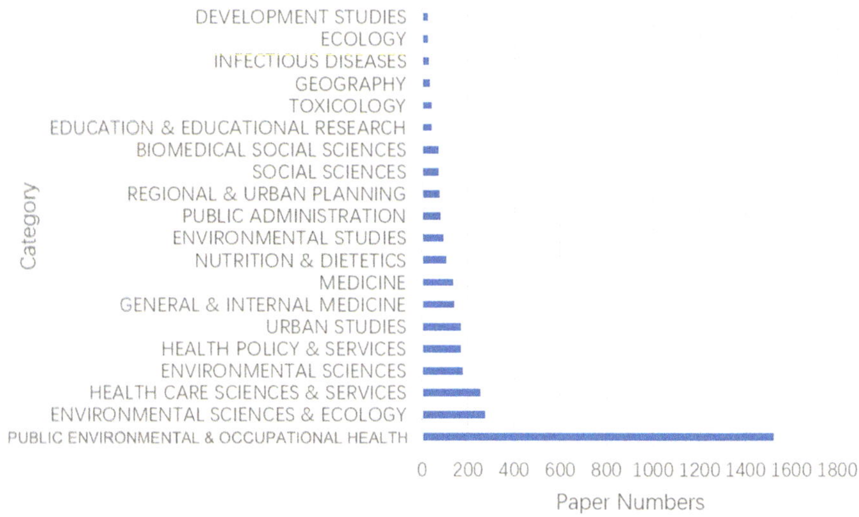

Fig. 2.2 Number of relevant research documents in major disciplines (cumulative value 1980–2019). (Source: drawn by authors)

studies (167 papers, ranking 6th) has become an important node of the interdisciplinary network, with the closest relationship with environmental studies and geography.

2.2.2.1 Research Hotspots

To make the diagrams clearer and more concise, the approach of displaying node labels by frequency and adjusting the threshold to 50 has been adopted. The result of the cluster analysis (Fig. 2.3) shows that the healthy cities research focusses mainly on nine research fields: behaviour, determinants of health, policy, ethnicity, lifestyle, health, infection, inner city and obesity. Topics in each field differ in both content and number. In terms of research keywords (Fig. 2.4), physical activity (ranking 2nd) and obesity (ranking 1st) are the focus of the research, while many other topics involve such issues as environment and association. The keyword network diagram shows that the research topics are relatively isolated except for smoking, which is highly connected with other topics. Since human health is often subject to the influence of a mixture of factors (e.g. environment and human behaviours have a joint impact on human health) (Frank et al. 2006), the confirmatory research based on assumed topic association is also worth exploring in the future.

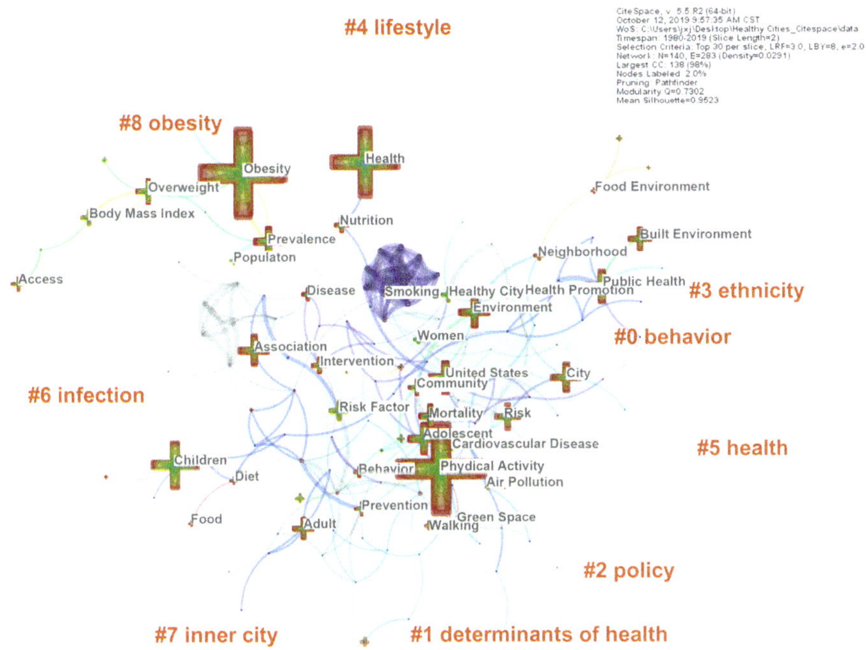

Fig. 2.3 Cluster analysis of keywords. (Source: drawn by author)

2.2.2.2 Research Evolution

Through the analysis of timeline and time-zone evolution (Figs. 2.5 and 2.6), this chapter includes the following findings: (a) from the perspective of the development timeline, healthy cities research enjoyed the highest degree of focus in the middle and later years of the 1990s and the 2000s; (b) in relation to the development continuity of the eight major clustering topics, the research on obesity, policy and ethnicity has survived to this day from the beginning of the 1990s, while the topics of lifestyle, health and prevalence were under the spotlight in the 1990s; (c) the keywords under research have undergone a shift from disease, risk impact and health promotion to environment, behaviours and nutrition, then to specific built environments (such as green spaces and food environment) and physical activities (such as walking), demonstrating a trend for the research objects to become increasingly detailed and micro.

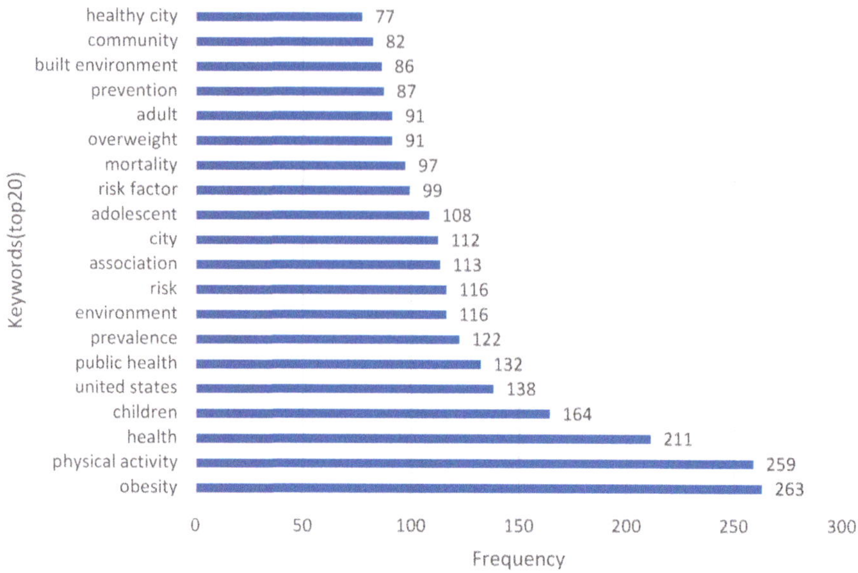

Fig. 2.4 Number of relevant research literatures of main keywords (cumulative value from 1980 to 2019). (Source: drawn by authors)

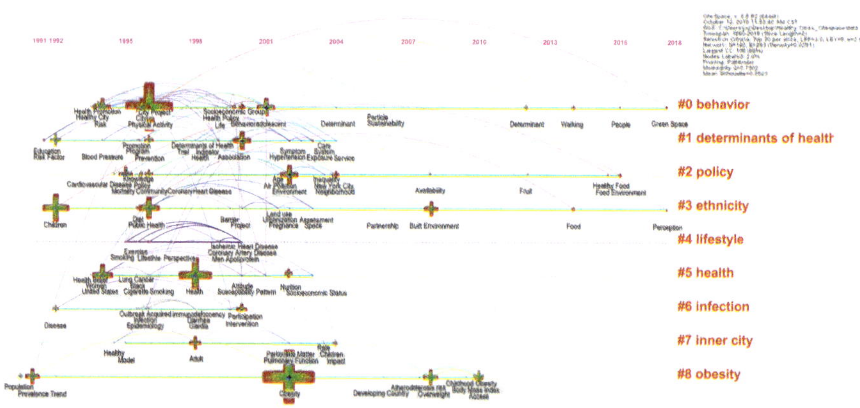

Fig. 2.5 Timeline analysis chart of research hotspots related to healthy cities. (Source: drawn by authors)

2.3 Healthy City Planning (HCP)

2.3.1 Urban Planning and Public Health

The impact of the built environment on public health has existed for a long time. Urban planning emerged as an important way for local governments to ensure urban

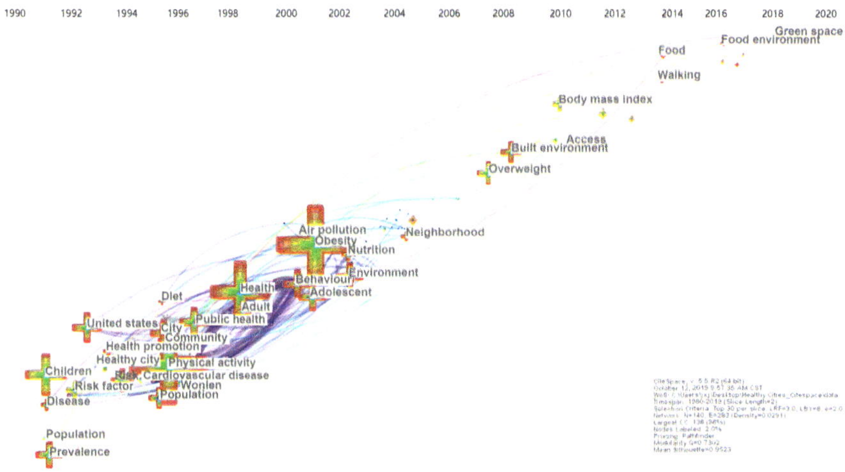

Fig. 2.6 Time-zone evolution of research hotspots related to healthy cities. (Source: drawn by authors)

public health. It is generally believed that modern urban planning in the UK originated from public health legislation in the 1840s.The formation of the Public Health Act directly related to urban problems emerging after the industrial revolution. Since then, the Act has begun to authorise the government to manage and operate urban affairs (Sun 2017). In the USA, in order to deal with the public health problem of infectious diseases caused by poor living conditions in the context of industrialisation, the government had promoted the improvement of built environments and public health through legislation and zoning resolutions, and large-scale infectious diseases had largely disappeared. By the 1940s, deaths caused by infectious diseases had fallen to 11% of the total (Cutler and Miller 2005; Jones 2005).

In this context, throughout the mid-twentieth century, consideration of the relationship between urban planning and public health waned until the 1980s when European and American cities, reflecting upon the prevalence of a range of chronic non-communicable diseases (NCDs), once again turned their attention to the interaction between the built environment and health. The rapid growth in the number of patients with NCDs is considered as "new outbreak" in cities around the world. By improving the quality of urban space and changing the way people live and work, urban planning may optimise the built environment and reduce the incidence of such diseases. Meanwhile, correlation between the increasing cases of mental illness and built environment is still under exploration, and the urban space that promotes social interaction is conducive to relieving psychological stress (Evans 2003). Therefore, optimisation of urban residents' physical and mental health is closely related to the built environment of cities. Urban planning, as a professional technology and a public policy for the arrangement and beautification of space, should take into account both traditional infectious diseases and new public health challenges, including new types of infectious, chronic non-infectious and psychological diseases.

2.3.2 Research Evolution of HCP

This chapter combs the number of papers per year over the past two decades and analyses the characteristics of related research in national cooperation, academic cooperation and institutional cooperation, respectively, in *Citespace 5.5.R* software. In order to clearly and indirectly display the visual analysis results, the software parameters are set as follows: (1) *Time Slicing* is from 1990 to 2019, and *Year Per Slice* is 2; (2) *Country*, *Institution* and *Keyword* are successively selected as *Node Types*; (3) *Links* uses the *Cosine algorithm*; (4) *TopN* selects the node data of 50; (5) The network does not use the clipping method.

Research literatures on global healthy cities show that the whole growth period increased flatly at first, then steeply, especially over the past 10 years. From Fig. 2.7, relevant studies began to emerge in 1991 and increase year by year, and the number of literatures significantly increased from 2007. In particular, a rapid growth occurred following a slight decline in 2008, and the number reached 177 in 2018. Combined with the development process of global healthy cities, the following conclusions can be reached: first, the annual growth of total number of literatures reflects the periodic characteristic. Statistics from 1991 to 2006, then 2006 to 2008, and the rapid increase since 2008, indicate that global research on healthy cities has become more and more significant. Secondly, the concept of "healthy city" was formally put forward in 1994, which greatly promoted the attention of researchers all over the world. In addition, multi-disciplinary participation and penetration, as

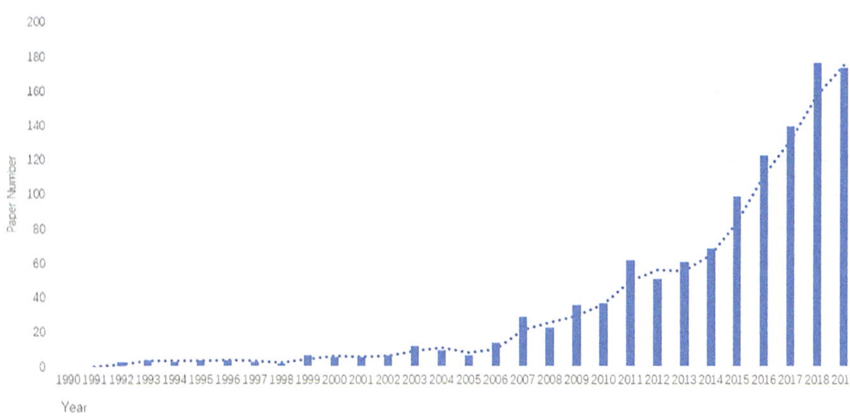

Fig. 2.7 Changes in the number of documents related to healthy city planning (1990–2019). (Source: drawn by authors)

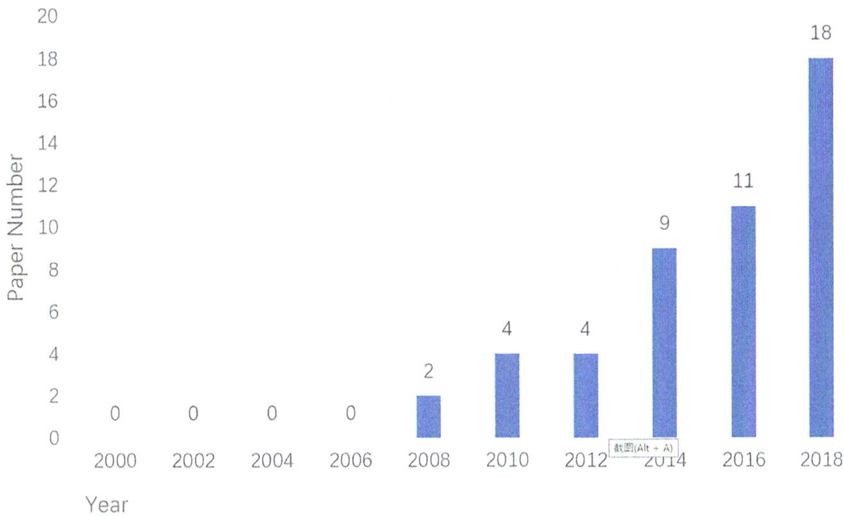

Fig. 2.8 Changes in the number of climate change related literature in healthy city planning (2000–2018). (Source: according to the results of CiteSpace analysis, the author collates and draws)

well as the promotion of "health in all policies", may have had an important impact on the rapid development since 2008.

Literature related to the theme of climate change in the field of healthy city planning began to appear in 2008, with rapid growth during the past 5 years (Fig. 2.8). Over the last 10 years, aggravation of global climate change and a series of corresponding health problems have prompted experts in the field of health and urban planning to pay more attention to the possible dual impact of urban planning on public health and climate change.

2.3.3 Research Hotspots of HCP

This chapter further uses the function of co-word analysis and cluster analysis in CiteSpace software to explore the important topics on which relevant researchers into healthy city planning have focussed. The basic principle of co-word analysis is to count the number of times that a group of words appear in the same group of literatures and to measure the degree of intimacy between them through co-occurrence (Li 2017). In this chapter, the collinear network is obtained by keyword co-occurrence analysis before generating knowledge clustering by map clustering algorithm and extracting labels from related citations by algorithm. The parameters set for analysis are as follows: (1) *Time Slicing* from 1990 to 2019, *Year Per Slice* is 2; (2) *Keyword* is selected as the *Node Types*; (3) *Links* use *Cosine algorithm*; (4) *TopN* selects 30 nodes data; (5) no network pruning method is used. Finally, knowledge clustering is named LSI (Latent Semantic Indexing) and LLR (Log-likelihood ratio), and five related research topic clusters are obtained: built environment (cluster 0#), atmospheric environment (cluster 1#), urban sustainability (cluster 2#), urban planning (cluster 3#) and the elderly (cluster 4#). Each cluster is mainly concentrated in 2004–2007, and the relevant research became mature 10 years ago.

As shown in the results (Fig. 2.11) of the keyword co-occurrence clustering spectrum, there are 123 network nodes, 680 connections, and the *Modularity Value* is 0.3766. The silhouette parameters of the five clusters are all greater than 0.65, indicating that the network clustering structure is significant and the clustering results are reasonable. The map network connection reflects that the built environment is important for undertaking research related to healthy city planning and has strong connections with the other four clusters.

The results of cluster analysis show that there are differences in the topic content and quantity of five research clusters related to healthy city planning. From the perspective of research keywords (Table 2.1), except some basic topics (health/healthy city/city), the research focuses are physical activity, built environment, walking, land use and urban design, etc. Climate change in the atmospheric environment cluster (1#, 48 papers) and its centrality is low (0.03). The cluster map (Fig. 2.9) shows that the built environment cluster (0#) has formed a highly networked research relationship, and other clusters have also formed a clear network pattern. The relationship between clusters is also relatively close.

2.3.3.1 Cluster 0#: Built Environment

Key words such as physical activity, urban form, transportation and walking are the main network nodes of cluster 0#. The main content of the study in cluster 0 # can be summarised as the impact of the built environment on residents' travel patterns, physical activities and health. In terms of residents' travel choices, Ewing and Cervero (2010) used a meta-analysis method to quantitatively analyse the relevant literature of "built environment–transportation travel" before 2009 and found that the planning and design of street networks is closely related to the accessibility of destination and mixed land use, and intersection density and destination density within walking range are significantly related to walking trips. Frank et al. (2008) analysed the relationship between built environment and physical activity, BMI, social and demographic characteristics in the Atlanta metropolitan area. Their results show better population density, better street connectivity, and that people who live in a higher degree of mixed-use are more likely to walk. Winters et al. (2010) used a multiple logistic regression model to analyse the relationship between cycling behaviour and the urban built environment and found that areas with fewer hills, higher intersection density, fewer expressways and urban arterial roads, more mixed land use, higher population density and more bicycle transportation facilities were more likely to have increased cycling behaviour. In terms of physical activity of residents, Lee and Moudon (2004) reviewed relevant literature in the field of health and found that physical exercise is related to the accessibility of entertainment facilities, neighbourhood safety and visual quality; and that walking and

Table 2.1 Co-occurrence analysis information of research key words related to healthy city planning (number of documents issued ≥40) (Source: CiteSpace analysis results)

Keyword	Paper number	Year	Centrality	Cluster#
Physical activity	345	2004	0.07	0
Built environment	212	2004	0.08	0
Walking	201	2004	0.08	0
Health	152	2000	0.08	0
Land use	123	2002	0.3	3
City	123	2006	0.05	2
Urban design	115	2008	0.05	0
Mortality	105	2003	0.08	1
Neighbourhood	101	2004	0.03	0
Urban planning	100	2008	0.06	2
Public health	95	2003	0.13	2
Impact	87	2006	0.12	1
Environment	77	2006	0.15	1
Walkability	75	2009	0.01	0
Association	71	2008	0.02	0
Obesity	70	2006	0.06	0
Design	65	2004	0.05	0
Transportation	62	2004	0.07	0
Green infrastructure	57	2007	0.06	3
Perception	55	2007	0.05	3
Accessibility	55	2016	0	2
Climate change	48	2009	0.03	1
Travel	47	2008	0.03	0
Air pollution	47	2008	0.03	1
Green space	45	2014	0.01	2
Transport	45	2008	0.01	0
Mental health	44	2012	0	2
Risk	43	2003	0.05	1

cycling are the main physical activity modes, mainly in the neighbourhood of streets and public facilities. Marshall et al. (2009) combined neighbourhood walkability with air pollution exposure, took Vancouver (Canada) as an analysis case, used the land-use regression model to estimate air pollutant concentration and analysed the overlap with neighbourhood walking ability. They found that there was a link between community income level, water pollution level and walking ability. In terms of residents' health, Wolch et al. (2014) emphasised that green space can promote public health, but it has become an environmental justice issue, requiring more sustainable policies to protect community green space.

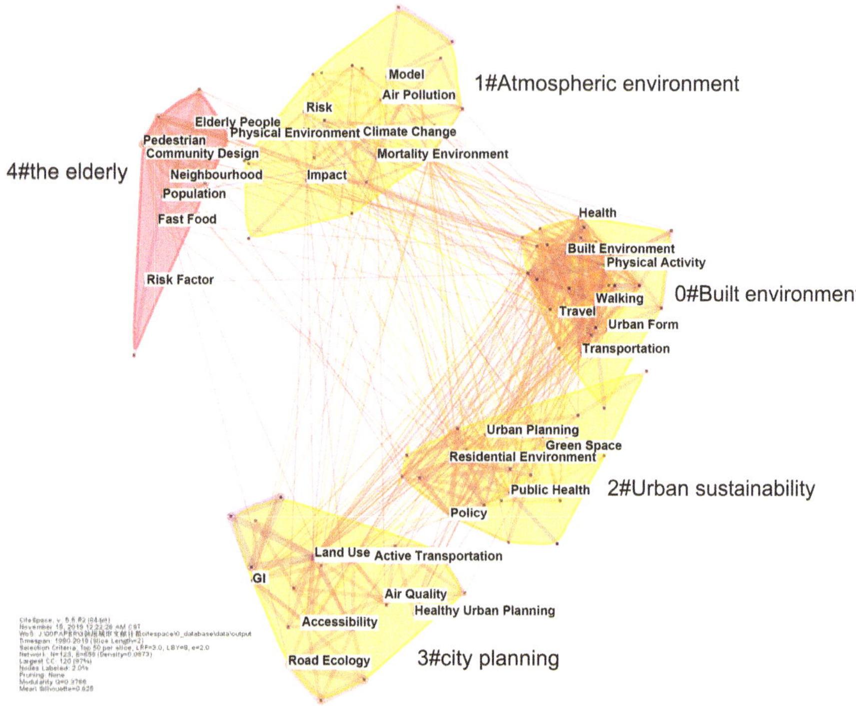

Fig. 2.9 Cluster map of key words co-occurrence in health city planning. (Source: CiteSpace analysis results sorting)

2.3.3.2 Cluster 1#: Atmospheric Environment

Mortality, impact, environment, climate change and risk are the main research hotspots of cluster 1#. Cluster 1# focusses on the atmospheric environment and discusses the role of urban planning and design in regulating the atmospheric environment and improving public health from two perspectives: thermal environment and air quality. Studies emphasise spatial planning and the importance of active intervention to improve the city's ability to adapt to the climate (Carter et al. 2015).

In terms of thermal environment, urban planning should consider both high-temperature and low-temperature. On one hand, Laaidi et al. (2012) conducted a case-control study in Paris and found that the risk of death was significantly related to the lowest average temperature on the day of death and the previous 6 days. On the other hand, Basagaña (2019) points out that the urban temperature is several degrees higher than the surrounding rural areas according to the global heat island effect records, and the increase of urban vegetation and use of cool materials play an important role in improving the urban temperature. Stone et al. (2010) studied urban morphology and high temperature data in the United States and found that the growth rate of extreme high temperature events in sprawling metropolitan areas is more than twice that of compact metropolitan areas. Urban parks and green spaces have the potential to provide thermal comfort and reduce the vulnerability of heat stress, so the design of climate response, based on evidence, can make the park more comfortable (Brown et al. 2015).

In terms of air quality, Xia et al. (2015) used the health impact assessment model to assess traffic pollution in Adelaide City and proposed the important role of promoting the change of transportation mode in urban planning decision-making and infrastructure construction to reduce death rates, disability cases and air pollution. Zou et al. (2009) constructed a population pollution air exposure model to assess population exposure in Dallas County, United States. It was found that the exposure was mainly due to vehicle pollution rather than industry, and that response strategies should be considered in urban planning.

2.3.3.3 Cluster 2#: Urban Sustainability

The main research hotspots of urban sustainable knowledge cluster 2# are cities, urban planning, public health, accessibility and green space, focussing on the role of transportation, urban ecosystems, open space and walking in promoting sustainable urban development and public health.

In urban transportation, public travel and auxiliary transportation should be an important part of the overall transportation structure. Hickman et al. (2013) pointed out that urban planning, as an important element, can play a vital role in improving the level of public transport, promoting walking and cycling, and thus promoting the transition to sustainable transport development. According to Sietchiping et al. (2012), the south area of Sahara Desert should integrate various commuting modes into urban infrastructure through urban planning and traffic infrastructure design to cope with increasingly severe urban traffic problems.

In terms of ecosystems and open space, Chawla (2015), focussing on children, highlights the positive impact of natural contact on their health and well-being, and points out the need for multi-scale integration of green elements from housing, schools, urban footpaths, greenways, parks and communities. Jackson (2003) argues that urban planning and design should enhance interdisciplinary cooperation and jointly promote the shaping of healthy material space, proposing that greening and visual and physical access to it are the key elements of health and should be considered from all aspects of neighbourhood design. Public open space has a positive impact on regional liveability and residents' health. Villanueva et al. (2015) provide a set of indicators to measure open space for urban planning and policymaking in Australia to help improve people's health.

Walking is beneficial to the health of residents. Walkability is an index used to describe and ensure a pedestrian-friendly area or road section. Moura et al. (2017) designed a methodological framework for measuring and evaluating the walkability of different populations, including seven key indicators, including connectivity, convenience, comfort, persistence, conspicuousness, coexistence and commitment, and carried out empirical research in Lisbon.

2.3.3.4 Cluster 3#: City Planning

Land use, perception, active traffic, accessibility and road ecology are the main research hotspots for cluster 3#. Literatures in cluster 3# focus on how urban planning and design methods can improve or promote residents' health and physical activity, safeguarding the supply of healthy food. Research findings reflect the importance of quantitative analysis and monitoring to propose planning or design interventions.

There is a close relationship between transportation and physical activity. Badland and Schofield (2005) considered physical activity and buildings attributed to transportation by reviewing existing evidence related to various urban design factors and physical activity behaviours. Density, age characteristics, street connectivity and mixed land use are significantly related. It is suggested that the relevant departments of urban design should cooperate with the health and transportation departments to use this basic evidence. Handy et al. (2002) propose that a combination of land-use patterns and transportation systems promoting walking and cycling can help to create active, healthy and liveable communities. Northridge et al. (2003) propose a conceptual framework for planning a healthy city, which includes three aspects: building built environmental monitoring indicators to assess the health impact of urban planning and policy decisions; comprehensively monitoring the possible health impacts of various activities of the public and private sectors; and analysing the possible health impacts of administrative management in "super-urban areas".

In addition, the food system has become a focus of researchers in the past 10 years. Eckert and Shetty (2011), through quantitative analysis of food accessibility by GIS, found that Toledo faced the challenge of food accessibility in communities and proposed specific strategies with which Toledo could deal with problems in food system planning. Páez et al. (2010) and Gordon et al. (2011) focussed on the impact of "food desert" on the health of residents (such as obesity). The former used the distance index to measure the cumulative measurement of accessibility of food facilities in Montreal (Canada), while the latter used the ratio index of accessible food supply facilities to evaluate the current situation of low-income communities in New York City.

2.3.3.5 Cluster 4#: Elderly

The node centrality of all hot words in Cluster 4# is low in the whole keyword co-occurrence network and the amount of related literature is small. Among them, the elderly group, neighbourhood, neighbourhood design and risk nodes are relatively important in the internal cluster network. Taking the elderly as the research object, the relevant literature of this cluster focuses on the impact of the urban built environment on the travel behaviour and physical activity of old people. Borst et al. (2008, 2009) focussed on the street characteristic environment and used the multiple linear regression model to analyse the attraction of micro street environments (such as slope/staircase, zebra crossing, trees, garden, bus station, and along street business) for the walking elderly. It was found that street cleanliness, landscape and along street vitality had significant attraction for them. Li et al. (2008) investigated and analysed the BMI, physical activity and built environment of the elderly in Portland, USA. It was found that the land mixing degree was negatively correlated with obesity, and fast-food restaurant density was the opposite; in addition, the land mixing degree, street connectivity and bus station density, to a certain extent, promoted the occurrence of physical activity.

2.3.4 Research Evolution

Analysis of the keyword evolution process (Fig. 2.10) shows that: (a) in 2000, relevant researches began to show a clear relationship of inheritance and continuity, all originating from the keyword "health"; (b) in the early 2000s, relevant researches focussed on walking, built environment, physical activity, land use, motorisation, etc. In the later stage, the research topics became more and more abundant (including more detailed issues), becoming the focus of scholars' research, such as obesity, children, climate change, the elderly, urban planning and design, walkability and the correlation between environment and health; (c) for nearly a decade, the related research has been further scattered and the research focusses are mainly on mental health, green space and accessibility issues; (d) overall, the inheritance of research hotspots shows a good continuity, and the process reflects the researchers' detailed and continuous thinking on relevant issues. For example, retrospective study of changes in the course of hot green space can be found, from health-built environment/physical activity to urban planning/urban design, and then focus on green

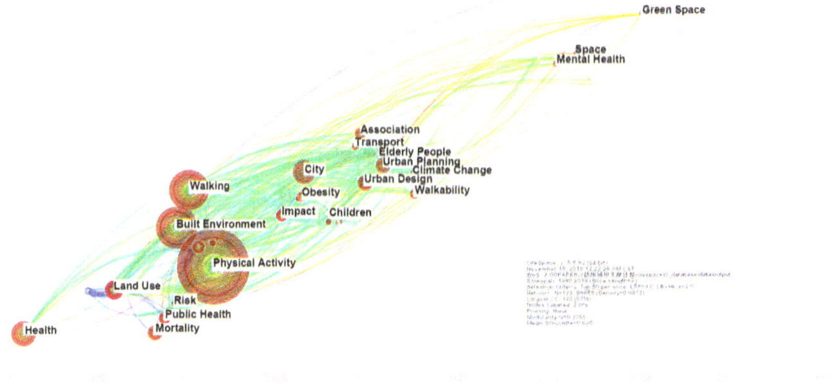

Fig. 2.10 Evolution of research hotspots related to healthy city planning. (Source: CiteSpace analysis results)

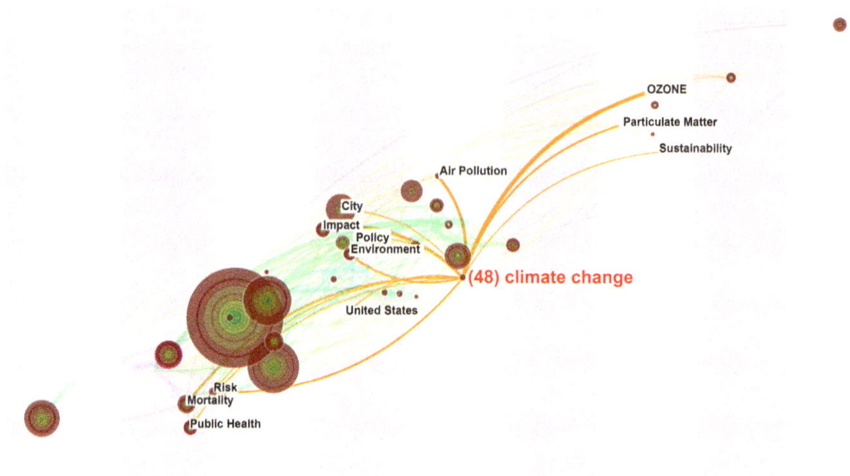

Fig. 2.11 "Climate change" research in the related research of healthy city planning. (Source: CiteSpace analysis results)

space; climate change research topics originated from public health, disease mortality and health risks, then being closely related to cities, health impacts, policies and the environment. The recent evolution of related climate change topics includes specific topics such as ozone, particulate matter and the concept of sustainability (Fig. 2.11).

2.4 Conclusion and Discussion

Based on the analysis of research evolution of health city and healthy city planning on the database of the Web of Science from 1980 to 2019, we discovered that five important research dimensions, including (1) built environment dimension focuses on the impact of built environment on travel models and physical activities of local residents and then their health outcome; (2) atmospheric environment dimension investigates the function of urban planning to adjust thermal environment and air quality, in which green space and urban form are included as major factors of built environment; (3) urban sustainability dimension is centralised with transportation system, ecosystem and open space; (4) city planning dimension explores how urban planning and design would promote physical activity, the accessibility of healthy food and healthy outcome; (5) elderly dimension studies the impact of built environment on travel behaviour and mental health of this specific population group. The literature investigating health city and its planning has rapidly increased since 1980s, with the themes becoming more comprehensive and multiple-focussed. The research related to climate change within the literature pay specific attention to ozone and particulate matter under the concept of sustainability. Further studies are

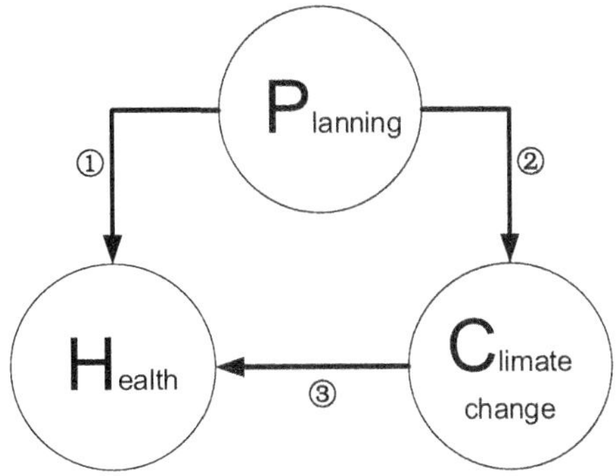

Fig. 2.12 Relationship between urban planning, health and climate change. (Source: drawn by authors)

needed to unravel the more specific mechanisms of city planning could influence climate change and public health.

Prior studies have confirmed that urban planning functions efficiently in achieving co-benefits of public health promotion and climate change mitigation. The relationship could be summarised as follows: (1) urban planning could promote public health by improving physical space (Fig. 2.12, ①); (2) urban planning could affect climate change through urbanisation and non-mobilisation (Fig. 2.12, ②); (3) the positive improvement of climate change brings health benefits on a large scale and for a long time (Fig. 2.12, ③). Urban planning affects public health though land use allocation, road systems, space form, green space and open space, as confirmed by existing studies. Through improvement of these built environment improvements, exposure to pollution can be reduced and physical activity and social interaction can be promoted. Healthy city planning can to a large extent achieve both positive public health and climate improvement, achieving co-benefits.

Although scholars have realised the potential connection among the three, further research is needed to enrich the understanding of these relationships. The analysis of keywords evolution process in this chapter shows the weak linkage between climate change and healthy city planning. It may be caused by the limited basic data and information resources, disciplinary barriers between analytical technologies, and lack of collaborative platforms and mechanisms. Diversification and enrichment of big data information sources, as well as the rapid increase in integration of related analysis technology, cooperation mode and platform, provide basic supports for future related research and planning practice. Healthy city planning needs to enhance its potential impact on climate change, which also deepens the understanding of planning effects on public health.

Acknowledgements This study has been carried out with financial support from the National Natural Science Foundation of China (41871359). In addition, we wish to thank Jie CAI, a masters student of urban planning, Tongji University, for her work in literature collection.

References

Anable J, Lane B, Kelay T (2006) An evidence base review of public attitudes to climate change and transport behaviour. The Department for Transport, London

Badland H, Schofield G (2005) Transport, urban design, and physical activity: an evidence-based update. Transp Res D: Transp Environ 10(3):177–196. https://doi.org/10.1016/j.trd.2004.12.001

Ballester J, Robine J-M, Herrmann FR, Rodó X (2011) Long-term projections and acclimatization scenarios of temperature-related mortality in Europe. Nat Commun 2(1):358. https://doi.org/10.1038/ncomms1360

Bambrick H, Kjellström T (2004) Good for your heart but bad for your baby? Med J Aust 181(2):61–62

Basagaña X (2019) Heat Islands/temperature in cities: urban and transport planning determinants and health in cities. In: Nieuwenhuijsen M, Khreis H (eds) Integrating human health into urban and transport planning: a framework. Springer International Publishing, Cham, pp 483–497. https://doi.org/10.1007/978-3-319-74983-9_23

Beggs PJ (2004) Impacts of climate change on aeroallergens: past and future. Clin Exp Allergy 34(10):1507–1513. https://doi.org/10.1111/j.1365-2222.2004.02061.x

Borst HC, Miedema HME, de Vries SI, Graham JMA, van Dongen JEF (2008) Relationships between street characteristics and perceived attractiveness for walking reported by elderly people. J Environ Psychol 28(4):353–361. https://doi.org/10.1016/j.jenvp.2008.02.010

Borst HC, de Vries SI, Graham JMA, van Dongen JEF, Bakker I, Miedema HME (2009) Influence of environmental street characteristics on walking route choice of elderly people. J Environ Psychol 29(4):477–484. https://doi.org/10.1016/j.jenvp.2009.08.002

Brown RD, Vanos J, Kenny N, Lenzholzer S (2015) Designing urban parks that ameliorate the effects of climate change. Landsc Urban Plan 138:118–131. https://doi.org/10.1016/j.landurbplan.2015.02.006

Caney S (2009) Justice and the distribution of greenhouse gas emissions. J Glob Ethics 5(2):125–146. https://doi.org/10.1080/17449620903110300

Carter JG, Cavan G, Connelly A, Guy S, Handley J, Kazmierczak A (2015) Climate change and the city: building capacity for urban adaptation. Prog Plann 95:1–66. https://doi.org/10.1016/j.progress.2013.08.001

Cervero R, Kockelman K (1997) Travel demand and the 3Ds: density, diversity, and design. Transp Res D: Transp Environ 2(3):199–219

Chawla L (2015) Benefits of nature contact for children. J Plan Lit 30(4):433–452. https://doi.org/10.1177/0885412215595441

Chinese Society for Urban Studies (2009) China's low carbon eco-city development strategy. China City Press Beijing

Costello A, Abbas M, Allen A, Ball S, Bell S, Bellamy R, Friel S, Groce N, Johnson A, Kett M, Lee M, Levy C, Maslin M, McCoy D, McGuire B, Montgomery H, Napier D, Pagel C, Patel J, de Oliveira JAP, Redclift N, Rees H, Rogger D, Scott J, Stephenson J, Twigg J, Wolff J, Patterson C (2009) Managing the health effects of climate change: Lancet and University College London Institute for Global Health Commission. Lancet 373(9676):1693–1733. https://doi.org/10.1016/S0140-6736(09)60935-1

Cutler D, Miller G (2005) The role of public health improvements in health advances: the twentieth-century United States. Demography 42(1):1–22. https://doi.org/10.1353/dem.2005.0002

Eckert J, Shetty S (2011) Food systems, planning and quantifying access: using GIS to plan for food retail. Appl Geogr 31(4):1216–1223. https://doi.org/10.1016/j.apgeog.2011.01.011

Evans GW (2003) The built environment and mental health. J Urban Health 80(4):536–555. https://doi.org/10.1093/jurban/jtg063

Ewing R, Cervero R (2010) Travel and the built environment. J Am Plan Assoc 76(3):265–294. https://doi.org/10.1080/01944361003766766

Frank LD, Sallis JF, Conway TL, Chapman JE, Saelens BE, Bachman W (2006) Many pathways from land use to health: associations between neighborhood walkability and active transportation, body mass index, and air quality. J Am Plan Assoc 72(1):75–87. https://doi.org/10.1080/01944360608976725

Frank LD, Kerr J, Sallis JF, Miles R, Chapman J (2008) A hierarchy of sociodemographic and environmental correlates of walking and obesity. Prev Med 47(2):172–178. https://doi.org/10.1016/j.ypmed.2008.04.004

Gordon C, Purciel-Hill M, Ghai NR, Kaufman L, Graham R, Van Wye G (2011) Measuring food deserts in New York City's low-income neighborhoods. Health Place 17(2):696–700. https://doi.org/10.1016/j.healthplace.2010.12.012

Gosling SN, McGregor GR, Lowe JA (2009) Climate change and heat-related mortality in six cities Part 2: climate model evaluation and projected impacts from changes in the mean and variability of temperature with climate change. Int J Biometeorol 53(1):31–51. https://doi.org/10.1007/s00484-008-0189-9

Handy SL, Boarnet MG, Ewing R, Killingsworth RE (2002) How the built environment affects physical activity: views from urban planning. Am J Prev Med 23(2, Supplement 1):64–73. https://doi.org/10.1016/S0749-3797(02)00475-0

Hickman R, Banister D (2014) Transport, climate change and the city. Routledge, London

Hickman R, Hall P, Banister D (2013) Planning more for sustainable mobility. J Transp Geogr 33:210–219. https://doi.org/10.1016/j.jtrangeo.2013.07.004

IPCC (2014) Climate change 2014: impacts, adaptation and vulnerability: Part B: regional aspects. Cambridge University Press, Cambridge

Jackson LE (2003) The relationship of urban design to human health and condition. Landsc Urban Plan 64(4):191–200. https://doi.org/10.1016/S0169-2046(02)00230-X

Jacobson MZ, Delucchi MA (2011) Providing all global energy with wind, water, and solar power, Part I: technologies, energy resources, quantities and areas of infrastructure, and materials. Energy Policy 39(3):1154–1169. https://doi.org/10.1016/j.enpol.2010.11.040

Jones MM (2005) Protecting public health in New York City: 200 years of leadership. New York Department of Health and Mental Hygiene (NYDOHMH), New York

Kennedy C, Corfee-Morlot J (2013) Past performance and future needs for low carbon climate resilient infrastructure—an investment perspective. Energy Policy 59:773–783. https://doi.org/10.1016/j.enpol.2013.04.031

Kennedy CA, Ibrahim N, Hoornweg D (2014) Low-carbon infrastructure strategies for cities. Nat Clim Chang 4(5):343–346. https://doi.org/10.1038/nclimate2160

Kwan SC, Hashim JH (2016) A review on co-benefits of mass public transportation in climate change mitigation. Sustain Cities Soc 22:11–18. https://doi.org/10.1016/j.scs.2016.01.004

Laaidi K, Zeghnoun A, Dousset B, Bretin P, Vandentorren S, Giraudet E, Beaudeau P (2012) The impact of heat islands on mortality in paris during the august 2003 heat wave. Environ Health Perspect 120(2):254–259. https://doi.org/10.1289/ehp.1103532

Lee C, Moudon AV (2004) Physical activity and environment research in the health field: implications for urban and transportation planning practice and research. J Plan Lit 19(2):147–181. https://doi.org/10.1177/0885412204267680

Li J (2017) Citespace: technology text mining and visualization, 2nd edn. Capital University of Economics and Business Press, Beijing

Li F, Harmer PA, Cardinal BJ, Bosworth M, Acock A, Johnson-Shelton D, Moore JM (2008) Built environment, adiposity, and physical activity in adults aged 50–75. Am J Prev Med 35(1):38–46. https://doi.org/10.1016/j.amepre.2008.03.021

Lindberg F, Thorsson S, Rayner D, Lau K (2016) The impact of urban planning strategies on heat stress in a climate-change perspective. Sustain Cities Soc 25:1–12. https://doi.org/10.1016/j.scs.2016.04.004

Marshall JD, Brauer M, Frank LD (2009) Healthy neighborhoods: walkability and air pollution. Environ Health Perspect 117(11):1752–1759. https://doi.org/10.1289/ehp.0900595

Moura F, Cambra P, Gonçalves AB (2017) Measuring walkability for distinct pedestrian groups with a participatory assessment method: a case study in Lisbon. Landsc Urban Plan 157:282–296. https://doi.org/10.1016/j.landurbplan.2016.07.002

Northridge ME, Sclar ED, Biswas P (2003) Sorting out the connections between the built environment and health: a conceptual framework for navigating pathways and planning healthy cities. J Urban Health 80(4):556–568. https://doi.org/10.1093/jurban/jtg064

Páez A, Gertes Mercado R, Farber S, Morency C, Roorda M (2010) Relative accessibility deprivation indicators for urban settings: definitions and application to food deserts in Montreal. Urban Stud 47(7):1415–1438. https://doi.org/10.1177/0042098009353626

Palinkas LA, Wong M (2020) Global climate change and mental health. Curr Opin Psychol 32:12–16. https://doi.org/10.1016/j.copsyc.2019.06.023

Raparthi K (2018) Assessing the role of urban planning policies in meeting climate change mitigation goals in Indian cities. J Urban Plan Dev 144(2):05018005. https://doi.org/10.1061/(ASCE)UP.1943-5444.0000440

Rissel CE (2009) Active travel: a climate change mitigation strategy with co-benefits for health. N S W Public Health Bull 20(2):10–13. https://doi.org/10.1071/NB08043

Shi J, Mu H-Z (2016) Sustainable development of large cities facing climate change—taking shanghai as an example. ResourEnviron Yangtze Basin 1:1

Sietchiping R, Permezel MJ, Ngomsi C (2012) Transport and mobility in sub-Saharan African cities: an overview of practices, lessons and options for improvements. Cities 29(3):183–189. https://doi.org/10.1016/j.cities.2011.11.005

Smoyer KE, Rainham DGC, Hewko JN (2000) Heat-stress-related mortality in five cities in Southern Ontario: 1980–1996. Int J Biometeorol 44(4):190–197. https://doi.org/10.1007/s004840000070

Stern N, Taylor C (2007) Climate change: risk, ethics, and the stern review. Science 317(5835):203–204. https://doi.org/10.1126/science.1142920

Stone B, Hess JJ, Frumkin H (2010) Urban form and extreme heat events: are sprawling cities more vulnerable to climate change than compact cities? Environ Health Perspect 118(10):1425–1428. https://doi.org/10.1289/ehp.0901879

Sturiale L, Scuderi A (2019) The role of green infrastructures in urban planning for climate change adaptation. Climate 7(10):119. https://doi.org/10.3390/cli7100119

Sun S (2017) Sustainable development and rational planning: 2017 annual national planning conference essays. China Construction Industry Press, China

Thomas F, Sabel CE, Morton K, Hiscock R, Depledge MH (2014) Extended impacts of climate change on health and wellbeing. Environ Sci Pol 44:271–278. https://doi.org/10.1016/j.envsci.2014.08.011

Villanueva K, Badland H, Hooper P, Koohsari MJ, Mavoa S, Davern M, Roberts R, Goldfeld S, Giles-Corti B (2015) Developing indicators of public open space to promote health and wellbeing in communities. Appl Geogr 57:112–119. https://doi.org/10.1016/j.apgeog.2014.12.003

Watts N, Adger WN, Agnolucci P, Blackstock J, Byass P, Cai W, Chaytor S, Colbourn T, Collins M, Cooper A, Cox PM, Depledge J, Drummond P, Ekins P, Galaz V, Grace D, Graham H, Grubb M, Haines A, Hamilton I, Hunter A, Jiang X, Li M, Kelman I, Liang L, Lott M, Lowe R, Luo Y, Mace G, Maslin M, Nilsson M, Oreszczyn T, Pye S, Quinn T, Svensdotter M, Venevsky S, Warner K, Xu B, Yang J, Yin Y, Yu C, Zhang Q, Gong P, Montgomery H, Costello A (2015) Health and climate change: policy responses to protect public health. Lancet 386(10006):1861–1914. https://doi.org/10.1016/S0140-6736(15)60854-6

WHO (1948) Basic documents. WHO, Geneva

WHO (1998) Health promotion glossary. WHO, Geneva

Winters M, Brauer M, Setton EM, Teschke K (2010) Built environment influences on healthy transportation choices: bicycling versus driving. J Urban Health 87(6):969–993. https://doi.org/10.1007/s11524-010-9509-6

Wolch JR, Byrne J, Newell JP (2014) Urban green space, public health, and environmental justice: the challenge of making cities 'just green enough'. Landsc Urban Plan 125:234–244. https://doi.org/10.1016/j.landurbplan.2014.01.017

Xia T, Nitschke M, Zhang Y, Shah P, Crabb S, Hansen A (2015) Traffic-related air pollution and health co-benefits of alternative transport in Adelaide, South Australia. Environ Int 74:281–290. https://doi.org/10.1016/j.envint.2014.10.004

Younger M, Morrow-Almeida HR, Vindigni SM, Dannenberg AL (2008) The built environment, climate change, and health: opportunities for co-benefits. Am J Prev Med 35(5):517–526. https://doi.org/10.1016/j.amepre.2008.08.017

Zou B, Wilson JG, Zhan FB, Zeng Y (2009) Spatially differentiated and source-specific population exposure to ambient urban air pollution. Atmos Environ 43(26):3981–3988. https://doi.org/10.1016/j.atmosenv.2009.05.022

Part II
Urban Climate Science in Action

Chapter 3
Embedding Climate Change in Urban Planning and Urban Design in New York City

Jeffrey Raven, Christian Braneon, and Cynthia Rosenzweig

Abstract Confronting the challenges of a rapidly urbanizing world threatened by climate change requires expanding the traditional influence and capabilities of urban planning and urban design, integrating climate science, natural systems and compact urban form to configure dynamic, desirable and healthy communities. Cost-effective planning and design measures that help mitigate emissions while bringing adaptive benefits should be prioritized. The chapter draws from the publication *Climate Change and Cities* (Cambridge University Press 2018) by the Urban Climate Change Research Network (UCCRN). The two-phase New York City case study by the Urban Design Climate Lab at the New York Institute of Technology and a team of international urban design climate experts illustrates how this emerging expertise can be replicated and implemented worldwide. Its focus on configuring people-centered public spaces that enhance energy efficiency and improve public health draws from four urban climate factors: improving efficiency of urban systems, both in energy and transportation; optimizing the form and layout of urban districts to enhance ventilation; promoting appropriate building materials with high reflectivity; and increasing green and blue urban infrastructure. The chapter highlights a set of tools and methods to measure success.

Keywords Urban climate · Mitigation · Adaptation · Urban planning · Urban design · Resiliency

---

J. Raven (✉)
School of Architecture & Design, New York Institute of Technology (NYIT-Manhattan), New York, NY, USA
e-mail: jraven@nyit.edu

C. Braneon · C. Rosenzweig
NASA Goddard Institute for Space Studies, New York, NY, USA
e-mail: christian.v.braneon@nasa.gov; cynthia.rosenzweig@nasa.gov

C. Ren, G. McGregor (eds.), *Urban Climate Science for Planning Healthy Cities*, Biometeorology 5, https://doi.org/10.1007/978-3-030-87598-5_3

Key Messages

- Cities should immediately implement climate action. Land-use patterns incorporate climate change impact assessments in urban areas.
- Climate change mitigation and adaptation strategies should form a core element in urban planning and urban design, taking into account local conditions. Investing in mitigation strategies that yield concurrent adaptive benefits should be prioritized. Integrating mitigation and adaptation can reduce counterproductive infrastructure and urban policies.
- Important strategies used by urban planners and urban designers to facilitate integrated mitigation and adaptation in cities include: (1) reducing waste heat and greenhouse gas emissions through energy efficiency, transit access, and walkability; (2) modifying the form and layout of buildings and urban districts; (3) use of heat-resistant construction materials and reflective surface coatings; and (4) increasing vegetative cover.
- Connections between climate change, infrastructure, and public health should be made clear to city planners, policymakers, and to the general public. The needs of the most disadvantaged and vulnerable citizens should be addressed in climate change planning and action.
- Climate risk assessments and climate action plans co-generated with the full range of stakeholders and scientists are most effective. Processes that are inclusive, transparent, participatory, multisectoral, multijurisdictional, and interdisciplinary are the most robust because of the inherent complexity and multiscale nature of climate resilient urban planning and design.
- The New York City Panel on Climate Change (NPCC) is a key resource for the City of New York as it responds to climate challenges by developing a broad range of climate resiliency policies and programs, as well as the knowledge base to support them. While specific to New York City and its metropolitan region, the approaches developed by the NPCC can contribute to efforts to enhance resilience as they are undertaken across governmental scales as well as in other locations.

3.1 Introduction

Developing the New York Metropolitan area in a denser, more compact form that mixes land use and supports mass transit reduces the regional carbon footprint. The resulting dense urban districts can be configured to reduce the impact of heat waves and storms while enhancing quality of life. Climate change mitigation and adaptation strategies should form a core element in urban planning and urban design, taking into account local conditions.

Investing in mitigation strategies that yield concurrent adaptive benefits should be prioritized; actions that reduce greenhouse gas (GHG) emissions while increasing resilience provide win-win outcomes. Integrating mitigation and adaptation can help New York and other cities avoid locking themselves into counterproductive infrastructure and policies. Decisions on urban form have long-term consequences and affect the city's capacity to reduce GHG emissions and respond to climate hazards.

Urban planning[1] and urban design[2] encompass multiple disciplines, providing critical input to inform systems, management, and governance for sustainability and resilience to climate change. Connections between climate change, infrastructure, and public health should be made clear to city planners, policymakers, and urban designers. Collaborative efforts, focused, for example, on the identification of vulnerable residents and resources, have been recognized as effective for enhancing community resilience during extreme events (Barata et al. 2018; Raven et al. 2018).

Section 3.1 introduces the four urban climate factors that can be used by urban planners and urban designers to facilitate integrated mitigation and adaptation in cities.

Section 3.2 provides an overview of climate hazards impacting public health in New York City.

Section 3.3 describes climate science-informed policy-making and participatory processes in New York City.

Section 3.4 presents a NYC Case Study demonstrating application of the urban climate factors. The Urban Design Climate Workshop planning process envisions urban design shaping transformative climate action in New York City, whose value proposition is based on positive public health and economic outcomes.

Section 3.5 focuses on the value of cross-sectoral collaboration for urban climate transformation.

3.2 The Four Climate Factors

There is a growing consensus around integrating urban planning and urban design, climate science, and policy to bring about desirable microclimates that address both mitigation and adaptation. It is widely accepted that to date there has been limited integration of climatic science into urban planning and design practice (Georgescu et al. 2014; Romero-Lankao et al. 2018). Many urban microclimate studies have documented that urban microclimate can affect human morbidity and mortality as

[1] Urban planning is a field of practice that helps city leaders to transform a sustainable development vision into reality using space as a key resource for development and engaging a wide variety of stakeholders in the process (UN-Habitat 2013).

[2] Urban design involves the arrangement and design of buildings, public spaces, transport systems, services, and amenities. Urban design is derived from but transcends planning, transportation policy, architectural design, development economics, engineering, and landscape.

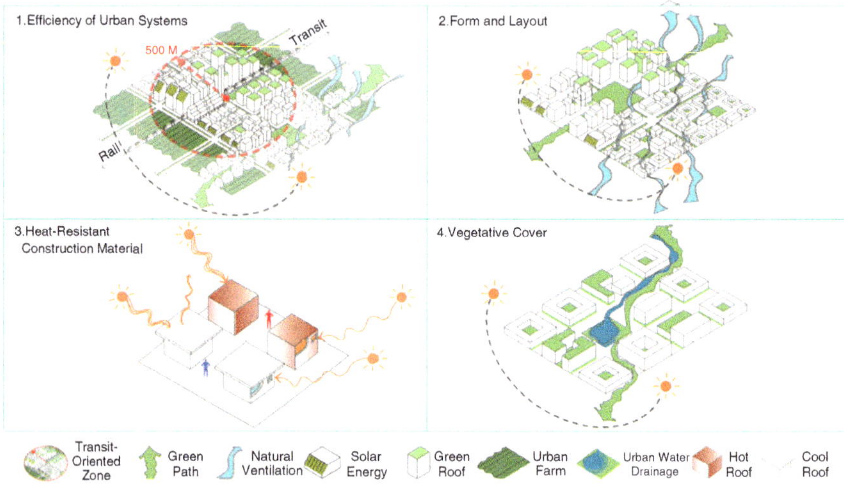

Fig. 3.1 The four urban climate factors: (1) reducing waste heat and greenhouse gas emissions through energy efficiency, transit access, and walkability; (2) modifying form and layout of buildings and urban districts; (3) use of heat-resistant construction materials and reflective surface coatings; and (4) increasing vegetative cover (Source: Raven 2016)

well as building energy performance and thermal comfort (Bader et al. 2018; Barata et al. 2018). The Urban Planning and Urban Design chapter of *Climate Change and Cities*: Second Assessment Report (ARC3.2) of the Urban Climate Change Research Network (UCCRN) focuses on urban design's central role in greenhouse gas emission (GHG) mitigation, stormwater management and urban heat stress adaptation and resilience. This research is producing new knowledge, expanding on the traditional influence and capabilities of urban planning and urban design to configure dynamic, desirable, and healthy communities. ARC3.2 identified four urban climate factors (Fig. 3.1) to guide decisions on urban form and urban function (Raven et al. 2018) that inform UCCRN's Urban Design Climate Workshop[3] (UDCW) series.

1. **Efficiency of urban systems**. Reducing urban waste heat and greenhouse gas emissions from infrastructure and industry through improvements in the efficiency of urban systems.

   Since the middle of the twentieth century, global urban development patterns have tended to increase outward from central cities. The infrastructure[4] network needed to maintain this sprawling, low-density development pattern has resulted

[3] The Urban Climate Change Research Network (UCCRN) established a multi-city, district-level Urban Design Climate Workshop (UDCW) program. The UCCRN UDCW sessions allow stakeholders to integrate and scale up mitigation and adaptation principles by reducing energy consumption in the built environment, strengthening climate resilience, and enhancing human quality of life.

[4] Infrastructure is typically expected to remain functional durable, and safe for 50 to more than 100 years.

in increased reliance on motor vehicles and networks of roads and parking land uses. This reliance on motor vehicles has consequently led to a significant increase in vehicle miles (or kilometers) travelled, a concomitant increase in GHG emissions, and an amplification of the urban heat island (UHI) effect through increased imperviousness, reduced green cover, and enhanced waste heat emissions (Stone et al. 2010).

Much of the current evidence indicates that densely occupied cities are more efficient in their energy use and generate less waste heat as a consequence (Resch et al. 2016). The evidence is especially strong for transport energy, which is largely based on cities relying on effective mass transit systems (Newman and Kenworthy 1989). Increasing urban population density through policies that co-manage land-use and transport networks is an important strategy for reducing urban GHG emissions (Dulal et al. 2011). In a study focused on urban spatial structure and extreme heat across more than 50 large US cities over the period of 1956–2005, Stone et al. (2010) found the annual frequency of extreme heat events to be rising more slowly in the most compact metropolitan regions; the most sprawling cities experienced a rate of increase in extreme heat events that was more than double that of the most compact cities. Furthermore, rates of deforestation were twice as high in the most sprawling metropolitan regions between 1992 and 2001.

2. **Form and layout**. Modifying the form and layout of buildings and urban districts can provide cooling and ventilation that reduces energy use and allow citizens to cope with higher temperatures and more intense runoff.

   Urban morphology is defined as the three-dimensional form and layout of the built environment and settlement pattern. It is responsible for varying the extent of airflow and solar impact in cities. Exploiting prevailing breezes is a key factor in implementing district-wide passive cooling strategies that can reduce energy loads in the built environment. At the neighborhood scale, streets oriented to the prevailing wind directions with staggered building arrangements together with street trees appear to offer the best possibility to deal with urban warming in warm humid cities (Smith and Levermore 2008).

   A comprehensive climate-based urban design supports developing and maintaining a network of green and blue infrastructure integrated with the built environment to conserve ecosystem functions and provide associated benefits to human populations (Raven et al. 2018). The location and form of green infrastructure should be determined in relation to the built environment and aligned in relation to natural systems, including water bodies,[5] solar impacts, and prevailing winds. Passive methods to increase comfort and reduce energy loads through solar design include orienting street and public space layout to reduce solar gain during hot months, shading through the configuration of adjacent vegetation, orienting neighborhood configurations to the sun's path, as well as vary-

[5] Localized water bodies can moderate temperature extremes through their high thermal storage capacity and through evaporative cooling.

ing building heights and breaks in the building line to control shadowing and increase solar access during cold months.

3. **Heat-resistant construction materials**. Selecting low heat capacity construction materials and reflective coatings can improve building performance by managing heat exchange at the surface.

   Increasing the surface reflectivity or albedo of urban materials is a well-established urban heat management strategy. An important advantage of albedo enhancement over other urban climate management strategies is its relatively low cost. In contrast to tree planting and other vegetative programs in which maximum cooling benefits are not realized until plants reach maturity, high-albedo coatings yield maximum benefits upon installation. However, their benefits diminish over time due to weathering, and as roofs become soiled.

   To explore the extent to which cool materials could reduce temperatures not only within the treated buildings themselves but also throughout the ambient urban environment, scientists at the Lawrence Berkeley Labs and the Columbia University *Center for Climate Systems Research* have modeled extensive albedo enhancement strategies (Rosenzweig et al. 2014). Industry analyses of these materials have found that the surface temperature of roofing materials can be reduced by as much as 50 °F/10 °C during periods of intense solar gain (Gaffin et al. 2012). In densely settled districts such as Manhattan, the potential to raise average albedos is great, but all cities can enhance their reflectivity through the use of higher albedo materials in routine resurfacing over time.

4. **Vegetative cover**. Increasing the vegetative cover in a city can simultaneously lower outdoor temperatures, building cooling demand, runoff, and pollution, while sequestering carbon.

   Evaporative cooling processes, fostered by the development of green spaces in cities, allow for sustainable management of the water cycle and a reduction of the UHI effect. Vegetation may be sited to maximize the absorption rate of solar radiation. Incorporating vegetation measures (green roofs and vegetative facades) reduce the amount of water flowing into sewage systems, mitigate temperature extremes, provide thermal insulation, and increase biodiversity in urban areas (Nolde et al. 2007; Schmidt 2009; Steffan et al. 2010; UNEP 2012; CIRIA 2019). Green streetscapes, including the use of permeable paving, provide shade and reduce thermal radiation.

3.3 New York City Climate Hazards

"Hurricane Sandy has focused attention on the significant effects that extreme climate events have on New York City" (Horton et al. 2015). "Although it is not possible to attribute any one extreme event such as Hurricane Sandy to climate change, sea level rise already occurring in the New York metropolitan region, in part due to climate change, increased the extent and magnitude of coastal flooding during the storm. This is an example of how long-term trends in climate variables can modify the risk of extremes" (Horton et al. 2015).

"Both temperature and precipitation extremes have significant impacts on New York City" (Horton et al. 2015). Summer (June, July, and August) temperatures are expected to increase in New York City throughout the twenty-first century (Horton et al. 2015), leading to more frequent and intense extreme heat events known as heat waves[6] (González et al. 2019). Heat waves increase energy demand (Sailor 2001; Schaeffer et al. 2012; Santamouris 2014) and mortality (Knowlton et al. 2007; Luber and McGeehin 2008; Anderson and Bell 2011; Rosenthal et al. 2014). "Moreover, higher temperatures associated with urbanization and the Urban Heat Island (UHI) (Oke 1982) exacerbate the impacts of extreme heat events" (Li and Bou-Zeid 2013; Ramamurthy and Bou-Zeid 2017; Ramamurthy et al. 2017; Ortiz et al. 2018; González et al. 2019).

3.3.1 New York City Climate

"New York City experiences significant precipitation throughout the year, with relatively little variation from month to month in the typical year" (Horton et al. 2015). "The primary large weather systems that affect New York City are cyclones. Cyclones refer to low-pressure regions where air converges and causes uplift. Cyclones can include extratropical cyclones, caused by mid-latitude weather fronts (e.g., Nor'Easters) and tropical cyclones, which originate in the tropical oceans (e.g., hurricanes). The tracks for both types of cyclones are used to associate extreme precipitation with storms" (González et al. 2019).

Annual average precipitation ranges between approximately 109 and 127 cm (43 and 50 inches), depending on the location within the city. "Summers in New York City are warm, with cool winters" (Horton et al. 2015). Annual mean air temperature in New York City[7] was approximately 54 °F/12.2 °C from 1971 to 2000 (Horton et al. 2015).

"Historical trends of daily maximum summer temperature in New York may be assessed using Central Park weather station, John F. Kennedy (JFK), and LaGuardia Airports during June, July, and August" (see Fig. 3.2) (González et al. 2019). "Central Park has the longest historical record, dating back to 1900, where the average annual daily maximum summer temperature has been rising at an average of 0.2 °F per decade from 1900 to 2013. JFK and LaGuardia weather stations go back to 1970, where average annual daily maximum summer temperatures have been

[6] Here, we follow the definition of heat waves according to the National Weather Service (NWS), that is, an interval of 3 (or more) consecutive days with temperatures of at least 90 °F (32.2 °C).

[7] Data from the Central Park weather station; temperatures from the meteorological station in Central Park tend to be lower than those in some other parts of New York City. This is due to the close proximity of the weather station to extensive vegetation.

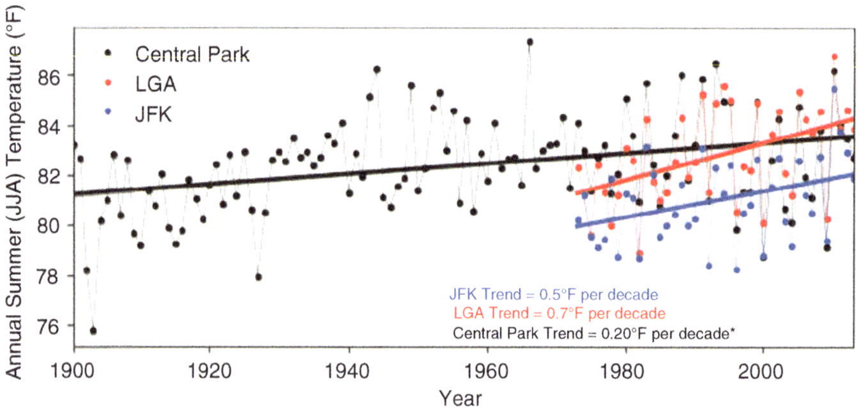

Fig. 3.2 Annual average daily maximum summer temperatures (June, July, August) in Central Park from 1900 to 2013, LaGuardia (LGA) Airport from 1970 to 2013, and John F. Kennedy (JFK) International Airport from 1970 to 2013. Solid lines represent linear trends for each station. Station records were obtained from the U.S. Historical Climatology Network (USHCN) Version 2.5 (Menne et al. 2013; González et al. 2019)

increasing at a rate of 0.5 °F per decade and 0.7 °F per decade, respectively" (González et al. 2019).

"The distance between these weather stations provides insights into processes that affect temperatures near the surface, such as sea breezes and the UHI. Sea breeze effects appear in stations located close to Long Island's southern shore (e.g., JFK), with lower daily maximum temperatures compared to their in-land counterparts" (see Fig. 3.2) (González et al. 2019). "Sea breeze impacts on temperatures show that geospatial heterogeneity of the urban landscape plays a role in near-surface temperatures, and therefore impact occurrences of extreme heat. The weather station located at JFK, which experiences afternoon sea breezes, has a mean summer maximum temperature of 80.6°F/27°C, whereas the other stations have a mean value of 82.7°F/28.2°C, which is 2.1°F/1.2°C higher. This is consistent with climatological studies (e.g., Gedzelman et al. 2003) of the UHI in the region, which have found that afternoon summer sea breezes may shift the center of the urban heat island west and north, toward New Jersey and The Bronx" (González et al. 2019). "Sea breezes form due to temperature differences between the air over land and ocean and are a feature of New York City coastal areas" (Haurwitz 1947; Childs and Raman 2005; González et al. 2019).

3.3.2 *Extreme Events*[8]

"The number of extreme events in a given year is highly variable" (Horton et al. 2015). For example, New York City recently recorded three consecutive years (2010–2012) with at least one day with maximum temperatures at or above 100 °F/37.8 °C. "Prior to 2010, the last day at or above 100°F was in 2001, and there has only been one other time on record (1952–1955) where New York City experienced more than two years in a row with maximum temperatures at or above 100°F" (Horton et al. 2015). From 1971 to 2000, New York City averaged 18 days per year with maximum temperatures at or above 90 °F/32.2 °C, 0.4 days per year at or above 100 °F, and two heat waves per year (Horton et al. 2015).

"Extreme precipitation timescales are highly asymmetrical" with heavy precipitation events generally ranging from less than an hour to a few days, whereas "meteorological droughts can range from months to years" (Horton et al. 2015). Between 1971 and 2000, New York City averaged 13 days per year with 2.54 cm (1 inch) or more of rain, 3 days per year with 5.08 cm (2 inches) or more of rain, and 0.3 days per year with 10.16 cm (4 inches) or more of rain. As with extreme temperatures, year-to-year variations in extreme precipitation events are large (Horton et al. 2015).

"There has been a small but not statistically significant trend toward more extreme precipitation events in New York City since 1900. For example, the four years with the greatest number of events with 2 inches or more of rain have all occurred since 1980 (1983, 1989, 2007, and 2011). Because extreme precipitation events tend to occur relatively infrequently, long time-series of measurements over large areas are needed to identify trends; there is a relatively large burden of proof required to distinguish a significant trend from random variability" (Horton et al. 2015). Over the larger Northeast region of the United States, intense precipitation events (defined here as the heaviest 1% of all daily events) have increased by approximately 70% over the period from 1958 to 2011 (Horton et al. 2014).

3.3.3 *Projections for the New York Metropolitan Region*

Climate change[9] is extremely likely to bring higher temperatures to the New York metropolitan region. Notably, heat waves and intense rainstorms are very likely to increase in magnitude and frequency. "The temperature trend since 1900 for the New York metropolitan region is broadly similar to the trend for the northeast

[8] "When a single climate variable or combinations of variables approach the tails of their distribution, this is referred to as an extreme event" (Horton et al. 2015).

[9] Climate change refers to a significant change in the state of the climate that can be identified from changes in the average state or the variability of weather and that persists for an extended time period, typically decades to centuries or longer. Climate change can refer to the effects of (1) persistent anthropogenic or human-caused changes in the composition of the atmosphere and/or land use, or (2) natural processes such as volcanic eruptions and Earth's orbital variations (IPCC 2013).

United States" (Horton et al. 2015). This trend cannot be explained by the urban heat island effect alone as it is present in both rural and urban weather stations (Horton et al. 2015).

Precipitation[1] has increased at a rate of approximately 2 cm (0.8 inches) per decade from 1900 to 2013. Year-to-year (and multiyear) variability of precipitation has also become more pronounced, especially since the 1970s. The standard deviation, a measure of variability, increased from 15.5 cm (6.1 inches) from 1900 to 1956 to 26.2 cm (10.3 inches) from 1957 to 2013 (Horton et al. 2015).

Heavy downpours in the New York metropolitan region are very likely to increase by the 2080s (Horton et al. 2015; González et al. 2019). Increase in precipitation, both in terms of the mean and extremes for the region (Ning and Bradley 2014). These precipitation changes are expected to occur in both winter and summer seasons (Fan et al. 2014; González et al. 2019).

Projected days above 90 °F/32.2 °C are expected to become more likely as summer temperatures increase. By the 2080s, projections show 24 (tenth percentile) to 75 (90th percentile) days above 90 °F/32.2 °C compared to the 1971–2000 baseline (10 days) (González et al. 2019).

"Increases in specific humidity combined with increasing temperatures might lead to a higher heat index, which has major consequences for human health and is a driver of peak energy demand for space cooling, as air conditioning systems remove sensible (temperature-related) and latent (moisture-related) heat from buildings. To assess combined air moisture and temperature impacts, concurrent hourly values must be used, rather than daily outputs from the model ensemble" (González et al. 2019). "Climate model results can be statistically or dynamically downscaled (e.g., using regional models embedded within global models), but some processes may not be captured by existing downscaling techniques. Examples include changes in land–sea breezes and the urban heat island effect on a warming planet" (Horton et al. 2015).

"Recent studies have projected future rainfall intensity–duration–frequency (IDF) curves for New York State. Rainfall extremes for durations longer than 1 h were projected using a set of downscaled GCMs and RCMs (DeGaetano and Castellano 2015, 2017; Castellano and DeGaetano 2017). Future IDF curves were developed by using change factors calculated as change in rainfall between past observations and downscaled climate model projections of future rainfall. These projections, which can be found online at http://ny-idf-projections.nrcc.cornell.edu/, are more certain for rainfall durations of more than 24 h than for those under 24 h. These are peer reviewed, local, sub-daily rainfall projections available for New York City, used in development of the New York City Climate Resiliency Design Guidelines (http://www.nyc.gov/resiliency)" (NYC Mayor's Office of Recovery and Resiliency 2018; González et al. 2019).

3.4 New York City Resilience and Infrastructure Plans

If future cities are to be sustainable and resilient, they must develop the physical and institutional capacities to respond to constant change and uncertainty. This will require strategies for long-term commitments across multiple electoral cycles and often among many political jurisdictions that constitute functional metropolitan areas (Romero-Lankao et al. 2018).

Climate science informed urban planning for resiliency formally started in New York City in 2008 when Mayor Bloomberg established the New York City Panel on Climate Change (NPCC). A collaboration of climate experts from around the New York metropolitan region, the NPCC was convened with the objective of providing the city with the most up-to-date climate science information relevant for the region that could be used in preparing for the impacts of climate change. The NPCC examines historical trends and develops future projections for risks specific to New York City around changes in temperature, precipitation, and sea level rise using state-of-the-art climate models. The goal is to make New York City and the surrounding metropolitan region more resilient to mean changes in climate and to future extreme events (e.g., Lempert et al. 1996; Kunreuther et al. 2013) (Horton et al. 2015). NPCC has created three reports to date—the first was released in 2010 (NPCC1), the second in 2015 (NPCC2), and the third (i.e., NPCC3) in 2019 (NPCC 2010, 2015, 2019).

For New York City-specific climate projections for the twenty-first century, quantitative global climate model-based projections are provided for means and extremes of temperature and precipitation (Horton et al. 2015). The NPCC has sought to present climate uncertainties clearly in order to facilitate risk-based decision-making for the use of policy tools such as incentives, regulations, and insurance. Among other risks, NPCC has analyzed future changes in extreme heat days and heat waves, and uses that inform information policies that aim to prepare residents, infrastructure, and buildings for these changes.

> "New methodologies for projections of heat wave characteristics for the New York metropolitan region were tested in NPCC3 using bias-corrected climate model projections. For the early part of the century (2020s), these results are consistent with those of NPCC2. In the later part of the century (2050s and 2080s), the NPCC3 results display the potential for more intense heat events with longer durations. Results show large changes across all heat wave metrics throughout the twenty-first century. The high-emissions scenario (RCP8.5) projects, in many cases, several times larger effects than the medium-emissions scenario. The uncertainty of the projections increases through time. The new NPCC3 methods include humidity, which is projected to increase by more than 30% from baseline values. These increases in atmospheric humidity with extreme temperatures are likely to have large societal implications reflected in public health and energy demands" (González et al. 2019).

> "NPCC3 confirms the NPCC2 projections for heat waves, hot days, and cold days as those of record for New York City in planning for the impacts of climate change and recommends the incorporation of the new methodologies into revised projections of record in NPCC4. Future work in projecting extreme heat and humidity for the NPCC should be directed to incorporating the spatial distribution of these extreme heat events to account for coastal

influence and UHI effects (e.g., sea breeze effects). This may require using regional climate models (RCMs) to dynamically downscale projections to finer spatial scales within the New York metropolitan region" (González et al. 2019).

3.4.1 Historical and Future Extreme Heat in New York City

Establishing the baseline is key to understanding how extreme heat and heatwaves have changed over time, especially as the climate change signal becomes more pronounced. A historical baseline and trend was established for the metrics described above using station data from the Central Park weather station. Between 1971 and 2000, New York City saw an average of 14 days over 90 °F (32 °C), 0.4 days over 100 °F (38 °C), and two heat waves a year lasting about four consecutive days each, with the number of events in a year being highly variable.

Extreme heat projections for these metrics were then developed utilizing the same methodology from the IPCC, using 35 global climate models (GCMs) and 2 greenhouse gas emissions scenarios—RCP[10] 4.5, a middle-emissions scenario, and RCP 8.5, a high-emissions scenario. Results of the analysis are shown in Table 3.1.

3.4.2 Stakeholder Engagement and Planning for Extreme Heat

NPCC engages directly with the New York City Mayor's Office of Climate Resiliency (MOCR) for discussion on research needs and how to best present climate information for use in policy in planning. MOCR works to connect NPCC members with additional relevant City agencies as needed, such as the NYC Department of City Planning, NYC Department of Environmental Protection, and the Mayor's Office of Climate and Sustainability. In addition, NPCC scientists connect with community members and infrastructure managers across the City to understand their unique risks.

The climate science information developed by NPCC has already been in use by City decision-makers, community members, and infrastructure managers in planning for the impacts of extreme heat and other changes in climate. While not a complete list of all initiatives, these activities include some of the following:

- The New York City Climate Resiliency Design Guidelines outline how to incorporate forward looking climate science from the NPCC into the design of all New York City capital projects (NYC Mayor's Office of Recovery and Resiliency

[10] Representative concentration pathways (RCPs) are sets of trajectories of concentrations of greenhouse gas emissions, aerosols, and land-use changes developed for climate models as a basis for long-term and near-term climate-modeling experiments (Moss et al. 2010).

Table 3.1 Projections of extreme heat days and heat waves in New York City in the 2020s, 2050s, and 2080s compared to the 1971–2000 baseline (Source: NPCC 2015)

Extreme heat	Baseline (1971–2000)	Low estimate (tenth percentile)	Middle range (25th to 75th percentile)	High estimate (90th percentile)
2050s				
Number of heat waves per year	2	4	5–7	7
Average heat wave duration (days)	4	5	5–6	6
Days per year with maximum temperatures at or above 90 °F	18	32	39–52	57
Days per year with maximum temperatures at or above 100 °F	0.4	2	3–5	7
2080s				
Number of heat waves per year	2	5	6–9	9
Average heat wave duration (days)	4	5	5–7	8
Days per year with maximum temperatures at or above 90 °F	18	38	44–76	87
Days per year with maximum temperatures at or above 100 °F	0.4	2	4–14	20

2018). The guidelines provide a consistent methodology for engineers, architects, and planners to design resilient facilities.

- With a core vision on City resilience, the OneNYC plan outlines major goals in preparing neighborhoods, buildings, infrastructure, and coastal defenses for the effects of climate change using NPCC analysis and projections (City of New York 2018).
- The New York City Climate Change Adaptation Task Force is a group of infrastructure stakeholders convened by the Mayor's Office that works to incorporate NPCC science information and share strategies for how key infrastructure systems can prepare for the effects of climate change. These infrastructure groups include energy, transportation, telecommunications, social infrastructure including hospitals and parks, as well as water, waste, and sewer systems.
- Developed by WE ACT for Environmental Justice, a community-based resilience group in Harlem, the Northern Manhattan Climate Action Plan is a guide for creating resiliency in the face of the disproportionate impacts of climate change on poor and working-class communities (WE ACT for Environmental Justice 2015).
- Cool Neighborhoods NYC is a comprehensive approach to keep communities safe in extreme heat that provides $106 million to help keep New Yorkers safe

during hot weather, mitigate the drivers of the urban heat island effect, and protect against the worst impacts of rising temperatures from climate change (City of New York 2017).

Stakeholder engagement is an ongoing, iterative process between NPCC scientists, city planners, and community members. Based upon stakeholder feedback utilizing the climate projections developed, the NPCC3 report published in 2019 examined a new methodology for assessing extreme heat, including heat wave intensity, distribution of extreme heat across New York City neighborhoods, and the effects of combined extreme heat and humidity events. These conversations between scientists and stakeholders continually push the NPCC to provide policy-relevant science that can inform responses to climate change over time.

While the NPCC presents a scientist-stakeholder engagement model that will hopefully be replicated around the world, there are still tremendous opportunities for NYC to develop a more integrated approach to adaptation and mitigation that supports district scale planning.

3.5 Urban Planning and Urban Design Case Study: Sunnyside Yard, NYC

New York City's population growth over the past decades to 8.5 million has placed stress on the city and region's core infrastructure. Recognizing the resulting land scarcity the public sector is reaching for innovative solutions to enhance transit infrastructure while meeting housing needs and job opportunities for NYC residents. Sunnyside Yard, a large rail yard in Western Queens, remains one of New York City's last great opportunities to solve these overlapping challenges in one district. This 180-acre site houses key rail operations for Amtrak, the MTA, and NJ Transit. The enormous rail infrastructure has divided neighboring communities in Queens. In 2015, the City of New York launched a feasibility analysis to build a deck over Sunnyside Yard to develop a planned high-density district in adjacent to the rapidly developing Long Island City district all while allowing rail operations to continue underneath (Fig. 3.3). With this ambitious initiative underway, impact studies have evaluated alternative futures for Sunnyside Yard, including urban climate and public health (Fig. 3.4).

3.5.1 Urban Design Climate Workshop (UDCW) Planning Process

The initial phase (i.e., Phase 1) evaluating alternative futures for the Sunnyside Yard district was led by the Urban Design Climate Lab studio at the New York Institute of Technology. Phase 2 was a UCCRN Urban Design Climate Workshop (UDCW)

Fig. 3.3 Sunnyside Yard, Queens; Manhattan in background

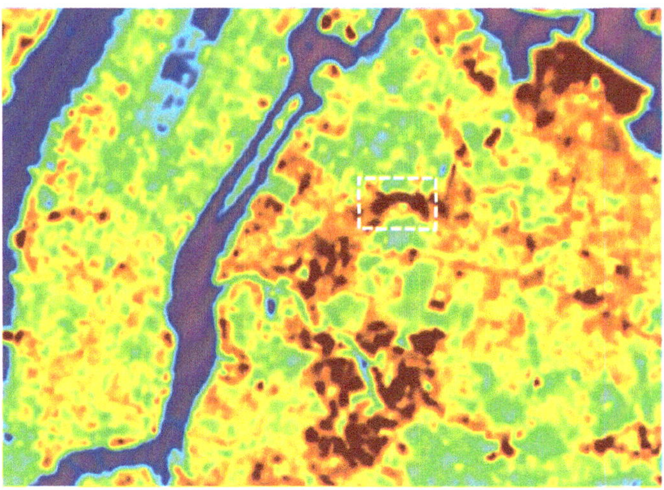

Fig. 3.4 Sunnyside Yard, Land Surface Temperature Scenario (Data Source: NASA, Landsat 8)

comprising urban designers, urban planners, urban climate scientists, policymakers, stakeholders, and graduate students working side by side to explore climate-driven urban design interventions for Sunnyside Yard in Queens, NY.

The UDCW used off-the-shelf Geographic Information Systems (GIS) mapping software and digital Parametric 3-D modeling tools (Rhinoceros, Grasshopper–with Ladybug plug-ins), drawing from Weather Data (EPW) files. The urban planning and design interventions were developed by considering outputs from three district development scenarios: Baseline (current condition), 2050 Baseline (business-as-usual), and 2050 Best Practice (climate-driven urban design) scenarios (Figs. 3.5, 3.6 and 3.7). Prior to the intervention, the design research includes (1) modeling and

Fig. 3.5 Baseline (current condition): Sunnyside Yard site in Queens, New York, as it is in present day, with the rail yard and surrounding areas

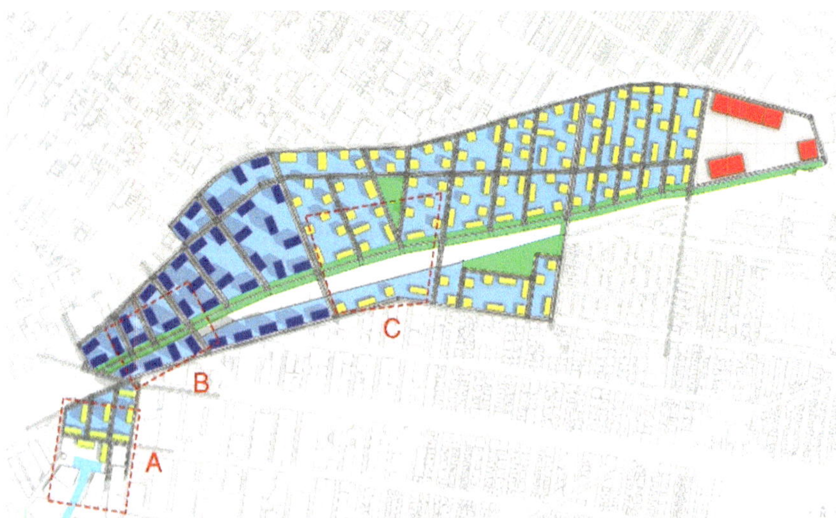

Fig. 3.6 Baseline (business-as-usual) 2050: Hypothetical scenario based on typical "market driven" decisions

Fig. 3.7 Best Practice (climate-driven urban design) 2050: Based on climate adaptive development, considering the 4 climate factors (efficiency of urban systems, form and layout, heat resistant construction materials and vegetative coverage); developed by the NYIT Urban Climate Lab

analyzing comparable districts of high density and low thermal stress; (2) developing evidence-based urban design guidelines that include climate factor diagrams and physical configurations; (3) identifying expected urban microclimate outcomes from peer-reviewed research. Participants also made assumptions pertaining to density increase, based on projected population, planned housing and infrastructure development, and rezoning information obtained from city agencies. The final urban design intervention has been configured through the articulation of design goals responding to the complexity of context, spatial scales and systems; pointing to coherent urban design strategies and outcomes. The preparation leading to the urban design workshop tested and refined the site intervention through microclimate modeling at district scale for current condition and 2050 scenarios.

The preparation leading to the Urban Design Climate Workshop (UDCW) tested and refined the urban design site intervention through the lens of Earth observations, microclimate models, and qualitative best practices; elements of the qualitative guidelines were applied to the Sunnyside Yard site. This urban design work was accomplished in a collaboration between the NYIT Graduate Urban Design students and the HOK[11] NYC office Urban Design Group.

[11] HOK, formerly Hellmuth, Obata + Kassabaum, is an American worldwide design, architecture, engineering and urban planning firm, founded in 1955.

This Baseline (business-as-usual) and Best Practices (climate-driven urban design) approach would provide compelling evidence to NYC policymakers (and a wider audience) on the value proposition (financial/health/public realm co-benefits) from our evidence-based strategies.

3.5.2 *Urban Microclimate Comparisons*

At the district scale, urban climate experts developed surface temperature distribution scenarios derived from satellite data; summertime Landsat 8 imagery from 2017 to 2018 was used to model the land surface temperature (LST) at the subdistricts of interest for comparison. The land surface temperature data generated is a proxy for the urban heat island and urban microclimate that helps identify district-level "hot spots," zones of elevated heat intensity.

Next, detailed urban climate models were used to identify hot spots at a neighborhood level. The goal of this detailed "hot spot" analysis at the city block level was to evaluate human comfort, and better understand the potential physiological stress on local population based on their vulnerability. For evaluating urban outdoor thermal comfort, several metrics are available, including the Universal Thermal Climate Index (UTCI).

3.5.3 *Urban Climate Scenarios*

There are two main categories of microclimate analysis approaches—observational approaches (e.g., techniques relying on thermal remote sensing) and simulation approaches (e.g., Energy Balance Models, Computational Fluid Dynamics). The main advantage of the numerical simulation approaches lies in their flexibility for comparative analyses based on different scenarios (Tsoka et al. 2018).

Three different Sunnyside Yard scenarios were modeled using 3 urban microclimate modeling approaches:

1. Earth Observations; land surface temperature modeling driven by thermal imagery
2. ENVI-met[12]; microclimate modeling applied to assess of human comfort

[12] The ENVI-met microclimate tool has been applied in a great number of microclimate studies concerned with the urban heat island and heat stress mitigation strategies (Bruse and Fleer 1998; Tsoka et al. 2018).

3. SOLWEIG[13]; modeling spatial variation of 3D radiation fluxes and mean radiant temperature[14]

3.5.4 *Testing Strategic Locations: Baseline vs. Best Practice*

Once 3D morphological models were generated, the climate modeling teams identified key subdistricts in the Sunnyside Yard development to test more detailed/granular outcomes; with human comfort outcomes specific to each.

In order to assess the microclimate outcomes under the 2050 Baseline (business-as-usual) and Best Practice (climate adaptive urban design) scenarios, ENVI-met 4.0 was used to model the air temperature and surface temperature distribution at Sunnyside Yard. The model simulates surface-plant-air interactions and consists of a 1D boundary layer and three sub-models for atmosphere, soil and plants (Fig. 3.8).

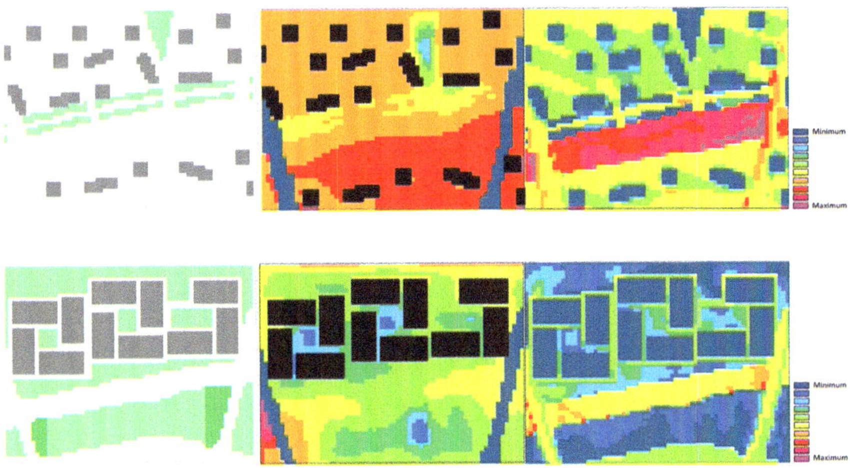

Fig. 3.8 A strategic subdistrict (See location C in Fig. 3.6) under the 2050 Baseline (top row) and 2050 Best Practice (bottom row) scenarios is illustrated by the ENVI-met grid cells in columns I, II, and III. The microclimate benefits of 2050 Best Practice urban design configuration is illustrated by these ENVI-met simulations modeling conditions at 6 pm in July:

- **Column I: Illustrates the variation of vegetative cover across the sub-district.**
- **Column II: Illustrates the variation in air temperature distribution across the sub-district.**
- **Column III: Shows the variation in surface temperature across the sub-district. Simulated surface temperatures range from 27 to 43 °C with significant surface cooling.**

---

[13] SOLWEIG is a component of the Urban Multi-scale Environmental Predictor (UMEP), which is freely available.

[14] Mean radiant temperature is one of the key meteorological variables governing human energy balance and human thermal comfort.

With average annual temperatures projected to rise by 4.1–5.7 °F by the 2050s as a result of climate change, New York City can expect more frequent extreme heat conditions like those experienced in the summers of 2017, 2018, and 2019. The primary impact of urban design at a neighborhood scale is to modify the local climate, which is governed by a hierarchy of atmospheric influences. During heatwave events, strong solar irradiance, high air temperatures and low wind-speed creates thermal discomfort, which can be alleviated by a design that does not obstruct airflow, generates shadows and uses cool surface materials.

In order to further examine the microclimatic impact of different urban designs, the SOLWEIG[15] model was used to compare alternative 2050 Sunnyside Yard designs (i.e., Baseline and Best Practice). SOLWEIG uses a three-dimensional description of the urban landscape and of land cover to derive key drivers of climate at the ground. These drivers include the spatial distributions of direct solar radiation, shadow patterns, and the sky view factor. Together, these can be used to estimate the distribution of the mean radiant temperature (MRT) of the urban surface and human comfort levels at selected locations. For the purposes of the Urban Design Climate Workshop, the model was run using the 2050 designs (i.e., Baseline and Best Practice) and data for 1 June 2018, which corresponds to a heatwave event in New York. In terms of the distribution of the MRT and human comfort values, the best practice design creates a more diverse shadow pattern and a greater area of relative comfort (by reducing MRT values).

3.6 UDCW Cross-Sectoral Collaborations

UCCRN Urban Design Climate Workshops (UDCW) are underway worldwide with urban designers, urban planners, climate scientists, policymakers, stakeholders, and graduate students working side by side. A primary source of collaboration for the UDCW are the urban design graduate students from the Urban Design Climate Lab at the New York Institute of Technology (NYIT). They worked in close collaboration throughout the Sunnyside Yard UDCW process, developing design scenarios and a large physical site model at the disposal of the UDCW. The goal of the Urban Design Climate Lab design studio at NYIT is to explore integrated, urban design and planning strategies in order to create sustainable and resilient communities. There are opportunities for some communities to adapt and thrive in changing global conditions, meet carbon-reduction goals, and sustain urban populations in more compact settings. In this urban design studio students explore how these compact communities can mitigate climate change by reducing GHG emissions through spatial efficiencies, pedestrian access to public transportation and preservation of open space and habitat. Sprawl is associated with ozone exceedances, poor water

[15] See: Lindberg et al. 2018. International Association for Urban Climate IAUC-University College Dublin provided this modeling contribution.

quality, and adverse health outcomes from obesity due to decreased physical activity as well as fatal road traffic injuries (Stone et al. 2010).

Another key UDCW collaborator is the urban design practitioner community. In the case of the Sunnyside UDCW, the global architecture firm HOK provided important case study scenario input (e.g., 3D morphological development support). Members of the Planning and Urban Design Committee of the American Institute of Architects NY Chapter (AIA New York) contributed expertise to the process. The UDCW in NYC drew its global experts from two simultaneous events: (1) The New York City launch by the Urban Climate Change Research Network (UCCRN) of the publication Climate Change and Cities: Second Assessment Report of the Urban Climate Change Research Network (ARC3.2), Cambridge University Press 2018; (2) The International Association for Urban Climate (IAUC) conference in NYC. This *New York City Urban Design Climate Workshop* (UDCW) was held on 11 August in New York City at the New York Institute of Technology.

The primary focus of the UDCW event was a roundtable discussion seeking common language (Fig. 3.9); common strategies between Urban Climate Scientists, Urban Designers, and Policymakers (Fig. 3.10). Participants explored urban design approaches, methods, and tools to strengthen community resilience through a systemic, interconnected public realm and green infrastructure to achieve reduced

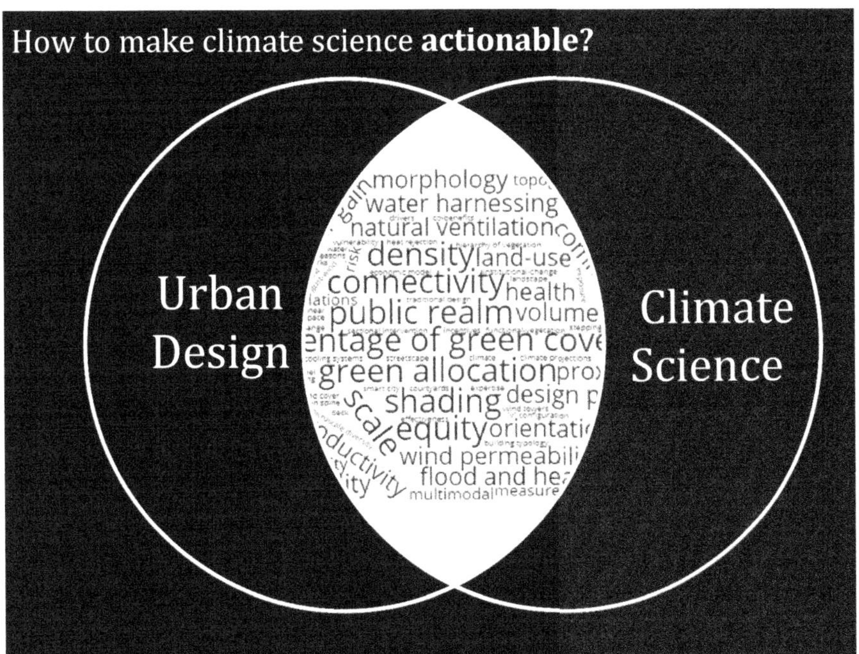

Fig. 3.9 "Word map" illustrating shared language and approaches between urban climate and urban design experts, from Urban Design Climate Workshop, NYC 2018

Fig. 3.10 Urban Design Climate Workshop with urban climate and urban design experts exchanging approaches

energy loads, cleaner air, and enhanced civic life. Urban design and climate experts analyzed prototype case study frameworks and methodologies in both local and international contexts.

The configuration of the four primary sub-zones (*Central Business District; Intermodal Transit Hub; Residential; Manufacturing*) was analyzed by the assembled group of experts through the lens of the four urban climate factors.

Participants confronted the district's energy efficiency challenges in responding simultaneously to strong winds, flood risks, and high temperature and humidity. They analyzed the configuration of the compact sub-districts; whether they successfully exploited natural ventilation. The UDCW's focus on energy-efficient design solutions drew from the assembled experts in energy, infrastructure planning, climate science, landscape architecture, and green building design. The workshop evaluated prescriptive measures and performance standards for a climate-resilient public realm, including urban ventilation, green infrastructure, and solar design.

The assembled experts discussed methodologies for identifying, monitoring, and prioritizing the Urban Climate Factors to achieve the best possible quality of life outcomes. The experts highlighted the challenges in prioritizing urban climate strategies, because they often defy simple measurement due to cascading impacts across sectors and spatial scales.

The UDCW concluded with the recognition that poorly designed dense urban environments illustrated by the Baseline 2050 (business as usual) produce a poor local climate. The Baseline 2050 also illustrates how the impervious surfaces of our "concrete jungles" cannot absorb storm water, causing greater risk of flooding. The Best Practice 2050 Sunnyside Yard urban design approach configured the densely occupied district so as to offset undesirable local outcomes through climate-sensitive urban design. The future scenarios modeled at these Urban Design Climate Workshops (UDCWs) effectively challenge decision-makers to configure dense urban districts to reduce the impact of increased urban heat and storms due to the changing climate while enhancing a high-quality, low-carbon lifestyle. UDCWs rapidly build capacity across multiple stakeholder groups to implement mitigation, adaptation, and transformation actions to respond to climate change.

3.7 Conclusion

Research shows that cost-effective planning and design measures that help mitigate GHG emissions while bringing adaptive benefits should be prioritized. Evidence-based strategies demonstrate how integrating climate science, natural systems, and compact urban form can configure dynamic, desirable, and healthy communities. The challenge is to configure these dense urban districts to reduce the impact of increased urban heat and storms due to the changing climate while enhancing a high-quality, low-carbon lifestyle. Urban designers and urban planners have a portfolio of climate change strategies that guide decisions on urban form and function in response to urban heat and climate mitigation and adaptation. To bridge between climate science and urgent climate action, scientists, engineers, and urban designers are organizing Urban Design Climate Workshops worldwide. This planning and design process effectively identifies and evaluates urban climate factors and then rapidly deploy tools and methods to configure sustainable and resilient districts at a local scale. The timing is critical, the science is clear, and urgent action is needed.

References

Anderson GB, Bell ML (2011) Heat waves in the United States: mortality risk during heat waves and effect modification by heat wave characteristics in 43 US communities. Environ Health Perspect 119(2):210–218

Bader DA, Blake RA, Grimm A, Hamdi R, Kim Y, Horton RM, Rosenzweig C, Alverson K, Gaffin SR, Crane S (2018) Urban climate science. In: Rosenzweig C, Solecki W, Romero-Lankao P, Mehrotra S, Dhakal S, Ali Ibrahim S (eds) Climate change and cities: second assessment report of the urban climate change research network (ARC3.2). Cambridge University Press, pp 27–60

Barata MML, Kinney PL, Dear K, Ligeti E, Ebi KL, Hess J, Dickinson T, Quinn AK, Obermaier M, Silva Sousa D, Jack D (2018) In: Rosenzweig C, Solecki W, Romero-Lankao P, Mehrotra S, Dhakal S, Ali Ibrahim S (eds) Climate change and cities: second assessment report of the urban climate change research network (ARC3.2). Cambridge University Press, pp 363–398

Bruse M, Fleer H (1998) Simulating surface–plant–air interactions inside urban environments with a three dimensional numerical model. Environ Model Softw 13(3–4):373–384. https://doi.org/10.1016/S1364-8152(98)00042-5

Castellano CM, DeGaetano AT (2017) Downscaling extreme precipitation from CMIP5 simulations using historical analogs. J Appl Meteorol Climatol 56(9):2421–2439. https://doi.org/10.1175/JAMC-D-16-0250.1

Childs PP, Raman S (2005) Observations and numerical simulations of urban heat island and sea breeze circulations over New York City. Pure Appl Geophys 162(10):1955–1980

CIRIA (2019) Benefits Estimation Tool (B£ST). https://www.susdrain.org/resources/best.html. Accessed 30 Mar 2020

City of New York (2017) Cool neighborhoods NYC: a comprehensive approach to keep communities safe in extreme heat. https://www1.nyc.gov/assets/orr/pdf/Cool_Neighborhoods_NYC_Report.pdf. Accessed 30 Mar 2020

City of New York (2018) Progress report: OneNYC 2018. April 2018. https://onenyc.cityofnewyork.us/wp-content/uploads/2018/05/OneNYC_Progress_2018.pdf. Accessed 30 Mar 2020

DeGaetano AT, Castellano CM (2015) Downscaled projections of extreme rainfall in New York State. Northeast Regional Climate Center, New York

DeGaetano AT, Castellano CM (2017) Future projections of extreme precipitation intensity-duration-frequency curves for climate adaptation planning in New York State. Clim Serv 5:23–35

Dulal HB, Brodnig G, Onoriose CG (2011) Climate change mitigation in the transport sector through urban planning: a review. Habitat Int 35(3):494–500. https://doi.org/10.1016/j.habitatint.2011.02.001

Fan F, Bradley RS, Rawlins MA (2014) Climate change in the northeastern US: regional climate model validation and climate change projections. Clim Dyn 43(1):145–161. https://doi.org/10.1007/s00382-014-2198-1

Gaffin SR, Imhoff M, Rosenzweig C, Khanbilvardi R, Pasqualini A, Kong AYY, Grillo D, Freed A, Hillel D, Hartung E (2012) Bright is the new black—multi-year performance of high-albedo roofs in an urban climate. Environ Res Lett 7(1):014029. https://doi.org/10.1088/1748-9326/7/1/014029

Gedzelman SD, Austin S, Cermak R, Stefano N, Partridge S, Quesenberry S, Robinson DA (2003) Mesoscale aspects of the Urban Heat Island around New York City. Theor Appl Clim 75(1):29–42. https://doi.org/10.1007/s00704-002-0724-2

Georgescu M, Morefield PE, Bierwagen BG, Weaver CP (2014) Urban adaptation can roll back warming of emerging megapolitan regions. Proc Natl Acad Sci 111(8):2909–2914. https://doi.org/10.1073/pnas.1322280111

González JE, Ortiz L, Smith BK, Devineni N, Colle B, Booth JF, Ravindranath A, Rivera L, Horton R, Towey K, Kushnir Y, Manley D, Bader D, Rosenzweig C (2019) New York City panel on climate change 2019 report chapter 2: new methods for assessing extreme temperatures, heavy downpours, and drought. Ann N Y Acad Sci 1439(1):30–70. https://doi.org/10.1111/nyas.14007

Haurwitz B (1947) Comments on the sea-breeze circulation. J Meteorol 4(1):1–8

Horton R, Yohe G, Easterling W, Kates R, Ruth M, Sussman E, Whelchel A, Wolfe D, Lipschultz F (2014) Chapter 16: Northeast. In: Melillo JM, Richmond TC, Yohe GW (eds) Climate change impacts in the United States: the third national climate assessment. U.S. Global Change Research Program, pp 371–395

Horton R, Bader D, Kushnir Y, Little C, Blake R, Rosenzweig C (2015) New York City panel on climate change 2015 report. Chapter 1: climate observations and projections. Ann N Y Acad Sci 1336:18–35. https://doi.org/10.1111/nyas.12586

IPCC (2013) Climate change 2013: the physical science basis. Contribution of working group i to the fifth assessment report of the intergovernmental panel on climate change. Cambridge University Press, Cambridge

Knowlton K, Lynn B, Goldberg RA, Rosenzweig C, Hogrefe C, Rosenthal JK, Kinney PL (2007) Projecting heat-related mortality impacts under a changing climate in the New York City Region. Am J Public Health 97(11):2028–2034. https://doi.org/10.2105/AJPH.2006.102947

Kunreuther H, Heal G, Allen M, Edenhofer O, Field CB, Yohe G (2013) Risk management and climate change. Nat Clim Chang 3(5):447–450. https://doi.org/10.1038/nclimate1740

Lempert RJ, Schlesinger ME, Bankes SC (1996) When we don't know the costs or the benefits: adaptive strategies for abating climate change. RAND Corporation, Santa Monica, CA. https://www.rand.org/pubs/reprints/RP557.html. Accessed 30 Mar 2020

Li D, Bou-Zeid E (2013) Synergistic interactions between urban heat islands and heat waves: the impact in cities is larger than the sum of its parts. J Appl Meteorol Climatol 52(9):2051–2064. https://doi.org/10.1175/JAMC-D-13-02.1

Lindberg F, Grimmond CSB, Gabey A, Huang B, Kent CW, Sun T, Theeuwes NE, Järvi L, Ward HC, Capel-Timms I, Chang Y, Jonsson P, Krave N, Liu D, Meyer D, Olofson KFG, Tan J, Wästberg D, Xue L, Zhang Z (2018) Urban Multi-scale Environmental Predictor (UMEP): an integrated tool for city-based climate services. Environ Model Softw 99:70–87. https://doi.org/10.1016/j.envsoft.2017.09.020

Luber G, McGeehin M (2008) Climate change and extreme heat events. Am J Prev Med 35(5):429–435. https://doi.org/10.1016/j.amepre.2008.08.021

Menne MJ, Williams Jr CN, Vose RS (2013) United States Historical Climatology Network (USHCN) version 2.5 serial monthly dataset

Moss RH, Edmonds JA, Hibbard KA, Manning MR, Rose SK, van Vuuren DP, Carter TR, Emori S, Kainuma M, Kram T, Meehl GA, Mitchell JFB, Nakicenovic N, Riahi K, Smith SJ, Stouffer RJ, Thomson AM, Weyant JP, Wilbanks TJ (2010) The next generation of scenarios for climate change research and assessment. Nature 463(7282):747–756. https://doi.org/10.1038/nature08823

Newman PG, Kenworthy JR (1989) Cities and automobile dependence: an international sourcebook. Gower Technical

Ning L, Bradley RS (2014) Winter precipitation variability and corresponding teleconnections over the northeastern United States. J Geophys Res Atmos 119(13):7931–7945. https://doi.org/10.1002/2014jd021591

Nolde E, Vansbotter B, Rüden H, König KW (2007) Innovative water concepts. Service water utilization in buildings. Berlin Senate Department for Urban Development, Berlin. https://www.stadtentwicklung.berlin.de/bauen/oekologisches_bauen/download/modellvorhaben/betriebswasser_englisch2007.pdf. Accessed 30 Mar 2020

NPCC (2010) Climate change adaptation in New York City: building a risk management response. Ann N Y Acad Sci 1196:1–354

NPCC (2015) Building the knowledge base for climate resiliency: New York City panel on climate change 2015 report. Ann N Y Acad Sci 1336:1–150

NPCC (2019) New York City panel on climate change 2019 report. Chapter 1: Introduction. Rosenzweig, C., and W. Solecki. Ann N Y Acad Sci 1439:22–29. https://doi.org/10.1111/nyas.14004

NYC Mayor's Office of Recovery & Resiliency (2018) Climate resiliency design guidelines. Version 2.0. NYC ORR, New York. https://www1.nyc.gov/assets/orr/pdf/NYC_Climate_Resiliency_Design_Guidelines_v2-0.pdf

Oke TR (1982) The energetic basis of the urban heat island. Q J R Meteorol Soc 108(455):1–24. https://doi.org/10.1002/qj.49710845502

Ortiz LE, Gonzalez JE, Wu W, Schoonen M, Tongue J, Bornstein R (2018) New York City impacts on a regional heat wave. J Appl Meteorol Climatol 57(4):837–851. https://doi.org/10.1175/jamc-d-17-0125.1

Ramamurthy P, Bou-Zeid E (2017) Heatwaves and urban heat islands: a comparative analysis of multiple cities. J Geophys Res Atmos 122(1):168–178. https://doi.org/10.1002/2016jd025357

Ramamurthy P, González J, Ortiz L, Arend M, Moshary F (2017) Impact of heatwave on a megacity: an observational analysis of New York City during July 2016. Environ Res Lett 12(5):054011. https://doi.org/10.1088/1748-9326/aa6e59

Raven J (2016) Urban Climate Factors. In Raven, J., et al. (2018) Urban planning and urban design. In: Rosenzweig C, Solecki W, Romero-Lankao P, Mehrotra S, Dhakal S, Ali Ibrahim S (eds) Climate change and cities: second assessment report of the urban climate change research network. Cambridge University Press

Raven J, Stone B, Mills G, Towers J, Katzschner L, Leone M, Gaborit P, Georgescu M, Hariri M (2018) Urban planning and urban design. In: Rosenzweig C, Solecki W, Romero-Lankao P, Mehrotra S, Dhakal S, Ali Ibrahim S (eds) Climate change and cities: second assessment report of the urban climate change research network. Cambridge University Press, Cambridge, pp 139–172

Resch E, Bohne RA, Kvamsdal T, Lohne J (2016) Impact of urban density and building height on energy use in cities. Energy Procedia 96:800–814. https://doi.org/10.1016/j.egypro.2016.09.142

Romero-Lankao P, Burch S, Hughes S, Auty K, Aylett A, Krellenberg K, Nakano R, Simon D, Ziervogel G (2018) Governance and policy. In: Rosenzweig C, Solecki W, Romero-Lankao P, Mehrotra S, Dhakal S, Ali Ibrahim S (eds) Climate change and cities: second assessment

report of the urban climate change research network. Cambridge University Press, Cambridge, pp 585–606
Rosenthal JK, Kinney PL, Metzger KB (2014) Intra-urban vulnerability to heat-related mortality in New York City, 1997–2006. Health Place 30:45–60. https://doi.org/10.1016/j.healthplace.2014.07.014
Rosenzweig C, Horton RM, Bader DA, Brown ME, DeYoung R, Dominguez O, Fellows M, Friedl L, Graham W, Hall C, Higuchi S, Iraci L, Jedlovec G, Kaye J, Loewenstein M, Mace T, Milesi C, Patzert W, Stackhouse PW Jr, Toufectis K (2014) Enhancing climate resilience at NASA Centers: a collaboration between science and stewardship. Bull Am Meteorol Soc 95(9):1351–1363. https://doi.org/10.1175/bams-d-12-00169.1
Sailor DJ (2001) Relating residential and commercial sector electricity loads to climate—evaluating state level sensitivities and vulnerabilities. Energy 26(7):645–657. https://doi.org/10.1016/S0360-5442(01)00023-8
Santamouris M (2014) On the energy impact of urban heat island and global warming on buildings. Energ Buildings 82:100–113. https://doi.org/10.1016/j.enbuild.2014.07.022
Schaeffer R, Szklo AS, Pereira de Lucena AF, Moreira Cesar Borba BS, Pupo Nogueira LP, Fleming FP, Troccoli A, Harrison M, Boulahya MS (2012) Energy sector vulnerability to climate change: a review. Energy 38(1):1–12. https://doi.org/10.1016/j.energy.2011.11.056
Schmidt M (2009) Rainwater harvesting for mitigating local and global warming. In: Fifth urban research symposium, vol 28.6, p 9
Smith C, Levermore G (2008) Designing urban spaces and buildings to improve sustainability and quality of life in a warmer world. Energy Policy 36(12):4558–4562. https://doi.org/10.1016/j.enpol.2008.09.011
Steffan C, Schmidt M, Köhler M, Hübner I, Reichmann B (2010) Rainwater management concepts, greening buildings, cooling buildings planning, construction, operation and maintenance guidelines. Berlin Senate Department for Urban Development, Berlin. http://www.stadtentwicklung.berlin.de/bauen/oekologisches_bauen/download/SenStadt_Regenwasser_engl_bfrei_final.pdf. Accessed 30 Mar 2020
Stone B, Hess JJ, Frumkin H (2010) Urban form and extreme heat events: are sprawling cities more vulnerable to climate change than compact cities? Environ Health Perspect 118(10):1425–1428. https://doi.org/10.1289/ehp.0901879
Tsoka S, Tsikaloudaki A, Theodosiou T (2018) Analyzing the ENVI-met microclimate model's performance and assessing cool materials and urban vegetation applications—a review. Sustain Cities Soc 43:55–76. https://doi.org/10.1016/j.scs.2018.08.009
UN-Habitat (2013) Urban planning for city leaders. UN-Habitat
United Nations Environment Programme (UNEP) (2012) Using ecosystems to address climate change: ecosystem based adaptation regional seas program report. UNEP
WE ACT for Environmental Justice (2015) NMCA—Northern Manhattan climate Action Plan. http://weact.nyc/climate/nmca-201508.pdf. Accessed 30 Mar 2020

Chapter 4
Cool Cities by Design: Shaping a Healthy and Equitable London in a Warming Climate

Anna Mavrogianni, Jonathon Taylor, Phil Symonds, Eleni Oikonomou, Helen Pineo, Nici Zimmermann, and Mike Davies

Abstract This chapter examines the concept of urban heat resilience in the context of the current research and policy trend towards the creation of healthy cities. This is achieved through the lens of research work carried out at the UCL Institute for Environmental Design & Engineering that focuses on urban climate risk, heat vulnerability mapping and co-creation with policy stakeholders in London. The chapter first sets out the challenges associated with healthy urban environments in a warming climate. It then proceeds with outlining the opportunities that arise from scientific advances in this area and the potential for knowledge exchange and co-design between academia, policy and citizens. The chapter closes with an exploration of solutions and pathways forward identified in past and ongoing research work.

Keywords Cities · Health · Well-being · Urban Heat Island · Climate change · Overheating · Heatwave · Vulnerability

A. Mavrogianni (✉) · J. Taylor · P. Symonds · H. Pineo · N. Zimmermann · M. Davies
Institute for Environmental Design and Engineering, The Bartlett, University College London, London, UK
e-mail: a.mavrogianni@ucl.ac.uk; jonathon.taylor@tuni.fi; p.symonds@ucl.ac.uk; helen.pineo@ucl.ac.uk; n.zimmermann@ucl.ac.uk; michael.davies@ucl.ac.uk

E. Oikonomou
UCL Energy Institute, The Bartlett, University College London, London, UK
e-mail: e.oikonomou@ucl.ac.uk

C. Ren, G. McGregor (eds.), *Urban Climate Science for Planning Healthy Cities*, Biometeorology 5, https://doi.org/10.1007/978-3-030-87598-5_4

Key Messages

- It is broadly recognised that the design and planning of buildings and cities are key factors for human health and well-being.
- Future cities will have to be resilient to urban heat risk, which will be exacerbated by the combined effects of climate change, the urban heat island and population ageing.
- Temporal-spatially explicit models of urban heat risk, often combined with mapping tools, can be used nowadays by planners and public health policymakers to identify urban areas where environmental, personal and social determinants of heat risk coincide across a city.
- A participatory, systems-based approach could be instrumental in addressing health inequities related to heat risk in urban environments.
- Enabling co-creation of solutions and effective communication between academics, planners, building construction practitioners, policymakers and communities is of fundamental importance for the mitigation of urban heat risk.

4.1 Introduction: Healthy City Design in a Climate of Change

The recent rise in the political economy of "wellness" and the well-being agenda among building designers and urban planners has emerged from a powerful shift in health policy that focuses on preventing ill health rather than curing disease. Urban design and planning is fast becoming a key instrument in promoting health and well-being in cities, and tackling the social determinants of health while achieving current climate change mitigation targets (Northridge et al. 2003; PHE 2018; Pineo et al. 2018b). This was reflected in the 2009 Zagreb Declaration for Healthy Cities that set healthy urban environment and design as one of the three core themes of Phase V of the World Health Organization (WHO) European Healthy Cities Network, alongside creating caring and supportive environments, and healthy living (WHO Regional Office for Europe 2009). Initiatives such as the WHO Healthy Cities programme have greatly contributed to a paradigm shift in academic research, urban planning and policy by highlighting the role that built environment design plays in forging healthy, sustainable and people centred cities (WHO 2018).

The climate change induced increase in the frequency and magnitude of hot episodes is expected to lead to a wide range of adverse effects on human health and well-being (IPCC 2014). These range from thermal discomfort, sleep disruption, reduced human performance, heat stroke and hyperthermia, to the exacerbation of certain chronic medical conditions and even death. According to WHO (2011), these effects are largely preventable. Following the deadly 2003 and 2006 heatwaves, European public health protection networks have greater preparedness to minimise heat risk than a few decades ago (Menne and Matthies 2009). Although recommended immediate response actions commonly triggered by extreme heat events feature meteorological early warning systems and public health advice, emphasis is now placed on building design and urban planning, such as the provision of shading and the improvement of green and blue infrastructure (Santamouris and Kolokotsa 2013).

As climate change, urban design and well-being are closely connected, the aim of this chapter is to place urban heat resilience in the context of the global Healthy Cities movement using London as a case study. It is divided into three parts:

1. Challenges
2. Opportunities
3. Solutions

The first part establishes the background guiding initiatives that aim to create climate resilient, sustainable, healthy and equitable urban environments. It outlines current urban transformation trends, including climate change, the Urban Heat Island (UHI) phenomenon, urbanisation, demographic change and population ageing, building stock changes, public health awareness and governance issues, and the root causes underlying the production of climate injustices in urban settings. The second part identifies key opportunity areas across research and policy arenas, including advances in urban climate science, the assessment of positive effects of urban warming (such as potential reductions in cold-related mortality and increases in physical activity), the development and mainstreaming of urban heat vulnerability mapping tools, "choice architecture", systems thinking, and knowledge exchange between academia, policy and citizens. The third part offers a synopsis of past and ongoing London focused academic research projects developed in close collaboration with city level and local government policymakers. Particular emphasis is given on successful applications of built environment and climate-related knowledge to the formulation of healthy urban design policies and solutions through co-creation processes that foster interdisciplinary and inter-sectoral partnerships.

4.2 Current and Future Challenges to Cities

Anthropogenic global climate change is now a well-established and unequivocal fact. According to the Intergovernmental Panel on Climate Change (IPCC 2013, 2018), extreme weather events, such as heatwaves, will become both more frequent and severe in the future, with potentially catastrophic impacts to public health and the economy. The UK experienced a 10-day heatwave in August 2003, which resulted in more than 2000 excess deaths in England and Wales, with similar thermal conditions being observed in July 2006 (Kovats and Hajat 2008) and, more recently, in the summer of 2018. Heat events of this magnitude will have become the norm by the middle of the century (Christidis et al. 2015). According to the UK Climate Projections (UKCP 2018), under the Representative Concentration Pathway 4.5 (RCP4.5), the projected 50th percentile change of mean summer temperature in the UK is 1.6 °C by the 2050s compared to the 1961–1990 baseline, with heatwave periods similar to 2018 likely to occur during half of the summers by the middle of the century. There is, as a result, increasing concern that ongoing and future warming trends pose unprecedented challenges to our communities. A recent inquiry by the Environmental Audit Committee (EAC 2018) highlighted the need for Government leadership and cross-departmental collaboration to prepare for heatwaves at multiple scales.

The negative effects of regional warming are exacerbated in urban areas due to the UHI phenomenon, an inadvertent modification of the local urban climate linked to urbanisation (Grimmond 2007). Heat islands are reported when temperatures are higher in an urban area relative to the surrounding rural environment as a result of more heat absorbed and trapped within urban environments due to high built densities, the substitution of natural surfaces with humanmade surfaces, and human activities generating additional heat; generally, the UHI effect is more pronounced during clear and calm nights (Oke 1982, 1987). London's UHI intensity markedly increases during heatwave events, with differences between central London and surrounding rural areas reaching 8–9 °C during the 2003 heatwave (MOL 2006) when, out of the around 2000 heatwave attributed deaths recorded nationwide, more than 600 were recorded in the London region (Kovats and Hajat 2008).

The urban heat risk increment is a pertinent issue in the context of global urbanisation trends. According to the most recent World Urbanization Prospects by the United Nations (UN DESA 2018), 55% of the world's population lived in urban areas in 2018 and this proportion is projected to increase to 68% by the middle of the century compared to only 34% in 1960. Urban population increases and the creation of new megacities will be predominantly concentrated in low income and lower middle income settings, mainly in Asia and Africa. In such settings, as well as in higher income global cities, such as London, urbanisation is likely to escalate heat stress risks among the most vulnerable segments of the population that will not have the means to adapt to a warming climate, thus jeopardising urban inclusivity and magnifying health inequities (Heaviside et al. 2017).

Epidemiological studies of heat-related health risk conclusively indicate that the most heat-vulnerable group is older people, who tend to be most susceptible to heat-related illnesses, such as heat stroke and exhaustion, and heat-related death (Kovats and Hajat 2008). The annual population growth rate of people aged 60 or older is around 3% and it is expected that by 2050 all regions in the world apart from Africa will nearly have a quarter of their population aged 60 or above (UN DESA 2017). Other heat risk factors include pre-existing illnesses, in particular cardiovascular and respiratory diseases, and social isolation, both of which are likely to increase with older age. The combined phenomena of climate change, the UHI, urbanisation and an ageing world will, therefore, result in an increasing number of people being exposed and vulnerable to climate change impacts in cities in the future. This will be especially relevant in major cities that have large numbers of older people (Smith et al. 2014; WHO 2017).

Despite the fact that people vulnerable to heat are likely to spend more time indoors compared to the rest of the population, the role of buildings as modifiers of exposure to high temperatures has remained an under-researched area to date. Epidemiological and built environment studies in Europe and the UK suggest that indoor heat exposure may be higher for people living in residential or care settings with one or more of the following characteristics (Vandentorren et al. 2006; Hajat et al. 2007; Kovats and Hajat 2008; DCLG 2012; ZCH 2015):

- Top floor
- Purpose built
- Single aspect flats
- With restrictors or other barriers to window opening
- No shading
- Low ceilings
- Large glazing areas (especially if they are south or west facing)
- High internal heat gains
- Mechanical Ventilation with Heat Recovery (MVHR) with no summer bypass
- Community heating systems with high heat gains through corridors
- Lack of air conditioning or other active cooling systems

This highlights the contribution of building characteristics to aggravating heat risk. In climates where wintertime heating is required, such as the UK, the focus of building stock transformation agendas has understandably been on winter energy efficiency, which aims to achieve two targets of fundamental importance: (1) climate change mitigation through the reduction of carbon emissions mainly related to space heating and (2) the elimination of fuel poverty. But as building fabrics become more insulated and airtight, the risk of overheating increases if appropriate passive cooling measures are not included at an early design and retrofit stage (Mavrogianni et al. 2012; Fosas et al. 2018).

Another challenge in heat risk mitigation in the UK is the social and cultural lack of knowledge of how to tackle heat-related hazards. Recent studies have suggested that vulnerable individuals perceive neither heat risks nor cold risks as potentially threatening to them and, due to the unprecedented nature of hot spells, they often do not know how to operate their homes during an extreme heat episode (Abrahamson et al. 2009; Wolf et al. 2010; White-Newsome et al. 2012). This underscores the importance of both public health initiatives aiming to increase risk awareness but also social cohesion and local knowledge bases that could help identify heat-vulnerable individuals and promote best practice advice during heat events.

The discussion above raises questions about how external stresses, such as urban warming and demographic trends, convert into a decline in population well-being and health. Many have argued that extreme weather events, such as heatwaves, often reveal underlying wider inequities in the distribution of vulnerability (Friel et al. 2008). The distribution of climate-related disadvantage varies geographically, as a function of the interplay of policy, sociodemographic and environmental factors (Macintyre et al. 2018). More focus should, therefore, be given to the study of unequal geographical distributions of climate risk and heat-related vulnerability and disadvantage in cities. According to Lindley et al. (2011), although significant research efforts have been made in recent years to identify the patterns of climate hazard exposures, the social, personal and environmental determinants of climate-related risk are less well understood or recognised in adaptation policy. Social and personal factors include age, income and education levels but also access to social networks and public health infrastructure that will have a protective effect during an extreme weather event. Environmental factors include the characteristics of the buildings and neighbourhood an individual inhabits, such as the presence and operationality of cooling building strategies, the surrounding microclimate, the type of amount and quality of accessible greenspace and other public spaces, the levels of noise, air pollution and perceived risk of crime that may inhibit natural ventilation through window opening in the summer. Hajat and Kosatky (2010) argue that an enhanced understanding of the determinants that explain heat risk heterogeneity is a fundamental prerequisite for the development of targeted climate change adaptation responses in cities. In a large and diverse city such as London, attention needs to be paid to the full gradient of inequality and its impact on the spatial variation of population vulnerability to heat risk. Socioeconomic status has not been shown to be a risk factor for heat mortality in the UK due to the low penetration of air conditioning across the building stock. However, Bundle et al. (2018) argue that it is plausible that there is a social gradient in indoor overheating risk as a result of a correlation between deprivation and heat risk factors, such as prevalence of long-term health conditions and neighbourhood crime rates, which may inhibit night cooling through window opening.

There is significant need to integrate overheating solutions into built environment policies for the planning and design of new development and infrastructure (including green spaces), building regulations and regeneration programmes. There are two categories of policy approaches to climate change adaptation: (1) "mainstreaming" of adaptation solutions into existing policy domains and programmes and (2) "stand-alone" adaptation policies (Runhaar et al. 2018). Mainstreaming solutions benefit from existing resources and the potential to develop co-benefits with other policy objectives, while stand-alone options may receive more attention and dedicated funding. Research by Cortekar et al. (2016) notes that adaptation policy efforts often start with a vulnerabilities assessment using locally specific climate impact resources, such as the UKCIP Adaptation Wizard, followed by the development of appropriate policy responses. Cortekar and colleagues noted significant challenges for adaptation policies related to unsuitability of a "one-size-fits-all" solution due to local variations, human and financial resource constraints, and lack of a systems approach to account for interconnections across climate change impacts and adaptation measures.

Such challenges tend to persist because urban environments are dynamic complex systems in a state of constant flux. They are a "drama in time" (Geddes 1904). The determinants of urban heat risk are commonly interconnected and counterintuitive phenomena. For example, if people use air conditioning to protect themselves from heat, at an aggregated level their decisions increase energy consumption and, if cooling systems are powered by fossil fuels, global warming. Such energy behaviour also contributes to the UHI because air conditioning generates heat directly outside the buildings with air conditioning units. In addition, if policymakers and planners take measures to build more densely, they may also increase the UHI effects locally. Mainstreaming of adaptation solutions and stand-alone adaptation policies rarely have merely technical dimensions and often require that people also change their behaviour. Yet, it is a true challenge to anticipate whole-system effects of potential policies, people's decisions and behaviour. Consequently, the study of urban environments and heat risks requires a whole-systems perspective for a broader understanding of effects in space and time.

4.3 Research Directions and Opportunities for Change

Local urban climate modification is a well-documented phenomenon worldwide (Oke 1987; Arnfield 2003). UHI studies have traditionally been observational and increasingly include remote sensing but advances in computing and big data analytics in recent years have also allowed the generation of prediction modelling frameworks. Modelling studies encompass a wide range of methods, including both deterministic and stochastic modelling approaches (Santamouris 2007; Mirzaei and Haghighat 2010; Stewart 2011; Mirzaei 2015). Deterministic approaches mainly include modelling of urban canyon temperature regimes, modelling of the urban canopy layer and dynamic numerical modelling. Stochastic approaches include statistical correlation analysis and Artificial Neural Network (ANN) modelling.

The development of these models paves the way for the evaluation of the impact of urban planning decisions on urban heat risk, energy use and health outcomes, hence facilitating informed policymaking (Wolf et al. 2015). For example, temporal-spatially explicit models of urban heat risk could enable planners and public health policymakers to identify "hotspots", i.e. areas where environmental, personal and social determinants of heat risk coincide across a city (Wilhelmi and Hayden 2010; Preston et al. 2011; Romero-Lankao et al. 2012). The fine scale mapping of local urban climate variables and heat risk factors is underpinned by the wide use of Geographic Information Systems (GIS) in research and practice (Wilson et al. 2010; Tomlinson et al. 2011; Merbitz et al. 2012; Aubrecht and Özceylan 2013; Harlan et al. 2013; Wolf and McGregor 2013; Wolf et al. 2014; Boumans et al. 2014; Dugord et al. 2014; Taylor et al. 2015; van der Hoeven and Wandl 2015). Maps can be a powerful communication tool when engaging with policymakers and other stakeholders with the aim to identify the ways built environment design impacts on health and well-being, for example by allowing the quick ranking of risk factors for an individual, i.e. age, housing type and location. Nonetheless, there are inherent challenges related to the difficulty to convey nuances or limitations in the underlying data, which further pinpoints the need for effective communication between academic researchers and policy stakeholders. Additional challenges may often emerge due to data-scale issues given that many attempts to map fine scale variations in vulnerability and risk encounter ethical issues related to data acquisition.

Despite the current focus on its physical determinants, urban heat risk is a socio-technical problem, which necessitates the adoption of an interdisciplinary approach in order to tackle it. For instance, the role of built environment design is indeed acknowledged by recent developments in the behavioural sciences. There is increasing recognition that human behaviour is often driven not by conscious, deliberative cognitive processes but instead by non-conscious, automatic responses that may be triggered by characteristics of, or cues within, our surrounding environment (Marteau et al. 2012; Hollands et al. 2017). Appropriate design strategies at different scales or, in other words, interventions to alter "choice architecture"—ranging from changes to the microenvironment of indoor spaces to the neighbourhood level could be employed to encourage healthy behaviours and social practices, thus reducing urban climate-related risks. This could include, for example, integrating passive cooling strategies, such as natural ventilation and shading, and green and blue infrastructure in the design of buildings and cities in such a way that they are intuitive and easy to implement, thus limiting or avoiding the use of active cooling.

This is very much echoed in the recent trend to adopt systems thinking and participatory research in built environment and urban health research (Homer and Hirsch 2006; Macmillan et al. 2016; Eker et al. 2018; Pineo et al. 2018b; Shrubsole et al. 2019). This trend is illustrated, for instance, by recent citizen science attempts to collect data related to urban heat risk factors, such as urban microclimate data (Rajagopalan et al. 2017) and building construction age (CASA 2019). Mapping interactions between various heat risk factors at the urban scale in collaboration with stakeholders will allow the identification of feedback interaction between drivers, barriers, and behaviours in space and time. It will also enable the identification of possible unintended consequences through this process and the potential reduction or elimination of negative consequences through the application of systems knowledge.

A participatory, systems-based approach could be an invaluable tool to addressing health inequities related to heat risk by enabling solution co-creation processes between academics, planners, building construction practitioners, policymakers and communities as described by Raven et al. (see Chap. 3). In the past, research project outputs were mainly published in academic outlets, hence limiting their accessibility to a wider audience. A two-way dialogue has thankfully been initiated in recent years between researchers and decision makers, whereby scientific evidence on the factors that may contribute to heat vulnerability is gradually but steadily being translated into policy action and industry best practice. The next section will offer some examples of research projects aiming to improve London's urban heat risk resilience with strong impact pathway plans.

4.4 Solution Pathways

The challenges and opportunities outlined in the previous sections have triggered a series of past and ongoing interdisciplinary London-focused research projects undertaken by University College London's (UCL) Institute for Environmental Design and Engineering (IEDE) in conjunction with other institutions. These projects have tackled various factors of urban heat risk, including the role of the UHI, building fabric characteristics, and social determinants of heat vulnerability, with a particular focus on older and ageing populations. While the main emphasis has been placed on residential environments, current research aims to expand existing methodological frameworks to assess the heat vulnerability of care settings. Translational work carried out in collaboration with London boroughs and the Greater London Authority (GLA) are presented at the end of the section.

4.4.1 London's Local Urban Climate

London's local urban climate and its impacts were the focus of the 4-year research project "The development of a Local Urban Climate model and its application to the Intelligent Design of cities" (LUCID) (Mavrogianni et al. 2011). This project improved our understanding of London's UHI through the development of a suite of local urban climate models spanning three scales: (1) citywide, (2) neighbourhood and (3) street level (Fig. 4.1). Data outputs from the local urban climate models were imported into a second suite of models aiming to assess the impacts of London's UHI on (1) energy use, (2) thermal comfort and (3) health.

Previous monitoring studies demonstrated that London experiences a significant UHI (Watkins et al. 2002). Modelling work carried out as part of the LUCID project allowed the exploration of the spatio-temporal variation of local temperatures across the city, including the effect of variables such as urban materials, land use and anthropogenic heat. The citywide London Unified Model (LondUM) was constructed as part of LUCID and is based on the MetOffice Unified Model (MetUM),

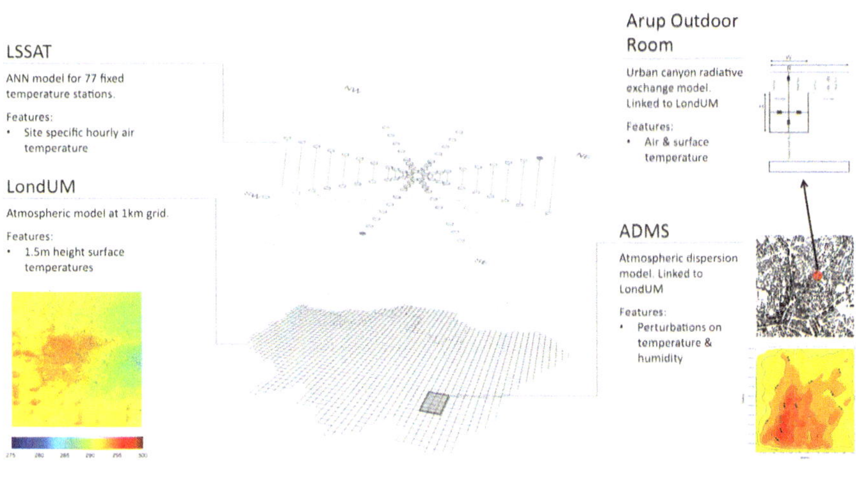

Fig. 4.1 Urban climate models developed within the context of the LUCID project. (Source: EPSRC LUCID project)

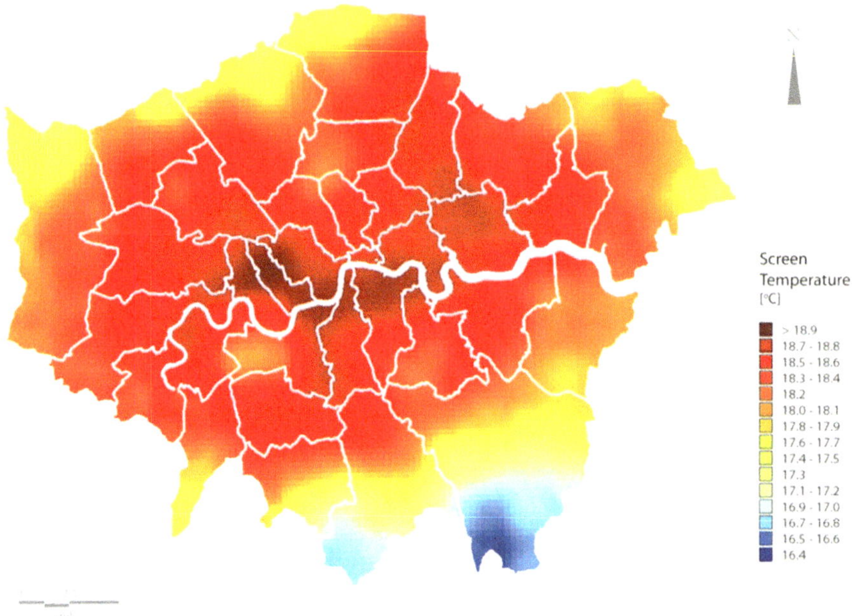

Fig. 4.2 Screen level temperature at 1.5 m height, 26 May–19 July 2006 (LondUM simulation). (Source: EPSRC LUCID project)

a numerical weather prediction model (Bohnenstengel et al. 2011). LondUM represents the Earth's surface using a variation of vegetated and non-vegetated surface properties (such as thermal reflectivity, emissivity and capacity), thus estimating the effect that the surface properties have on the local climate.

Figure 4.2 illustrates an example of LondUM's outputs. It was found that a heat island intensity (i.e. a temperature difference between core urban and rural locations) of 5 °C during the hours after sunset occurs frequently. The intensity peaks on nights with low winds and clear skies. A considerable variation of local urban temperatures over short distances was also observed; external air temperatures may vary by 3–4 °C within only a kilometre. This is attributed to the heterogeneity of the local microclimatic characteristics, including land use and surface properties as well as urban morphology. Although on average the thermal centre of London's UHI is located around the British Museum, the advection of cool rural air changes its pattern on windier days and redistributes heat across London. Modelling also demonstrated that the spatial distribution of London's green infrastructure has a significant cooling effect for London, reducing citywide night-time temperatures by up to 2–3 °C compared to a hypothetical scenario in which all London green surfaces were replaced by built up areas. Anthropogenic heat, on the other hand, generated from traffic, industry, space heating and other human activities, was shown to increase the night-time heat island intensity by up to 2 °C on average over a year. Such temperature differences are important in a health impact context taking into account that an increase in relative risk of heat-related death of 3.8% per °C above a 24.8 °C mean maximum ambient daily outdoor temperature threshold has been estimated for London (Armstrong et al. 2011).

According to LUCID's impact assessment models, the UHI currently has an appreciable net energy benefit for London resulting in around a 13% reduction in residential space heating loads (Mavrogianni et al. 2009). However, these benefits might be offset by increased cooling loads if there is a significant uptake of air conditioning in the future. Although the vast majority of London homes are currently naturally ventilated, temperature monitoring studies have shown that bedrooms in almost half of the homes in the monitored sample overheated even during a mild summer (Pathan et al. 2017). The heat island also has a substantial impact on the burden of heat-related mortality. An epidemiological study using LondUM outputs estimated that the proportion of heat-related deaths attributable to the London's heat island increased by 9.5% as one moves from outer to central London (Milojevic et al. 2011). It was demonstrated that heat-related mortality risk increases in London areas with high average building height, potentially due to the high prevalence of overheating in top floor flats which are prone to heat exposure from radiation loading in these areas. This is in agreement with previous studies (Mavrogianni et al. 2009). In winter, nevertheless, the heat island has a protective effect as it results in decreased cold-related mortality (Milojevic et al. 2011).

Embedding passive cooling strategies in urban design is important and measures such as the improvement and expansion of London's green infrastructure will have co-benefits for health and well-being (Fairbrass et al. 2018). However, Oikonomou et al. (2012) highlighted the importance of high thermal quality envelopes coupled with solar shading as the most effective way of reducing indoor heat exposure in London. The study indicated that the relative contribution of the thermal quality of domestic buildings on internal temperatures is higher than that of the building's location within the heat island.

4.4.2 London Buildings as Modifiers of Heat Exposure

There has been increasing recognition in the last few decades of the indoor environment's impact on human health and well-being. Building envelopes can indeed act as modifiers of human exposure to heat, light, noise and air pollution. Nevertheless, in contrast to the wealth of epidemiological research focusing on outdoor climate and air pollution and their influence on health outcomes, indoor environmental quality remains a relatively under-researched field to date. Taking into account that the UK population spends around 90% of their time indoors on average, it is imperative that we estimate the modifying effect of buildings on human exposure to harmful environmental conditions (Vardoulakis et al. 2015).

This has been the remit of the "Air Pollution and WEather-related Health Impacts: Methodological Study based On spatio-temporally disaggregated Multi-pollutant models for present day and future" (AWESOME) project, which developed a building physics based indoor environment model of the UK housing stock (Taylor et al.

2016) to examine the contribution of the building envelope as a modifier of exposure to outdoor pollutants and heat, and to provide an integrated assessment of the associated health risks. These estimates were based on a large number of simulations using the building physics tool EnergyPlus, with input data derived from national housing surveys and energy efficiency retrofit databases (Mavrogianni et al. 2012; Oikonomou et al. 2012). The model outputs were applied to a geographically referenced housing stock model and aggregated at the postcode level. The National Institute for Health Research (NIHR) Health Protection Research Unit (HPRU) in Environmental Change and Health built on this work with the aim to evaluate more broadly the impact of urban stock heat adaptation transformations, housing energy retrofit strategies and urban greening on indoor environmental quality and health. For the HPRU project, a meta-modelling framework based on the outputs of multiple EnergyPlus runs was developed using Artificial Neural Networks (ANN) (Symonds et al. 2016) (Fig. 4.3). This model framework significantly reduces the computing time required for the calculation of occupant exposures to indoor overheating or pollutants with a reduced number of housing inputs,

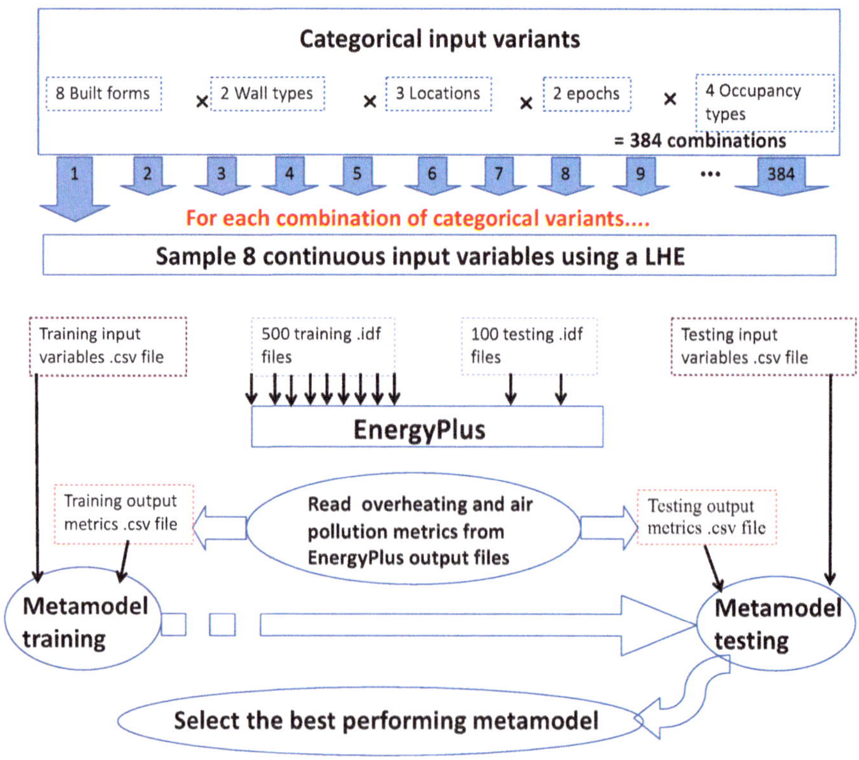

Fig. 4.3 The meta-modelling framework underpinning the HPRU housing stock model. (Source: Symonds et al. 2016)

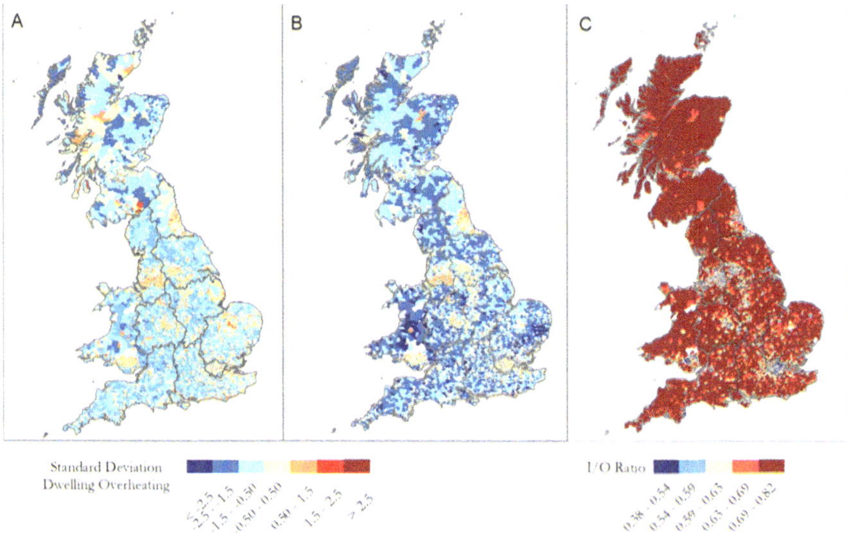

Fig. 4.4 (**a**) Mean Maximum Daytime living room Temperature (MMDT, °C), (**b**) Mean Maximum Night-time bedroom Temperature (MMNT, °C), and (**c**) Indoor/Outdoor ratios for $PM_{2.5}$, for the pensioners' occupancy assumption aggregated at the Lower Super Output Area level. (Source: NERC AWESOME project)

facilitating rapid estimation of exposure at the housing stock or population level under a range of different climate and adaptation scenarios.

The relative internal heat and air contaminant exposure levels across the UK housing stock, as derived from the AWESOME project, are mapped in Fig. 4.4 (Taylor et al. 2016). It was shown that people living in flats and newly built properties are exposed to lower indoor air pollution from outdoor sources. This could possibly be attributed to the fact that the majority of newer, purpose built flats are characterised by higher levels of building air tightness and feature smaller exposed surface areas compared to older houses. Additional work by Taylor et al. (2016) has, however, indicated that they are also exposed to higher air pollution from indoor sources. In terms of indoor overheating risk, flats, bungalows and more recently built, and highly airtight dwellings were among the building types that were most at risk of overheating. As a result, people living in urban flats, such as London, are on average exposed to higher levels of overheating and indoor air pollutants due to the higher number of flats in urban areas.

4.4.3 Identifying Heat-Vulnerable Older Londoners

The importance of an individual's age for heat vulnerability is well established in epidemiological literature (Kovats and Hajat 2008). Although a plethora of existing studies have evaluated the impacts of climate change on urban environments, and population susceptibility to adverse health effects, there are relatively few studies currently that specifically focus on older people and ageing populations in cities. In a similar manner, climate adaptation strategies are rarely tailored to older citizens. The "Seasonal Health and Resilience for Ageing Populations and Urban Environments" (SHARPER) 2-year project aimed to fill this knowledge gap. The project studied three global cities: London, New York and Shanghai (Arup and Partners 2016).

The heat vulnerability maps and assessment tools developed for SHARPER were heavily based on the London urban heat vulnerability index developed by Wolf and McGregor (Wolf and McGregor 2013; Wolf et al. 2014). The index used a Principal Components Analysis (PCA) for socioeconomic determinants of heat risk to develop a Heat Vulnerability Index (HVI). This includes population density, and health status (including mobility levels). The HVI was subsequently overlaid on top of UHI satellite data to identify "hotspots" of heat risk. For the purposes of the SHARPER project, the index was expanded and revised to include updated Census data as shown in Fig. 4.5. This allowed the identification of areas where the population is most at risk from heat, and provided estimates of the resilience of ageing and older populations in each city at a high spatial resolution (Arup and Partners 2016).

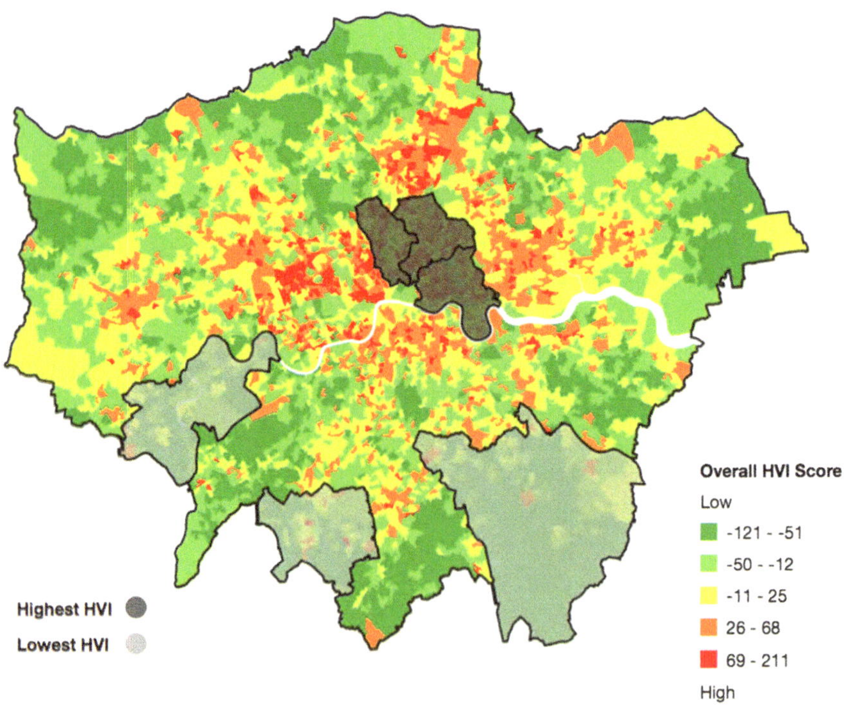

Fig. 4.5 Heat vulnerability index for London. (Source: Arup and Partners 2016, © Dr. Tanja Wolf/King's College London)

The study found that inner city people aged 65+ are at high risk from a rise in the frequency of heatwaves. The London Boroughs of Hackney, Islington and Tower Hamlets, in particular, are characterised by the highest number of heat vulnerability areas. This is due to a combination of heat risk factors such as high population density, and a greater proportion of older people with health and mobility issues in these boroughs. Although the borough of Hammersmith and Fulham has high land surface temperatures, the socioeconomic heat vulnerability factors are lower here. Richmond upon Thames, Bromley and Sutton contained the lowest concentration of populations at heat risk.

Similarly to the assessment of urban heat vulnerability, the design of adaptation measures often focus on the general population and only a small body of literature is concerned with the development of targeted responses for urban older and ageing populations. Many authors have argued that the identification of urban design and planning strategies to reduce the heat exposure and associated risks for older people in cities should become a research priority (Gamble et al. 2013). The SHARPER project went a step further by outlining a range of practical interventions targeted to older and ageing populations in cities aiming to alleviate heat vulnerability. The interventions were identified and evaluated through detailed stakeholder consultation. It was recommended that "win-win" measures that improve the resilience of older people to both extreme heat and cold, as well as contributing to the wider agenda of healthy and sustainable building design, should be prioritised.

Examples of such measures include:

- Drawing upon the wealth of knowledge and skills that older citizens have, identifying and using existing community networks, and empowering citizens by involving them in the development of heat resilience plans.
- Supporting and expanding the role of social care services in providing access to energy, food, water, sanitation and medicines during and after a heatwave.
- Enhancing existing and creating new green and blue infrastructure across the city.
- Implementing passive cooling strategies at the microclimatic level through the introduction of shading structures, trees, ventilation pathways through buildings, courtyards, etc.
- Climate proofing new build as well as retrofitted buildings by introducing passive cooling building design strategies at an early design stage.
- Improving existing heatwave alert systems and streamlining communication channels to ensure that timely information and advice reaches vulnerable people during a hot episode.

4.4.4 London's Triple Heat Jeopardy

A limitation of the Wolf and McGregor and SHARPER heat vulnerability indices is that the role of building characteristics as modifiers of heat exposure was not taken into account fully. The "Triple Heat Jeopardy" mapping framework was developed

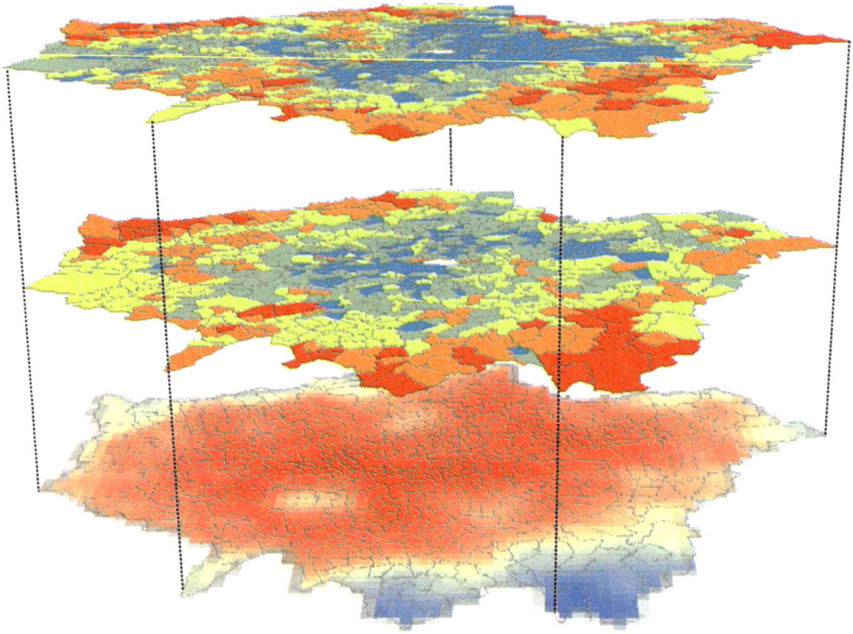

Fig. 4.6 The London Triple Heat Jeopardy mapping framework combining (from top to bottom) age-related relative risk of heat mortality, dwelling temperature anomaly, and local urban temperature data. (Source: NERC AWESOME project)

by Taylor et al. (2015) with input from GLA policymakers, who have been at the forefront of climate change adaptation at an urban scale. To identify locations at greatest risk of heat-related mortality, the Triple Jeopardy Index combines three data layers: (1) population age, (2) local urban temperature and (3) indoor temperature estimates for over 2.6 million residential addresses in London (Fig. 4.6). An epidemiologically based mathematical relationship was then used to estimate heat deaths during a hot episode based on an individual in the population's age and age-specific heat mortality risk, and the UHI and housing components of heat exposure. The methodological approaches taken by Wolf and McGregor and Taylor et al. are fundamentally different. The models include different inputs, for example: (1) Taylor et al. do not include multiple social determinants of heat risk like Wolf and McGregor, and (2) Taylor et al. include more detailed estimates of heat exposure, both from housing and the UHI. Calculation outputs are also different, with the Taylor et al. model estimating population mortality risk, while the Wolf and McGregor model estimates a vulnerability index, which was subsequently demonstrated to be more aligned to ambulance callouts than mortality (Wolf et al. 2014).

Interestingly, the findings by Taylor et al. indicate that once building properties are factored into the evaluation of urban heat risk, heat-related mortality is likely to be higher in the outskirts of London rather than the centre of the city despite the increased external temperatures (Fig. 4.7). This is mainly attributed to the high proportion of

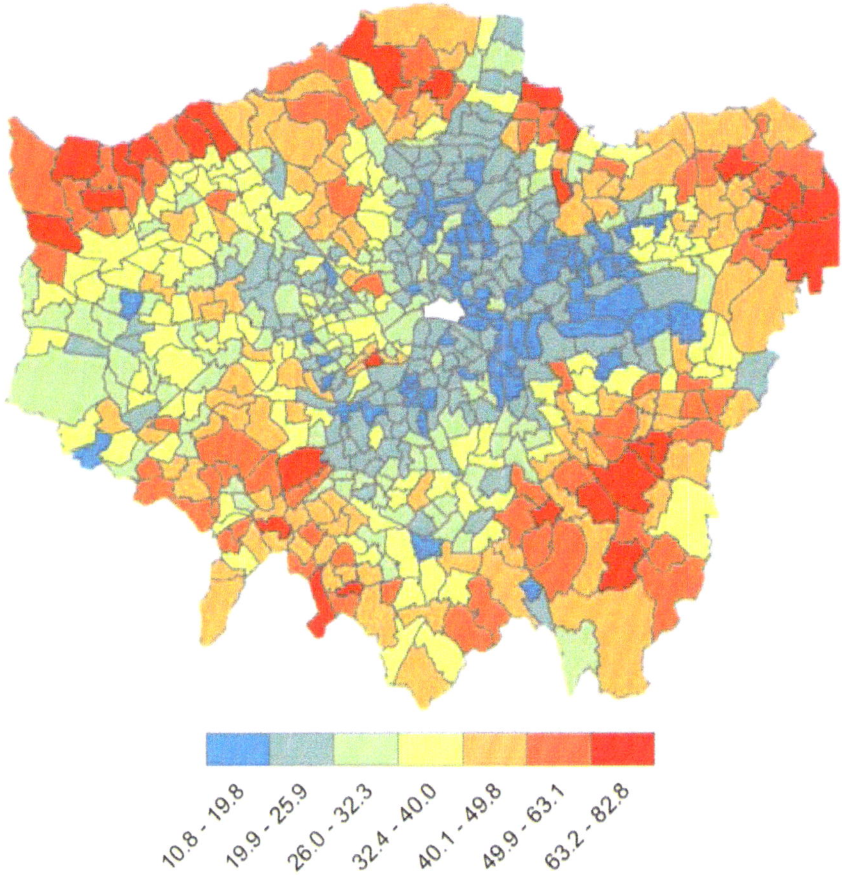

Fig. 4.7 Estimated mortality per million population during a warm 55-day period in London. (Source: Taylor et al. 2015)

older people in the outer boroughs, as well as the makeup of their housing stock, which includes poorly insulated bungalows. The contradictions between the Taylor et al. and SHARPER indices should not be seen as a weakness of either approach but rather as an indication of the complexity of urban heat vulnerability assessment. The comparison of varying approaches is a valuable step towards the integration of multiple heat risk determinant factors in GIS-based risk assessment frameworks. Such tools can become a great asset for urban planners and policymakers aiming to quantify the reduction in heat-related health risks resulting from climate adaptation policies and compare their relative advantages in cost benefit analyses.

4.4.5 Climate Proofing London Care Homes

In recent years, there has been an increasing interest in the climate resilience of the care and extra care sector in the UK. This discussion is underpinned by the findings of a recent project commissioned by the Joseph Rowntree Foundation (JRF) (Gupta et al. 2016; Gupta and Gregg 2017). The study suggested that the health risks associated with heat exposure are not adequately recognised by social care practitioners, potentially due to ingrained habits, norms and the prevalent "culture of warmth" that dictates that older people need to be kept warm at all times. In addition, poor building design and the lack of effective heat management in care and extra care settings may contribute to increased indoor heat exposure that could have detrimental impacts for the most vulnerable residents. This was further reflected in the UK Government's 2017 Climate Change Risk Assessment (CCRA) produced by the Committee on Climate Change (CCC) (Kovats and Osborn 2016), which identified overheating risk in care homes as a key risk and research priority for the health and social care system.

The interdisciplinary pilot project "*Mapping Climate Disadvantage for London's Care Provision*" was an 1-year pilot project (Mavrogianni et al. 2018a; Oikonomou et al. 2018). Its overarching aim was to accelerate the development of equitable responses to climate change for care provision in urban environments, with a focus on overheating risk using Greater London as a case study. A meta-analysis of outputs of existing housing research using building physics-based models was carried out specifically focusing, for the first time, on care settings. This quantified the impact of building geometry and physical properties on the current and future overheating risk of London care home typologies. The location of London care homes was then mapped and information layers of the spatial variability of climate disadvantage in these areas was overlaid on them using spatial analysis techniques. Preliminary mapping suggests that 20% of London care homes are located in areas of "extremely high" or "acute" socio-spatial vulnerability (Fig. 4.8), defined by the ClimateJust framework as the result of "an equally weighted combination of neighbourhood level scores for indicators including population sensitivity to heat, exposure, ability to prepare, respond and recover" (Climate Just 2018).

The Natural Environment Research Council (NERC) recently funded the ClimaCare project, which is currently underway and aims to develop methods that will support the UK's care provision in preparing for rising heat stress under climate change. It is envisaged that ongoing and future work in this area will have important implications for building designers, social care providers and managers by helping making suggestions on how to modify existing inspection processes and management practices in order to identify and combat overheating risks for care home residents.

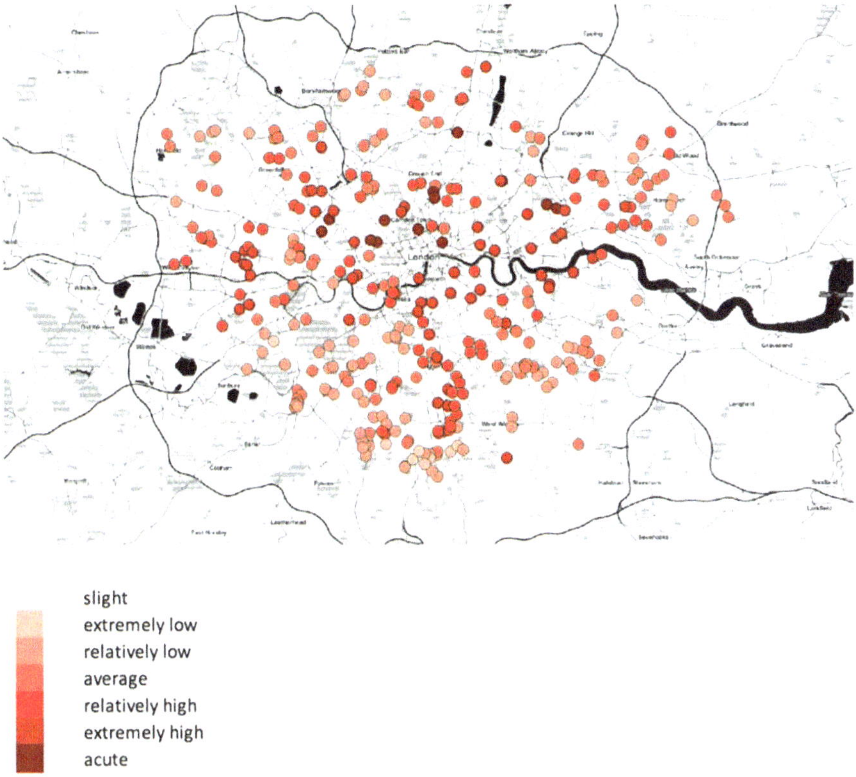

Fig. 4.8 Location of London care homes in relation to socio-spatial heat vulnerability indicators. (Source: Mavrogianni et al. 2018a; Oikonomou et al. 2018)

4.4.6 Working with London Local Authorities

As discussed earlier, the promotion of collaborations between academics, policy-makers and the wider public and the operationalisation of research outcomes have a pivotal role in the development of effective urban heat risk mitigation plans. Two examples of such successful translational studies are the UCL Institute for Environmental Design & Engineering (IEDE) projects carried out in collaboration with two London Local Authorities, the London Boroughs of Islington and Hounslow.

The "Climate Resilience South Islington Project" (CRISP) was led by the Borough of Islington and assessed levels of awareness, attitudes and habits of older social housing residents in the south of Islington using a questionnaire survey (Kolm-Murray et al. 2013). As a follow-up to this study (Mavrogianni et al. 2015; Makantasi and Mavrogianni 2016), the UCL team undertook monitoring of indoor temperatures in a small sample of flats across three estates (a 1900s low-rise, a 1950s mid-rise and a 1960s high-rise building). It was found that vulnerable social housing tenants already experience indoor overheating even during a not particularly warm summer, with the flats located in the 1960s high-rise buildings being most prone to high temperatures. Building performance simulation was subsequently used to test the effectiveness of a range of retrofit interventions under current and future climate, thus generating useful guidance for the Borough's housing team. One key finding was that future urban warming could inhibit the cooling potential of natural ventilation beyond a certain time point. In other words, natural ventilation alone may not be able to cool dwellings in the future, with noise and air pollution concerns further limiting its feasibility.

The collaborative project with Hounslow Council (Taylor et al. 2017; Mavrogianni et al. 2018b) disseminated research outputs of the HPRU project, described above, and tailored them in such a way so as to facilitate their integration into the Borough's existing emergency management systems. This included modelled overheating risk indicators produced at individual physical address level for a range of external weather conditions. 2D and 3D visualisations and animations of the spatio-temporal variation in dwelling overheating risk and UHI intensities during hot spells were also produced (Fig. 4.9).

The broader lessons learnt during these two collaborative projects on urban heat risk can be summarised as follows:

- It is of paramount importance to engage with all key stakeholders from the early stages of the project; a co-design element is essential.
- Flexibility is required on the researchers' end to allow for work to be performed within the time and personnel resource constraints experienced by the Borough.

Fig. 4.9 A 3D visualisation of the area around Worton Way, Hounslow, showing the overheating risk of dwellings at 28 °C outdoor temperature (blue dwellings represent the coolest, while red are the hottest). (Source: Taylor et al. 2017)

- It is vital to achieve a good understanding of the Borough's datasets (including GIS), procedures and emergency planning tools at an early stage to allow for the effective integration of new data.

4.4.7 Complex Urban Systems

To improve the situation locally, it is important to not only integrate the research on the spatial and social characteristics of climate and heat vulnerability, respectively, but to also incorporate a whole systems perspective and to collaborate with local decision makers and stakeholders. The "Complex Urban Systems for Sustainability and Health" (CUSSH) project is a 4-year Wellcome Trust funded project that aims to develop critical evidence on how to achieve the far-reaching transformation of cities needed to address vital environmental imperatives for population and planetary health. Only launched in 2018, it has started to use cutting edge science and systems-based participatory methods to articulate visions of development, help shape policy decisions, and accelerate the implementation of transformational changes for health and sustainability in different income settings (Belesova et al. 2018; Zimmermann et al. 2018a). Across the six partner cities (London, Rennes, Nairobi, Kisumu, Beijing and Ningbo), the CUSSH project will be supporting city decision makers in developing ambitious urban sustainability and health policies. The project assembles a multidisciplinary team of experts in participatory system dynamics, behaviour change, microsimulation, air pollution modelling, and urban systems complexity and public engagement. In London, CUSSH is initially focussing on green infrastructure with its connections to environmental quality, sustainability, population health and health equity.

Although green infrastructure has the potential to attenuate the UHI effect, it is important to establish potential synergies and unintended consequences of this approach as well as understanding how such green infrastructure can be provided and managed. In collaboration with GLA policymakers, a systems approach was adopted using participatory system dynamics modelling. An initial causal diagram of potential causal effects and feedback relationships associated with the provision, maintenance and health impacts of green infrastructure in London was developed. Figure 4.10 shows a selected number of these relationships; it reveals that the

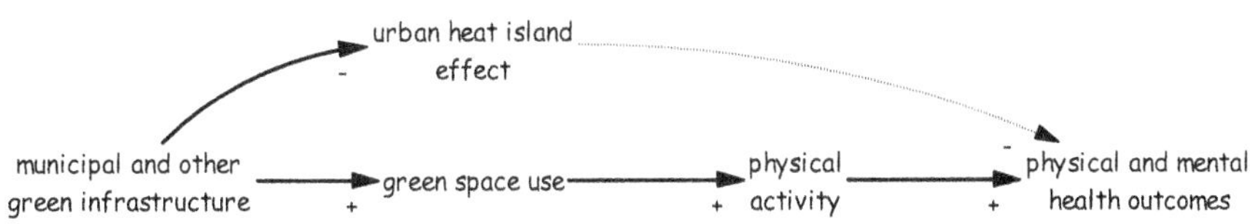

Fig. 4.10 Selected causal effects of adding municipal and other green infrastructure. A plus (minus) next to an arrow indicates a same (inversely) directed causal relationship. A dashed arrow indicates the high ambiguity of the adverse summer and useful winter effects of the UHI. (Source: Zimmermann et al. 2018b)

existence of municipal and other green infrastructure tends to increase their use, with positive effects on physical activity and health. At the same time, they reduce the UHI effect, with beneficial health effects, at least under warm conditions (Zimmermann et al. 2018b).

This approach revealed explicitly that green infrastructure and the UHI are interconnected in a complex system. It brought together the multiple perspectives of the participating stakeholders. The visual nature of this approach allowed policymakers to see the interconnection of variables and their dependencies, including unintended consequences and synergies (Black 2013).

A key finding arising from the CUSSH work is that researchers can provide policymakers with an evidence-based assessment of heat vulnerability, enabling policymakers to direct scarce resources towards design and retrofit measures that have the best chance of reducing the health impacts of extreme heat events. Such collaboration would benefit from researchers gaining greater understanding of policymakers' constraints and opportunities. Ongoing research in IEDE shows that there is an increasing number of urban health indicator tools displaying data about built environment exposures at low spatial scales (Pineo et al. 2018a), and climate vulnerability tools were among the indicator tools identified. Research into local decision makers' use of evidence in the UK found that participants were interested in qualitative data (alongside quantitative data), data displayed on maps, case studies and indicators that could help build a business case for particular interventions (McGill et al. 2015). The modelling results from the various projects discussed in this section are an example of research outputs that might be useful for practitioners as they include maps, images and locally specific data.

4.5 Summary

The role of building and urban design in public health promotion has been elevated in recent decades. This chapter set out to discuss a series of UCL IEDE research projects focusing on London's urban heat resilience in the context of an urban health research and policy agenda. The current and future challenges to global megacities, such as London, were explored. This was followed by a discussion of how advancements in a wide range of scientific fields, such as climate science, built environment studies, GIS, behavioural psychology, and systems thinking could be harnessed to develop evidence based and solution driven decision frameworks for policymakers working in the area of urban climate resilience. It is widely recognised that we live in a warming, ageing and increasingly urbanised world and that certain segments of the population will be hardest hit by the effects of these major changes to our urban environments. In London, academic research has been centred around modelling the external environment and the modifying effect of the building stock on indoor environment exposures, and mapping the environmental, social and personal determinants of climate-related risk. This work is co-created with public health bodies, city level and local government policymakers. Further translational work is needed to strengthen the collaborations between academic teams and stakeholders, thus bridging the fields of climatology, building and urban design, and public health in order to increase climate resilience, erase climate injustice and promote health equity in urban settings.

Acknowledgements The research work presented in this chapter was funded by the Engineering and Physical Sciences Research Council (EPSRC) [LUCID project, EP/E016375/1, EP/E016308/1 and EP/E016448/1], the Natural Environment Research Council (NERC) [AWESOME project NE/I007938/1, SHARPER project NE/M021149/1], the Wellcome Trust [CUSSH project, 209387/Z/17/Z], the Arup Global Research Challenge, the UCL Bartlett Synergy Grant, the Adaptation and Resilience in a Context of Change (ARCC) Network, the Islington Council and the National Institute for Health Research Health Protection Research Unit (NIHR HPRU) in Environmental Change and Health [HPRU-2012-10016] at the London School of Hygiene and Tropical Medicine in partnership with Public Health England (PHE), and in collaboration with the University of Exeter, University College London, and the MetOffice. The views expressed are those of the authors and not necessarily those of the NHS, the NIHR, the Department of Health or Public Health England.

References

Abrahamson V, Wolf J, Lorenzoni I, Fenn B, Kovats S, Wilkinson P, Adger WN, Raine R (2009) Perceptions of heatwave risks to health: interview-based study of older people in London and Norwich, UK. J Public Health 31(1):119–126

Armstrong BG, Chalabi Z, Fenn B, Hajat S, Kovats S, Milojevic A, Wilkinson P (2011) Association of mortality with high temperatures in a temperate climate: England and Wales. J Epidemiol Community Health 65(4):340–345

Arnfield AJ (2003) Two decades of urban climate research: a review of turbulence, exchanges of energy and water, and the urban heat island. Int J Climatol 23(1):1–26

Arup & Partners (2016) Seasonal health and resilience for ageing urban populations and environments. London, UK

Aubrecht C, Özceylan D (2013) Identification of heat risk patterns in the US National Capital Region by integrating heat stress and related vulnerability. Environ Int 56:65–77

Belesova K, Hale J, Aplin J, Baghaei Lakeh A, Davies M, Dianati K, Mberu B, Michie S, Osrin D, Symonds P, Taylor J, Wilkinson P, Zimmermann N (2018) Complex urban systems for sustainability and health: a structured approach to support the development and implementation of city policies for population and planetary health. In: World Health Organisation (WHO) international healthy cities conference Belfast 2018, changing cities to change the world, Belfast, UK, 1–4 October 2018

Black LJ (2013) When visuals are boundary objects in system dynamics work. Syst Dyn Rev 29(2):70–86

Bohnenstengel S, Evans S, Clark PA, Belcher S (2011) Simulations of the London urban heat island. Q J R Meteorol Soc 137(659):1625–1640

Boumans RJ, Phillips DL, Victery W, Fontaine TD (2014) Developing a model for effects of climate change on human health and health–environment interactions: heat stress in Austin, Texas. Urban Clim 8:78–99

Bundle N, O'Connell E, O'Connor N, Bone A (2018) A public health needs assessment for domestic indoor overheating. Public Health 161:147–153

CASA (2019) Colouring London. http://colouringlondon.org. Accessed 20 May 2019

Christidis N, Jones GS, Stott PA (2015) Dramatically increasing chance of extremely hot summers since the 2003 European heatwave. Nat Clim Chang 5(1):46–50
Climate Just (2018) Climate Just web tool, London
Cortekar J, Bender S, Brune M, Groth M (2016) Why climate change adaptation in cities needs customised and flexible climate services. Climate Services 4:42–51
DCLG (2012) Investigation into overheating in homes, Literature review. Department for Communities and Local Government (DCLG), London
Dugord P-A, Lauf S, Schuster C, Kleinschmit B (2014) Land use patterns, temperature distribution, and potential heat stress risk–the case study Berlin, Germany. Comput Environ Urban Syst 48:86–98
EAC (2018) Heatwaves: adapting to climate change, London
Eker S, Zimmermann N, Carnohan S, Davies M (2018) Participatory system dynamics modelling for housing, energy and wellbeing interactions. Build Res Inf 46(7):738–754
Fairbrass A, Jones K, McIntosh A, Yao Z, Malki-Epshtein L, Bell S (2018) Green infrastructure for London: a review of the evidence. The Engineering Exchange and the London Sustainability Exchange, London
Fosas D, Coley DA, Natarajan S, Herrera M, de Pando MF, Ramallo-Gonzalez A (2018) Mitigation versus adaptation: does insulating dwellings increase overheating risk? Build Environ 143:740–759
Friel S, Marmot M, McMichael AJ, Kjellstrom T, Vågerö D (2008) Global health equity and climate stabilisation: a common agenda. Lancet 372(9650):1677–1683
Gamble JL, Hurley BJ, Schultz PA, Jaglom WS, Krishnan N, Harris M (2013) Climate change and older Americans: state of the science. Environ Health Perspect 121(1):15–22
Geddes P (1904) Civics: as Applied Sociology. Sociological Papers 1:103–118
Grimmond S (2007) Urbanization and global environmental change: local effects of urban warming. Geogr J 173(1):83–88
Gupta R, Gregg M (2017) Care provision fit for a warming climate. Archit Sci Rev 60(4):275–285
Gupta R, Walker G, Lewis A, Barnfield L, Gregg M, Neven L (2016) Care provision fit for a future climate. York
Hajat S, Kosatky T (2010) Heat-related mortality: a review and exploration of heterogeneity. J Epidemiol Community Health 64(9):753–760
Hajat S, Kovats RS, Lachowycz K (2007) Heat-related and cold-related deaths in England and Wales: who is at risk? Occup Environ Med 64(2):93–100
Harlan SL, Declet-Barreto JH, Stefanov WL, Petitti DB (2013) Neighborhood effects on heat deaths: social and environmental predictors of vulnerability in Maricopa County, Arizona. Environ Health Perspect 121(2):197–204
Heaviside C, Macintyre H, Vardoulakis S (2017) The urban heat island: implications for health in a changing environment. Curr Environ Health Rep 4(3):296–305
Hollands GJ, Bignardi G, Johnston M, Kelly MP, Ogilvie D, Petticrew M, Prestwich A, Shemilt I, Sutton S, Marteau TM (2017) The TIPPME intervention typology for changing environments to change behaviour. Nat Hum Behav 1(8):1–9. https://doi.org/10.1038/s41562-017-0140
Homer JB, Hirsch GB (2006) System dynamics modeling for public health: background and opportunities. Am J Public Health 96(3):452–458
IPCC (2013) Climate change 2013: the physical science basis. In: Contribution of Working Group I to the fifth assessment report of the intergovernmental panel on climate change. Cambridge University Press, Intergovernmental Panel on Climate Change (IPCC), Cambridge, New York. https://www.ipcc.ch/report/ar5/wg1/. Accessed 6 Mar 2020
IPCC (2014) Climate change 2014, synthesis report, Summary for Policymakers, Geneva. https://www.ipcc.ch/report/ar5/syr. Accessed 6 Mar 2020
IPCC (2018) Global warming of 1.5°C, Summary for policymakers, Geneva.. https://www.ipcc.ch/sr15. Accessed 6 Mar 2020
Kolm-Murray J, Smith A, Clarke C (2013) Individual and community resilience to extreme weather events amongst older people in south Islington: attitudes, barriers and adaptive capacity. Climate Resilience Islington South Project (CRISP)-Final Report, London

Kovats RS, Hajat S (2008) Heat stress and public health: a critical review. Annu Rev Public Health 29:41–55

Kovats RS, Osborn D (2016) UK climate change risk assessment evidence report: Chapter 5, People and the built environment. Contributing authors: Humphrey K, Thompson D, Johns D, Ayres J, Bates P, Baylis M, Bell S, Church A, Curtis S, Davies M, Depledge M, Houston D, Vardoulakis S, Reynard N, Watson J, Mavrogianni A, Shrubsole C, Taylor J, Whitman G. Report prepared for the Adaptation Sub-Committee of the Committee on Climate Change, London

Lindley S, O'Neill J, Kandeh J, Lawson N, Christian R, O'Neill M (2011) Climate change, justice and vulnerability. Joseph Rowntree Foundation, York. https://www.jrf.org.uk/report/climate-change-justice-and-vulnerability. Accessed 6 March 2020

Macintyre H, Heaviside C, Taylor J, Picetti R, Symonds P, Cai X-M, Vardoulakis S (2018) Assessing urban population vulnerability and environmental risks across an urban area during heatwaves–Implications for health protection. Sci Total Environ 610:678–690

Macmillan A, Davies M, Shrubsole C, Luxford N, May N, Chiu LF, Trutnevyte E, Bobrova Y, Chalabi Z (2016) Integrated decision-making about housing, energy and wellbeing: a qualitative system dynamics model. Environ Health 15(1):S37. https://doi.org/10.1186/s12940-016-0098-z

Makantasi A-M, Mavrogianni A (2016) Adaptation of London's social housing to climate change through retrofit: a holistic evaluation approach. Adv Build Energy Res 10(1):99–124. https://doi.org/10.1080/17512549.2015.1040071

Marteau TM, Hollands GJ, Fletcher PC (2012) Changing human behavior to prevent disease: the importance of targeting automatic processes. Science 337(6101):1492–1495. https://doi.org/10.1126/science.1226918

Mavrogianni A, Davies M, Chalabi Z, Wilkinson P, Kolokotroni M, Milner J (2009) Space heating demand and heatwave vulnerability: London domestic stock. Build Res Inf 37(5–6):583–597

Mavrogianni A, Davies M, Batty M, Belcher S, Bohnenstengel S, Carruthers D, Chalabi Z, Croxford B, Demanuele C, Evans S, Giridharan R, Hacker J, Hamilton I, Hogg C, Hunt J, Kolokotroni M, Martin C, Milner J, Rajapaksha I, Ridley I, Steadman J, Stocker J, Wilkinson P, Ye Z (2011) The comfort, energy and health implications of London's urban heat island. Build Serv Eng Res Technol 32(1):35–52. https://doi.org/10.1177/0143624410394530

Mavrogianni A, Wilkinson P, Davies M, Biddulph P, Oikonomou E (2012) Building characteristics as determinants of propensity to high indoor summer temperatures in London dwellings. Build Environ 55:117–130

Mavrogianni A, Taylor J, Davies M, Thoua C, Kolm-Murray J (2015) Urban social housing resilience to excess summer heat. Build Res Inf 43(3):316–333

Mavrogianni A, Oikonomou E, Raslan R, Dennett A (2018a) Mapping climate disadvantage for London's care provision. In: Healthy city design 2018 international congress, Royal College of Physicians, London, 15–16 October 2018

Mavrogianni A, Taylor J, Shrubsole C, Symonds P, Chalabi Z, Davies M, Wilkinson P, Palmer T, O'Connell E, Thompson R, Bone A, Figueiredo A (2018b) Urban heat vulnerability mapping: working with a London Borough to translate research outputs. In: World Health Organisation (WHO) International healthy cities conference Belfast 2018, changing cities to change the world, Belfast, UK, 1–4 October 2018

McGill E, Egan M, Petticrew M, Mountford L, Milton S, Whitehead M, Lock K (2015) Trading quality for relevance: non-health decision-makers' use of evidence on the social determinants of health. BMJ Open 5(4)

Menne B, Matthies F (2009) Improving public health responses to extreme weather/heat-waves–EuroHEAT. World Health Organization, Copenhagen, Denmark. http://www.euro.who.int/__data/assets/pdf_file/0010/95914/E92474.pdf. Accessed 6 Mar 2020

Merbitz H, Buttstädt M, Michael S, Dott W, Schneider C (2012) GIS-based identification of spatial variables enhancing heat and poor air quality in urban areas. Appl Geogr 33:94–106

Milojevic A, Wilkinson P, Armstrong B, Davis M, Mavrogianni A, Bohnenstengel S, Belcher S (2011) Impact of London's urban heat island on heat-related mortality. Epidemiology 22(1):S182–S183

Mirzaei PA (2015) Recent challenges in modeling of urban heat island. Sustain Cities Soc 19:200–206
Mirzaei PA, Haghighat F (2010) Approaches to study urban heat island–abilities and limitations. Build Environ 45(10):2192–2201
MOL (2006) London's urban heat island: a summary for decision makers, London
Northridge ME, Sclar ED, Biswas P (2003) Sorting out the connections between the built environment and health: a conceptual framework for navigating pathways and planning healthy cities. J Urban Health 80(4):556–568
Oikonomou E, Davies M, Mavrogianni A, Biddulph P, Wilkinson P, Kolokotroni M (2012) Modelling the relative importance of the urban heat island and the thermal quality of dwellings for overheating in London. Build Environ 57:223–238
Oikonomou E, Mavrogianni A, Raslan R, Dennett A (2018) Mapping climate disadvantage for care provision in London, UK: a sociospatial heat vulnerability assessment. In: Public health science 2018, The Lancet, 23 November 2018, Queen's University, Belfast
Oke TR (1982) The energetic basis of the urban heat island. Q J R Meteorol Soc 108(455):1–24
Oke TR (1987) Boundary layer climates, 2nd edn. Routledge, Abingdon
Pathan A, Mavrogianni A, Summerfield A, Oreszczyn T, Davies M (2017) Monitoring summer indoor overheating in the London housing stock. Energ Buildings 141:361–378. https://doi.org/10.1016/j.enbuild.2017.02.049
PHE (2018) Healthy people, healthy places: building a healthy future, London
Pineo H, Glonti K, Rutter H, Zimmermann N, Wilkinson P, Davies M (2018a) Urban health indicator tools of the physical environment: a systematic review. J Urban Health 95(5):613–646
Pineo H, Zimmermann N, Cosgrave E, Aldridge RW, Acuto M, Rutter H (2018b) Promoting a healthy cities agenda through indicators: development of a global urban environment and health index. Cities & health 2(1):27–45
Preston BL, Yuen EJ, Westaway RM (2011) Putting vulnerability to climate change on the map: a review of approaches, benefits, and risks. Sustain Sci 6(2):177–202
Rajagopalan P, Santamouris M, Andamon M (2017) Public engagement in urban microclimate research—an overview of a citizen science project. In: Schnabel MA (ed) Proceedings of the 51st international conference of the Architectural Science Association (ANZAScA 2017), Wellington, New Zealand, 29 November–2 December 2017. Architectural Science Association Adelaide, Australia, pp 703–712
Romero-Lankao P, Qin H, Dickinson K (2012) Urban vulnerability to temperature-related hazards: a meta-analysis and meta-knowledge approach. Glob Environ Chang 22(3):670–683
Runhaar H, Wilk B, Persson Å, Uittenbroek C, Wamsler C (2018) Mainstreaming climate adaptation: taking stock about "what works" from empirical research worldwide. Reg Environ Chang 18(4):1201–1210
Santamouris M (2007) Heat island research in Europe: the state of the art. Adv Build Energy Res 1(1):123–150
Santamouris M, Kolokotsa D (2013) Passive cooling dissipation techniques for buildings and other structures: the state of the art. Energ Buildings 57:74–94
Shrubsole C, Hamilton I, Zimmermann N, Papachristos G, Broyd T, Burman E, Mumovic D, Zhu Y, Lin B, Davies M (2019) Bridging the gap: the need for a systems thinking approach in understanding and addressing energy and environmental performance in buildings. Indoor Built Environ 28(1):100–117. https://doi.org/10.1177/1420326X17753513
Smith KR, Woodward A, Campbell-Lendrum D, Chadee DD, Honda Y, Liu Q, Olwoch JM, Revich B, Sauerborn R (2014) Human health: impacts, adaptation, and co-benefits. In: Field CB, Barros VR, Dokken DJ, Mach KJ, Mastrandrea MD, Bilir TE, Chatterjee M, Ebi KL, Estrada YO, Genova RC, Girma B, Kissel ES, Levy AN, MacCracken S, Mastrandrea PR, White LL (eds) Climate change 2014: impacts, adaptation, and vulnerability, Part A: Global and sectoral aspects, Contribution of Working Group II to the Fifth Assessment Report of the Intergovernmental Panel on Climate Change. Cambridge University Press, Cambridge, New York, pp 709–754. https://www.ipcc.ch/site/assets/uploads/2018/02/WGIIAR5-Chap11_FINAL.pdf. Accessed 6 Mar 2020

Stewart ID (2011) A systematic review and scientific critique of methodology in modern urban heat island literature. Int J Climatol 31(2):200–217

Symonds P, Taylor J, Chalabi Z, Mavrogianni A, Davies M, Hamilton I, Vardoulakis S, Heaviside C, Macintyre H (2016) Development of an England-wide indoor overheating and air pollution model using artificial neural networks. J Build Perform Simul 9(6):606–619. https://doi.org/10.1080/19401493.2016.1166265

Taylor J, Wilkinson P, Davies M, Armstrong B, Chalabi Z, Mavrogianni A, Symonds P, Oikonomou E, Bohnenstengel SI (2015) Mapping the effects of urban heat island, housing, and age on excess heat-related mortality in London. Urban Clim 14:517–528. https://doi.org/10.1016/j.uclim.2015.08.001

Taylor J, Davies M, Mavrogianni A, Shrubsole C, Hamilton I, Das P, Jones B, Oikonomou E, Biddulph P (2016) Mapping indoor overheating and air pollution risk modification across Great Britain: a modelling study. Build Environ 99:1–12. https://doi.org/10.1016/j.buildenv.2016.01.010

Taylor J, Shrubsole C, Mavrogianni A (2017) Urban heat vulnerability mapping team placement. Hounslow London Borough Council, London

Tomlinson CJ, Chapman L, Thornes JE, Baker CJ (2011) Including the urban heat island in spatial heat health risk assessment strategies: a case study for Birmingham, UK. Int J Health Geogr 10(1):42. https://doi.org/10.1186/1476-072X-10-42

UKCP (2018) Maps & key findings, Key findings summary spreadsheet, Bristol

UN DESA (2017) World population prospects. World population prospects, Volume II: Demographic profiles, New York

UN DESA (2018). World urbanization prospects: the 2018 revision, Key facts, New York

van der Hoeven F, Wandl A (2015) Amsterwarm: mapping the landuse, health and energy-efficiency implications of the Amsterdam urban heat island. Build Serv Eng Res Technol 36(1):67–88

Vandentorren S, Bretin P, Zeghnoun A, Mandereau-Bruno L, Croisier A, Cochet C, Ribéron J, Siberan I, Declercq B, Ledrans M (2006) August 2003 heat wave in France: risk factors for death of elderly people living at home. Eur J Public Health 16(6):583–591

Vardoulakis S, Dimitroulopoulou C, Thornes J, Lai K-M, Taylor J, Myers I, Heaviside C, Mavrogianni A, Shrubsole C, Chalabi Z, Davies M, Wilkinson P (2015) Impact of climate change on the domestic indoor environment and associated health risks in the UK. Environ Int 85:299–313. https://doi.org/10.1016/j.envint.2015.09.010

Watkins R, Palmer J, Kolokotroni M, Littlefair P (2002) The London Heat Island: results from summertime monitoring. Build Serv Eng Res Technol 23(2):97–106. https://doi.org/10.1191/0143624402bt031oa

White-Newsome JL, Sánchez BN, Jolliet O, Zhang Z, Parker EA, Timothy Dvonch J, O'Neill MS (2012) Climate change and health: indoor heat exposure in vulnerable populations. Environ Res 112:20–27. https://doi.org/10.1016/j.envres.2011.10.008

WHO (2011) Public health advice on preventing health effects of heat. New and updated information for different audiences. WHO, Copenhagen

WHO (2017) Climate change and health, World Health Organization (WHO). http://www.who.int/mediacentre/factsheets/fs266/en. Assessed 6 March 2020

WHO (2018) Types of healthy settings, Healthy cities. https://www.who.int/healthy_settings/types/cities/en. Assessed 6 March 2020

WHO Regional Office for Europe (2009) Zagreb declaration for healthy cities: health and health equity in all local policies. WHO, Copenhagen

Wilhelmi OV, Hayden MH (2010) Connecting people and place: a new framework for reducing urban vulnerability to extreme heat. Environ Res Lett 5(1):014021. https://doi.org/10.1088/1748-9326/5/1/014021

Wilson SM, Richard R, Joseph L, Williams E (2010) Climate change, environmental justice, and vulnerability: an exploratory spatial analysis. Environ Justice 3(1):13–19

Wolf T, McGregor G (2013) The development of a heat wave vulnerability index for London, United Kingdom. Weather and Clim Extrem 1:59–68

Wolf J, Adger WN, Lorenzoni I, Abrahamson V, Raine R (2010) Social capital, individual responses to heat waves and climate change adaptation: an empirical study of two UK cities. Glob Environ Chang 20(1):44–52
Wolf T, McGregor G, Analitis A (2014) Performance assessment of a heat wave vulnerability index for greater London, United Kingdom. Weather Clim Society 6(1):32–46
Wolf T, Chuang W-C, McGregor G (2015) On the science-policy bridge: do spatial heat vulnerability assessment studies influence policy? Int J Env Res Public Health 12(10):13321–13349
ZCH (2015) Overheating in Homes, The big picture, Full report. London
Zimmermann N, Symonds P, Davies M, Wilkinson P, Dianati K, Taylor J, Milner J (2018a) Urban systems complexity in sustainability and health: an interdisciplinary modelling study. The Lancet Planetary Health 2:S21. https://doi.org/10.1016/S2542-5196(18)30106-2
Zimmermann N, Symonds P, Dianati K, Davies M, Wilkinson P. (2018b) Applying participatory system dynamics in planning for green infrastructure and health. In: Healthy City Design 2018 International Congress, Royal College of Physicians, London, 15–16 October 2018

Chapter 5
Manchester: The Role of Urban Domestic Gardens in Climate Adaptation and Resilience

Gina Cavan, Fraser Baker, Konstantinos Tzoulas, and Claire L. Smith

Abstract Urban green infrastructure (GI) is a well-known solution to enhance resilience to climate hazards, particularly through the climate regulating functions of green spaces including cooling the city and moderating runoff, in addition to improving health and wellbeing. Domestic gardens provide a valuable contribution to the larger GI network in cities, however, there is less detailed knowledge about these areas and their role in urban climate regulation, in comparison to publicly accessible spaces such as parks. Environmental models require detailed information relating to GI type, structure, and height, since the functionality of different GI types (e.g. mown grass, rough grass, shrubs, trees) is variable in its influence upon urban microclimate. Whilst remote sensing classification techniques can distinguish broad vegetated and non-vegetated classes, they cannot identify the diverse range of surface types within gardens.

This chapter presents a novel approach to collecting fine-scale high quality information on urban domestic gardens to produce new climate information for planning healthy cities, within the case study city of Manchester, UK. The approach combines citizen science data with high resolution areal imagery and environmental modelling, to quantify and map cooling potential and runoff attenuation under different scenarios. This found that greening gardens could be the solution to reducing future risk of heat wave events and surface water flooding, particularly in neighbourhoods with a high proportion of gardens.

This new climate information was presented in an action plan, co-developed with project partner organisations, embedded within the existing policy framework. Engagement of citizens directly led to positive climate resilient actions within

---

The original version of this chapter was revised. The correction to this chapter is available at https://doi.org/10.1007/978-3-030-87598-5_19

G. Cavan (✉) · F. Baker · K. Tzoulas
Department of Natural Sciences, Manchester Metropolitan University, Manchester, UK
e-mail: g.cavan@mmu.ac.uk; F.Baker@mmu.ac.uk; K.Tzoulas@mmu.ac.uk

C. L. Smith
The Open University, Milton Keynes, UK
e-mail: C.smith@open.ac.uk

C. Ren, G. McGregor (eds.), *Urban Climate Science for Planning Healthy Cities*, Biometeorology 5, https://doi.org/10.1007/978-3-030-87598-5_5

Manchester as residents pledged and implemented climate adaptation solutions within their gardens, such as replacing impervious surfaces with vegetation. This project demonstrates how cross-disciplinary collaboration among scientists, planners, government, non-governmental organizations, the private sector and the general public helps to address the challenges that lie ahead for cities as healthy environments.

Keywords Urban green infrastructure · Heatwaves · Flooding · Citizen science · Gardens · Urban resilience

Key Messages

- Urban green infrastructure is a well-known solution to enhance resilience to climate hazards and improve health and well-being.
- A significant proportion of the green infrastructure in cities is contained within private domestic gardens, but the lack of detailed data means that its contribution to a city's climate resilience is unknown, hindering effective planning and policymaking.
- Novel methods combining citizen science and high-resolution image analysis were applied to investigate domestic gardens in the case study of Manchester, UK. It was discovered that gardens are on average only 50.23% green infrastructure, thus reducing the city's total green infrastructure by almost 10% and reducing their effectiveness for building climate resilience.
- Environmental modelling was conducted to quantify and map the cooling potential and runoff attenuation under different scenarios finding that greening gardens could be the solution to reducing future risk of heat wave events and surface water flooding, particularly in neighbourhoods with a high proportion of gardens.
- This new climate information can aid planning for building healthy cities. An action plan was co-developed with project partner organisations and embedded within the existing policy framework. Engagement of citizens directly led to positive climate resilient action as residents implemented changes within their gardens, such as replacing impervious surfaces with vegetation.
- Cross-disciplinary collaborations among scientists, planners, government, non-governmental organisations, the private sector, and the general public are required to address the challenges that lie ahead for cities as healthy environments.

5.1 Introduction

Cities are particularly vulnerable to warming temperatures and the increasing incidence of extreme weather events, prompting action by governments and policymakers to deliver solutions to improve urban resilience to climate impacts and protect the health and well-being of city dwellers (Rosenzweig et al. 2018). It is well known that urban green infrastructure (GI)—the interconnected network of vegetation and water in cities—contributes to the health of a city and influences urban climate. Urban GI can help to moderate microclimate through cooling via evapotranspiration, shading, and storing less heat than built surfaces, and mitigate flooding through intercepting and capturing rainfall, reducing the rate and volume of surface water runoff (Gill et al. 2007; Bowler et al. 2010). Urban GI therefore offers potential to aid urban resilience in adapting cities to climate change. Accordingly, it has received increasing interest in environmental planning and policymaking (Carter et al. 2015). GI provides a wide range of other ecosystem services (ES) (a concept linking the functioning of ecosystems to human benefits), in addition to climate regulation. Importantly, there is a strong association between GI and health—since GI not only reduces climate-related health hazards such as heat-related deaths, but also improves physical and mental health and well-being (Lindley et al. 2019).

Domestic gardens offer a valuable source of GI within an urbanised environment (Gaston et al. 2005; Davies et al. 2009; Cameron et al. 2012). They are important patches of greenspace that can provide connectivity between larger areas of green spaces (parks, recreation grounds, etc.), therefore improving provision of urban ES. Whilst, individually, one domestic garden may appear insignificant, collectively domestic gardens can comprise over one third of a city's surface area (Mathieu et al. 2007), and become more important as a GI resource as urbanisation increases (Loram et al. 2011). Domestic gardens provide diverse ES, which benefit urban residents through both active and passive engagement (Lin et al. 2017). Health and well-being benefits, such as restorative benefits, depend both upon the resident interacting with the physical space and characteristics of the garden, in terms of vegetation, design, size, etc. (Lin et al. 2017). Some ecosystem services are also gained via passive means, such as climate regulation, whereby benefits extend beyond the immediate garden parcel to influencing climate at the neighbourhood scale. Differences in both form and management of domestic gardens will significantly affect the ES provision (Cameron et al. 2012).

Despite their importance as an urban GI resource, domestic gardens have received comparably little attention to public green spaces, and thus there is a lack of data and reliable information about the quantity and quality of GI provision by domestic gardens (Cameron et al. 2012), not least regarding their contribution to a city's climate resilience. Furthermore, the lack of regulation of domestic gardens means that homeowners can alter garden composition how they choose, with local government having little control over existing gardens. As a result, policymakers and planners perceive domestic gardens as a particularly challenging issue to influence. Changes to garden composition can have significant implications for ES provision. For example, conversion of front gardens to hard surfaces, housing extensions, infilling, and adding impervious surfaces such as paths, whilst a gradual process, could result in significant loss of total garden resource. Such changes in garden composition reduce resilience to climate change, which highlights the need to engage citizens, in order to stimulate positive outcomes that aid climate resilience. With growing urban densification, domestic gardens are perceived increasingly as a luxury and thus there is a need to better understand private gardens and their role in ES provision (Cameron et al. 2012) to support effective environmental policy and planning practice.

This chapter presents the outcomes of a co-produced project at the science–policy–practice interface that aimed to improve understanding of domestic gardens in a large city and the ES they provide, with a particular focus upon the benefits that enhance climate resilience. The research aimed to embed new knowledge in user organisations to illustrate the benefits of urban gardens for reducing climate risks in order to facilitate improvements in spatial planning and installation of GI. Whilst the project's aim primarily focused upon generating new knowledge to inform GI planning, the need to involve citizens was clear in order to implement adaptation solutions. Accordingly, the methodology incorporated a citizen science element to both gather data and as an educational tool. The project was carried out in Manchester, UK, led by Manchester Metropolitan University and University of Leicester with public (Manchester City Council), private (Southway Housing Trust) and third sector (Manchester City of Trees and Lancashire Wildlife Trust) organisations as partners.

5.2 Study Area

Manchester, UK, is a post-industrial city with a growing population of 541,319 people (2016 mid-year estimate) within an area of 115.6 km^2 (Manchester City Council 2018). Manchester experiences a warm temperate fully humid climate, with warm summers (Cfb) and mild winter temperatures (Kottek et al. 2006). Manchester's daytime air temperature UHI is around 3 °C (night-time, 5 °C) in summer (Smith et al. 2011), and the city is especially vulnerable to extreme weather events, including flooding and warm temperatures (Kaźmierczak and Cavan 2011; Smith et al. 2011, 2015; Carter et al. 2015). The Indices of Deprivation in 2015 ranked Manchester as 5 out of 326 local authorities in England, where 1 is the most deprived, confirming that it has one of the highest proportions of deprived neighbourhoods in England (DCLG 2015). This profile of high deprivation increases vulnerability of populations to extreme weather events and affects the capacity of the city's population to prepare for, respond to, and recover from climate-related extreme events (Kaźmierczak and Cavan 2011).

Green and blue infrastructure is a core component of Manchester City Council's long-term plan for delivering a green, sustainable, healthy and liveable city throughout the twenty-first century. The Manchester Strategy, overseen by stakeholders and developed through extensive consultation with residents, businesses and partner organisations, sets out a long-term vision for Manchester's future, with five key priorities to realise: a thriving and sustainable city; a highly skilled city; a progressive and equitable city; a liveable and low-carbon city; and a connected city (Manchester City Council 2016). Several key plans support this strategy, relating to themes such as local development, transport, ageing, and poverty. Of key interest here is the Manchester Green and Blue Infrastructure Strategy (MGBIS) 2015–2025, the policy document for guiding GI planning in the city. The MGBIS and accompanying Implementation Plan outlines how stakeholders will work together to improve Manchester's GI (Manchester City Council 2015). This recognises the valuable role of GI in delivering environmental and social benefits to the city, creating attractive, successful neighbourhoods, and encouraging business investment (Manchester City Council 2015).

The MGBIS undertook an initial stocktake of Manchester's GI resources and estimated that GI covers 58% of the land area of the city, including five river valleys, three canals, over 160 parks, street trees, woodland, private gardens and other areas of natural environment (Manchester City Council 2015). This assessment, based upon OS Mastermap datasets, was limited in its representation of domestic gardens, since it was unable to distinguish between land surface cover types within a garden, and assumed gardens are completely GI. Since land surface cover in urban areas is a major driver of the provision of climate regulating and other ecosystem services, and gardens comprise 20% of the land area in the city, resulting environmental models that estimate the provision of, for example, cooling potential and runoff retention, will be inaccurate. This data gap will overestimate the amount of GI in a given area; thus, the need to enhance existing GI will be underestimated, affecting delivery of GI solutions. Furthermore, since the distribution of gardens varies across Manchester (Fig. 5.1), environmental model outputs will not have a uniform error across the city.

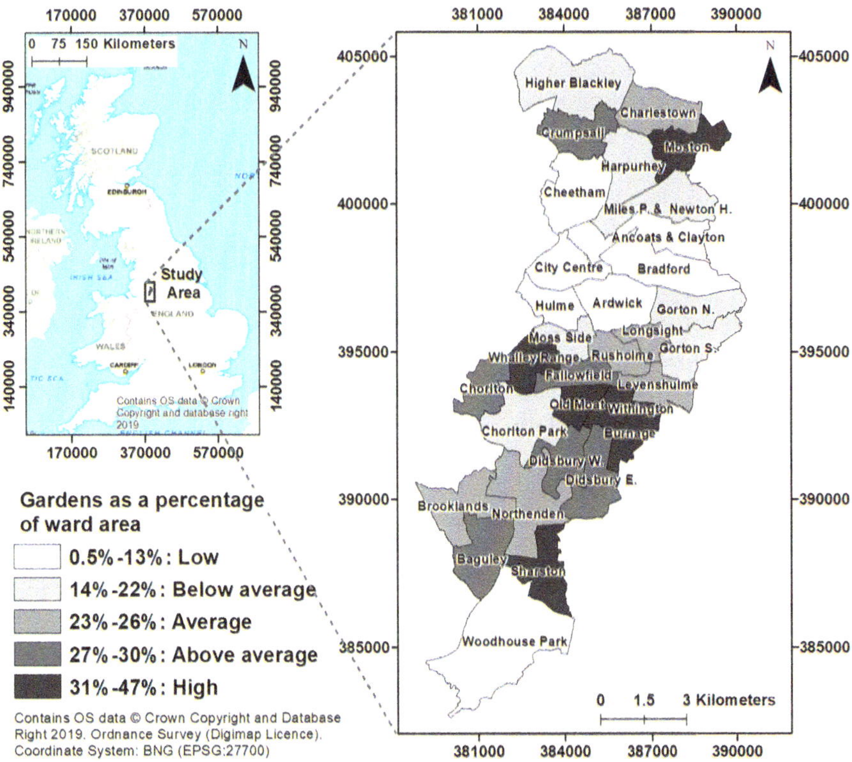

Fig. 5.1 The case study area of Manchesters, UK, showing gardens as a percentage of ward area

Manchester comprises 226,640 households of which approximately 68% is private stock (owner occupied and private rented), with the remainder owned by Local Authority or Registered Social Landlords (Housing Associations) (Manchester City Council 2018). Manchester has a large proportion of apartments (26.6%) and terrace housing (36.0%), in comparison to the national averages, and a low amount of detached housing (4.3%) (AGMA 2010). There are 156,573 gardens, which comprise over one fifth (20.4%, 23.6 km^2) of the total land area of Manchester, and garden coverage (as a proportion of total land cover) varies considerably across the city, between 0.5% in the City Centre ward and 47% in suburban areas (Fig. 5.1). Given the extensive terrace housing, many gardens in the city are small spaces, with 22.52% of gardens having a footprint of less than 50 m^2.

5.3 Establishing the Baseline: Fine Resolution Mapping of Garden Green Infrastructure

Vegetation has an important influence on urban climate. At the microscale, vegetation type and structure are important for influencing climate regulating services. For example, grasses mitigate runoff completely, and trees reduce runoff by 60%, in comparison to tarmac surfaces (Armson et al. 2013). Regarding cooling functions, mature parkland trees reduce radiant temperatures by 5–7 °C, whilst grasses have very little effect on reducing radiant temperatures (Armson et al. 2012). An understanding of the mix of land surface cover types within gardens is therefore needed. This lack of data on the land surface cover in gardens necessitated the collection of a detailed dataset in order to understand the contribution of gardens to climate resilience on a city scale.

Obtaining detailed land surface cover information for gardens is challenging due to the small scale and heterogeneity in land cover across gardens. Local knowledge from citizens can be a useful resource in crowdsourcing information. This has been found to be valuable in previous studies related to garden green infrastructure, for example, to gather data on garden use and maintenance (Loram et al. 2011; Lin et al. 2017) as well as in other disciplinary areas, including climate and atmospheric sciences (Muller et al. 2015). However, using crowd-sourced information for accurately characterising garden composition has not been attempted previously. Further benefits of working with local communities to create a garden database include the exchange of knowledge between local residents, scientists, and stakeholders.

To obtain fine-scale high-quality information on urban gardens, a novel methodology was employed, which combined citizen science with high-resolution image analysis. This involved two steps: firstly, designing and implementing an online citizen science survey (CSS) tool to capture quantitative data relating to garden composition, and secondly, classification of high-resolution optical imagery to verify and extend the CSS results. This combined approach enabled an assessment of the quality of citizen science data. Presented below is a summary of the methodology, with full details provided in Baker et al. (2018).

The CSS tool, promoted as *My Back Yard*, was co-designed with the project partners, and open for completion for 6 months (July–December 2016). The CSS required respondents to estimate the proportional cover of ten common garden land surface cover types (buildings, hard impervious, hard pervious, bare soil, trees, shrubs, mown grass, cut grass, cultivated and water), required for environmental modelling in order to account for the ecological processes of varying land cover types. In addition, the CSS performed as a public engagement tool to enable learning about the role of urban gardens in climate regulation. On completion of the survey, respondents received feedback on their results and how they compare with other responses from their neighbourhood. Specific guidance on ways to improve their garden to aid climate resilience was provided and respondents could then pledge their support to implementing adaptation solutions, committing to make positive changes to their own garden. The survey was open to anyone with a UK residential postcode; however, results presented are responses collected for Manchester residents only. Over one thousand useable responses to the survey were received, with a good diversity of gardens captured, in terms of both size and amount of garden GI (Baker et al. 2018).

Image analysis was required to validate and extend the CSS database, by categorising broad surface cover types for all digitised garden parcels in Manchester. Validation was important to provide an accuracy of CSS garden land surface cover estimations, and extension was needed to quantify garden surface coverage beyond the gardens reported by the CSS respondents. Validation of the CSS land surface estimations revealed a mean accuracy of 76.63% (s = 15.24%), demonstrating that citizens are able to provide valid estimates of garden surface coverage proportions. High-Resolution True-colour Aerial Imagery (TAI) (spatial scale 0.125 m) was used for the object-based image analysis, to classify seven land surface cover categories for all garden parcels in the city, which achieved an overall classification accuracy of 82%. CSS land surface estimations were then extrapolated, to provide the proportional coverage of ten garden land surface cover types (buildings, hard impervious surfaces, hard pervious surfaces, bare soil, trees, shrubs, mown grass, rough grass, cultivated land, water) for every garden parcel in the city.

The resulting dataset quantifies the proportion of ten land surface cover types in all gardens across the city. This reveals that the proportion of GI in an average Manchester garden is 50.23% (Table 5.1). Average garden land surface cover proportions vary considerably between wards, with GI in an average garden varying between 15.44% (lowest, Hulme ward) and 52.74% (highest, Burnage ward) (Fig. 5.2). Furthermore, it is evident that there are differences in the provision of different types of GI, with tree cover in an average garden varying greatly between 2.15% (lowest, Moss Side ward) and 25.52% (highest, Didsbury East), highlighting the increased quality of GI in some wards, since trees provide increased climate regulation benefits (Fig. 5.2).

Combining garden GI with public green space datasets (from Ordnance Survey) reveals that Manchester has a total of 49.0% GI—almost 10% less than previously estimated, since gardens were assumed to be completely comprised of green space. This means that over one fifth (20.94%) of Manchester's GI is contained within domestic gardens, for which city residents are responsible for maintaining and protecting.

Table 5.1 Average proportional land surface cover across Manchester's gardens

Garden land surface cover type	Proportion (%)
Buildings	1.32
Hard Impervious	33.77
Hard Pervious	9.53
Bare Soil	5.15
Trees	16.54
Shrubs	9.19
Mown Grass	14.46
Rough Grass	1.97
Cultivated	7.62
Water	0.45
Total green infrastructure (GI)	50.23

Total GI is the sum of trees, shrubs, mown grass, rough grass, cultivated and water

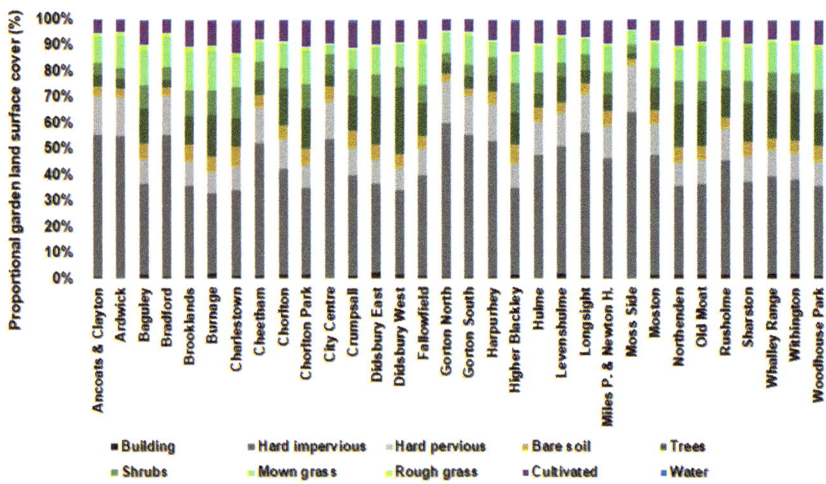

Fig. 5.2 Proportional garden land surface cover for the average garden in each ward

Spatial variation in garden GI between wards is also evident, varying between 0.2% (lowest, City Centre) and 27% (highest, Burnage) (Fig. 5.3a). Results reveal that whilst the wards with the greatest proportion of garden space to ward area (including Old Moat, Withington, and Burnage, Fig. 5.1) still have the highest proportion of garden GI, they experience a significant reduction. For example, in Burnage, domestic gardens cover 47.2% of the ward area, but garden GI cover is only 27% of the ward area, with the remaining proportion (20.2%) comprising other land surface cover types including buildings, hard surfaces and bare soil. This demonstrates that gardens are not performing as well as they could be in delivering GI.

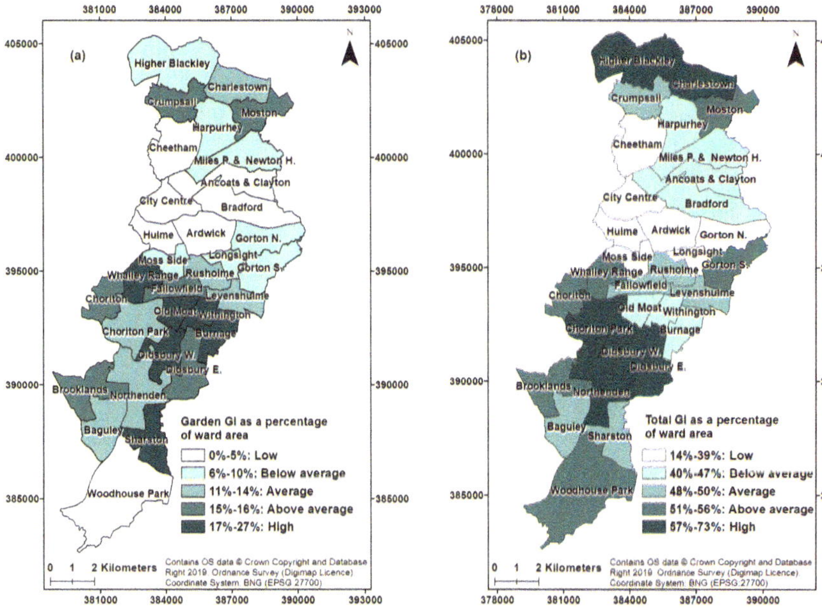

Fig. 5.3 (**a**) Garden GI as a percentage of ward area, (**b**) total GI as a percentage of ward area

Combining the new garden GI data with all other GI in the ward (e.g. public parks) reveals that total ward GI varies between 13.65% (lowest, City Centre) and 73.36% (highest, Higher Blackley) (Fig. 5.3b). Of the total GI in a ward, domestic gardens contribute between 1.34% (lowest, City Centre) and 62.41% (highest, Burnage) of the GI. These large spatial variations in GI quantity and type will affect the spatial provision of climate regulating services, as discussed in Sect. 5.4.

5.4 The Contribution of Domestic Gardens to Climate Regulation

Environmental modelling was undertaken to estimate and map the cooling potential and runoff attenuation functions of GI, to indicate benefits related to reduced risks from heatwaves and flooding. Risks of heatwaves and flooding were identified as priority climate risks by the project partners and a wider group of stakeholders forming the Manchester Green and Blue Infrastructure Strategy Group.

The cooling potential of domestic gardens was assessed using an urban energy exchange model, which produces outputs of maximum surface temperature expected during the two hottest days in summer (98th percentile) for a specified study area (Gill et al. 2007; Cavan et al. 2014; Carter et al. 2015). Runoff attenuation was assessed using a model based upon the US Soil Conservation Service approach

(Gill et al. 2007), providing the percentage of rainfall that becomes surface runoff for a given amount of daily rainfall, in this case the 99th percentile daily winter rainfall (assuming normal soil moisture conditions). Both models were available within an online decision-support tool (the "STAR Tools"), which have been applied to cities in the UK, Europe and Africa, and used by policymakers, researchers, and green infrastructure professionals to guide planning and policy (The Mersey Forest and University of Manchester 2011).

The STAR tools were run for the electoral wards administrative unit (32 in total) to demonstrate how land cover change influences climate risks across the city. Whilst ward boundaries do not reflect climatic or environmental processes, these were the preferred spatial unit for the project partners, since electoral wards are the spatial units used to elect local government councillors, and thus have importance for local planning policy. Wards vary in size and population, with the national average around 5500 people per ward. Baseline and future climate projections data were available from the UK Climate Projections 2009, the most up-to-date high-resolution climate information for the UK at the time. Climate data for 1961–1990 baseline time period and 2050s High emissions scenario (2050H) 90% probability level (to represent the maximum amount of change) were selected, where the high emissions is equivalent to the IPCC A1FI scenario (Murphy et al. 2009). Four scenarios were run to demonstrate the contribution of gardens to climate regulation (Table 5.2).

Results demonstrate that because gardens are comprised of heterogeneous land surfaces, wards experience higher maximum surface temperatures (between 0.1 and 4 °C; Fig. 5.4) and higher surface runoff (between 0 and 6%; Fig. 5.5) than they would if gardens were wholly green space. Differences are most pronounced in the wards that have a higher proportion of gardens to ward area, in particular, Burnage, Old Moat and Withington (with garden areas comprising 47.26%, 46.49%, and 39.62% of the ward, respectively). As previously demonstrated, domestic gardens contain a significant amount of the city's green space, and thus make an important contribution to climate regulation.

Climate projections were run to demonstrate potential future change and the impact of greening gardens on climate regulation. Figures 5.6 and 5.7 show ward level results for the modelled maximum surface temperatures and surface runoff, respectively. This highlights the impact of garden greenspace in the future, as the difference between Future I scenario (future climate with accurate garden greenspace) and Future II scenario (future climate with gardens wholly greenspace). Greening

Table 5.2 Description of the four scenarios for assessing cooling potential and runoff attenuation

Scenario	Description
Baseline I	1961–90 climate with the assumption that gardens are completely green space
Baseline II	1961–90 climate with more accurate garden greenspace data
Future I	2050s high emissions scenario (90% probability level) with more accurate garden greenspace data
Future II	2050s high emissions scenario (90% probability level) with the assumption that gardens are completely green space

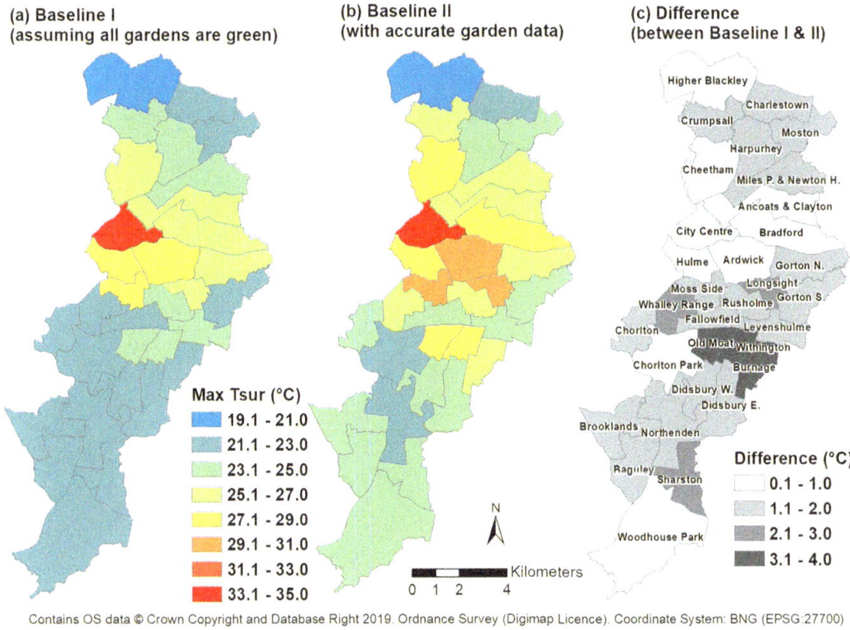

Fig. 5.4 Modelled maximum surface temperatures for (**a**) Baseline I scenario, (**b**) Baseline II scenario, (**c**) Difference between Baseline I and Baseline II scenarios

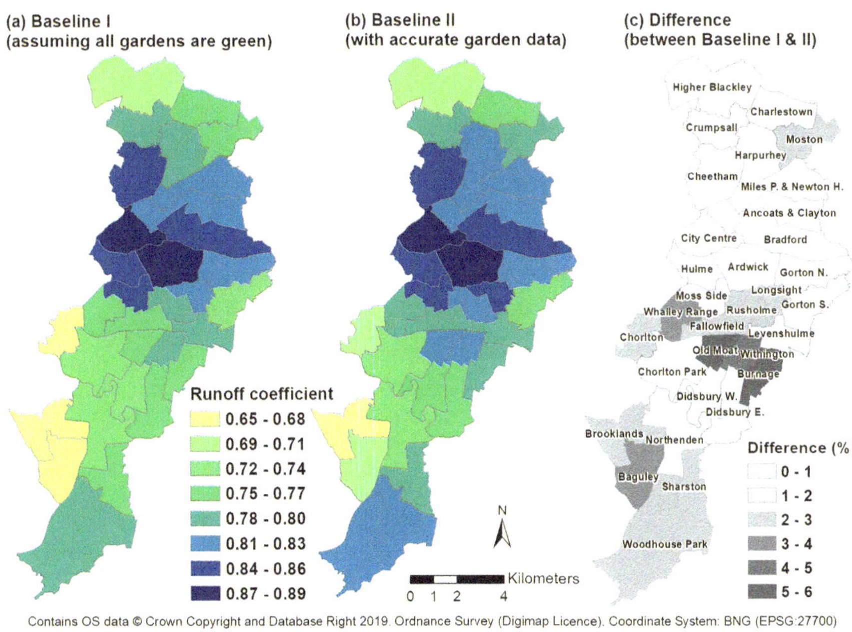

Fig. 5.5 Modelled surface runoff coefficient (where a coefficient of 1 is complete runoff) for (**a**) Baseline I scenario, (**b**) Baseline II scenario, (**c**) Difference between Baseline I and Baseline II scenarios

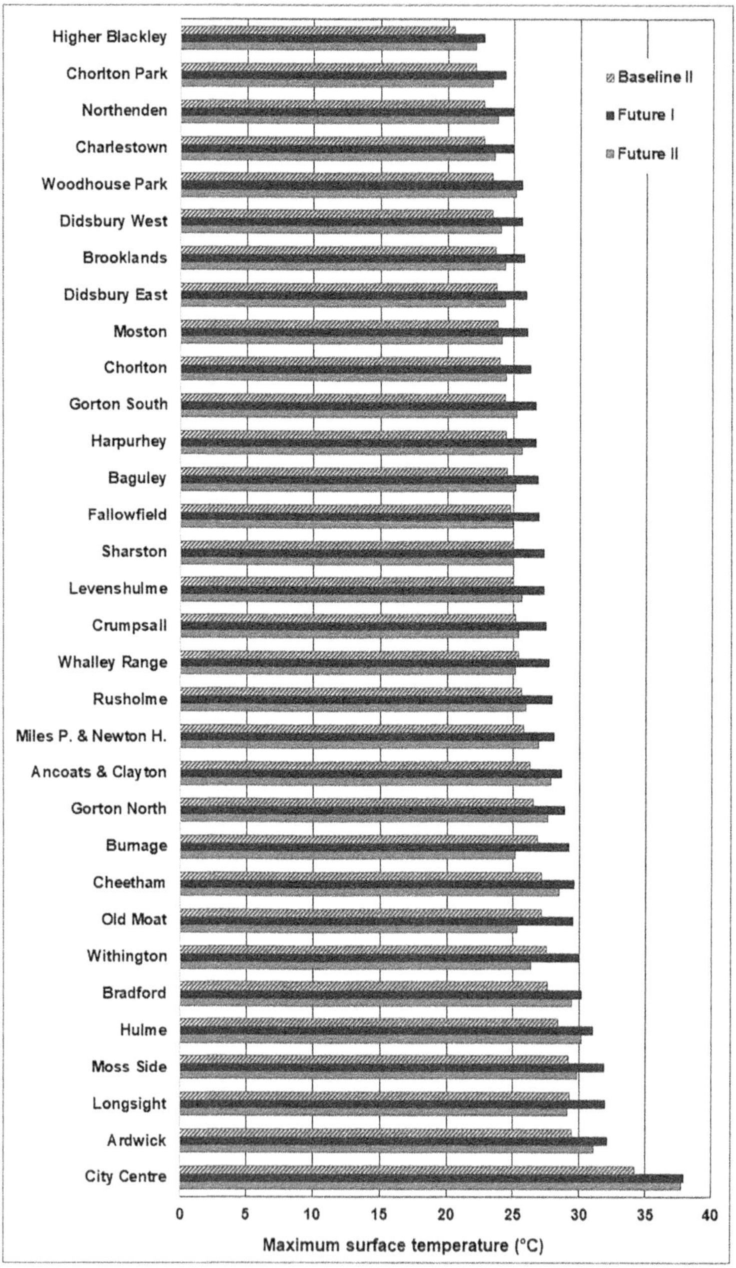

Fig. 5.6 Ward level modelled maximum surface temperature for baseline and future scenarios

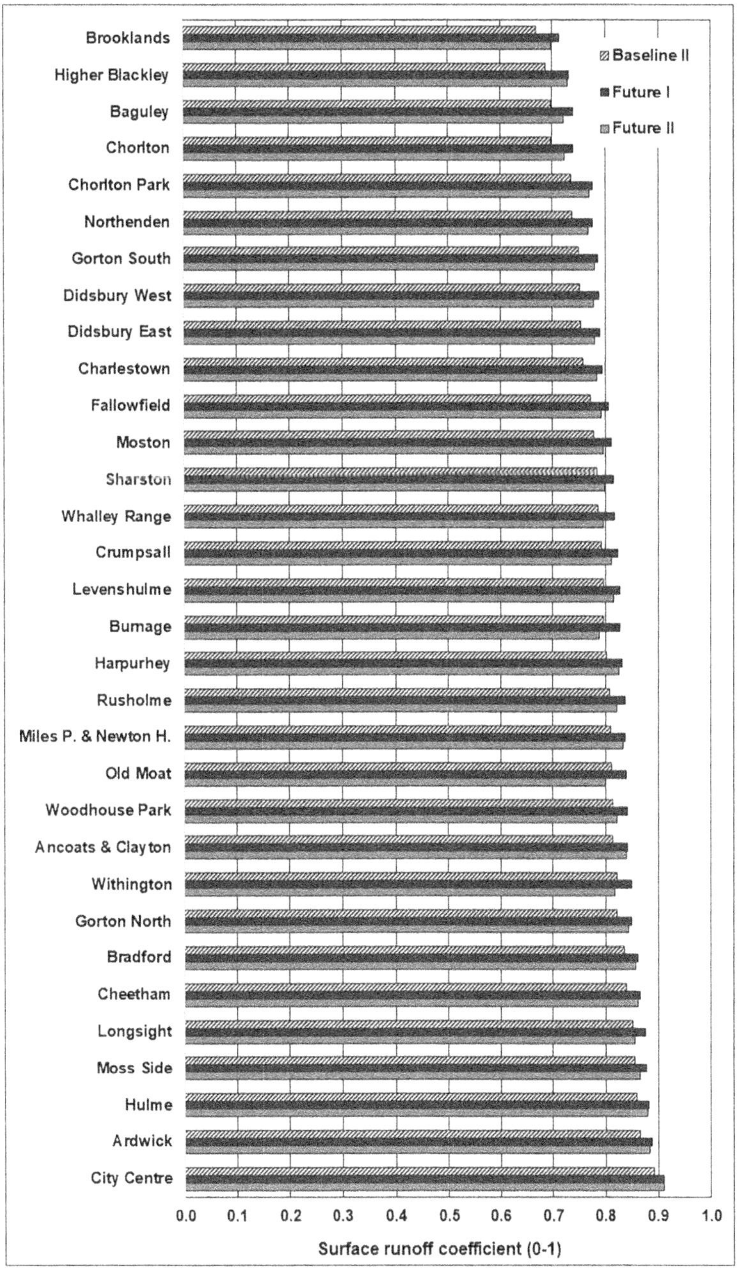

Fig. 5.7 Ward level surface runoff for baseline and future scenarios

gardens would reduce ward surface temperatures in future by between −0.1 and −4.2 °C ($M = -1.7$ °C; $s = 1.0$ °C). It is evident that for six wards (Burnage, Whalley Range, Sharston, Withington, Old Moat and Longsight), greening gardens could mitigate the increased risk of heatwave events due to climate change on the neighbourhood, since the temperatures under the Future II scenario falls below Baseline II scenario (Fig. 5.6). These wards have a high proportion of gardens to ward area (between 24.93 and 47.26%); therefore, complete greening of gardens in these wards has a significant impact upon reducing future heatwave risks. Greening gardens could reduce future ward surface runoff by up to −4.0% ($M = -1.0$%; $s = 1.0$%), and in three wards (Withington, Old Moat, Burnage) this could mitigate the increased risk of surface runoff due to future climate change on the neighbourhood (Fig. 5.7). These three wards have the highest proportion of gardens to ward area (over 39%, Fig. 5.1). These reductions are of significant value, since UK meteorological records show that the 10 warmest years have all occurred since 2002 and extreme rainfall events have increased over the past 50 years (Kendon et al. 2019). This highlights the need for urgent action to protect vulnerable GI (which includes gardens) in urban areas.

5.5 Engaging Citizens to Encourage Positive Action in Gardens

Whilst the primary focus of the CSS was to gather data on the proportion of garden land surface cover types across the city, this also acted as a useful educational tool to engage citizens about gardens and their importance for delivering ecosystem services, in particular, climate regulation and biodiversity. As outlined above, over 20% of Manchester's GI is contained within domestic gardens, which city residents are responsible for maintaining. Since these are private spaces subject to limited planning control, engagement of city residents is key to halting loss of garden GI, and positive action is needed to increase garden GI. These changes (either positive or negative) will have a direct and measurable effect on the city's climate resilience.

On completion of the CSS, respondents received simple information about why their gardens are important for enhancing the city's climate resilience and biodiversity, and were given guidance on actions that could improve their garden. Respondents could then pledge their support to implementing adaptation solutions, committing to make positive changes to their own garden. Over 700 pledges were received, including:

- 252 respondents pledged to plant a variety of plant types in their garden to enhance wildlife.
- 228 respondents pledged to use drought resistant plants in their garden and to collect water.
- 150 respondents pledged to plant trees for shade to reduce heatwave impacts and improve air quality.
- 119 respondents pledged to replace the hard surfaces in their garden with green space to reduce surface water flooding.

A follow-up survey was emailed to the CSS respondents 1 year later to understand if the pledges had translated into action. For those who pledged to implement an action, completion rates were high, varying between 84.5 and 94.8%, depending upon the specific pledge and associated action. For example, completion rates for the action to replace hard surfaces with green space were lower, likely reflecting that such changes are more challenging and costly to implement. Further research into the barriers and enablers to action is ongoing to support delivery of the action plan, presented in Sect. 5.6.

5.6 Influencing Policy and Practice: An Action Plan to Increase Green Space in Manchester's Gardens

An action plan was co-developed with the four project partner organisations—Manchester City Council, Lancashire Wildlife Trust, Southway Housing Trust, and City of Trees. These organisations all have an interest in gardens and are able to influence planning policy and practical projects in green spaces across Manchester. The action plan was created using the Delphi technique (Curtis 2004; Okoli and Pawlowski 2004), an iterative process where partners discuss issues and arrive at a consensus, culminating in a shared action plan that could be embedded into existing policy and delivery mechanisms. Three iterative discussions took place to co-develop the action plan; these involved:

1. Project partners identifying and discussing their existing priorities relating to GI interventions and associated delivery mechanisms in order to reach a consensus on their mutual and positively reinforcing priorities and delivery mechanisms for GI interventions.
2. Project partners reaching a consensus on the potential of the GI spatial database to facilitate integration of priorities and delivery mechanisms and on a common evaluation of the potential to pool mechanisms for delivering mutual priorities in GI interventions.
3. Project partners agreeing a common action plan for GI interventions that strengthen the functionality of ecosystems in areas of GI need.

Through the three iterative discussions, the final aim of the action plan was agreed to *encourage a cultural change in the way that people appreciate the benefits of their gardens*. Project partners committed to achieving this aim by (1) working in partnership; (2) pooling delivery mechanisms; and (3) delivering the actions for improving green space and wildlife in gardens.

The action plan included key actions for each organisation, which bring together and complement existing activities that the project partners undertake to improve green space in gardens as well as creating the groundwork for future garden projects. Each action included a delivery mechanism, targeted areas where the action was needed (based on the spatial analysis presented), the partners involved (including collaborations beyond the project), timescales for delivery, and delivery

Table 5.3 An overview of the Action Plan, outlining the six key themed actions, delivery mechanisms, and specific examples

Types of actions	Delivery mechanisms	Examples
Promote actions that increase green space and wildlife in gardens	Citizen science Data collection and analysis Events Information packs Lobbying On the ground projects Public promotion	The Wildlife Trust campaigns including "My Wild City" and "My Wild Garden" Engaging Manchester in national campaigns including Britain in Bloom and Grow Wild Awards City of Trees' Heritage Trees project, an online citizen science database of peoples' stories, memories and photographs of local heritage trees
Undertake practical garden related improvement projects	Events On the ground projects	Community Greening projects, food growing projects such as "Real Food Wythenshawe", and tree planting initiatives
Engage in garden related policy development	Lobbying	Update the Manchester GI strategy in light of new garden data Establish RPs green infrastructure group
Undertake research on gardens	Data collection and analysis Citizen science On the ground projects	Ongoing collaborative green infrastructure research Spatial mapping of the connectivity of garden corridors
Provide training and practical skills relating to gardens	On the ground projects Events Information packs	The "green streets" projects involving grants for groups or organisations for innovative urban forestry projects, training on tree planting and tree management skills Training on wildlife identification and recording in gardens
Promote the value of gardens	Citizen science Events Information packs	A species recording campaign led by Lancashire Wildlife Trust

mechanisms. There were six key themed actions, shown in Table 5.3, together with some specific examples and delivery mechanisms from the action plan.

The action plan was published as a short briefing document for a general audience (Fig. 5.8), supported by an online spatial database of the key mapped outputs, and has already stimulated a follow-on campaign, My Wild City, a collaboration between Lancashire Wildlife Trust and Manchester City Council funded by Esme Fairburn Foundation. My Wild City is a 4-year project that aims to inspire and engage city residents into getting involved in practical actions in their gardens, parks and wildlife corridors, and develop a new 10-year biodiversity strategy for the city.

To maximise impact of the action plan, partners agreed that it should be embedded within the existing local policy context. Whilst there is no specific planning policy for domestic gardens, they are included in the Manchester Green and Blue Infrastructure Strategy (MGBIS), which complements strategic objectives of the

Fig. 5.8 The MBY Action Plan, illustrating Lancashire Wildlife Trust's actions and delivery mechanisms, available to download at www.mmu.ac.uk/mybackyard

Manchester Strategy and Manchester Climate Change Strategy. The action plan will be delivered through the Manchester Green Infrastructure Strategy Group, an informal partnership of public, private, and third sector organisations, of which all the project team are members. The Strategy Group meets four times per year to provide updates and review progress on delivery. The MGBIS is due to be updated in 2020/21, which will include the updated evidence documented here, together with priority actions and delivery mechanisms highlighted. Embedding this new evidence within the existing policy framework will ensure that progress on actions is appropriately monitored and reported.

5.7 Conclusions

Urban green infrastructure (GI) is a well-known solution to enhance resilience to climate hazards. In this context the quantity and spatial distribution of GI has been researched in relation to creating cool islands, moderating urban heat islands and reducing surface water runoff, mitigating flooding. In order to understand the impact of GI in moderating climatic extreme events, environmental models require detailed information relating to urban GI type, structure, and height, since the functionality of different GI types (e.g. mown grass, rough grass, shrubs, trees) is variable in its influence upon urban microclimate. Whilst comprehensive geospatial datasets exist for GI in public spaces, there is less information about privately owned domestic gardens, which make up a large area of cities, and assumed to contain a significant proportion of GI. The lack of data regarding the land surface cover in gardens necessitates the collection of a detailed dataset in order to enhance understanding of the contribution of gardens to climate resilience on a city scale, and provide knowledge to inform spatial planning and policymaking. This project applied a novel approach, combining citizen science with remote sensing analysis, within a collaborative project partnership of universities, private and third sector organisations, which culminated in an action plan to increase green space and enhance wildlife in domestic gardens across Manchester.

In Manchester, domestic gardens comprise over one fifth (20.4%, 23.6 km^2) of the total land area of the city. Gardens, however, are not comprised wholly of green space, and the proportion of GI in an average Manchester garden is 50.23%. At the city scale, this means that Manchester has a total of 49.0% GI—almost 10% less than previous estimations which assumed gardens were completely comprised of green space.

Environmental modelling was conducted to quantify and map climate regulation services including cooling potential and runoff attenuation under different scenarios. This demonstrates that greening gardens could be the solution to reducing future risk of heatwave events and surface water flooding, particularly in neighbourhoods with a high proportion of gardens to ward area. Since city residents are responsible for maintaining and protecting private domestic gardens, such actions necessitate the engagement of citizens within a receptive policy environment. To begin to achieve this ambition, an action plan was co-developed with project partner organisations who are able to influence planning policy and practical projects in green spaces across Manchester. Embedding the action plan within the existing policy framework is beneficial since this will ensure that progress on actions is appropriately monitored and reported in the future.

The online survey tool and results are transferable to other cities and demonstrates how citizen science within a science–policy–practice project framework can provide the basis for shared learning to improve the resilience of the urban environment to climate change. Finally, it is clear that cross-disciplinary collaborations among scientists, planners, government, non-governmental organisations, the private sector, and the general public are required to address the challenges that lie ahead for cities as healthy environments.

Acknowledgements This research was carried out as part of the project "*Assessing the contribution of domestic gardens to urban ecosystem services*" (2016–2018), funded by Natural Environment Research Council NE/N017374/1 and NE/N017374/2. We gratefully acknowledge support from the project partner organisations, and would like to warmly thank Dave Barlow, Philippa Reece, Pete Stringer, and James Hall. We would also like to give a special thank you to the late Tom Butlin who developed the My Back Yard tool.

References

AGMA (2010) Association of Greater Manchester Authorities, Greater Manchester Strategic Housing Market Assessment http://www.manchester.gov.uk/download/downloads/id/14074/gm_strategic_housing_market_assessment_shma_update_may_2010.pdf. Accessed 4 Oct 2019

Armson D, Stringer P, Ennos AR (2012) The effect of tree shade and grass on surface and globe temperatures in an urban area. Urban For Urban Gree 11(3):245–255. https://doi.org/10.1016/j.ufug.2012.05.002

Armson D, Stringer P, Ennos AR (2013) The effect of street trees and amenity grass on urban surface water runoff in Manchester, UK. Urban For Urban Gree 12(3):282–286. https://doi.org/10.1016/j.ufug.2013.04.001

Baker F, Smith CL, Cavan G (2018) A combined approach to classifying land surface cover of urban domestic gardens using citizen science data and high resolution image analysis. Remote Sens 10(4):537

Bowler DE, Buyung-Ali L, Knight TM, Pullin AS (2010) Urban greening to cool towns and cities: a systematic review of the empirical evidence. Landsc Urban Plan 97(3):147–155. https://doi.org/10.1016/j.landurbplan.2010.05.006

Cameron RWF, Blanuša T, Taylor JE, Salisbury A, Halstead AJ, Henricot B, Thompson K (2012) The domestic garden—its contribution to urban green infrastructure. Urban For Urban Gree 11(2):129–137. https://doi.org/10.1016/j.ufug.2012.01.002

Carter JG, Cavan G, Connelly A, Guy S, Handley J, Kazmierczak A (2015) Climate change and the city: building capacity for urban adaptation. Prog Plann 95:1–66. https://doi.org/10.1016/j.progress.2013.08.001

Cavan G, Lindley S, Jalayer F, Yeshitela K, Pauleit S, Renner F, Gill S, Capuano P, Nebebe A, Woldegerima T, Kibassa D, Shemdoe R (2014) Urban morphological determinants of temperature regulating ecosystem services in two African cities. Ecol Indic 42:43–57. https://doi.org/10.1016/j.ecolind.2014.01.025

Curtis IA (2004) Valuing ecosystem goods and services: a new approach using a surrogate market and the combination of a multiple criteria analysis and a Delphi panel to assign weights to the attributes. Ecol Econ 50(3):163–194. https://doi.org/10.1016/j.ecolecon.2004.02.003

Davies ZG, Fuller RA, Loram A, Irvine KN, Sims V, Gaston KJ (2009) A national scale inventory of resource provision for biodiversity within domestic gardens. Biol Conserv 142(4):761–771. https://doi.org/10.1016/j.biocon.2008.12.016

DCLG (2015) The English indices of deprivation 2015. Department for Communities and Local Government, London

Gaston KJ, Warren PH, Thompson K, Smith RM (2005) Urban domestic gardens (IV): the extent of the resource and its associated features. Biodivers Conserv 14(14):3327–3349. https://doi.org/10.1007/s10531-004-9513-9

Gill SE, Handley JF, Ennos AR, Pauleit S (2007) Adapting cities for climate change: the role of the green infrastructure. Built Environ 33(1):115–133

Kaźmierczak A, Cavan G (2011) Surface water flooding risk to urban communities: analysis of vulnerability, hazard and exposure. Landsc Urban Plan 103(2):185–197. https://doi.org/10.1016/j.landurbplan.2011.07.008

Kendon M, McCarthy M, Jevrejeva S, Matthews A, Legg T (2019) State of the UK climate 2018. Int J Climatol 39(S1):1–55. https://doi.org/10.1002/joc.6213

Kottek M, Grieser J, Beck C, Rudolf B, Rubel F (2006) World map of the Köppen-Geiger climate classification updated. Meteorol Z 15(3):259–263

Lin BB, Gaston KJ, Fuller RA, Wu D, Bush R, Shanahan DF (2017) How green is your garden? Urban form and socio-demographic factors influence yard vegetation, visitation, and ecosystem service benefits. Landsc Urban Plan 157:239–246. https://doi.org/10.1016/j.landurbplan.2016.07.007

Lindley SJ, Cook PA, Dennis M, Gilchrist A (2019) Biodiversity, physical health and climate change: a synthesis of recent evidence. In: Marselle MR, Stadler J, Korn H, Irvine KN, Bonn A (eds) Biodiversity and health in the face of climate change. Springer International Publishing, Cham, pp 17–46. https://doi.org/10.1007/978-3-030-02318-8_2

Loram A, Warren P, Thompson K, Gaston K (2011) Urban domestic gardens: the effects of human interventions on garden composition. Environ Manag 48(4):808. https://doi.org/10.1007/s00267-011-9723-3

Manchester City Council (2015) Manchester's great outdoors: a green and blue infrastructure strategy for Manchester 2015–25. MCC, Manchester

Manchester City Council (2016) Our Manchester: the manchester strategy. MCC, Manchester

Manchester City Council (2018) Intelligence Hub (Manchester Statistics) Mid-year population estimates for Manchester, ONS, Crown Copyright. https://dashboards.instantatlas.com/viewer/report?appid=9853f1a1b412404b806dab4d1918def7. Accessed 12 Feb 2018

Mathieu R, Freeman C, Aryal J (2007) Mapping private gardens in urban areas using object-oriented techniques and very high-resolution satellite imagery. Landsc Urban Plan 81(3):179–192. https://doi.org/10.1016/j.landurbplan.2006.11.009

Muller CL, Chapman L, Johnston S, Kidd C, Illingworth S, Foody G, Overeem A, Leigh RR (2015) Crowdsourcing for climate and atmospheric sciences: current status and future potential. Int J Climatol 35(11):3185–3203. https://doi.org/10.1002/joc.4210

Murphy JM, Sexton DMH, Jenkins GJ, Boorman PM, Booth BBB, Brown CC, Clark RT, Collins M, Harris GR, Kendon EJ, Betts RA, Brown SJ, Howard TP, Humphrey KA, McCarthy MP, McDonald RE, Stephens A, Wallace C, Warren R, Wilby R, Wood RA (2009) UK climate projections science report: climate change projections. Met Office Hadley Centre, Exeter

Okoli C, Pawlowski SD (2004) The Delphi method as a research tool: an example, design considerations and applications. Inf Manag 42(1):15–29. https://doi.org/10.1016/j.im.2003.11.002

Rosenzweig C, Solecki WD, Romero-Lankao P, Mehrotra S, Dhakal S, Ibrahim SA (2018) Climate change and cities: second assessment report of the urban climate change research network. Cambridge University Press

Smith CL, Webb A, Levermore GJ, Lindley SJ, Beswick K (2011) Fine-scale spatial temperature patterns across a UK conurbation. Clim Chang 109(3):269–286. https://doi.org/10.1007/s10584-011-0021-0

Smith CL, Cavan G, Lindley S (2015) Urban climatic map studies in UK: greater Manchester. In: Ng E, Ren C (eds) The urban climatic map: a methodology for sustainable urban planning. Routledge, London, pp 313–326

The Mersey Forest and University of Manchester (2011) STAR tools: surface temperature and runoff tools for assessing the potential of green infrastructure in adapting urban areas to climate change. https://maps.merseyforest.org.uk/grabs. Accessed 4 Oct 2019

Chapter 6
Tokyo Heat Island Adaptation Measures

Yasunobu Ashie

Abstract "It is difficult to improve the quality of life because Tokyo is already so developed. Tokyo has come about without any urban planning," said former Tokyo Governor Shintaro Ishihara. The heat island effect in Tokyo, which is one of the most densely populated areas in Asia, has been intensifying in recent years. In order to solve this problem, urban and regional planning must be used to devise the layout of buildings and land cover, and appropriate countermeasure need to be taken to improve the environmental. This chapter describes recent efforts in Tokyo to secure the flow of wind through urban spaces and actual trials for cooling outdoor apparent temperatures. Section 6.1 outlines the current state of the heat island effect in Tokyo. It illustrates the number of tropical nights in Tokyo and temperature conditions, and explains the effect on the heat island phenomenon, people and the urban environment. In Sect. 6.2, the sensible temperature in urban spaces and the analyzing methods are described. A person standing in urban spaces receives various heat from the sun, soil surface, building walls and atmosphere. We developed a numerical model of radiant heat exchange in urban spaces considering the complex 3D geometries formed by buildings and the location of surface materials to estimate the district plans that lower sensible temperatures. This analysis method was applied to a marathon course in the Ginza district, and the thermal impact on people through various cooling measures such as planting trees, water sprinklers and road materials were discussed. Another topic is the case of a bus stop in a semi-open outdoor space equipped with a cooling bench using groundwater. It is unique in that physiological and psychological factors such as heart rates and thermal physiological effects were surveyed with thermal conditions. Section 6.3 is about wind paths. The Tokyo wind path mapping method featuring sea wind is explained and actual urban projects were introduced. Additionally, a quantitative analysis was performed to find a relation between the wind path securement method and sensible temperatures.

Keywords Tokyo · Urban heat islands · Adaptation · Sensible temperature · Wind path

Y. Ashie (✉)
Building Research Institute, Ibaraki, Japan
e-mail: ashie@kenken.go.jp

C. Ren, G. McGregor (eds.), *Urban Climate Science for Planning Healthy Cities*, Biometeorology 5, https://doi.org/10.1007/978-3-030-87598-5_6

Key Messages

- Tokyoites feel the heat.
- Numerical prediction of sensible temperature.
- How to create the wind path.
- Toward thermally comfortable city.

6.1 Tokyo Heat Island Phenomenon and Environmental Impact

6.1.1 Situation of the Heat Island Phenomenon in Tokyo

It is believed that the heat island phenomenon (high temperature phenomenon in cities) was discovered in nineteenth-century London. When the heat island phenomenon was later confirmed by observations around the world, extensive research to clarify the mechanism of its occurrence was undertaken (Landsberg 1981). It was finally possible to grasp the heat island phenomenon as the object of countermeasures after the start of the present century. Especially in Japan, urban heat island countermeasure has been one of the most important social issues.

Figure 6.1 compares change over time of annual average temperature with 1900 as the base year in Tokyo, as the Japan average, and the global average (Ministry of the Environment 2013). Tokyo's air temperature change over the 100 years of the previous century compared with the Japan average and global average which were about 1 °C was higher at 3 °C. This 2 °C difference in the temperature increase is viewed as an effect of heat islands and implementing measures to control this phenomenon is an urgent task.

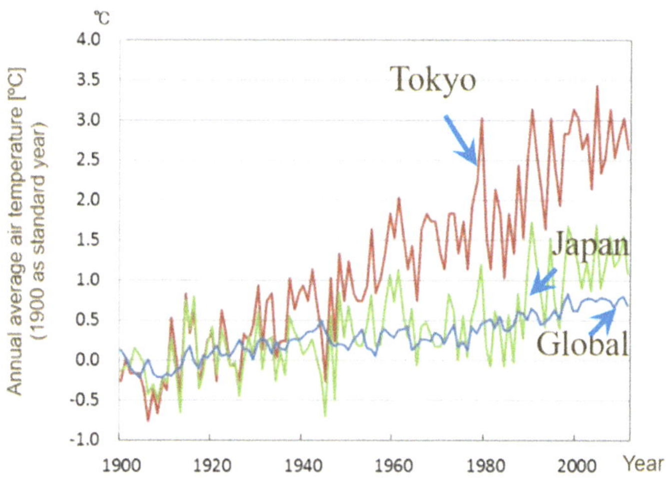

Fig. 6.1 Change of annual average air temperature in Tokyo, Japan, and Worldwide (Ministry of the Environment 2013) from 1900 to 2010. Based on JMA observation data

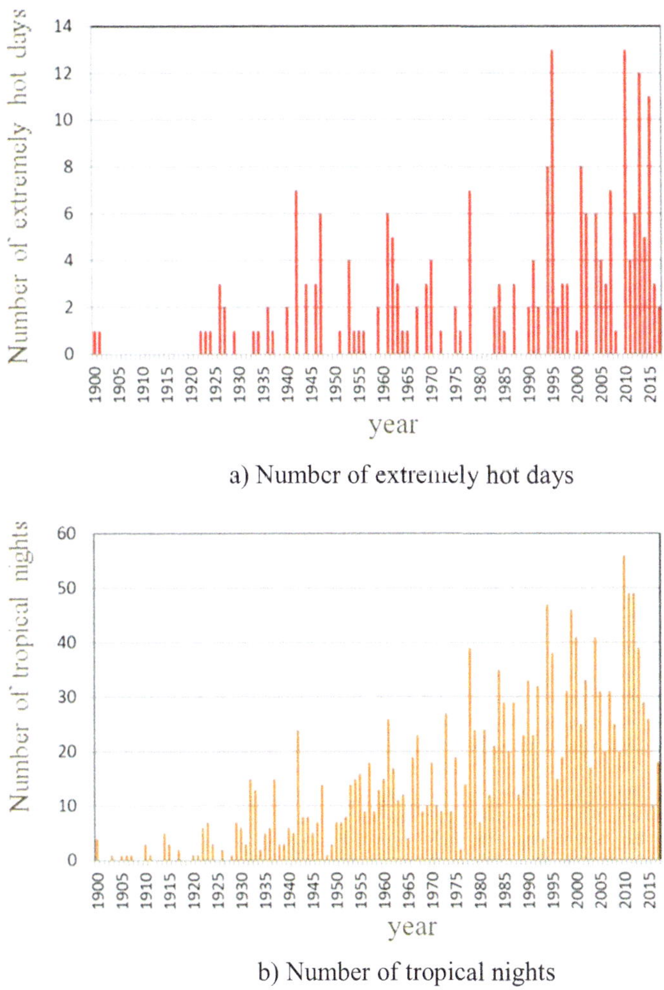

Fig. 6.2 Change in number of extremely hot days and tropical nights (Ministry of the Environment 2018) from 1900 to 2015. Based on JMA observation data. In December 2014, the observation point moved from Otemachi, Chiyoda-ku to Kitanomaru Park. (**a**) Number of extremely hot days. (**b**) Number of tropical nights

Figure 6.2 shows the change over years in the number of extremely hot days and tropical nights in Tokyo (Ministry of the Environment 2018). Extremely hot days and tropical nights are forecast terms used by the Japan Meteorological Agency in weather forecasts. Extremely hot days are days when the high temperature is recorded as 35 °C or higher, and tropical nights is when the low temperature is 25 °C or higher. The number of hot days and tropical nights are increasing over time, and there are concerns about the impact on urban residents such as heat stroke and nighttime sleeplessness.

6.1.2 Environmental Impact of Heat Island Phenomenon

As the classic example of heat risk, Fig. 6.3 shows change over time of the number of deaths from illnesses caused by heat using demographic statistics by the Ministry of Health, Labour and Welfare (National Institute for Environmental Studies, Japan 2009). It reveals that the numbers of deaths of both men and women caused by heat-related illness have increased sharply in recent years. Because it shows a strong positive correlation between the number of people transported in ambulances suffering from heat-related illnesses with the daily maximum air temperature, it can be stated that the occurrence of extreme weather in recent years threatens the lives and health of people. And elderly people are susceptible to heat environments, so it is necessary to pay attention to the fact that the arrival of the superannuated society will lead to a rise in heat risk. The percentage of Japan's population 65 years of age and older was 9.1% in 1980, but increased to 17.4% by 2000. It predicted that this will pass 30% in 2030, so in the future it will continue to be impossible to ignore the problem of heat risk from the perspective of the aging of the population.

Table 6.1 is a list of environmental impacts caused by the heat island phenomenon based on the results of collected past literature and expert interviews (Ministry of the Environment 2004). From this, it can be seen that the environmental impact of the heat island phenomenon is not limited to thermal illnesses, but affects various factors such as ecosystems and energy consumption, and the way in which the

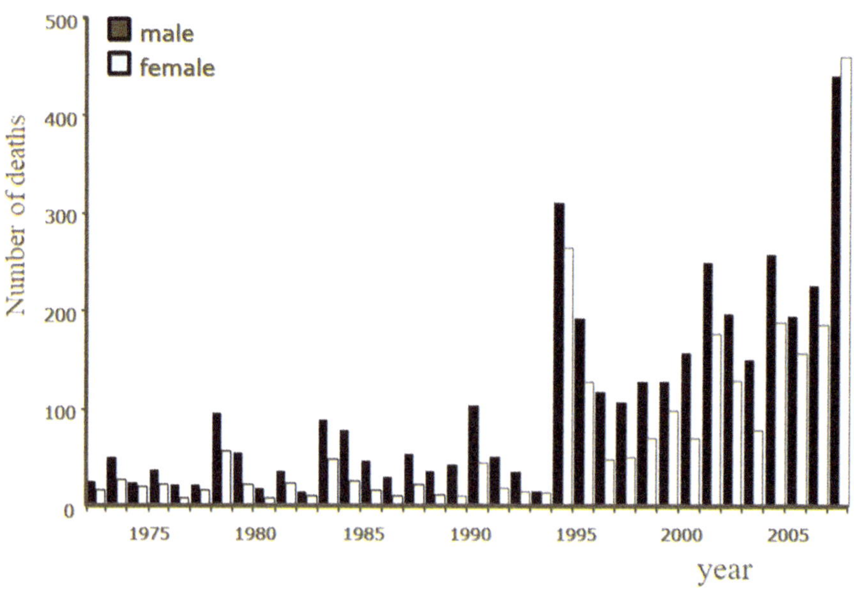

Fig. 6.3 Number of deaths caused by heat-related illnesses (National Institute for Environmental Studies, Japan 2009) from 1970 to 2010. Based on the demographic statistics by MHLW

Table 6.1 Environmental impact of heat island phenomenon (Ministry of the Environment 2004)

Target		Impact index	Phenomenon index	Index application examples
Human	Heat stroke	Number of occurrences	High temperature	The temperature at which the number of occurrences increases rapidly differs depending on the age group. (Working-age population (15–64 years old): 35 °C, elderly (65 years old or older): 33 °C)
			WBGT	Heat stroke occurrence increases rapidly when wet-bulb globe temperature (WBGT) is 28 °C or higher
	Cardiovascular disease	Death rate	High temperature	Mortality due to heart disease in the elderly (65 years and older) increases at 33 °C or higher
	Fatigue	Subjective responses	Indoor/outdoor traffic frequency	Fatigue increases with more frequent traffic between cooled interiors and outdoors. In general, temperature differences 5 °C or greater affect body temperature regulation
	Sleeping disorder	–	Room temperature	–
	Virus infection	Expansion of vector mosquito habitat	Annual average temperature	The distribution of Aedes albopictus goes further north every year, and the distribution is similar to areas with an average annual temperature of 11 °C or higher
		Increase in population density of vector mosquitoes	Daily average temperature	Accumulated warm days with an average temperature of 11 °C or higher (Accumulated warm days per generation of Aedes albopictus 350 °C day)
Ecosystem	Biota	Change in population	(change over years)	Change over years of species inhabiting certain area
	Plant growth	Photosynthesis speed	Air temperature/humidity	The photosynthesis rate decreases at the optimum temperature (25 °C for cedar)
Air pollution	NOx, etc.	Increased contamination concentration	–	Air pollution concentration rises in the mixolimnion in the winter, and on the entire sea breeze front in the summer
	Photochemical oxidant	Increased contamination concentration	High temperature	Photochemical smog is stimulated over the city (in the mixolimnion)
Weather	Torrential rain	Torrential rain frequency	–	The heat island phenomenon may be involved in the generation of cumulonimbus clouds in Tokyo
Energy	Consumption	Increased consumption	High temperature (summer)/ Low temperature (winter)	Increased power consumption
		Reduced consumption	Low temperature (summer)/ High temperature (winter)	Reduction in power consumption

impact occurs differs between the summer and winter. For example, it has been pointed out that summer heat islanding increases the risk of heat stroke, while winter heat islanding can lead to the northward movement of virus-carrying mosquitoes. Regarding building air-conditioning, summer heat islanding cooling energy increases (disadvantageous for humans), and winter heat islanding affects heating energy decreases (advantageous for humans). When planning heat island countermeasures, we should collect the information for making decisions to consider these factors comprehensively.

6.1.3 Mitigation and Adaptation Measures

The Paris Agreement (adopted on 12 December 2015) which is the international framework concerning climate change since the Kyoto Protocol was adopted in 1997 stipulates that the signatory states fulfill their obligation to submit reduction targets at the same time as it stipulates "mitigation" which means taking measures to reduce emissions of or absorb greenhouse effect gases and "adaption" to the impacts of greenhouse effect gasses which have already begun to occur. The reason why adaptation has been emphasized is that it is considered that even if each country's reduction targets are achieved in total, it will be difficult to hold down the advance of global warming to less than 2 °C.

In Japan, in order for the government to comprehensively and systematically promote overall consistent efforts to deal with the various aspects of climate change, the Climate Change Adaptation Plan was enacted by the Cabinet on November 27, 2018, based on the Climate Change Adaptation Act. This plan clearly presented "Impact of hot weather on life" in the fields of lives of the people and city life from among seven areas which are the object of climate change adaptation.

The following is the concept of the relationship of mitigation measures with adaptive measures. Table 6.2 organizes the relationship between mitigation

Table 6.2 Relationship of mitigation countermeasures with adaptive countermeasures in global warming countermeasures and heat island countermeasures

	Global warming countermeasures	Heat island countermeasures
Mitigation measures	• Restricting emissions of CO_2 (low carbon) • Restricting emissions of GHG and other fluorocarbons • Sequestering carbon dioxide (CCS)	• Reducing anthropogenic heat • *Improving ground surface covering* • Improving the urban morphology (ventilation)
Adaptive measures	• Strengthening dikes on rivers and seacoasts • Increasing the water storage capacity of dams and buildings • *Urban greening and cool spots* • Improving air-conditioning systems against future heats	• Promoting heatstroke prediction • Encouraging appropriate use of cooling equipment • Promoting appropriate rehydration • *Urban greening and cool spots*

measures and adaptive measures in global warming countermeasures and heat island countermeasures. Urban greening and cool spot measures are understood to be one kind of adaptive measure considered from the perspective of global warming countermeasures. Mitigation measures among heat island measures correspond to the contents clearly stated in the General Principles for Heat Island Countermeasures, which were reduction of anthropogenic heat, improvement of ground surface cover (including urban greening and cool spots), and improvement of the urban morphology. Adaptive measures among heat island measures are intended to curtail the impacts to human being by human activities (not going out, evacuating to cool places, drinking water, etc.), but not to directly reduce the atmospheric temperature. Planting trees to create cool spots is a kind of adaptive heat island countermeasure in the sense that they are places to take refuge from the heat.

6.2 Sensible Temperature and Heat Countermeasures

6.2.1 Sensible Temperature in Urban Spaces

People say "it's hot today" in daily conversation, but there are many ways of roughly expressing thermal environment for human being using numerical values of air temperature. However, the sensible temperature which people feel is, strictly speaking, not the air temperature, exists against a complex physical and physiological background which is decided by the action on the heat balance of human bodies of humidity, wind speed, irradiation, and so on.

The Building Research Institute has worked to develop numerical fluid analysis technologies in order to evaluate heat island countermeasures. This study virtually reproduces the elevations, ground cover, building layout, etc. of a city in a super computer to predict the flow of wind, temperature distribution, etc. inside a city, based on CFD (computation fluid dynamics) (Ashie and Kono 2011). And to simulate the sensible temperature, in addition to fluid analysis, it is necessary to apply the results of numerical analysis of radiation fields in urban space to obtain the mean radiation temperature (MRT) in a model of the heat balance of a human body.

The local surfaces are divided into surface elements to calculate the view factor of the surface elements based on the geometrical shape of urban space to compute the convergence of the reflection and absorption of short and long wave radiation. Figure 6.4 shows the relationship of the CFD grid with surface elements. Surface elements exist in the CFD grid in optional orientations and sizes. In other words, the CFD grid and surface elements are independent, and the boundaries of the CFD grid need not agree with the positions of the surface elements. Accordingly, it can reproduce the complex shape of buildings and undulations of the ground surface with greater precision than the conventional staged surface division method.

The mean radiation temperature (MRT) of the human body is calculated by aggregating the radiance of the short waves and long waves based on the angular relationship with the surface elements around the human body. And from the CFD

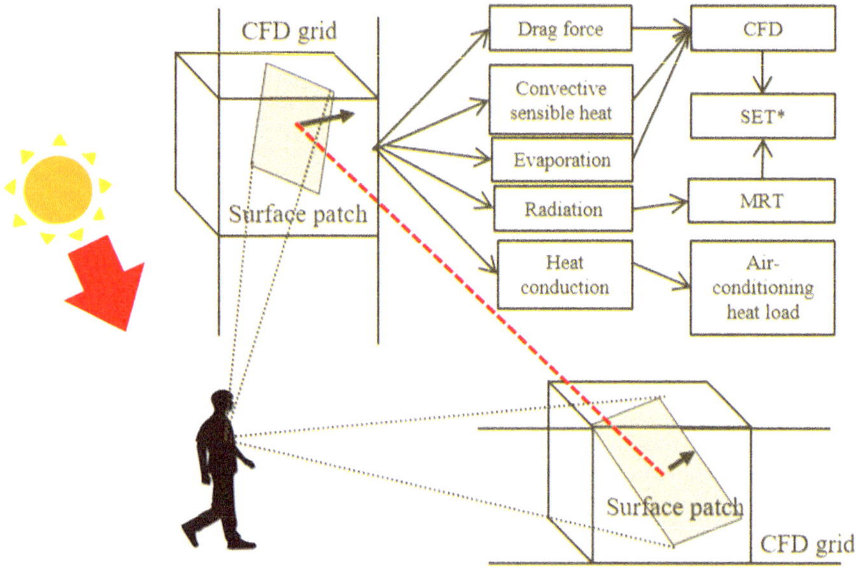

Fig. 6.4 Evaluation of the sensible temperature by combining the CFD grid with surface elements

analysis results, it is possible to find the sensible temperature (new standard effective temperature, SET*) by substituting information about the air temperature, humidity, and air current velocity into the thermal physiology model.

6.2.2 *Radiant Temperature*

A thermal imaging camera is used to measure the surface temperature, but the temperatures displayed by the thermal imaging camera are, strictly speaking, radiant temperatures (equivalent surface temperature that assumed the surface of the measured object is black). It is difficult to compute the emissivity of the surface of the object measured on site and the impact of reflection from the surroundings, so the actual surface temperature is almost never reverse analyzed from the radiant temperature. The numerical analysis, on the other hand, can directly obtain the surface temperature, also obtaining information about long wavelength luminance of the surface elements including the self-radiation of the surface elements and the impact of reflection from the surroundings. So, the results of computing the long wavelength luminance of surface elements hypothesizing emissivity of 1 are converted to temperature, expressing the distribution of radiant temperature in urban space. The following is the temperature conversion equation.

$$T^* = \sqrt[4]{q_L / \sigma} - T_0 \tag{6.1}$$

where T^*: radiant temperature, q_L: long wavelength radiation luminance, σ: Stefan–Boltzmann constant, T_0: absolute zero.

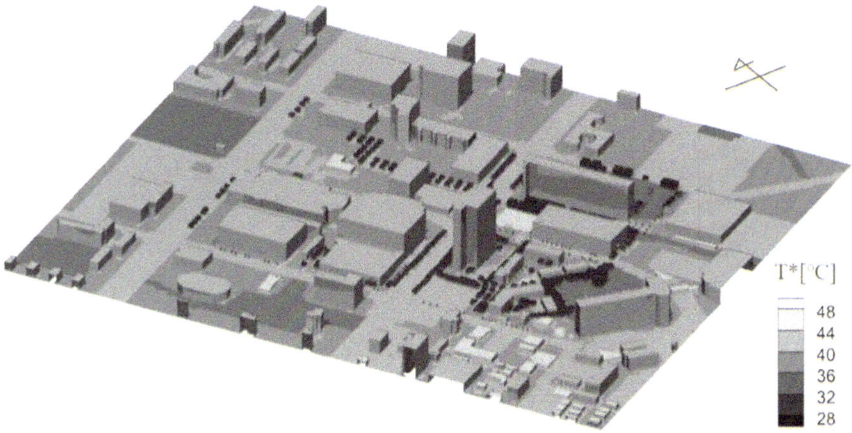

Fig. 6.5 Sample analysis of radiant temperature (Tsukuba Science City)

Figure 6.5 shows an example of analysis of radiant temperature. The direction of the building wall surface faithfully reproduces the actual layout, allowing us to figure out the impact of the ground surface cover (asphalt paving, greenery), materials of buildings (medium to high rise concrete, low rise wooden construction), and sunshine (sunny place, shady place) from the radiant temperature distribution.

6.2.3 Climograph

The Tokyo Olympic and Paralympic Games (Tokyo 2020) will be held from the end of July to the beginning of September. The previous Tokyo Olympics was held in October 1964, but this time it will be held in midsummer, so measures to protect the audience and athletes from the heat are an important concern. Figure 6.6 shows a climograph for Tokyo and London. During the summer in Tokyo, the temperature is 10 °C higher than in London, and it is predicted that the outdoor heat environment during the daytime will be severe. Since the climate conditions are very different from those at the 2012 London Olympics, it is necessary to bear measures against heat stroke during outdoor athletic events in mind.

Also, due to the COVID-19 pandemic, the Tokyo Olympic and Paralympic Games have been postponed to 2021. In addition, it was hastily decided to move the location of the marathon event from Tokyo to Sapporo in consideration of the heat environment. These major changes and final decisions have occurred since the beginning of 2020. The heat countermeasures to be extended in Sect. 6.2.4 for the marathon course have been surveyed in the Tokyo area before these major changes in the situation and final decisions were made. Although the marathon event will likely not be held in Tokyo, the scientific examination itself of heat countermeasures is meaningful, so they will be shown in the next section.

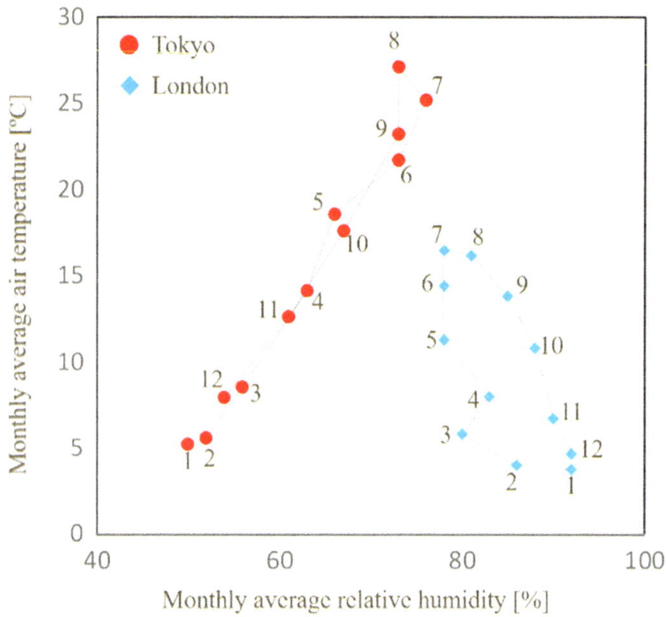

Fig. 6.6 Climographs

6.2.4 Numerical Simulation of Heat Countermeasures on the Marathon Course

Marathon course heat countermeasures are attracting the public's attention as a result of the 2020 Tokyo Olympics. Heat countermeasures considered include high albedo pavement, water-retentive pavement, spraying water and mists, and shading sheets, and a numerical simulation was done to study the effects of countermeasures in a case where these were implemented in a downtown section (1 km^2) shown in Fig. 6.7. The marathon course is expected to include Chuodori Street, where priority heat countermeasures will be taken. In addition, assuming that the spectators will move from major stations, heat countermeasures are planned for surrounding streets. The analysis is done by the procedure explained below.

1. The Meteorological Agency performs numerical simulations to predict the weather 24 h a day 365 days a year, and climate information such as potential temperature, wind, and atmospheric radiation are abstracted based on horizontal 500 m meshes of this archive data.
2. Based on the Meteorological Agency's archive data, climate information is spatio-temporally interpolated at 500 m which is the top edge of the CFD analysis range to build a 24-h boundary condition data base on CFD analysis dates.

Fig. 6.7 Marathon course heat countermeasures

3. Boundary conditions at the 500 m height of the top edge of the CFD analysis range are used to perform coupled unsteady analysis of radiation and CFD, obtaining the air temperature and humidity in urban space, short wave and long wave radiance, and 24-h value of wind speed. The short wave and long wave radiances are converted to mean radiant temperature (MRT) of the body considering the angular relationship.
4. Based on radiation and CFD analysis results on a typical summer day, August 10, 2013, air temperature, humidity, air flow velocity, and MRT information are substituted in the thermal physiology model to obtain the new standard effective temperature SET*.

Figure 6.8 shows the simulated air temperature distribution during a summer day. It shows that areas where the temperature exceeds 34 °C exist in urban spaces, and that their heat environments are severe. The prevailing wind blows south-east from the ocean and the air temperature on Showadori Street where ventilation is poor is particularly high. The temperature is low on Harumidori Street where ventilation is relatively better.

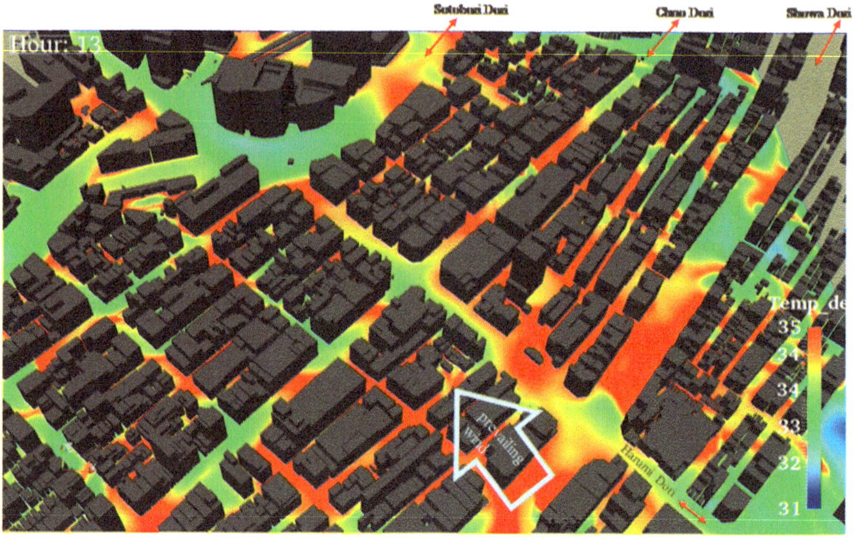

Fig. 6.8 Example of air temperature analysis (13:00 on August 10, 2013)

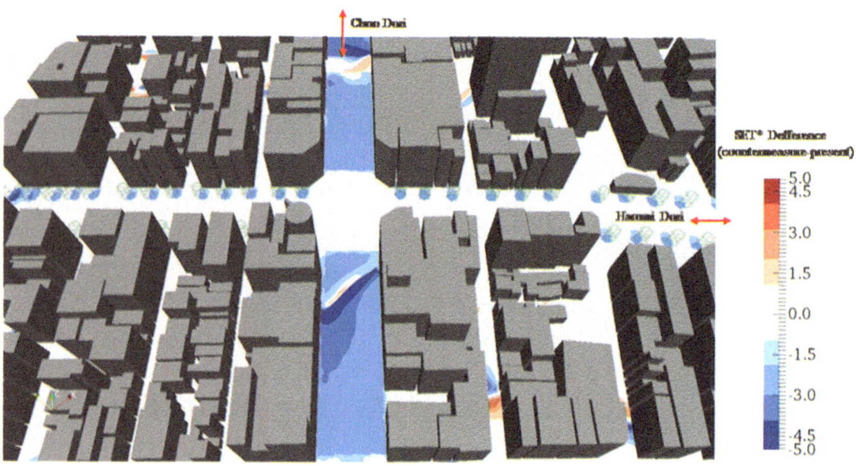

Fig. 6.9 Effects of heat countermeasures (13:00 on August 10, 2013)

SET* taking the heat countermeasures is calculated and it is shown in Fig. 6.9 based on finite difference of with and without the heat countermeasures. On Chuodori which is on the marathon course, SET* was lowered by about 2–3 °C, confirming the effectiveness of the heat countermeasures. Even on Harumidori, mitigation effects are clearly seen around trees, but the effect is small in areas without coverage by the tree crowns, so increasing the number of trees, introducing trees with good branch shapes and other supplementary measures are beneficial.

6.2.5 *Demonstration of Heat Countermeasures in Bus Stop Waiting Room*

Promoting the use of public transit is advocated as way to build a low carbon society. But there are many bus stops without surrounding shade, and it is feared that the heat environment around such places will be severe in the summer months. It is necessary to consider not only air-conditioning on the bus, but also reducing the heat risk to users of bus stops in outdoor spaces. So, a waiting room of a bus stop was built experimentally and used to test a variety of heat countermeasures.

The waiting room of the bus stop was designed as a semi-outdoor space, so instead of refrigerating machine, passive cooling methods such as ensuring shady space and using cold heat and heat of vaporization of ground water were considered. Figure 6.10 shows heat countermeasures in the bus stop waiting room (Ministry of the Environment 2017). It was designed by combining screening solar radiation with highly reflectance awnings, and western sunshade evaporative cooling louvers and regulating the rise of surface temperatures with water retentive blocks, and body surface temperature reduction by water cooled benches. As shown by Fig. 6.11, building a waterscape inside the bus stop waiting room can visually create coolness.

Figure 6.12 shows the results of a survey taken on a summer day, performed by Ministry of the Environment (Ministry of the Environment 2017). Figure 6.12a compares the air temperature as a thermal environment and the results of measurements of SET* outside and inside the bus stop waiting room. The heat environment inside the bus stop waiting room is better than it is outside, at 0.5 °C lower than the air temperature and 6 °C lower at SET*. Figure 6.12b shows the results of a survey of physiology/psychology. In this case too, the environment tends to be better in the bus stop waiting room than it is outside, as the heart rate increase rate falls by 3.5%, the heat sensation is improved by 4 ranks, and the comfort sensation is improved by 2.5 ranks. The temperature of the ground water is 17 °C immediately after it is pumped up, but in fact, the temperature of the seat surface is about 20 °C when the groundwater is circulated through the cooling bench. The cooling effects of the cooling bench is not directly reflected in the evaluation of SET* using the globe

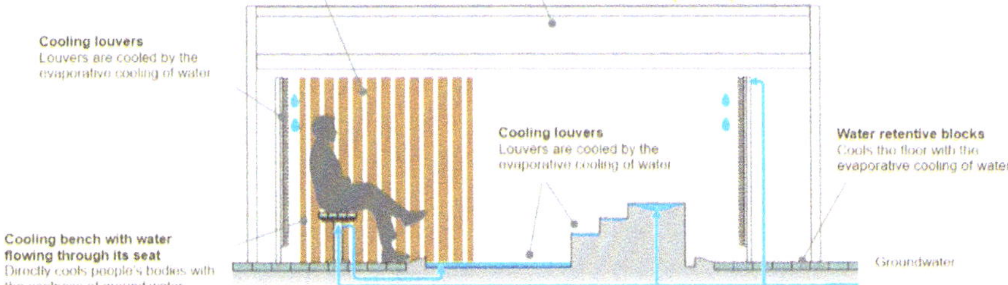

Fig. 6.10 Measures against heat in Bus station waiting room (Ministry of the Environment 2017)

Fig. 6.11 View of the interior of the bus stop waiting room (Ministry of the Environment 2017)

temperature. In an experiment using test subjects on the other hand, the test subject declares sitting on the cooling bench, so it is presumed that the effects of the contact temperature are reflected in the declaration results.

6.3 Wind Paths

6.3.1 Tokyo Coastal Wind and Heat Island

Due to the difference in specific heat between the sea and land, the land tends to warm up more during the day while the sea does not. Updrafts are generated on the land warmed by solar radiation in the daytime, and the atmospheric pressure near the surface becomes low. To compensate for this, a cool breeze blows from the sea to the land during the day. When there is a mountain range in the area behind, the sea breeze moves deep inland, coupled with the pressure effect caused by the altitude difference. Looking from the sea breeze flowing across the surface with a thickness of several hundreds of meters, urban buildings are only somewhat fine roughness, and it is not enough to transform the natural wind system including the powerful driving source of the temperature difference between the Japanese Alps and Tokyo Bay.

However, when zooming up near the ground surface, it is clear that the presence of individual buildings greatly affects the flow of local wind. In recent years, the image of the high-rise buildings in the Tokyo coastal area is called "Tokyo Wall," and there are concerns about the environmental impact, including that the ventilation of the urban space is hindered by the wall of buildings blocking the sea breeze, promoting the heat island phenomenon. Even if there is a building or group of

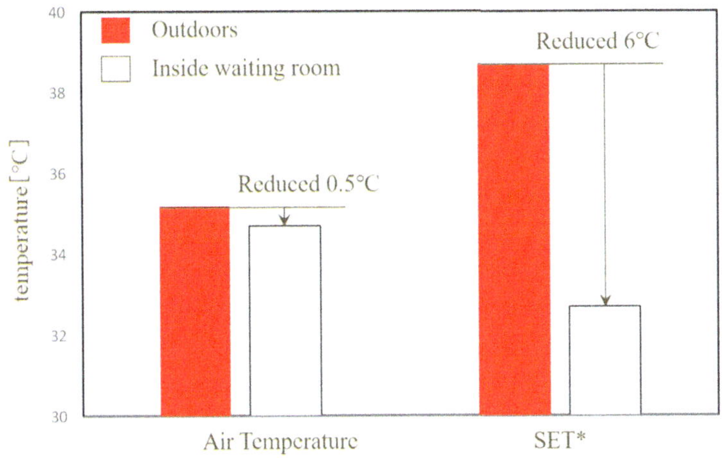

(a) Thermal environment

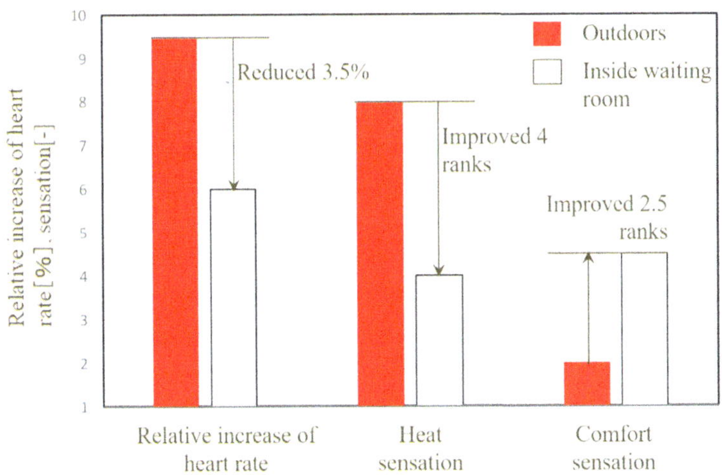

(b) Physiology/psychology

Relative increase of heart rate: percentage increase of heart rate with heart rate when quiet as the standard (%)
Heat sensation: subjective declaration in 9 levels (1: cold, 9:hot)
Comfortable sensation: subjective declaration in 7 levels (1: uncomfortable, 7: comfortable)

Fig. 6.12 Results of survey on a summer day (14:00 to 15:00 on August 5, 2016). Figures are created using the survey data performed by Ministry of the Environment (Ministry of the Environment 2017). (**a**) Thermal environment. (**b**) Physiology/psychology. Relative increase of heart rate: percentage increase of heart rate with heart rate when quiet as the standard (%). Heat sensation: subjective declaration in 9 levels (1: cold, 9: hot). Comfortable sensation: subjective declaration in 7 levels (1: uncomfortable, 7: comfortable)

buildings, the wind bypasses its surroundings, so the wind flow is not interrupted at all. The area behind a building or group of buildings will certainly have reduced wind speed, but it will be pulled by the wind that flows over the city, and the wind speed will recover as it travels down a certain horizontal distance. However, it is also true that insufficient wind has become chronic for urban residents. This is because there are too many weak winds in large cities where buildings are densely packed. Therefore, wind paths that bring cool breezes from the sea that occurs during the daytime in the summer into urban spaces has become a countermeasure for heat islands. The government's Heat Island Countermeasures Principle clearly states in "3. Improving the urban form" "securing wind paths from green spaces and water surfaces" (Heat Island Countermeasure Promotion Council 2004, revised in 2013). According to the Ministry of Land, Infrastructure, Transport and Tourism guidelines, wind paths that contribute to heat island countermeasures are: "continuous, open spaces such as rivers, green spaces and streets, and are effective for ventilation in urban spaces" (Ministry of Land, Infrastructure, Transport and Tourism 2013).

6.3.2 *Existence of Wind Paths and Evaluation Level*

It is necessary to measure the wind locally in order to check whether the wind is actually flowing along the river or street. Therefore, an anemometer was installed along rivers and streets in Tokyo, and field observations were conducted. The survey results are shown in Fig. 6.13. The figure also shows the results of wind direction

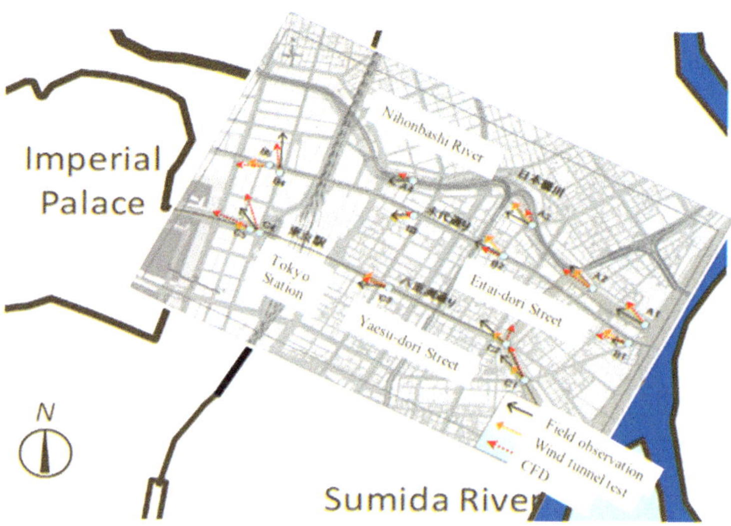

Fig. 6.13 Wind conditions in urban spaces (July 31, 2005 at 12:00)

and wind speed from wind tunnel experiments (using 1/750 model) and numerical calculations (CFD simulation was performed). Looking at the results of these three methods, it is apparent that the wind direction is formed along each route of the Nihonbashi River, Eitai-dori Street, and Yaesu-dori Street. The three methods are similar to each other. The survey results confirm that the wind from the sea that blows during the day flows through rivers and streets in urban spaces.

The wind flow in urban areas is complicated in three dimensions, but it is easier to understand if the evaluation height is specified and classified. Figure 6.14 compares the height of the wind path with the height of the building. There are three possible levels of wind path evaluation: 2 m, 10 m, and 50 m above ground with consideration to building form.

2 m above ground is equivalent to the air above a person standing on the ground and is lower than the height of a detached house. The height of a detached house is stipulated as 10 m or 12 m by the Building Standards Act, but generally there are many less than 10 m, e.g., 7 m. At an evaluation height of 2 m above ground, winds flowing through residential sites, narrow streets, and small parks are mainly targeted, and the horizontal width of the wind is several meters.

10 m above ground is equivalent to the height of a detached house and is lower than that of a general office building. In Japan, the Urban Building Act, which is the predecessor of the current Building Standards Act, has a 100 Shaku limit on the height of buildings. "Shaku" is the traditional length measures in Japan, 100 shaku is nearly equal to 31 m and the height of general office buildings in existing urban areas concentrates often 31 m or less. At an evaluation height of 10 m above ground, winds flowing through rivers, main roads, and large-scale green spaces are mainly targeted, and the horizontal width of the wind is tens of meters.

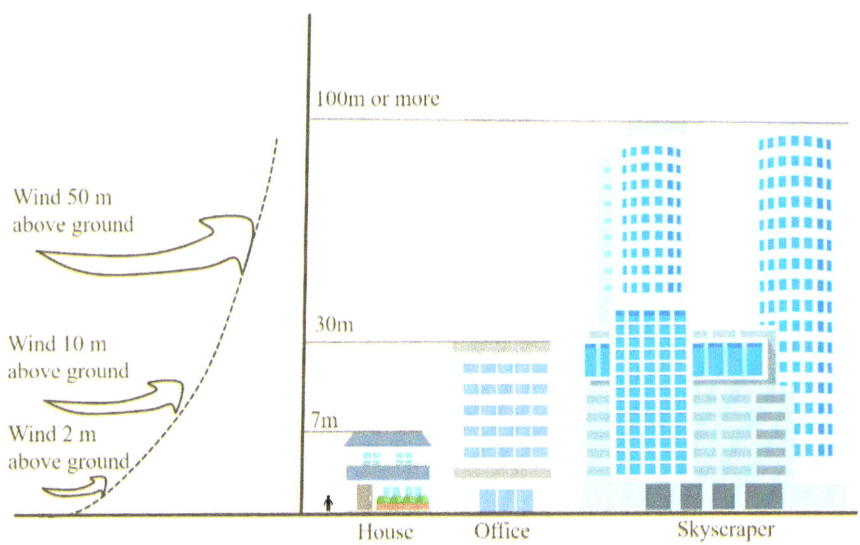

Fig. 6.14 Wind paths and evaluation level

The height of 50 m above ground is equivalent to the air above general office buildings, which is lower than the height of skyscrapers. There is no clear definition for skyscrapers, but in general, buildings with a height of 100 m or more are often called skyscrapers. At an evaluation height of 50 m above ground, the main target is the wind that bypasses the skyscrapers and flows through areas that are not highly developed, with a horizontal wind width of 100 m or more.

6.3.3 Heat Island Countermeasure Map and Wind Paths

The heat island countermeasure map outlines heat island countermeasure policies on a map based on the current state of heat islands such as urban wind flow and temperature distribution, ground cover, artificial exhaust heat and urban form, and consists of the current situation diagram and the countermeasure policy diagram. An overview of the heat island countermeasure map is shown in Fig. 6.15 (Ministry of Land, Infrastructure, Transport and Tourism 2013). The heat island countermeasure map can be of various scales from prefectures, municipalities, and specific districts, and is basically created for each jurisdiction of local governments. Based on the consideration and securing of "wind paths" in city planning, it is best for the city department of the local public body to be the principle department, and for the heat island countermeasure map to be created while coordinating with other departments such as the environmental department. Based on the countermeasure policy diagram of the heat island countermeasure map created by local governments, it is necessary for businesses to organize considerations for heat island countermeasures in individual plans/projects and seek consensus through discussions with local governments.

In preparing the countermeasure policy diagram "wind path" of the heat island countermeasure map, the following classifications were considered for the wind flowing in urban spaces.

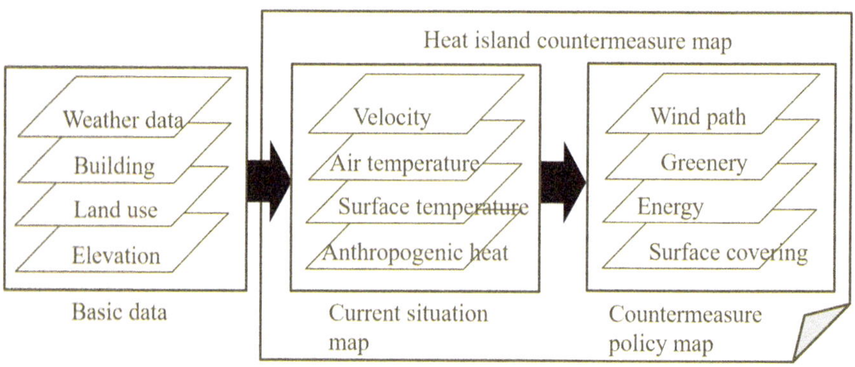

Fig. 6.15 Heat island countermeasure map (Ministry of Land, Infrastructure, Transport and Tourism 2013)

- Classification I—type in which the sea breeze flows into the urban space from the coast along the ground surface through streets, rivers, and green spaces
- Classification II—type in which the sea breeze flowing in the sky is guided by buildings along streets and rivers and flows near the ground surface
- Category III—Continuous open spaces such as rivers, green spaces, and streets for guiding land breezes into urban spaces (often the same space as wind paths through which daytime sea breezes flow)
- Classification IV—Continuous gaps such as streets, rivers, and open spaces between buildings to guide the advection of cold air from urban green spaces to surrounding urban areas

Furthermore, it arranges as follows according to the gap scale.

- Class 1 Path (Classification I)—Wind flow using land and river winds across prefectures
- Class 2 Path (Classification I, II, III)—Wind flow using open spaces in roads and parks
- Class 3 Path (Classification IV)—Local wind flow from green spaces

Using wide-area CFD analysis technology, large-scale calculations of about 50 cases were carried out by changing the prevailing wind and time period. Here, based on the calculation results of wind speed 10 m above ground, Fig. 6.16 shows wind flow in urban spaces in Tokyo in one map (Ministry of Land, Infrastructure, Transport and Tourism 2013). When local governments make environmental plans, it is first necessary to understand local environmental information. The countermeasure policy diagram "wind path" specifies in advance public spaces such as local rivers, streets, and green spaces as wind paths, and does not suppress private rights. The main purpose is to use the urban building plan considering the wind flowing through public spaces as data for urban development companies to consider.

6.3.4 Urban Development Examples Considering Wind Paths

6.3.4.1 Nihonbashi/Daimaruyu District: Removal of Station Buildings and Elevated Roads

In the Nihonbashi/Daimaruyu area of Tokyo (the abbreviation for the Otemachi, Marunouchi, and Yurakucho areas), the town planning council of Nihonbashi and the Daimaruyu area is the main body, and experts on heat island countermeasures and related local governments (Tokyo Prefecture, Chiyoda City, Chuo City) and the country (Ministry of Land, Infrastructure, Transport and Tourism, Ministry of the Environment) participated and planned urban redevelopment. At that time, various studies such as wind tunnel experiments and numerical simulations were conducted from the viewpoint of wind paths regarding redevelopment with the removal of the station building at Tokyo Station and the Metropolitan Expressway along the

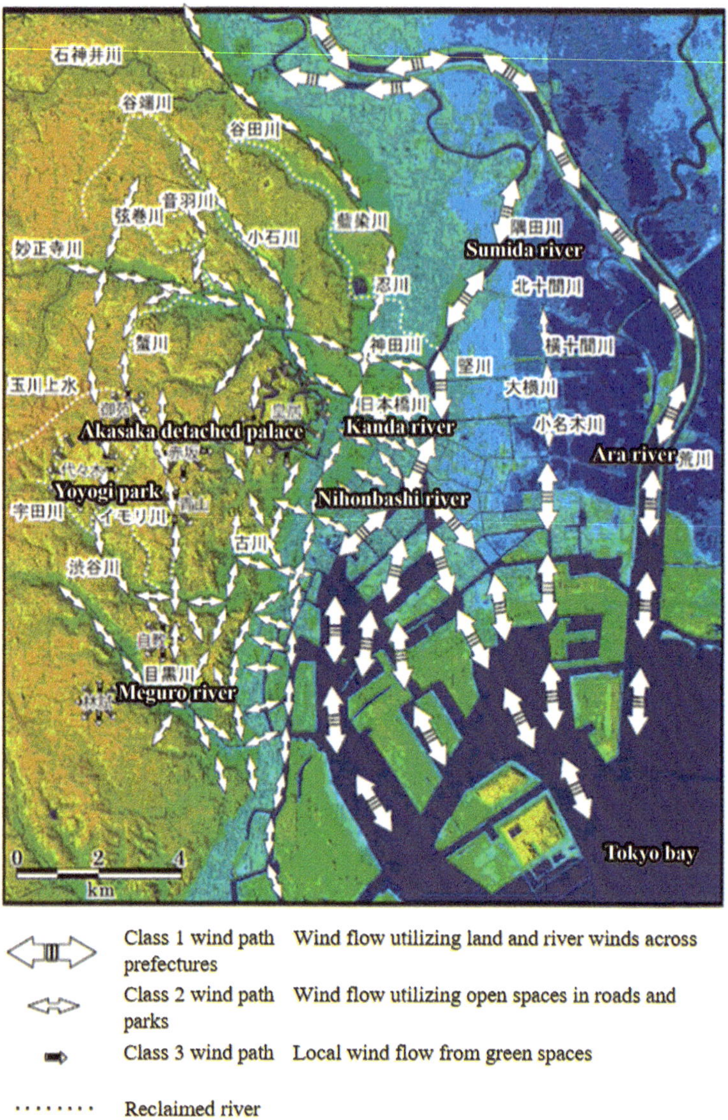

Fig. 6.16 Tokyo Wind path map: countermeasure policy map "Wind Path" (Ministry of Land, Infrastructure, Transport and Tourism 2013)

Nihonbashi River. Figure 6.17 shows the urban model (1/750) used for the wind tunnel experiment in the redevelopment around Tokyo Station. The redevelopment plan removes the station building that blocks the wind blowing toward the station on the main street and builds new twin towers on both sides. The plan was implemented, the twin towers opened in 2007, and the station building was removed after relocation of station building tenants was completed.

(a) Before redevelopment

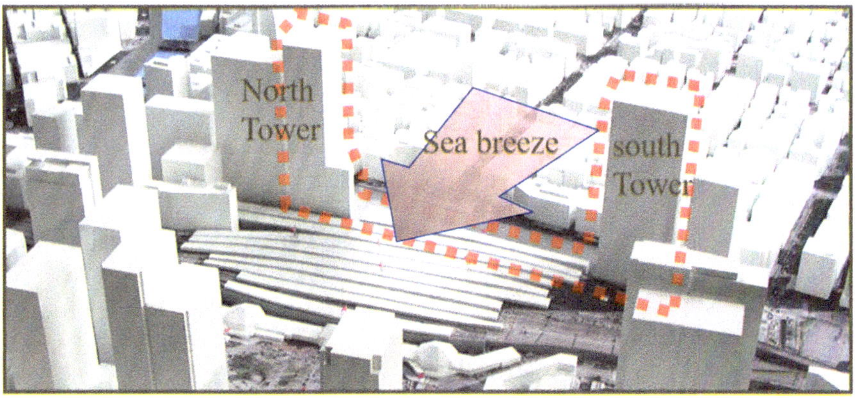

(b) After redevelopment

Fig. 6.17 Redevelopment around Tokyo Station. (**a**) Before redevelopment. (**b**) After redevelopment

Another redevelopment is the redevelopment around the Nihonbashi River, Tokyo. The Nihonbashi section of the highway runs on the Nihonbashi River with foundation piles driven into the Nihonbashi River, and follows the path of the river. In 2018, it was decided that the elevated road with foundation piles would be removed and the 1.8 km highway from Kandabashi JCT to Edobashi JCT would be targeted for placement underground. When this road project is finished, the surface of the Nihonbashi River, which was previously covered by the elevated road, and Nihonbashi Bridge, the starting point of all national roads in Japan, will be revitalized as a local landscape. A local NPO is considering a development plan to improve the walking environment around the river by setting back and lowering the buildings along the river. Figure 6.18 shows the exterior of the development plan.

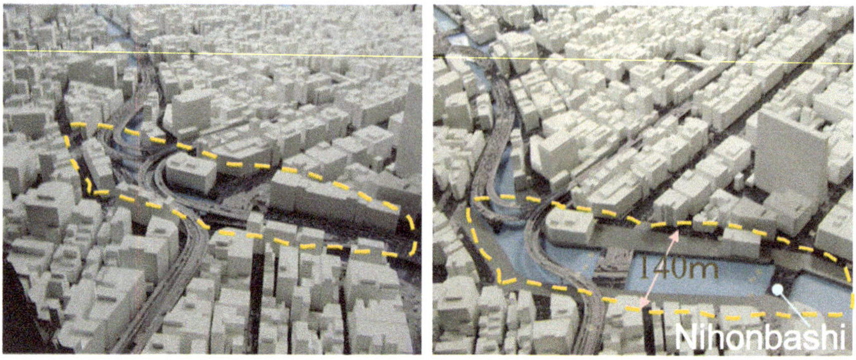

(a) Before redevelopment (b) After redevelopment

Fig. 6.18 Redevelopment around Nihonbashi River. The redevelopment area is indicated by the dotted line. Elevated roads will be removed and the height and volume along the river reduced. (**a**) Before redevelopment. (**b**) After redevelopment

6.3.4.2 Regional Urban Regeneration Vision Around Osaki Station: V-Shaped Building Layout Around Meguro River

The area around Osaki Station in Shinagawa City, Tokyo, was designated as an urgent urban renewal area in July 2002, and the Osaki Urban Renaissance Vision was compiled by the local liaison committee in November 2004.

In this area, global warming countermeasures and heat island countermeasures are positioned as important issues, and in order to form a network of water, greenery and wind, it is securing a cohesive green area, reducing ground cover temperature by greening areas along the Meguro River, roads and pedestrian spaces, adopting pavement materials to improve the thermal environment of areas such as roads and adopting exterior materials that suppress radiant heat from walls, devising the shape and arrangement of buildings, and securing wind paths for wind to blow through area by arranging roads and streets so they encourage wind to pass.

In particular, in order to secure the wind path, the buildings are arranged in a V-shape toward the river along the current street so that the wind from the Meguro River can be taken in. By arranging the buildings in a V-shape, the wind flowing through the river effectively spreads in the city area in the wind direction, while hardly allows infiltration from the opposite direction. With such a building arrangement, it is believed that cold sea breezes from the south during the summer day are actively guided into the urban area, while the infiltration of seasonal wind from the north in the winter can be suppressed to the fullest extent. Figure 6.19 shows the actual building layout around the Meguro River in the area around Osaki Station.

Fig. 6.19 Redevelopment around Meguro River. The arrows were added to the building layout created by Nihon Sekkei Inc. to show ideal summertime wind flow

6.3.4.3 Shinagawa Station/Tamachi Station Area Town Development Guidelines: Restrictions on Building Height

The area around Shinagawa Station and Tamachi Station is located to the south of the central Tokyo, and convenient travel to the location has increased with the opening of the Tokaido Shinkansen at Shinagawa Station and the internationalization of Haneda Airport. Furthermore, in recent years, various developments have been promoted mainly on the seaside from Shinagawa Station to Tamachi Station, and since there are many large-scale low and unused areas, considerable development is expected in the future. Under these circumstances, the Tokyo Metropolitan Government formulated a basic plan to improve the urban and living environment around Shinagawa Station in September 2006, and released a technical guideline for securing wind paths (Tokyo Metropolitan Government 2006b, revised in 2014).

The guideline set forth the most characteristic slogan for creating an environmental model city, contributes to environmental improvements such as the introduction of wind into the city and a countermeasure against the heating of urban spaces, and aims to guide wind to not only this area but also inland. Under the conditions of a south-southeast aerial wind, the main wind direction for this region in the summer, fluid simulations that take into account the topography and building arrangement were performed, and areas where the wind speed is 4 m/s or more are defined as "major wind paths." Figure 6.20 shows "major wind paths" on a land use map with arrows. Buildings to be constructed in "main wind paths" will be limited to 50 m or less in principle, and if existing buildings are to be rebuilt at a height of more than 50 m, they are basically limited or restricted to the current height or less.

平成13年度土地利用現況調査等を基に作成

Fig. 6.20 Location of wind paths (Tokyo Metropolitan Government 2006b, revised 2014). Created based on 2001 land use survey

6.3.5 *Wind Paths and Heat Countermeasures*

6.3.5.1 Urban Development and Environmental Impact

In urban development, various physical environmental loads that cause the heat island phenomenon occur. For example, convection heat to the atmosphere due to the high temperature of features, artificial exhaust heat (sensible heat, latent heat) from cars and building facilities, and attenuation of wind kinetic energy due to buildings built together in large numbers. These physical environmental loads not only change the temperature and wind conditions in the development area, but also affect the temperature and wind conditions around the development area. Figure 6.21 shows the relationship between the development area and the surrounding area. Uniform land use is assumed throughout the whole area (specifically, land use in the windward and leeward areas). When a development area is newly created, a physical environmental load different from the conventional one occurs, so that environmental influences occur in the surrounding area.

6.3.5.2 Environmental Analysis of Development Area and Leeward Area

There are cases where local residents point out local weakening in wind speed to local governments and urban development companies. What kind of damage will local weakening cause? One thing that can be said is that the wind is weak and heat dissipation of the human body is hindered, so local residents will feel hotter. In past studies, there were cases in which the heat dissipation characteristics of human body model were examined in detail by CFD (Ono et al. 2006). The close relationship is understood because wind speed and airflow turbulence affect the convective heat transfer coefficient of the human body. However, since the leeward area exists

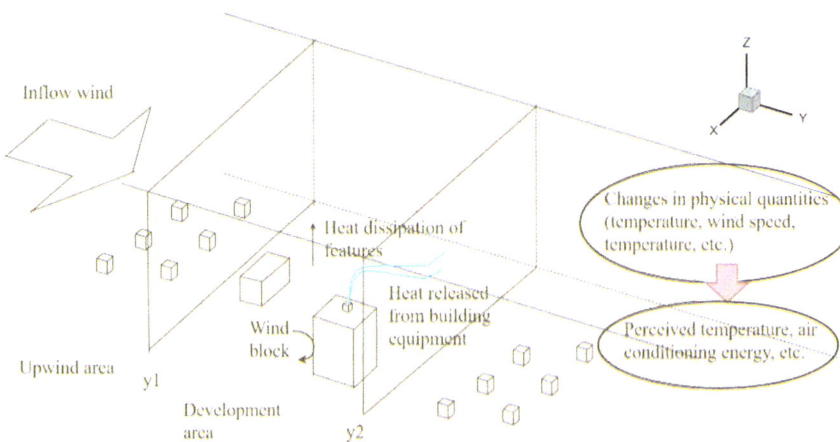

Fig. 6.21 Development area and surrounding area

in a wide range, which is said to be several times the height of the building, it is clearly technically burdensome to directly and wholly understand the temperature increase when the leeward area is weakened by urban development. Therefore, by conducting a fluid simulation on the effects of the following buildings on the wind environment in the leeward area, we first attempt a quantitative evaluation limited to the wind.

- The evaluation range of the leeward area is the horizontal range of D × W starting from the leeward of the development area (site) including buildings (D: twice the maximum height of planned buildings, W: maximum horizontal width of the development area that goes directly to the wind direction).
- The prevailing wind is, for example, 5.1 m/s at 44 m above ground, the wind direction is south-southeast, and the standard value of the evaluation wind speed is 4 m/s. These conditions vary from region to region.
- The rate of occurrence of the evaluation wind speed shall be the ratio of the area that is equal to or greater than the standard value of the evaluation wind speed to the area of the evaluation range of the leeward area, and the rate of occurrence of the evaluation wind speed shall be 0.4 to 0.5 or more.

Next, using the simulation results of the wind environment, the deterioration of the thermal environment due to the reduction in wind speed in the leeward evaluation area is considered. Regarding wind speed above the leeward evaluation area, the state of the wind environment before and after development is grasped by numerical calculation. Therefore, it is assumed that the sensible temperature is damaged in the mesh below the standard wind speed in the leeward evaluation area, and the thermal environment deterioration R_{HI} due to wind speed reduction in the leeward evaluation area is expressed using the following equation.

$$R_{HI,A2,t1}(t) = W_{Q,A2,t1}(t) \times \eta_1 \times \eta_2 \times \text{standard value of evaluation wind speed} - u_{A2,t1}(x, y, z(=50\,\text{m}), t) \times V_{\text{below standard},A2,t1} \tag{6.2a}$$

$$R_{HI,A2,t2}(t) = W_{Q,A2,t2}(t) \times \eta_1 \times \eta_2 \times \text{standard value of evaluation wind speed} - u_{A2,t2}(x, y, z(=50\,\text{m}), t) \times V_{\text{below standard},A2,t2} \tag{6.2b}$$

Here, $W_Q(t)$ is a weighting factor representing the probability that a person is in the outdoor space, η_1 is the increase in sensible temperature due to the decrease in wind speed [°C/(m/s)], η_2 is the wind speed adjustment factor, $\langle du \rangle$ is the spatial average value in the evaluation area (meshes that exceed the reference value of the evaluation wind speed are accumulated at 0), $V_{\text{below standard}}$ is the sum of the mesh area below the standard value of the evaluation wind speed, $A2$ is the leeward evaluation area, and $t1$ and $t2$ indicate before and after development.

For the increase in sensible temperature due to the decrease in wind speed η_1, during standard summer daytime weather conditions in Tokyo, the following relationship is obtained when a sensitivity analysis is performed for SET* (Japan Sustainable Building Consortium 2017).

$$\eta_1 = \frac{\partial \text{SET}^*}{\partial(-u)} \cong 2\left[\frac{^\circ\text{C}}{\text{m/s}}\right] \tag{6.3}$$

For the wind speed adjustment factor η_2, the following relationship is firstly obtained when the power law (power 0.25) is applied.

$$\eta_2 = \frac{u(z = 1.5\,\text{m})}{u(z = 50\,\text{m})} = \left(\frac{1.5}{50}\right)^{0.25} \cong 0.42 \tag{6.4}$$

However, wind situation around buildings is very complicated, and we will set $\eta_2 = 1$ for safety.

In this way, the deterioration of the thermal environment in the leeward evaluation area could be easily estimated from the numerical simulation results of the wind speed without performing full-fledged temperature calculations. Considering that it is practically difficult to directly control the sensible temperature in the leeward area, securing the wind path leads to rational maintenance of the thermal environment in the leeward area.

There are no problems if the wind path is sufficiently secured, but if this is not the case, it will be an important perspective to consider what kind of response will be required from developers regarding environmental degradation.

Since the leeward area is an asset (site) of others, it is practically difficult for a developer to develop concrete measures such as watering and greening. In addition, seeking direct compensation increases the financial burden of urban development and is not always useful. On the other hand, it is easy for businesses to work on designs and countermeasures against heat in the development area. The place where environmental improvements will be made will move from the leeward area to the development area, but if the development is a highly public place, there will be benefits for local residents, and if measures such as tree planting are promoted in the development area, a cooling effect is also expected due to cool air advancing to the leeward area. Therefore, countermeasures against heat in the development area and securing wind paths in the leeward area will be evaluated by using the sensible temperature as a common index, and the balance between them will be examined.

However, since it is necessary to secure a certain amount of wind paths, the allowable reference value 0.4 is set as the rate of occurrence of the evaluation wind speed (ratio of the area that is equal to or greater than the standard value of the evaluation wind speed to the area of the evaluation range of the leeward area). If the rate of occurrence of the evaluation wind speed is less than 0.4, the development plan cannot be approved by the government regardless of the degree of heat countermeasures in the development area.

The reduction effect of sensible temperature SET* when heat countermeasures are implemented in the development area is defined as follows.

$$Q_{HI,A1,t1}(t) = W_{Q,A1,t1}(t) \times \text{allowable upper limit up SET}^* - \text{SET}^*_{A1,t1}(x,y,z,t) \times V_{\text{allowance},A1,t1} \quad (6.5a)$$

$$Q_{HI,A1,t2}(t) = W_{Q,A1,t2}(t) \times \text{allowable upper limit of SET}^* - \text{SET}^*_{A1,t2}(x,y,z,t) \times V_{\text{allowance},A1,t2} \quad (6.5b)$$

Here, Q_{HI} is the index representing the effect of reducing the sensible temperature, $\langle f(d\text{SET}^*)\rangle$ is the spatial average value in the evaluation area (meshes that exceed the allowable upper limit of SET* are accumulated at 0.), $V_{\text{allowance}}$ is the sum of the mesh area not exceeding the allowable upper limit of SET*, and subscript A1 represents the development area.

Since the development area is smaller in area than the leeward area, it is considered that a detailed numerical simulation including radiation calculations can be performed on the sensible temperature when heat countermeasures are implemented. However, detailed numerical simulations including radiation calculations may be difficult for some developers. In such cases, a method may be used in which only the determination of the sun and shade is performed without performing radiation calculations, and the surface temperature for each covering is tabulated to give thermal boundary conditions to the thermal airflow simulation. It is simply evaluated with the follow equation by changing the air temperature T in the development area obtained through thermal airflow simulations in this way to the reduction in sensible temperature SET*.

$$Q_{HI,A1,t1}(t) = W_{Q,A1,t1}(t) \times \eta_3 \times \text{allowable upper limit of } T - T_{A1,t1}(x,y,z,t) \times V'_{\text{allowance},A1,t1} \quad (6.6a)$$

$$Q_{HI,A1,t1}(t) = W_{Q,A1,t1}(t) \times \eta_3 \times \text{allowable upper limit of } T - T_{A1,t1}(x,y,z,t) \times V'_{\text{allowance},A1,t1} \quad (6.6b)$$

Here, η_3 is the reduction in sensible temperature (°C/°C), and $\langle f(dT)\rangle$ represents the spatial average value in the evaluation area (mesh exceeding the allowable upper limit value of T integrated as 0). $V'_{\text{allowance}}$ is the sum of the mesh area not exceeding the allowable upper limit of T. For the reduction in sensible temperature due to lowered temperatures, during standard summer daytime weather conditions in Tokyo, the following relationship is obtained when a sensitivity analysis is performed for SET* (Japan Sustainable Building Consortium 2017).

$$\eta_3 = \frac{\partial \text{SET}^*}{\partial T} \cong 1\left[\frac{^\circ\text{C}}{^\circ\text{C}}\right] \quad (6.7)$$

The effect of improving the thermal environment in the development area can be estimated by multiplying the evaluation factor k by the deterioration of the thermal environment due to wind speed reduction in the leeward evaluation area. For development companies, this is a target value that should be referenced when planning measures to improve the thermal environment in the development area. The possibilities of deregulation on the building height can be discussed using this relation.

$$\text{affordable}\, Q_{HI,A1,t2}(t) - Q_{HI,A2,t1}(t) \geq k \times \left(R_{HI,A2,t2}(t) - R_{HI,A2,t1}(t)\right) \quad (6.8a)$$

$$\text{unaffordable}\, Q_{HI,A1,t2}(t) - Q_{HI,A2,t1}(t) < k \times \left(R_{HI,A2,t2}(t) - R_{HI,A2,t1}(t)\right) \quad (6.8b)$$

6.3.5.3 Trial calculations

Consider a development area and evaluation area as shown in Fig. 6.22. First, the improvement effect of heat countermeasures is calculated as follows by decreasing SET* by 1 °C from 30 °C to 29 °C, then performing $W_{Q,AI}(t) = 5$, $V_{\text{allowance}} = 5000$.

$$\Delta Q = Q_{HI,A1,t2}(t) - Q_{HI,A1,t1}(t) = 5 \times (40 - 29) \times 5000 - 5 \times (40 - 30) \times 5000 = 25{,}000$$

Due to the development, when the leeward evaluation area ($W_Q = 1$) generation rate changes from 100% to 40%, the average wind speed of the mesh below the standard value of the evaluation wind speed is 3 m/s (decrease of 1 m/s), and $\eta_1 = 2$, $\eta_2 = 1$, $W_{Q,A2}(t) = 1$, if the deterioration of the thermal environment due to the wind speed reduction in the leeward evaluation area is

$$\Delta R = R_{HI,A2,t2}(t) - R_{HI,A2,t1}(t) = 1 \times 2 \times 1 \times (4 - 3) \times 12{,}000 \times 0.6 - 0 = 14{,}400$$

And the evaluation factor is $k = 1$,

$$\Delta Q > \Delta R$$

In this case, the outdoor living density in the development area $W_{Q,A1}(t)$ and outdoor density in the leeward area $W_{Q,A2}(t)$ are roughly set to 5, 1, respectively, but the

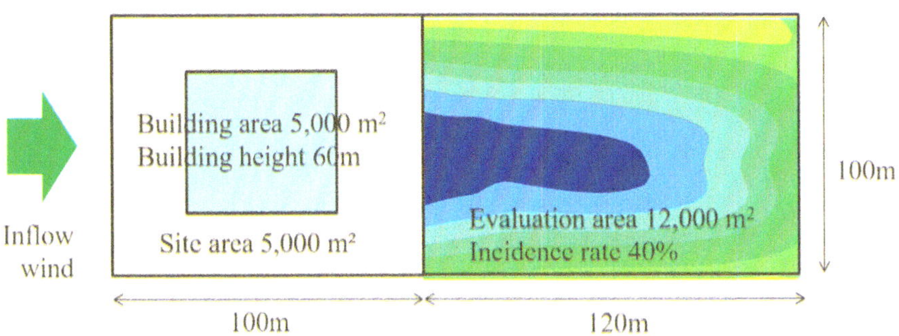

Fig. 6.22 Trial calculations case

setting of the numerical values differs between areas where a large number of unspecified users visit such as public institutions and shopping centers, and general residential areas. In addition, the evaluation factor k is a coefficient that represents the balance between the development area and the leeward area, and it is desirable to consider the overall adjustment, including the opinions of the region and main members, when determining it.

6.4 Closing Remarks

After publication of the General Principles for Heat Island Countermeasures, the national government and local public bodies enacted guidelines one after another: Building Design Guideline to the Mitigation of the Heat Island Phenomenon (Ministry of Land, Infrastructure, Transport and Tourism 2004), Osaka Prefecture Heat Island Countermeasure Promotion Program (Osaka Prefecture 2004), Heat Island Countermeasures Guideline (Tokyo Metropolitan Government 2006a) and the Heat Island Countermeasure Guideline (Ministry of the Environment 2009). The preparation of each type of guideline was linked to many survey research studies, and each incorporated the newest development tools such as urban canopy model or urban CFD. The advance of computer technology, digitalization of urban environment, etc. are factors which support highly realistic heat island analysis. The heat island countermeasures are primarily intended to first achieve overall reduction of the air temperature in cities, and various effects of greening of sites and

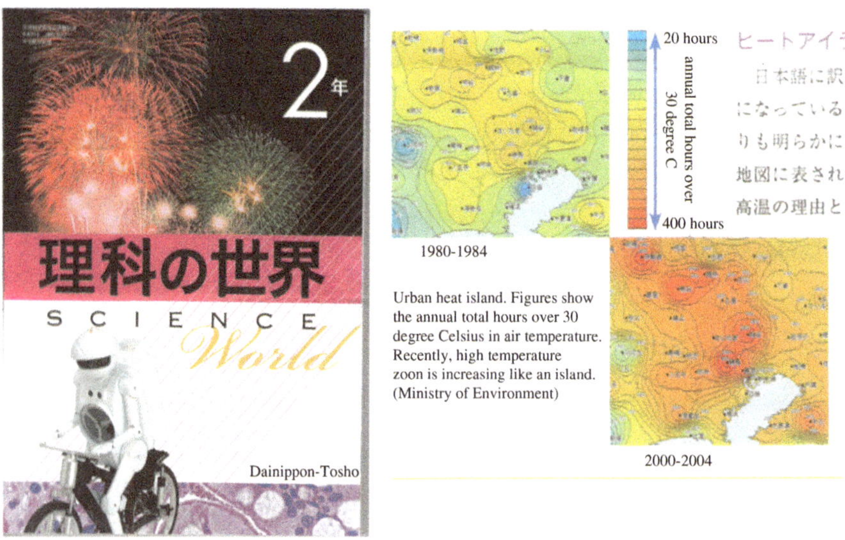

Fig. 6.23 Heat island phenomenon explained in a Japanese textbook (Dainippon-Tosho 2012)

buildings, water-retentive pavement, ventilation enhancements created by innovative building layouts, and anthropogenic heat reduction etc. have been verified based on numerical calculations.

Many special news reports have explained just what the heat island phenomenon is, stimulating concern with the heat island phenomenon and measures to counter it among the general public. Discussions of the heat island phenomenon have been taken up by educators in junior and senior high school textbooks (Fig. 6.23). At the present time, the word, "heat island" is generally widespread and has become a word used frequently in TV weather forecasts. For this reason, the major result of heat island countermeasures can be said to be the way the word "heat island" has generally penetrated society.

There is a concept called "Factor 4." It is the idea that if benefits are doubled and environment load halved by development, the benefits per environmental load increased 4 times. The Ministry of Land, Infrastructure, Transport and Tourism has introduced CASBEE-HI, which is a system that evaluates building plans from the perspective of heat island countermeasures. This system sets environmental efficiency which places the quantity of decline of sensible temperature inside a building site as the numerator as the benefits of the building plan, and places the environmental load which consists of heat, humidity and wind energy loss discharged outside the building site as the denominator as an evaluation yardstick. The results of evaluation by CASBEE predict inducement action according to rank linked with supplementary systems of the national government and local governments.

The adaptive measures among heat island countermeasures discussed above include measures to lower local sensible temperature. In the background is the fact that it would be very costly and time-consuming to achieve the overall reduction of the temperature of a city. Urban development on the other hand, is often forced to create environmental load, and accurate evaluation of external diseconomies and benefits have been long-term challenges. Although adaptive measures on the flow of environmental conservation and urban development occupy different positions, yardsticks that they focus on both conform with sensible temperature.

As an impact of urban heat island phenomenon, sensible temperature could be cited, and its relationship with adaptive measures among heat island countermeasures can be described. But the impacts of the heat island effects are extremely diverse including concentrated rainfall, atmospheric pollution and ecosystems other than sensible temperature and many points are unclear. If the scale of development is expanded to include multiple city blocks, further study may be necessary.

References

Ashie Y, Kono T (2011) Urban-scale CFD analysis in support of a climate-sensitive design for the Tokyo Bay area. Int J Climatol 31(2):174–188. https://doi.org/10.1002/joc.2226

Dainippon-Tosho (2012) Rika no Sekai 2: Cover. p 277

Heat Island Countermeasure Promotion Council (2004, revised in 2013) Heat Island Countermeasures Principle. http://www.env.go.jp/air/life/heat_island/taikou.html. Accessed 20 Jan 2020
Japan Sustainable Building Consortium (2017) CASBEE HI (Heat Island) comprehensive assessment system for built environment efficiency assessment manual. EDN 2017:75–87
Landsberg HE (1981) The urban climate. International geophysics series, vol 28. Academic Press, New York
Ministry of Land, Infrastructure, Transport and Tourism (2004) Architectural design guidelines for mitigating the heat island phenomenon. http://www.mlit.go.jp/kisha/kisha04/07/070716/02.pdf. Accessed 20 Jan 2020
Ministry of Land, Infrastructure, Transport and Tourism (2013) City Planning Guidelines for Mitigating the Heat Island Phenomenon. https://www.mlit.go.jp/common/001023246.pdf. Accessed 20 Jan 2020
Ministry of the Environment (2004) FY2003 Investigation and Study on Environmental Impact of Heat Island Phenomenon. http://www.env.go.jp/air/life/heat_island/reports/h15gaiyo.pdf. Accessed 20 Jan 2020
Ministry of the Environment (2009) Heat Island Countermeasure Guidelines. http://www.env.go.jp/air/life/heat_island/guideline/h20.html. Accessed 20 Jan 2020
Ministry of the Environment (2013) Heat Island Countermeasure Guidelines, rev edn. http://www.env.go.jp/air/life/heat_island/guideline/h24/chpt1.pdf. Accessed 20 Jan 2020
Ministry of the Environment (2017) FY2016 Investigation and Study on Low Carbon City Utilizing Surplus Groundwater. http://www.env.go.jp/air/report/h28-02/28.html. Accessed 20 Jan 2020
Ministry of the Environment (2018) Guidelines for Countermeasures against Heat in Towns, rev edn. http://www.env.go.jp/air/life/heat_island/guidelineH30/02_gaiyo.pdf. Accessed 20 Jan 2020
National Institute for Environmental Studies, Japan (2009) NIES Research Booklet, No. 32. https://www.nies.go.jp/kanko/kankyogi/32/10-11.html. Accessed 20 Jan 2020
Ono T, Murakami S, Ooka R, Takahashi T, Omori T, Saotome T (2006) Evaluation of mean convective heat transfer coefficient of a human body in outdoor environment: proposal of the formula for mean convective heat transfer coefficient of a human body by means of wind tunnel and CFD analysis. J Environ Eng (Transaction of AIJ) 601:9–14
Osaka Prefecture (2004) Osaka Prefecture Heat Island Countermeasure Promotion Program
Tokyo Metropolitan Government (2006a) Heat Island Countermeasure Guidelines
Tokyo Metropolitan Government (2006b, revised in 2014) Shinagawa Station/Tamachi Station Area Town Development Guidelines. https://www.toshiseibi.metro.tokyo.lg.jp/seisaku/guideline2014/pdf/siryo_3.pdf. Accessed 20 Jan 2020

Chapter 7
Hong Kong: Climatic Application in Urban Planning and Design at Multiple Scales for Creating a Healthy Living Environment

Emily Ying Yang Chan, William B. Goggins III, Zhe Huang, and Chi Shing Wong

Abstract Climate is an important topic for public health practice in twenty-first century, exacerbated by the extreme climate events brought about by climate change. As the frequency and severity of climate-related disasters increase, the need for understanding the health effects of extreme climate events is rapidly growing. Hong Kong is a densely populated city located in a subtropical region with high urbanization, heat island effect, population movement, and aging. All of these would aggravate the effects brought by extreme climate events. In recent years, a series of studies focusing on the associations between meteorological factors and all-cause or disease-specific mortality and hospital admissions have been published. These studies make clear that climate variations will affect various health outcomes, including those related to communicable and non-communicable diseases. Apart from mortality and hospital admissions, studies about discomfort and mild symptoms due to extreme climate events were also conducted to contribute to a better understanding of climate-related healthy living. This chapter highlights the example of the adverse health impact caused by urban heat island effect as a result of poor urban planning

E. Y. Y. Chan (✉)
Collaborating Centre for Oxford University and CUHK for Disaster and Medical Humanitarian Response (CCOUC), Faculty of Medicine, The Chinese University of Hong Kong, Shatin, Hong Kong SAR, China

GX Foundation, Shatin, Hong Kong SAR, China
e-mail: emily.chan@cuhk.edu.hk

Z. Huang · C. S. Wong
Collaborating Centre for Oxford University and CUHK for Disaster and Medical Humanitarian Response (CCOUC), Faculty of Medicine, The Chinese University of Hong Kong, Shatin, Hong Kong SAR, China
e-mail: huangzhe@cuhk.edu.hk; cswong@cuhk.edu.hk

W. B. Goggins III
JC School of Public Health and Primary Care, The Chinese University of Hong Kong, Shatin, Hong Kong SAR, China
e-mail: wgoggins@cuhk.edu.hk

C. Ren, G. McGregor (eds.), *Urban Climate Science for Planning Healthy Cities*, Biometeorology 5, https://doi.org/10.1007/978-3-030-87598-5_7

and design, and the example of more urban green space as a good urban planning and design practice that promotes urban healthy living environment. The evidence and the anticipated adverse health effects should provide suggestions and directions to policies and practices related to urban planning and design, in addition to setting priorities for planned response strategies, i.e., adaptation and mitigation, to offer more thorough protection for creating a healthy living environment for the urban population.

Keywords Communicable disease · Health effect measurement · Non-communicable diseases · Response strategies · Urban green space · Urban heat island effect

Key Messages

- Climate events like extreme temperatures, high humidity, and extreme wind speed can create serious threats to the health of urban residents. Urban planning and design should thus take these climate-related health threats into account in order to mitigate or adapt to the urban climate health impact to create a healthy urban living environment.
- Different indicators can be used to measure the health effects of urban climate, from the most serious deaths to hospital admissions, medical help--seeking, self-medication and self-care, and mild symptoms. In terms of measurability, health effects such as mortality and hospital admissions are often more readily observable. In contrast, it is more difficult to conduct data collection and assessments on the first, second, and third levels.
- In hot season, beyond the threshold of daily mean temperature of 28.2 °C, every 1 °C increase was found to be associated with an estimated 1.8% increase in mortality. In cold season, low temperature was found to have a lagged effect on health persisting up to 3 weeks. With a 1 °C drop in daily mean temperature, the subsequent 3 weeks had a 3.8% cumulative increase in mortality.
- In hot days (i.e., average temperatures above 29 °C), lower wind speeds were significantly associated with increased deaths, and the association was stronger in the region with high urban heat island indices (UHII). Urban micro heat islands, as a result of poor urban planning and design, thus exacerbate the negative health consequences of high temperatures and low wind speeds, including mortality.
- Urban green space has been linked to the health of surrounding population, because it not only supports the ecological integrity of the city, but also improves health and promote healthy behaviors. It can moderate temperatures by providing shade and cooling an area, reduce environmental pollution, and serve as sites for physical activities.
- When the daily mean temperature exceeded the threshold of 29 °C during hot season, a rise of every 1 °C increased hospital admissions by 4.5%, whereas for every 1 °C drop during cold season, admissions increased by 1.4%.

7.1 Introduction

Climate events like extreme temperatures, high humidity, and extreme wind speed can create serious threats to the health of urban residents. In recent years, the frequency of extreme events and disasters related to climate has risen. Climate change has also altered disease transmission patterns and distribution in urban cities. Urban planning and design should thus take these climate-related health threats into account in order to mitigate or adapt to the urban climate health impact to create a healthy urban living environment. This chapter will examine existing research on climate impact on health risks in Hong Kong, a highly developed and densely populated subtropical metropolis in southern China. It will also discuss adaptation and response strategies to address climate-related urban health risks (Chan 2019).

7.1.1 Health Effect Measurements

Different indicators can be used to measure the health effects of urban climate, from the most serious deaths to hospital admissions, medical help-seeking, self-medication and self-care, and mild symptoms (Fig. 7.1). Generally speaking, when a climate extreme event like heatwave happens, individuals may develop discomfort, mild symptoms, and subtle health effects. Among them, most can self-adjust through their own physical self-healing mechanism which relieves the discomfort. A small portion of them need to take over-the-counter medicine to get over the discomfort

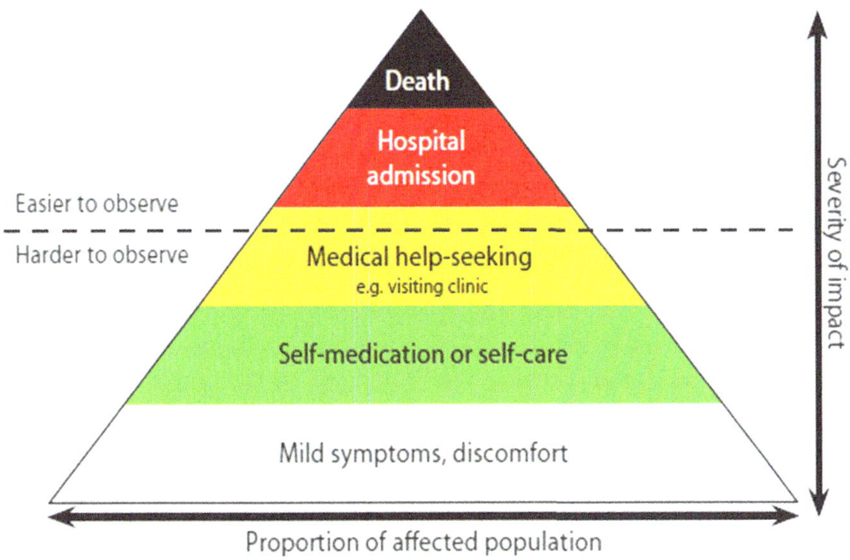

Fig. 7.1 Health effect pyramid

while a smaller number may seek primary healthcare. An even smaller proportion may need more intensive care and hospitalization due to their serious conditions. The most severe health impact is death, but this level constitutes the smallest proportion of the affected population when compared to the previous four levels.

In terms of measurability, health effects such as mortality and hospital admissions are often more readily observable. In contrast, it is more difficult to conduct data collection and assessments on the first, second, and third levels. For instance, patients may not seek the healthcare services provided by the government and they are not required to report to any authorities about their use of over-the-counter medication or self-care treatments. Subsequently, it is more difficult to quantify as impact of a climate event on human health at those levels. While most people may experience no symptoms or only mild symptoms, some may be severely affected by climate-related diseases and disorders, including communicable and non-communicable diseases (Chan 2017, 2019).

7.2 Health Risks of Urban Climate

7.2.1 Mortality

Using data from 1998 to 2006 from Hong Kong Observatory, Census and Statistics Department, Environmental Protection Department, and Centre for Health Protection, the relationships between mortality and temperature under both hot and cold seasons in Hong Kong were analyzed. In hot season, beyond the threshold of daily mean temperature of 28.2 °C, every 1 °C increase was found to be associated with an estimated 1.8% increase in mortality (Chan et al. 2012). Studies also found that cardiovascular and respiratory infection-related deaths were more sensitive to high temperature and people of lower socio-economic status were more vulnerable to heat-related mortality. In cold season, low temperature was found to have a lagged effect on health persisting up to 3 weeks (Goggins et al. 2013a). With a 1 °C drop in daily mean temperature, the subsequent 3 weeks had a 3.8% cumulative increase in mortality. When comparing to several intermittently hot and cold days, continuously very cold days led to higher mortality. Moreover, older people were found to be a high-risk group to the adverse health impact of cold weather.

7.2.1.1 Overall Urban Planning and Design and Mortality: Urban Heat Island Effect as a Negative Example

The pace of urbanization in a city may change the health outcomes of urban residents. In addition to exposure to extreme ambient temperatures, the densely built high-rises block the airflow and the heat emissions from air-conditioners, vehicles and industrial activities are trapped. In other words, the closeness of buildings reduces wind speeds and inhibits cooling by convection. At the same time, building materials like concrete absorb and retain some of the heat. All these poor planning and design contribute to the temperature rise in the urban area. When the urban cooling rates are slower than those in rural areas, the urban heat island effect develops. A study found that urban heat island effects were directly associated with health risks in Hong Kong (Goggins et al. 2012a). In hot days (i.e., average temperatures above 29 °C), lower wind speeds were significantly associated with increased deaths, and the association was stronger in the region with high urban heat island indices (UHII). Urban micro heat islands, as a result of poor urban planning and design, thus exacerbate the negative health consequences of high temperatures and low wind speeds, including mortality. Case Box 1 describes the potential health implication of urban green space.

Case Box 1 Urban Planning and Design and Mortality: Urban Green Space as a Positive Example

Since urbanization did exacerbate health risks posed by urban climate, urban heat island, and pollution (McMichael 2000; Frumkin 2002), it is important that urban planning and design should try to mitigate these problems caused by urbanization. For example, urban green space has been linked to the health of surrounding population, because it not only supports the ecological integrity of the city, but also improves health and promote healthy behaviors. It can moderate temperatures by providing shade and cooling an area (Nowak et al. 1998; Cummins and Jackson 2001; Wolch et al. 2014), reduce environmental pollution, and serve as sites for physical activities. A study suggested that a higher green space density was significantly associated with lower cardiovascular and diabetes mortality. Associations were stronger for males and low-income area residents (Xu et al. 2017). However, although the reasons vary, literature indicates that the distribution of green space in urban area may be disproportionately associated with affluent communities and "access to green space" is increasingly recognized as an environmental inequality issue.

7.2.2 Hospital Admissions

Other than affecting the mortality, urban climate has a direct impact on medical service utilization and health needs. As temperature changes, the relative humidity and air pollution level will also alter, which may further aggravate the impact of extreme temperatures on human health, leading to an increase in the incidence of related diseases. In Hong Kong, the public hospital database covers 83% of all hospitalizations and provides a reliable database for studying the relationship between urban climate and hospital admissions. Through the data collected, a number of associations between disease-specific hospital admission and meteorological factors in Hong Kong could be studied. Overall, when the daily mean temperature exceeded the threshold of 29 °C during hot season, a rise of every 1 °C increased hospital admissions by 4.5%, whereas for every 1 °C drop during cold season, admissions increased by 1.4% (Chan et al. 2013).

7.2.2.1 Communicable Diseases

Urban climate and environmental factors play a key role in spreading communicable diseases. Warmer air or seawater temperatures, increased precipitation, and urban planning and design factors such as urban overcrowding and deforestation create more favorable conditions for pathogens and vectors, contributing to the spread of water-, food-, and vector-borne diseases. In Hong Kong, with improved water, sanitation, and hygiene practices, communicable disease pattern had been stable since 1990s. However, a number of communicable diseases were found to be associated with climate factors in urban context.

Hand, Foot, and Mouth Disease

Hand, foot, and mouth disease symptomized by sores in the mouth and a rash on the hands and feet is a disease caused by enteroviruses, usually transmitted through person-to-person contact, respiratory droplets, or contaminated objects. More commonly found among children, the admission rate of hand, foot, and mouth disease is affected by temperature changes. At a daily mean temperature ranging from 8 °C and 20 °C or above 25 °C, a rising trend was observed for hand, foot, and mouth disease hospital admissions. High humidity created additional health risk. Admissions increased drastically on days with relative humidity higher than 80%. In addition, hospital admissions arising from the disease were found to be positively correlated with moderate rainfall, strong wind speed, and solar radiation (Wang et al. 2016).

Rotavirus and Norovirus Infection

Rotavirus and norovirus are two common viruses causing diarrheal illness, transmitted mainly through direct contact with an infected person, or touching objects where the viruses are present. In Hong Kong, short-term changes in rainfall, temperature, and humidity are significantly associated with hospitalization due to rotavirus and norovirus infections among the hospitalized patients younger than 5 years old (Wang et al. 2018b). Higher rainfall was found to be associated with fewer rotavirus, but more norovirus hospitalizations. These associations were stronger for rotavirus in winter and for norovirus in summer and the duration of association with rotavirus was notably longer than that of norovirus. Moreover, higher temperatures were found to be associated with fewer hospitalizations for both rotavirus and norovirus infection, while higher relative humidity was generally associated with more norovirus, but fewer rotavirus, hospitalizations.

Salmonellosis

Salmonellosis food poisoning is an infection usually caused by bacteria Salmonella, while Salmonella is commonly found in food animals, pets, and eggs. It can cause diarrhea, fever, abdominal cramps, and vomiting. High temperatures, high relative humidity, and light rainfall are found to be positively associated with Salmonella hospitalizations. A Hong Kong study indicated that higher temperature was strongly associated with more hospitalizations over the entire range of temperatures observed (Wang et al. 2018a). In addition to temperature, other meteorological factors were also found to be associated with the number of general infectious disease-related hospitalization. For example, positive associations were found for relative humidity above 60% and rainfall between 0 and 0.14 mm. Case Box 2 shows how SARS in 2003 might affect urban community and its implications on urban infrastructure.

Pneumonia

As one of the most common communicable respiratory disease, pneumonia is an infection targeting the air sacs in one or both lungs. In hot season, high temperature (>28 °C) and high relative humidity (>82%) were significantly associated with more pneumonia among older people (≥60 years old) (Lam et al. 2018a). The hospitalization rate for pneumonia among the elderly aged 75 or over increased along with higher wind speed and stronger solar radiation. During the cold season, the pneumonia admissions demonstrated U-shaped associations with temperature and relative humidity among adults over 60.

Case Box 2 Controlling the Emergence of New Infectious Diseases in Urban Area: Severe Acute Respiratory Syndrome (SARS) 2002–2003

In November 2002, it was reported that an unknown outbreak of infectious disease was emerging in Guangzhou, the capital city of Guangdong Province in mainland China. While Guangzhou is a geographic neighbor of Hong Kong, there was limited documentation of relevant communications or exchanges between the two health authorities. In February 2003, an infected visitor from Guangzhou visited Hong Kong. This visitor became the primary source of spreading the infection, later known as severe acute respiratory syndrome (SARS), first in Hong Kong, and later in other urban cities around the world. In the following months, the outbreak of SARS in the densely populated, vertically developed and poorly ventilated urban Hong Kong had infected 1755 people, among whom 299 eventually died. About 20% of the infected were healthcare personnel working in hospital wards where the SARS patients were treated, and six of them died eventually. The outbreak of SARS not only affected the health of the population, but also had a huge socio-economic impact on the city.Among many measures that were developed post-SARS in 2003, improved sewage pipe and ventilation designs for high-rise buildings were adopted to mitigate the health risk of similar infectious diseases, which are also relevant to health protection to community during extreme climate events.Sources: Lee (2014), World Health Organization (2004, April 21).

7.2.2.2 Non-communicable Diseases

Non-communicable diseases (NCDs) are diseases that do not spread from person to person and they usually appear to be chronic. Although some NCD patients may not be aware of their own disease status, they ultimately need long-term investment and protection from the healthcare system to maintain their health. According to World Health Organization (2018), more than 70% of the global mortality were due to NCDs. In Hong Kong, over half (55.2%) of 46,662 registered deaths in 2016 were attributed to five major NCDs, namely cancer, heart diseases, stroke, chronic lower respiratory diseases, and diabetes (Centre for Health Protection of the Department of Health 2018). Even though there are still some uncertainties, research in extreme climate events showed significant increase in the risk of certain NCDs.

Allergy

In a population-based cross-sectional study in Hong Kong conducted in 2017, 22.4% and 15.7% of the patients with skin and nasal allergies, respectively, claimed that their symptoms worsened during hot weather (Lam and Chan 2019). About 24% had reported an allergic history during the study period. Compared to people without a history of allergies, people with allergic symptoms claimed increased mucus secretions and oral ulcers, worsened sleep quality, and depression during hot weather. In summary, pre-existing allergic symptoms were found to be associated with more adverse health effects and worse quality of life during hot days.

Asthma

Asthma is a common respiratory disease characterized by chronic airway inflammation and high sensitivity to various triggers. In hot season, asthma hospitalizations in Hong Kong were lowest at 27 °C, rose to a peak at 30 °C, then plateaued between 30 °C and 32 °C, while in cold season, temperature is negatively associated with the number of asthma hospitalizations (Lam et al. 2016). Compared to those of other age groups, adult asthma admissions were most sensitive to temperatures in both seasons, while admissions among children aged 5 or below were least associated. Higher humidity and ozone levels in hot season, and low humidity in cold season were also associated with more asthma admissions.

Chronic Obstructive Pulmonary Disease (COPD)

Chronic obstructive pulmonary disease (COPD) is one of the most common causes of respiratory hospitalization among older people characterized by long-term breathing problems and poor airflow. During the cold season, consistent hockey-stick associations with temperature and relative humidity were found for COPD admissions among older people (Lam et al. 2018a). The lowest morbidity temperature and relative humidity were at about 21–22 °C and 82%, respectively. In summary, the low-temperature-admissions associations with COPD were stronger and were strongest among older people.

Acute Myocardial Infarction

Acute myocardial infarction, also known as heart attack, is caused by progressive narrowing of coronary arteries resulting in inadequate blood supply to the myocardium. While no significant association between high temperature and acute myocardial infarction was found, patients with cardiovascular disease had an increased risk of acute myocardial infarction hospital admissions in cold weather (Goggins et al. 2013b). In general, if the daily mean temperature of a 14-day period dropped below 24 °C, acute myocardial infarction admissions increased by 3.7% with every 1 °C drop in temperature.

Specifically, the acute myocardial infarction admissions among diabetic patients were found to be linearly and negatively associated with temperature in cold season (Lam et al. 2018b). The association between myocardial infraction admissions and temperature among non-diabetic patients was weaker and this association began only when the temperature was below 22 °C. In hot season, the acute myocardial infarction hospitalization among diabetic patients started increasing when temperature rose above 28.8 °C. In contrast, the rate of infarction admission among non-diabetic patients was not related to temperature. The difference in sensitivity to temperature between patients with and without diabetes was most apparent in the group below 75 years old and among first-admission cases in cold season.

Heart Failure

In Hong Kong, low temperature is closely related to increasing heart failure-related hospital admissions and mortality (Goggins and Chan 2017). When days with higher and lower temperatures were compared, it was observed that heart failure admissions on days with lower temperature were 2.63 times of those on higher temperature days. The association with cold weather was stronger among older age groups and for new hospitalizations compared to recurrent ones, after adjusting for the presence of co-morbidities. For relative humidity, both extremely high and extremely low relative humidity were modestly associated with more heart failure-related hospital admissions, i.e., a U-shaped association.

Stroke

Stroke is mainly divided into two types. Hemorrhagic stroke is caused by the bursting of a blood vessel inside the brain which cuts down the flow of blood, while ischemic stroke is caused by blood clots formed inside the vessels of brain. Within the daily mean temperature range of 8.2–31.8 °C, admissions due to hemorrhagic stroke increased with temperature drop (Goggins et al. 2012b). As the average 5-day mean temperature dropped by 1 °C, hemorrhagic stroke admissions increased by 2.7%. This association was stronger among older people and females. In contrast, the link between daily mean temperature and ischemic stroke was less apparent. If the mean temperature of a consecutive 14 days was below 22 °C, admissions due to ischemic stroke increased by 1.6% with a 1 °C drop.

Mental Disorders

A linear positive correlation between temperature and the number of mental disorder admissions was observed, with the strongest effect among older people (≥75 years old), and the effect lasted about 2 days (Chan et al. 2018b). The relative risk increased significantly when the temperature was above 19.4 °C.

In sub-disease analyses, transient mental disorders and episodic mood disorders showed positive associations with temperature while drug-related mental disorders demonstrated a positive association when temperature rose over a threshold of about 20 °C. Moreover, depressive disorder and other nonorganic psychoses showed a significantly lower risk at lower temperature. In contrast, no obvious association with temperature was observed for anxiety, dissociative and somatoform disorders, schizophrenic disorders, and alcohol-related mental disorders.

7.2.3 *Medical Help-Seeking, Self-Care, and Mild Symptoms*

Urban climate phenomena pose health impacts with varying degrees of severity. Apart from death or hospital admissions, people may be sick or have severe symptoms, which could require some medical care. In addition, discomfort and mild symptoms due to climate events should not be dismissed. Although related data tend to lead to underestimation of the real health impact due to the difficulty of data collection, they would help the respective government authorities apprehend the overall health risks during extreme climate events, which in turn contribute to establishing a more comprehensive warning system and contingency planning as well as better urban planning and design to mitigate and adapt to the adverse health effect.

A study focusing on the help-seeking behavior among older people in Hong Kong showed that when the highest temperature on that day exceeded 32 °C, the number of emergency help call increased significantly along with temperature rise (Chan et al. 2011). During hot season, almost half of the calls were made for explicit health-related reasons including dizziness, shortness of breath, and general pain. There is also a gender difference on responses to high temperature events. When the temperature climbed up to 30.5 °C, call frequency among females increased. As for call frequency among males, it started to increase at 33.5 °C.

Another territory-wide large-scale telephone survey was conducted immediately after the coldest day in Hong Kong since 1957 to investigate the impact of the cold spell on Hong Kong community (Chan et al. 2017a). Among all the participants, around 67% reported that the cold spell brought about physical discomfort. However, only 13% thought that they had to seek medical help. For those who thought they had the need to seek medical assistance, most (around half) went to private medical clinics, followed by public clinics and traditional Chinese medical practitioner. Around 13% of the interviewed participants bought over-the-counter drugs.

7.3 Response Strategies

Adaptation and mitigation are two main approaches to respond to extreme climate events. Adaptation refers to the process of adjustment made according to the actual and expected climate and its effects in order to moderate harm or exploit beneficial opportunities, while mitigation refers to basic prevention measures to relieve or stabilize climate change.

7.3.1 Adaptation Strategies

7.3.1.1 Hot Weather

In a Hong Kong-wide telephone survey study conducted in 2017, 45.3% of the interviewees were found to believe that hot weather would not affect their health. Among them, 28.8% were 65 years old or above (Chan et al. 2018a). About 87% of the respondents were aware of the hot weather warning issued by the Hong Kong Observatory during the study period. It was found that 37.2–97.5% of the interviewees adopted at least one personal heat protection measure. The most commonly adopted measure was to drink more water (97.5%), and the least was to use sunscreen (37.2%). Male participants were less likely to avoid sunlight and use sunscreen, while air-conditioners were less used by people with lower education levels. As the hot weather warning service could reach most parts of Hong Kong, most people could take personal precaution measures accordingly.

7.3.1.2 Cold Weather

According to a territory-wide telephone survey on the impact of the coldest day since 1957 in Hong Kong in 2016 on health risk, health risk perception and self-care behaviors among the Hong Kong community, 45.0% of the respondents were identified to have higher potential health risks (e.g., older people and those reporting a history of chronic disease, living alone and receiving low-income assistance from the government), but 63.7% among them underestimated their health risks in cold weather (Chan et al. in press). In contrast to the fact that the cold wave in 2016 was more severe, Hong Kong citizens took more protective measures against cold weather during the 2017 cold spell. Those who suffered adverse health effects during the 2016 cold spell tended to wear more clothes during the 2017 cold spell. It was also found that women, those who felt cold during the study period and older people aged 70 years old or above were more likely to use heating devices.

7.3.1.3 Typhoon

As the frequency and severity level of extreme weather events associated with climate change are increasing, understanding household preparedness is important for managing the relevant health risks. A study in Hong Kong in 2018 explored the relationships among the risk perception of disasters, family preparedness, and the immediate effects of typhoon Mangkhut (Chan et al. 2019). Among all respondents, 93.9% and 74.3%, respectively, reported of having taken routine and special emergency preparedness measures for typhoon Mangkhut. One-third of the respondents reported that Typhoon Mangkhut had some forms of influence on them. Respondents who perceived that their home were at high risk during the typhoon and had routine emergency preparedness were more likely to take special emergency preparedness for the typhoon. However, there is no significant relationship between the implementation of special measures and the impact of typhoon. The study shows that current household emergency preparedness measures may not be sufficient to respond to the effects of a super typhoon. Better urban planning and design may reduce the health impact of typhoon to make up for the insufficient household emergency preparedness.

7.3.2 Mitigation Strategies

The carbon-reduction behaviors as measures to tackle climate change focus on the positive health impact produced—in addition to the improvement in the environment—when adaptation and mitigation measures are introduced. For example, consuming less red meat reduces an individual's cholesterol level and at the same time helps reduce the greenhouse gases produced by livestock farming; urban planning and design like building more cycling paths and better public transportation systems can encourage healthy travel and reduce emission of greenhouse gas and poisonous pollutants.

Unlike increasing urban green space density and reducing urban heat island effect, some mitigation measures involving urban planning and design may not have direct benefit to human health but can benefit society, the economy and the environment and thus maintain health indirectly. For example, forest protection policies help biodiversity, maintain a balance in the ecosystem, and ensure the supply and quality of drinking water.

Nature reduces climate-related health risks by acting as a buffer against climate change, and urban planning measures like afforestation can strengthen both nature and human society. For instance, planting mangroves can protect coastal areas by reducing the erosion caused by floodwater, and building green cities and increasing the number of green spaces can reduce the mortality rate and the incidence of cardiovascular and respiratory diseases in addition to alleviating the urban heat island effect.

In Hong Kong, it was found that a higher proportion of the older age groups practiced carbon-reduction behaviors (Chan et al. 2017b). This is in contrast to the general conception that the younger generation adopt such environmental friendly behaviors more readily. In addition, a larger proportion of females and married individuals were found to practice carbon-reduction behaviors. The same study also investigated the practice of environmental friendly behaviors bearing health co-benefits in the urban population. Using less packaging and shopping bags was the most commonly practiced carbon-reduction behavior, with 70% of interviewees reported to do it every day, which may be attributed to the government's Environmental Levy Scheme on Plastic Shopping Bags. Another green scheme implemented by the Water Supplies Department encouraged people to shorten shower time to less than 5 min. Almost half of the survey respondents reported that they had never practiced nor considered this behavior.

As for waste management, around 60% of the interviewees indicated that they separated household waste at least once a week. Different housing types found varying patterns for the practice of waste separation. The ratio was higher among those living in subsidized self-own housing, including Home Ownership Scheme and Sandwich Class Housing Scheme, accounting for around 70%. The ratio was around 60% for private housing and it was the lowest for public housing, accounting for around 55%. This might reflect that residential facilities under the subsidized apartments came with a more comprehensive waste separation mechanism and facilities. Going vegetarian is another carbon-reduction behavior with health co-benefits. The study found that only 3.5% of the interviewees were vegetarians, while around half of the non-vegetarians had never thought of having at least one green meal a week. Bringing along one's own dining utensils is a green behavior at individual level. However, such practice was not commonly practiced. As high as 70% of the respondents have never considered bringing their own utensils.

7.4 Conclusion

This chapter highlights the linkage between climate events and health outcomes of various sub-groups as well as various response strategies to climate-related health risks in Hong Kong, a subtropical urban city. These macro-level studies should be translated into micro-level urban planning and design practices to help create a healthy urban living environment. Government and relevant parties should always take into consideration the specific health needs of the high-risk groups in climate events when undergoing urban planning and design to help offer a more thorough protection for the well-being of citizens in face of climate change. More urban planning and design studies at the micro and translational level are urgently needed in face of the worsening adverse impact of climate change.

In particular, it was found that sustained periods of high temperature and low wind speed aggravated the adverse health impact of urban heat island. More efforts should be paid on urban planning and design study and practice to minimize wall effect of crowded high-rises, ensure adequate air ventilation, and thereby lower urban temperature. Notable urban planning and design measures like building height restrictions, green buildings, and ensuring more green space may be effective ways to minimize urban heat island effect and its associated health risks. Concerted efforts of various stakeholders (including government, property developers, construction companies, community advocacy groups, researchers and citizens) should aim to integrate and apply climate-related health impact research findings in urban planning and design to create a healthy living environment for urban residents.

References

Centre for Health Protection of the Department of Health (2018) Non-communicable diseases watch. https://www.chp.gov.hk/files/pdf/ncd_watch_may_2018.pdf. Accessed 1 Apr 2020

Chan EYY (2017) Public health humanitarian responses to natural disasters. Routledge, London

Chan EYY (2019) Climate change and urban health: the case of Hong Kong as a Subtropical City. Routledge, London

Chan EYY, Goggins WB, Kim JJ, Griffiths S, Ma TKW (2011) Help-seeking behavior during elevated temperature in Chinese population. J Urban Health 88(4):637. https://doi.org/10.1007/s11524-011-9599-9

Chan EYY, Goggins WB, Kim JJ, Griffiths SM (2012) A study of intracity variation of temperature-related mortality and socioeconomic status among the Chinese population in Hong Kong. J Epidemiol Community Health 66(4):322–327. https://doi.org/10.1136/jech.2008.085167

Chan EY, Goggins WB, Yue JS, Lee P (2013) Hospital admissions as a function of temperature, other weather phenomena and pollution levels in an urban setting in China. Bull WHO 91(8):576–584

Chan EYY, Huang Z, Mark CKM, Guo C (2017a) Weather information acquisition and health significance during extreme cold weather in a subtropical city: a cross-sectional survey in Hong Kong. Int J Disaster Risk Sci 8(2):134–144. https://doi.org/10.1007/s13753-017-0127-8

Chan EYY, Wang SS, Ho JY-E, Huang Z, Liu S, Guo C (2017b) Socio-demographic predictors of health and environmental co-benefit behaviours for climate change mitigation in urban China. PLoS One 12(11):e0188661. https://doi.org/10.1371/journal.pone.0188661

Chan EYY, Man AYT, Lam HCY (2018a) Personal heat protective measures during the 2017 heatwave in Hong Kong: a telephone survey study. First Global Forum for Heat and Health, Hong Kong, December 2018

Chan EYY, Lam HC, So SH, Goggins WB, Ho JY, Liu S, Chung PP (2018b) Association between ambient temperatures and mental disorder hospitalizations in a subtropical city: a time-series study of Hong Kong special administrative region. Int J Environ Res Public Health 15(4):754

Chan EYY, Man AYT, Lam HCY, Chan GKW, Hall BJ, Hung KKC (2019) Is urban household emergency preparedness associated with short-term impact reduction after a super typhoon in subtropical city? Int J Environ Res Public Health 16(4):596

Chan EYY, Lam HCY, Huang Z, Liu S, Guo CL, Goggins WB (in press) Are risk perception and its accuracy associated with personal health protection behaviours against cold waves? BMC Public Health
Cummins SK, Jackson RJ (2001) The built environment and children's health. Pediatr Clin N Am 48(5):1241–1252. https://doi.org/10.1016/S0031-3955(05)70372-2
Frumkin H (2002) Urban sprawl and public health. Public Health Rep 117(3):201–217. https://doi.org/10.1093/phr/117.3.201
Goggins WB, Chan EYY (2017) A study of the short-term associations between hospital admissions and mortality from heart failure and meteorological variables in Hong Kong: weather and heart failure in Hong Kong. Int J Cardiol 228:537–542. https://doi.org/10.1016/j.ijcard.2016.11.106
Goggins WB, Chan EYY, Ng E, Ren C, Chen L (2012a) Effect modification of the association between short-term meteorological factors and mortality by urban heat islands in Hong Kong. PLoS One 7(6):e38551–e38551. https://doi.org/10.1371/journal.pone.0038551
Goggins WB, Woo J, Ho S, Chan EYY, Chau PH (2012b) Weather, season, and daily stroke admissions in Hong Kong. Int J Biometeorol 56(5):865–872. https://doi.org/10.1007/s00484-011-0491-9
Goggins WB, Chan EYY, Yang C, Chong M (2013a) Associations between mortality and meteorological and pollutant variables during the cool season in two Asian cities with sub-tropical climates: Hong Kong and Taipei. Environ Health 12(1):59. https://doi.org/10.1186/1476-069X-12-59
Goggins WB, Chan EYY, Yang C-Y (2013b) Weather, pollution, and acute myocardial infarction in Hong Kong and Taiwan. Int J Cardiol 168(1):243–249. https://doi.org/10.1016/j.ijcard.2012.09.087
Lam HCY, Chan EYY (2019) Effects of high temperature on existing allergic symptoms and the effect modification of allergic history on health outcomes during hot days among adults: an exploratory cross-sectional telephone survey study. Environ Res 175:142–147. https://doi.org/10.1016/j.envres.2019.05.017
Lam HCY, Li AM, Chan EYY, Goggins WB (2016) The short-term association between asthma hospitalisations, ambient temperature, other meteorological factors and air pollutants in Hong Kong: a time-series study. Thorax 71(12):1097–1109. https://doi.org/10.1136/thoraxjnl-2015-208054
Lam HCY, Chan EYY, Goggins WB (2018a) Comparison of short-term associations with meteorological variables between COPD and pneumonia hospitalization among the elderly in Hong Kong—a time-series study. Int J Biometeorol 62(8):1447–1460. https://doi.org/10.1007/s00484-018-1542-2
Lam HCY, Chan JCN, Luk AOY, Chan EYY, Goggins WB (2018b) Short-term association between ambient temperature and acute myocardial infarction hospitalizations for diabetes mellitus patients: a time series study. PLoS Med 15(7):e1002612. https://doi.org/10.1371/journal.pmed.1002612
Lee SH (2014) Historical perspectives in public health: experiences from Hong Kong. In: Griffiths SM, Tang JL, Yeoh EK (eds) Routledge handbook of global public health in Asia. Routledge, Oxford, pp 5–20
McMichael AJ (2000) The urban environment and health in a world of increasing globalization: issues for developing countries. Bull WHO 78:1117–1126
Nowak DJ, McHale PJ, Ibarra M, Crane DE, Stevens JC, Luley CJ (1998) Modeling the effects of urban vegetation on air pollution. In: Gryning S, Chaumerliac N (eds) Air pollution modeling and its application XII, vol 399–407. Plenum Press, New York
Wang P, Goggins WB, Chan EY (2016) Hand, foot and mouth disease in Hong Kong: a time-series analysis on its relationship with weather. PLoS One 11(8):e0161006. https://doi.org/10.1371/journal.pone.0161006
Wang P, Goggins WB, Chan EYY (2018a) Associations of Salmonella hospitalizations with ambient temperature, humidity and rainfall in Hong Kong. Environ Int 120:223–230. https://doi.org/10.1016/j.envint.2018.08.014

Wang P, Goggins WB, Chan EYY (2018b) A time-series study of the association of rainfall, relative humidity and ambient temperature with hospitalizations for rotavirus and norovirus infection among children in Hong Kong. Sci Total Environ 643:414–422. https://doi.org/10.1016/j.scitotenv.2018.06.189

Wolch JR, Byrne J, Newell JP (2014) Urban green space, public health, and environmental justice: the challenge of making cities 'just green enough'. Landsc Urban Plan 125:234–244. https://doi.org/10.1016/j.landurbplan.2014.01.017

World Health Organization (2004) Emergencies preparedness, response: summary of probable SARS cases with onset of illness from 1 November 2002 to 31 July 2003. http://www.who.int/csr/sars/country/table2004_04_21/en/. Accessed 1 Apr 2020

World Health Organization (2018) Noncommunicable diseases. https://www.who.int/news-room/fact-sheets/detail/noncommunicable-diseases. Accessed 1 Apr 2020

Xu L, Ren C, Yuan C, Nichol JE, Goggins WB (2017) An ecological study of the association between area-level green space and adult mortality in Hong Kong. Climate 5(3):55

Chapter 8
Air Pollution Prevention and Urban Planning Response: Building a Healthy Beijing

Xiaoyi Fang, Xin Li, Xiyan Ren, Jian He, Wupeng Du, Yajun Xiong, Yonghong Liu, Chen Cheng, and Shuo Zhang

Abstract Building a healthy city is our goal. Herein, we take Beijing as an example to elaborate the aspects of air pollution governance, air quality status and causes, and urban planning response. Especially after 2008, the Beijing Olympic experience transformed China's national policy for air pollution governance. Compared with 2013, the air pollution in Beijing improved significantly in 2018. In Beijing, the number of haze days in 2018 was 83 days, which is 38 days less than the 121 days on average in each of the past 4 years, a reduction of 31.4%. Beijing is surrounded by mountains on three sides, the arc-shaped mountains formed by the Taihang Mountains and the Yanshan Mountains lead to low air mobility in the warm sector of the mountain front and the accumulation of pollutants in front of the arc-shaped mountains. Local emissions account for two thirds of the main source of annual PM2.5 emissions in Beijing, and the other one third of the main sources is regional transport. As the pollution level increased, the regional transport contribution showed a significant upward trend. The Beijing Urban Master Plan (2016–2035) was approved in 2017, integrating the improvement of environmental air quality and

X. Fang (✉) · Y. Liu · C. Cheng · S. Zhang
Chinese Academy of Meteorological Sciences, Beijing, China
e-mail: fangxy@cma.gov.cn

X. Li
Counselors Office of Beijing Municipal People's Government, Beijing, China

X. Ren
Chinese Academy of Urban Planning & Design Beijing Planning& Design Consultants LTD, Beijing, China

J. He
Beijing Urban Planning and Design Institute, Beijing, China

W. Du
Beijing Municipal Climate Center, Beijing, China

Y. Xiong
Beijing-Tianjin-Hebei Prediction and Early Warning Center for Environmental Meteorology, Beijing, China

C. Ren, G. McGregor (eds.), *Urban Climate Science for Planning Healthy Cities*, Biometeorology 5, https://doi.org/10.1007/978-3-030-87598-5_8

adaptation to climate change into urban planning, and design is an important part of the implementation of strategic positioning.

Keywords Air pollution · Urban planning · Climate · Regional coordinated development · Health Beijing

Key MessagesBeijing is surrounded by mountains on three sides, the arc-shaped mountains lead to low air mobility, especially with the higher temperature under the global climate change.

- The rapid development and social progress in China are accompanied by environmental problems. Haze governance has risen to an unprecedented level of strategic consideration in China.
- After 2008, the Beijing Olympic experience transformed China's national policy for air pollution governance. Compared with 2013, the air pollution in Beijing improved significantly in 2018.
- The improvement of environmental air quality and adaptation to climate change in Beijing are spatially integrated with the Beijing–Tianjin–Hebei region and cannot be isolated. Coordinated governance is the only way.
- A case study (The Beijing Urban Master Plan (2016–2035)) is conducted to demonstrate how to protect and restore the natural ecosystem, optimize the ecological spatial pattern, strengthen the urban resilience, and mitigate and adapt to climate change in regional coordinated development.

8.1 Background

The news that Beijing and several other large and medium-sized cities in central and eastern China collectively became hard-hit areas by haze was selected by the Global Times as one of the top ten international news stories of 2013 (Zhao 2013), suggesting China's air pollution has become a global issue. According to statistics, in 2013, the average number of haze days in Beijing and six other major cities in China reached 161 days, and the frequent haze period was from late fall to winter (Wang and Zheng 2014). The rapid development and social progress in China are accompanied by severe environmental problems. The climatic change caused the decline of the capacity of atmospheric diffusion (Wang and Zheng 2013; Mao et al. 2018). Especially with the urbanization (Zhou et al. 2017), adverse meteorological conditions are the important reasons for heavy air pollution (Pan and Shan 2017; Zhu et al. 2018b).

In September 2013, the State Council of China publicly released the Air Pollution Prevention and Control Action Plan (referred to as the Action Plan hereafter) in response to the increasingly severe haze weather (General Office of the State Council of China 2013). Since then, haze governance has risen to an unprecedented level of strategic consideration in China. To protect the atmospheric environment, governments at central and local levels play a leading role (Yu and Tie 2015; Shi et al. 2016), and the environmental protection departments, scientific research institutes, enterprises and institutions, and various sectors of the society fight together (Ma and Zhang 2014; Yang et al. 2016).

The Action Plan mainly set forth 35 specific action plan requirements from ten aspects: to reduce emissions from pollution sources (e.g., industrial sources, non-point sources, mobile sources); to adjust and optimize the industrial structure, and promote industrial transformation and upgrading; to accelerate technological transformation of enterprises and improve the capability for innovation in science and technology; to accelerate the restructuring of the energy industry and increase the clean energy supply; to tighten the market access for energy conservation and environmental protection and optimize the spatial layout of industries; to better play the role of a market mechanism and improve environmental and economic policies; to improve the law and regulation system and exercise strict supervision and management in accordance with the system; to establish the regional cooperation mechanism and coordinate the regional environmental governance; to establish a monitoring, early warning, and emergency response system to properly cope with heavy pollution weather; and to clarify the responsibilities of the government, enterprises and society, and mobilize the public to participate in environmental protection. The targeted Action Plan was not established overnight but reflected the course of air pollution governance in China, fully demonstrating the formation of haze over decades of development and accumulation and necessitating medium- to long-term prevention and control strategies as well as scientific measures. Haze governance is not based on a single action but requires long-term and multi-faceted efforts.

Herein we take Beijing as an example to elaborate the aspects of air pollution governance, air quality status and causes, and urban planning response.

8.2 Air Pollution Governance in Beijing

The air pollution governance and management in Beijing since the twenty-first century can be divided into the following stages:

8.2.1 Forced Fulfillment of the Promise of the Beijing Olympics and Pioneering of Regional Coordinated Air Quality Assurance

When Beijing bid for the Olympics in 2001, it promised the world it would achieve good air quality. After the successful bidding, it became a key concern to ensure the air quality was up to standard during the period of the Olympics. As early as 1998, Beijing launched the Blue Sky Project to gradually implement various stages of air pollution control measures. Starting in 2000, Beijing also supported multiple projects such as the Study on Air Pollution Control Countermeasures and the Study on Air Quality Compliance Strategies in Beijing. Studies showed that although the air pollution governance measures adopted in the early 2000s made significant progress, the concentrations of atmospheric particulate matter and ozone in Beijing were still very high. PM2.5 accounted for more than 60% of PM10, leading to a highly oxidizing atmosphere and severe secondary pollution. The air pollution in Beijing and its surrounding areas has significant regional characteristics. Under certain meteorological conditions, pollutants discharged from surrounding provinces and cities were significantly transported to Beijing. The air pollution in Beijing was characterized as typical combined air pollution. Improving the air quality in Beijing required the joint efforts of surrounding provinces/autonomous regions/municipalities (Li 2013).

Therefore, based on the air quality problems faced during the 2008 Beijing Olympics and the research status, the Beijing Environmental Protection Bureau proposed a Study on the Transport and Transformation of Air Pollutants in Beijing and Surrounding Areas and the Goals of Air Quality in Beijing in 2005, which was funded by the Beijing Science and Technology Commission (Beijing Science and Technology Commission 2005). The project systematically analyzed the formation mechanism and characteristics of regional combined air pollution in Beijing and surrounding provinces and cities, the impact of regional economic and social development on air quality, regional air pollutant emission conditions, and key areas and pollution sources that have a major impact on air quality in Beijing. The Beijing Air Quality Assurance Program for the 29th Olympics and Measures for Provinces, Autonomous Region, and Municipality Surrounding Beijing was proposed. The different potentials of industrial sources for reduced emissions of SO_2, NO_x, and PM10 were enacted for Tianjin, Hebei, Inner Mongolia, and Shanxi. The annual emission reduction of CO, NO_x, PM, and HC can be realized in these regions by implementing various control measures for motor vehicles, and governance measures for the emission reduction of volatile organic compounds (VOCs) were implemented at gas stations.

In addition, to cope with extremely adverse weather conditions, the Emergency Measures for Air Pollution Control under Extremely Unfavorable Weather Conditions during the Beijing Olympics and Paralympic Games (Ministry of Environmental Protection of China 2008) were formulated. The formulation of the two measures provided a scientific basis for determining the specific projects and requirements for pollution and emission reduction in air quality assurance work during the Olympics. A multi-sectorial leading group for air quality monitoring, prediction, and forecasting was established to form a regional atmospheric stereoscopic monitoring network, which was based on conventional ground monitoring and supplemented by advanced technological means, such as tower observation, laser radar, and satellite remote sensing inversion, and multi-mode ensemble forecasting was adopted to establish and operate a regional air quality forecasting and early warning system, providing a scientific basis for timely initiation of air pollution control emergency supervision plan and scientific decision-making.

These measures were fully implemented during the 2008 Beijing Olympics. By adopting the joint mechanism of the six provinces and cities in northern China, the air quality was significantly better than during the same period in 2007 and all met the standards, and a record of first-class blue sky for a single month in summer was set, successfully ensuring the air quality during the Beijing Olympics (Greenpeace 2008).

8.2.2 The Olympic Experience Became a Useful Reference for Urban and Regional Control of Air Pollution and Transformed China's National Policy for Air Pollution Governance

The air quality assurance mission for the 2008 Olympics was the first attempt in China to promote air quality through major events, particularly through regional coordination to ensure the air quality during major events. It has become an effective attempt and classic case for inter-regional prevention and control of air pollution across provinces and cities in China, and offers the beneficial experience of promoting environmental governance and management in China, providing the world with new cases of improving regional environmental quality as well as promoting coordinated environmental and economic development in a large developing country.

8.2.2.1 Phase One of Joint Control of Air Pollution in Beijing, Tianjin, and Hebei and the Surrounding Areas After the Olympics

To solidify the successful experience of air quality assurance during the Olympics, Beijing and surrounding provinces and cities took the initiative to strengthen exchanges and cooperation through work such as jointly carrying out regional air pollution prevention and control measures, as well as conducting research and formulation of air pollutant emission standards, to achieve information sharing of atmospheric environmental resources and promote reduced pollution emissions. In 2010, the Guidelines of the General Office of the State Council on Promoting Joint Prevention and Control of Air Pollution to Improve Regional Air Quality were released, followed by the launching of the plan for joint prevention and control of air pollution in key regions, which was approved by the State Council in October 2012 (General Office of the State Council of China 2010).

Making full use of its advantages in scientific research, after the Olympics, Beijing launched the Project of Research and Demonstration on Formation Mechanism and Prevention and Control Measures of Combined Air Pollution in Beijing and Near Surrounding Areas (Beijing Science and Technology Commission 2009), which established five sub-projects including three-dimensional monitoring, ozone and precursor control, a pollution source list, model simulation and prediction, and joint prevention and control strategy research. The project completed the establishment of a regional atmospheric environmental observation network based on 25 ground observation stations, carried out real-time ground monitoring and passive monitoring of dry and wet deposition, established multiple VOC monitoring stations and carried out monitoring, developed CO/CH_4 analysis technology, developed and applied core technologies, such as data assimilation and pollution source inversion, improved the ensemble prediction system, and initially evaluated the economic development, pollution contribution, and regional cooperation status of the Beijing–Tianjin–Hebei region. The above work provided a scientific basis for further exploring the formation mechanism of air pollution and improving the joint prevention and control work in the Beijing–Tianjin–Hebei region.

8.2.2.2 Phase Two of the Joint Control of Air Pollution in Beijing, Tianjin, and Hebei and the Surrounding Areas After the Olympics

Since the 18th National Congress of CPC, China proposed a major national strategy for the coordinated development of Beijing, Tianjin, and Hebei. With joint prevention and control of air pollution as a priority field, and with optimization and upgrading of industrial structure and realization of innovation-driven development as the focus of cooperation, the air pollution prevention and control work in the Beijing–Tianjin–Hebei region and surrounding areas entered a new era, accelerating the process of air quality regionalization management.

To further increase the air pollution prevention and control efforts in Beijing–Tianjin–Hebei and surrounding areas, in accordance with China's Air Pollution Prevention and Control Action Plan and the Detailed Rules for the Implementation of the Air Pollution Prevention and Control Action Plan in Beijing–Tianjin–Hebei and Surrounding Areas (Ministry of Environmental Protection of China 2013), and with the approval of the State Council, Beijing led the establishment of a collaboration team for air pollution prevention and control in Beijing–Tianjin–Hebei and surrounding areas in October 2013, and established a work mechanism for joint prevention and control of regional air pollution, with members including seven provinces/municipalities/autonomous regions (Beijing, Tianjin, Hebei, Shanxi, Inner Mongolia, Shandong, and Henan), and eight central ministries and commissions (the National Development and Reform Commission, the Ministry of Industry and Information Technology, the Ministry of Finance, the Ministry of Environmental Protection, the Ministry of Housing and Urban–Rural Development, the Ministry of Communications, the National Energy Administration, and the China Meteorological Administration). The collaboration team had an office responsible for implementation, communication, and assurance. All members worked together to form a joint effort for pollution control, on the basis of air pollution prevention and control work in their respective administrative areas.

At the same time, a special information platform for regional air quality monitoring and pollution source supervision was established, an expert committee on regional air pollution prevention was set up, the medium- and long-term planning for joint prevention and control of regional air pollution was formulated, a regional air quality forecasting and early warning consultation platform was established, and a real-time video conference was realized. A Beijing–Tianjin–Hebei environmental law enforcement joint work mechanism was established to coordinate members to carry out special law enforcement in the jurisdictions of six provinces/autonomous regions/municipalities and to conduct non-scheduled joint law enforcement. In addition, China's first cross-regional regulatory coordination mechanism for motor vehicle emissions pollution was established.

Since 2013, the prevention and control of air pollution in the Beijing–Tianjin–Hebei region and surrounding areas have been comprehensively promoted, achieving remarkable results. In 2017, the average annual concentration of PM2.5 in 70 cities of the seven provinces/autonomous regions/municipalities consisting of the Beijing–Tianjin–Hebei region and surrounding areas was 55 $\mu g/m^3$, which was 11.5% and 37.2% lower than that in 2016 and 2013, respectively. Among them, the average annual concentration of PM2.5 in Beijing, Tianjin, and Hebei was 64 $\mu g/m^3$, which was 9.9% and 39.4% lower than that in 2016 and 2013, respectively. In addition, the annual average concentrations of PM10, SO_2, and NO_2 in the seven provinces/autonomous regions/municipalities of the region were 25.0%, 51.1%, and 4.0% lower than that in 2013, respectively. The average number of days with heavy air pollution in the 70 cities in the region decreased significantly, and the regional air quality continued to show an overall trend towards improvement (Beijing Municipal Environmental Protection Bureau 2018).

Under the leadership of governments at all levels, and after 5 years of arduous and unremitting efforts of the whole society (Zhu et al. 2018a), the air pollution in the Beijing–Tianjin–Hebei region improved significantly. The following is the situation in Beijing.

8.3 Current Status of Air Quality in Beijing

8.3.1 Haze Days

The average number of haze days in Beijing in 2018 was significantly lower than that in the past 4 years (from 2014 to 2017, the same hereinafter), and the severity of haze was significantly reduced. Taking the representative observatory station in Beijing as an example (as shown in Fig. 8.1), the number of haze days in 2018 was 83 days, which is 38 days less than the 121 days on average in each of the past 4 years, a reduction of 31.4% (Fig. 8.2). In terms of the degree of change, on average, the numbers of severe, moderate, and mild haze days decreased by 18, 23, and 21 days, respectively, while the number of slightly mild days increased by 24 days (Table 8.1).

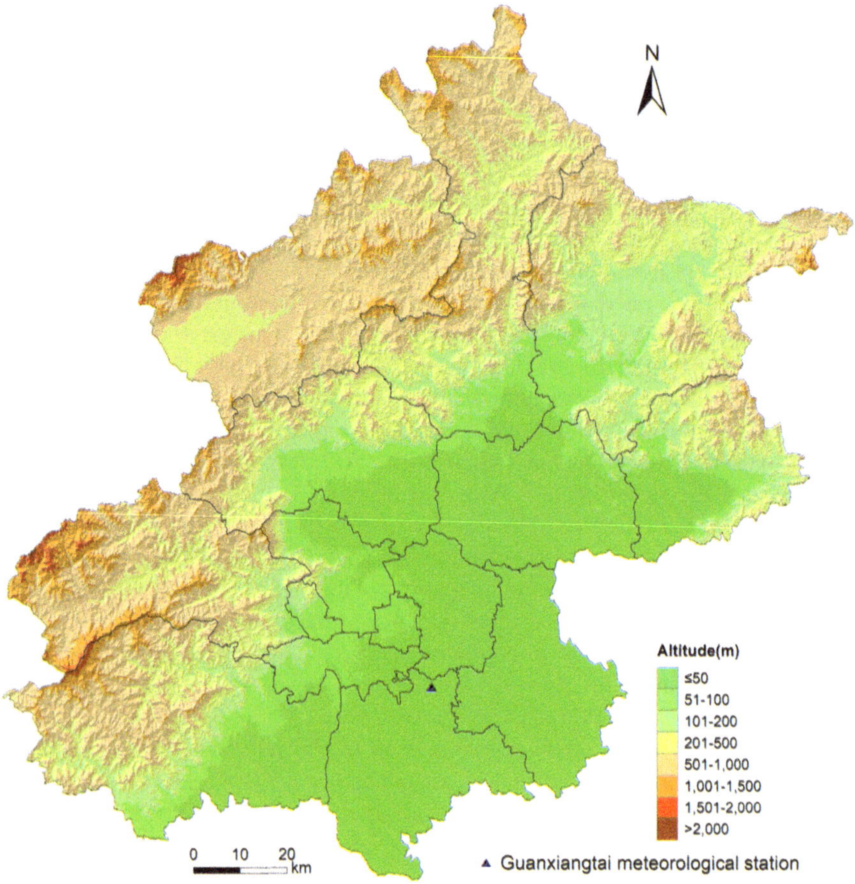

Fig. 8.1 Schematic diagram of the Beijing terrain and the location of the observatory. (Source: drawn by author)

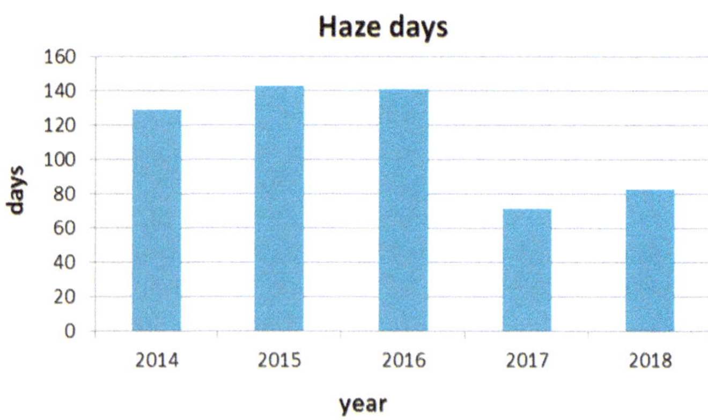

Fig. 8.2 Number of haze days as per the Beijing observatory from 2014 to 2018. (Source: drawn by author)

Table 8.1 Changes in the number of days with different haze levels as per Beijing observatory from 2014 to 2018 (Source: author)

Year	Slight haze (days)	Minor haze (days)	Moderate haze (days)	Severe haze (days)	Total (days)
2014	26	33	40	30	129
2015	38	41	42	22	143
2016	32	45	38	26	141
2017	21	22	16	13	72
2018	53	14	11	5	83

8.3.2 *Particulate Matter*

The mass concentration of particulate matter in Beijing in 2018 was significantly lower than that in the past 4 years. According to the data of the national control station of the Beijing Environmental Protection Monitoring Center (Beijing Municipal Environment Monitoring Center 2019), the annual average mass concentration of PM2.5 in Beijing in 2018 was 51 $\mu g/m^3$, which is 23 $\mu g/m^3$ lower than 74 $\mu g/m^3$ in the past 4 years, demonstrating a 31.1% decrease. The average annual mass concentration of PM10 was 78 $\mu g/m^3$, which is 21.2% lower than 99 $\mu g/m^3$ in the past 4 years. The concentration of particulate matter was significantly reduced, also proving that measures, such as the joint prevention and control of air pollution for the Beijing–Tianjin–Hebei region, are continuing to achieve satisfactory results.

8.3.3 *Ozone*

The number of days with excessive ozone levels in Beijing in 2018 was 58 days, which is 3 days more than the average number of days (55 days) over the past 4 years, suggesting an increase of 5.5%. This is one day more than the 57 days in 2017, suggesting an increase of 3.6%. The ozone concentration of most months in Beijing in 2018 was higher than the average over the past 4 years, whereas the ozone concentration in May and July–October was 6.1% and 10.7% lower, respectively, than that in the same period over the past 4 years. Compared with that in 2017, the ozone concentration in Beijing in May–July, September, and November 2018 was reduced by 8.5%, 25.2%, and 16.2%, respectively, whereas the ozone concentration in the other months of 2018 was higher than that in the same period of 2017. As the concentration of particulate matter continues to decrease, ozone has become the primary pollutant in summer (the data do not reflect the fact that ozone has declined for three consecutive years in Beijing).

8.3.4 *Number of Days Complying with the Air Quality Standard*

Compared with that in 2013, when the Action Plan was not implemented, the number of days with compliant air quality in Beijing in 2018 increased significantly, and the occurrence of heavy pollution decreased significantly. The number of compliant days in 2018 was 227 days, accounting for 62.2% of all days, suggesting an increase of 51 days from that in 2013. Among them, the number of first-class excellent days was 72, which is an increase of 31 days compared with that in 2013. In the period of 2013–2018, the number of days with heavy air pollution in Beijing decreased year by year from 58 days in 2013 to 15 days in 2018, i.e., a decrease of 43 days over 6 years. In 2018, there were 195 consecutive days with no heavy pollution of PM2.5, suggesting a significant decrease in the occurrence of heavy pollution.

8.4 Causes of Air Pollution in Beijing

8.4.1 *Geographical Terrain*

As can be seen geographically (Fig. 8.1), Beijing is surrounded by mountains on three sides, with the Taihang Mountains to the west and southwest, the Yanshan Mountains to the north and northeast, the Bohai Sea to the east, and the North China Plain to the south. The overall terrain is characterized as high in the west and north and flat in the south and east. The arc-shaped mountains formed by the Taihang Mountains and the Yanshan Mountains lead to less air mobility in the warm sector of the mountain front and the easy accumulation of pollutants in front of the arc-shaped mountains. Especially under weak southerly winds, local pollution and the transport of external pollutants are superimposed on each other, which likely causes heavy pollution.

8.4.2 *Local Pollution Sources*

According to the source analysis results released by the Beijing Municipal Environmental Protection Monitoring Center on 14 May 2018 (Han and Zhu 2018), local emissions accounted for two thirds of the main source of annual PM2.5 emissions in Beijing. Among the contributions of local sources, mobile sources, dust sources, industrial sources, living nonpoint sources, and coal sources accounted for 45%, 16%, 12%, 12%, and 3%, respectively, and other sources such as agricultural and natural sources accounted for 12%. The diesel vehicles traveling in Beijing contributed the most to the mobile sources, building construction and road dust were equally important in terms of the dust sources, the VOCs discharged by the petrochemical, automotive, and printing industries contributed prominently to the industrial sources, and the use of everyday solvents accounted for approximately 40% of the living sources.

The latest changes in local pollution sources can be summarized as "two increases, two decreases, and one prominence." The two increases refer to the increase in the rate of the contribution of mobile sources and dust sources. During different time periods and within spatial extents, mobile sources are the largest source of atmospheric PM2.5 in Beijing and accounted for as much as 45% of the local emissions, showing a significant increase from the past. The proportion of dust sources also increased from the past, reaching 16%. The two decreases refer to the decrease in the rate of contribution of coal sources and industrial sources. The one prominence refers to the contribution from living nonpoint sources being more prominent than before.

8.4.3 Regional Transport of Pollutants

According to the source analysis results (Han and Zhu 2018), the regional transport of PM2.5 accounted for one third of the main sources in Beijing and, as the pollution level increased, the regional transport contribution showed a significant upward trend. For example, in 2017, the regional transport contributed approximately 20 μg/m^3 to the average annual PM2.5 concentration of 58 μg/m^3. On an annual average basis, the regional transport contributed 34% ± 8% to the annual PM2.5.

Among the transport channels in the pollutant area, the southern and eastern transport channels contributed significantly. The southern transport channel was located along the Taihang Mountains. Under southerly winds, the average annual rate of contribution of regional transport was between 35 and 53%, whereas under easterly winds, the average annual rate of contribution of regional transport was in the range of 29–49%, both being much higher than the annual average. In particular, under southeastern winds, sulfate was the most prominent component of PM2.5, the coal sources accounted for the largest proportion of the sources, and the transport characteristics were relatively significant.

8.5 Urban Planning Response Based on Environmental Improvement

Based on the above analysis, Beijing has made great progress in air pollution prevention and control. However, problems still exist due to pollution emissions and regional coordinated governance. Regional coordinated development and rational planning/construction are closely related to pollution prevention and control, against which no city can be immune. Therefore, it is necessary to establish the concept of regional sustainable urbanization and rationally plan a high-quality city according to the characteristics and positioning of Beijing.

8.5.1 The Beijing Urban Master Plan (2016–2035) (BUMP)

The Beijing Urban Master Plan (2016–2035) (hereafter referred to as the Beijing Urban Master Plan) was approved on September 13, 2017. The plan clarifies the urban strategic orientation of Beijing as the national political center, cultural center, international exchange center, and science and technology innovation center (Beijing Municipal Commission of Planning and Natural Resources 2018a). It emphasizes that all work in Beijing must adhere to the city's strategic positioning of the "four centers." The most important external environmental condition for the construction of the four centers is excellent air quality. Integrating the improvement of environmental air quality and adaptation to climate change into urban planning and design is an important part in the implementation of strategic positioning.

In the overall planning, through the establishment of a meteorological wind field model, analysis of pollutant source and concentration compliance, analysis of annual as well as winter and summer dominant wind direction, combined with the simulation of the solar wind field environment, as well as on the basis of the topographical terrain, ecological resources and wind field environment of Beijing, the important components of the ecological conservation area in the northwestern part of the Beijing–Tianjin–Hebei coordinated development pattern were determined to include the Mentougou District, Pinggu, Huairou, Miyun, Yanqing, and the mountainous areas of Changping and Fangshan (Beijing Municipal Commission of Planning and Natural Resources 2018b), which are the large "oxygen bars" in Beijing and the key areas that guarantee the sustainable development of the capital, and should be protected and constructed as ecological conservation areas (Fig. 8.3).

8.5.2 The Layout of Urban Functions in BUMP

In terms of the layout of urban functions and control of air pollution sources, different functional positioning as well as construction goals and visions were proposed for the northwestern region, northeastern region, and southern region. At the same time, it was proposed that the city strictly forbid the development of the production and processing links of general manufacturing, resolutely withdraw from the general manufacturing industry, eliminate on-site industries and production processes with heavy pollution and high energy and water consumption, close metal and nonmetal mines, and conduct an orderly exit from high-risk hazardous chemical production and management enterprises. In terms of production methods, the following were proposed: adhere to green development, cyclic development, low-carbon development, and comprehensively implement green production modes such as

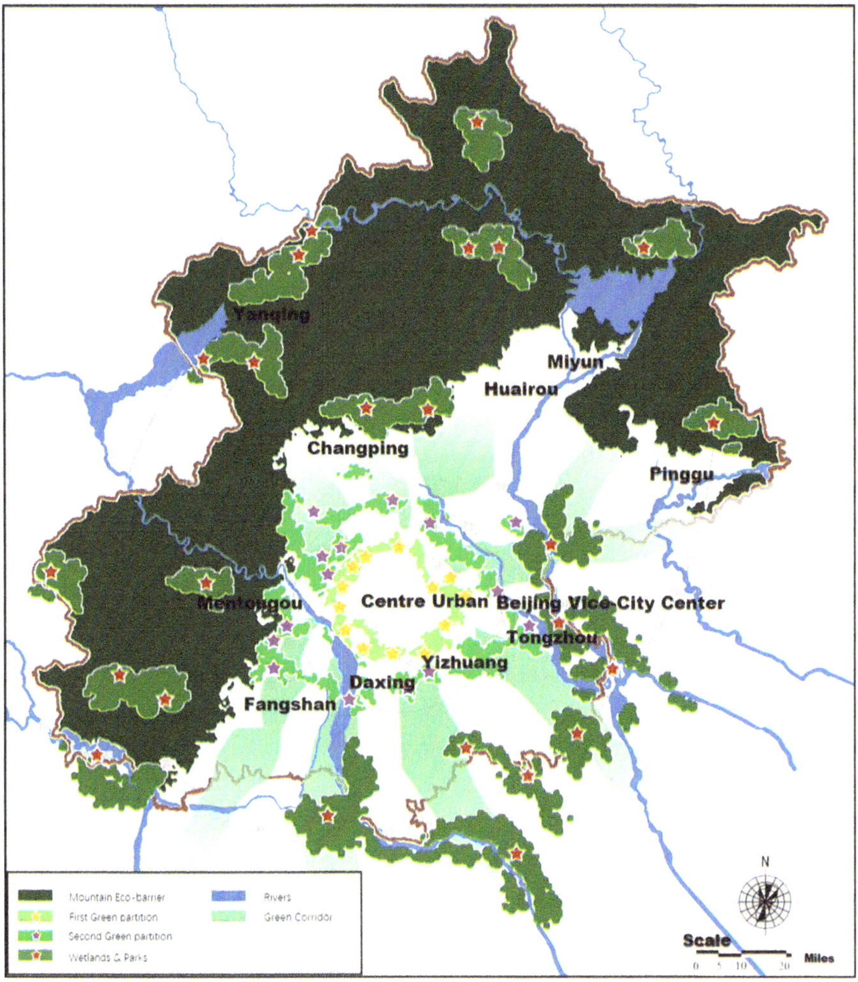

Fig. 8.3 Planning map of green spatial structure within the city territory of Beijing. (Source: drawn by author)

source reduction, process control, vertical extension, lateral coupling, and terminal regeneration. In addition, the following were proposed: promote cleaner production, develop a cyclic economy, and form a resource-saving, environmentally friendly, and economically efficient industrial development model.

In terms of urban functional dispersal and land use structure adjustment, the "leaving open space and increasing green space" policy for land use after dispersal has been strengthened, and a relatively balanced urban cold source with a certain scale has been constructed. While providing more recreational sites as well as improving the livelihood and service level for the people, it also creates the underlying surface conditions for the construction of ventilation corridors and improvement of urban ambient air quality.

8.5.3 The Ventilation Corridor System from Urban Plan to Urban Design in BUMP

At the urban design level, the building height is strictly controlled. The overall control of building height is strengthened, and the height and site layout of super high-rise buildings (above 100 m) are strictly controlled. The height control of buildings along the central axis and its extension line, as well as along Chang'an Avenue and its extension line, are strengthened to form a good urban space order. The control of the height of the buildings around mountains and on both sides of rivers is strengthened to create comfortable public spaces (Beijing Municipal Commission of Planning and Natural Resources 2018b). Scientific and reasonable urban design, as well as the construction of ventilation corridors in urban built-up areas guided by building height, can be carried out to effectively reduce the overall urban wind resistance roughness, lessen the impact of the underlying surface on urban wind fields, and form a better ventilation environment.

The Beijing Urban Master Plan attaches great importance to the creation of "blue-green interwoven" ecological space, which means that the water network and the green vegetation network are intertwined, and the work of ventilation corridor planning to enrich urban ecological products and improve ecological space quality. In the Master Plan, the ecological environmental assessment of natural resources and ecological spaces, such as green spaces, water systems, and wetlands, is carried out, and ecological restoration is conducted for the problem areas. The construction of the greenway system, the ventilation corridor system, and the blue network system means the water network is the focus. The ecological restoration support mechanism is researched and established to continuously improve the quality of ecological space. The construction of a multi-stage ventilation corridor system requires that the central urban ventilation corridor system be built and improved to enhance the overall air circulation of the built-up area. By 2035, five primary ventilation corridors with a width of more than 500 m and multiple secondary ventilation corridors with a width of more than 80 meters will be formed. In the long term, the ventilation corridor network system will be formed with strict control of the construction scale, gradually opening the key nodes that block the connectivity of the corridors (Du et al. 2016; Fang et al. 2018) (Fig. 8.4).

8.5.4 Other Management and Governance of the Control of Air Pollution in BUMP

The Beijing Urban Master Plan also attaches great importance to protecting and restoring the natural ecosystem, optimizing the ecological spatial pattern, strengthening the urban resilience, and mitigating and adapting to climate change. The Master Plan puts forward six goals, strategies, and measures to greatly improve the scale and quality of the ecology based on the management of the marking control

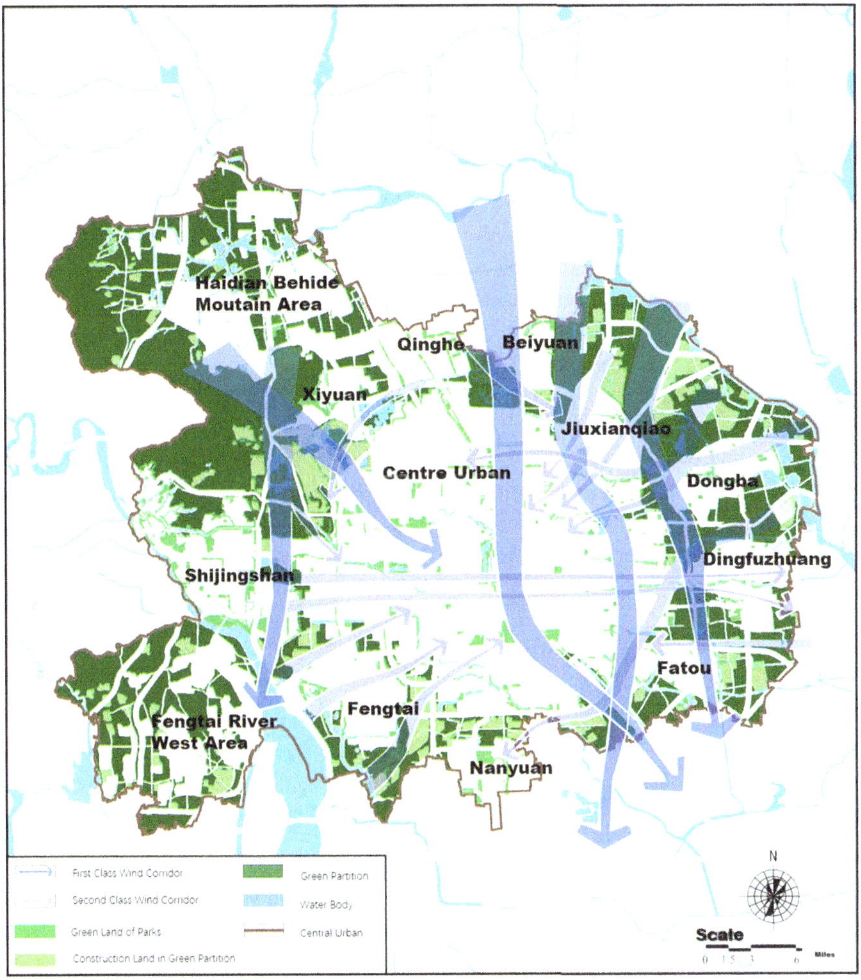

Fig. 8.4 Schematic diagram of the ventilation corridor in the central city area of Beijing. (Source: drawn by author)

and construction requirements. At the same time, it proposes to improve the green space structure, the forest city vision, and the recreational green spatial system of the city's green space system. Four measures for implementing the most stringent water resource management system are proposed, and the goal of building a sponge city guided by ensuring water security, preventing water pollution, and protecting aquatic ecology is proposed.

The Beijing Urban Master Plan adheres to the combination of source emission reduction, process management, and terminal governance in the control of air pollution and overall improvement of environmental quality. Multiple measures with multi-party joint and multi-pronged approaches are used simultaneously to promote industrial transformation and upgrading with an environmental forcing mechanism. In particular, the following rigid target is set: under normal meteorological conditions, the annual average concentration of fine particulate matter (PM2.5) in the atmosphere will drop from the current 80.6 μg/m^3 to approximately 56 μg/m^3 by 2020, and the air environmental quality will be fundamentally improved by 2035, reaching the international advanced level by 2050 (Beijing Municipal Commission of Planning and Natural Resources 2018c). In order to achieve this rigid goal, comprehensive measures are proposed to control the emission of coal-fired pollutants, promote pollution reduction in the transportation sector, reduce the total industrial pollution emissions, and strictly control dust and agricultural non-point source pollution.

In order to realize green development while achieving comprehensive emission reduction planning, measures are proposed to control the total energy consumption and optimize the energy structure. Based on the current total energy consumption of approximately 68.53 million tons of standard coal in Beijing, it is proposed that the total consumption be controlled at 76.5 million tons of standard coal by 2020 and hopefully to approximately 90 million tons of standard coal by 2035 (Beijing Municipal Commission of Planning and Natural Resources 2018c). The following are also proposed: build a diversified high-quality energy system, develop local new energy and renewable energy according to local conditions, actively introduce clean high-quality energy from other cities, provide a more stable and safe energy supply, and strive to build a high-quality energy system primarily based on electricity and natural gas, supplemented by geothermal energy, solar energy, and wind energy. The proportion of high-quality energy will increase from the current 86.3–95% by 2020, and to 99% by 2035. The proportion of new energy and renewable energy in total energy consumption will increase from the current 6.6% to more than 8% by 2020 and to 20% by 2035 (Beijing Municipal Commission of Planning and Natural Resources 2018c). According to the UN's sustainable development goals and China's commitment to climate change, and in accordance with the requirements of the National Development and Reform Commission and the Ministry of Housing and Urban–Rural Development to jointly develop climate-adapted cities, Beijing plans to improve the adaptability of the urban infrastructure and the carbon sink capacity of the urban system, and to enhance the ability of emergency response to extreme weather events, thereby building Beijing into a climate-smart demonstration city.

8.5.5 Regional Coordinated Development due to Regional Air Pollution Diffusion

From the perspective of the Beijing–Tianjin–Hebei region where Beijing is located, the improvement of environmental air quality and adaptation to climate change in Beijing are spatially integrated with the Beijing–Tianjin–Hebei region and cannot be isolated. The industrial layout should be arranged and adjusted according to the

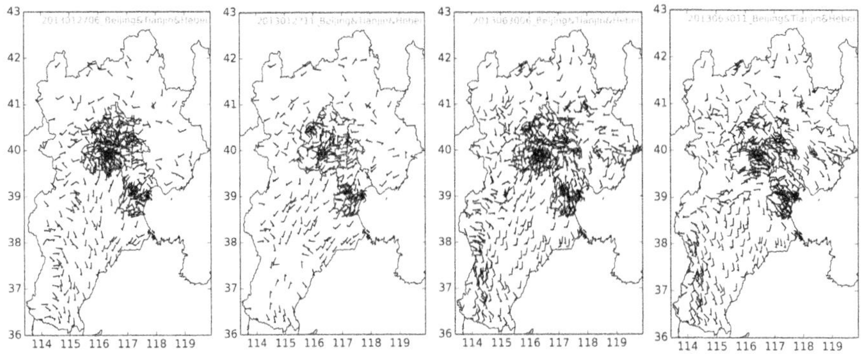

Fig. 8.5 Distribution of the wind direction during the pollution process on January 27, 2013 (left 2 panels), and June 30 (right 2 panels). (Source: drawn by author)

flow field pattern within the Beijing–Tianjin–Hebei region. According to the statistics of meteorological stations of the Beijing–Tianjin–Hebei region in the past 30 years, under the conditions of widespread haze and southerly winds in the Beijing–Tianjin–Hebei region in the fall–winter and in the summer, the Caofeidian area and the adjacent Leting Harbor Development Zone are dominated by sea–land breeze (mostly southwesterly winds), and the Huanghua area is dominated by land–sea breeze (mostly southwesterly winds). The two areas have sea and land diffusion capacity and, when the regional pollution process occurs in the Beijing–Tianjin–Hebei region, their wind field directions are inconsistent with the direction of the regional wind field. The transport of pollutants from the two areas to Beijing is relatively small, and hence the two areas are suitable as the site selected for the industrial agglomeration areas of the Beijing–Tianjin–Hebei region, especially as the preferred direction and region for the shifts of industries from Beijing. The mountain-front cities of Xingtai–Handan–Shijiazhuang–Baoding–Beijing are prone to the accumulation of pollutants due to topographic influences. This zone is also the aorta of railways and highway passages. Under southerly conditions in summer and winter, there exists a transport wind field belt from the southwest to Beijing in the above mountain-front city zone (Fig. 8.5), which therefore should not be retained for heavy chemical industry. If it must be retained, clean production technology innovation or strict pollution control needs to be carried out. The urban agglomeration in this zone should form a cooperation mechanism of joint prevention and control of pollution as soon as possible.

8.5.6 Planning Strategy Based on Regional Collaborative Governance

In view of this situation, the Beijing Urban Master Plan attaches great importance to the coordinated development of the Beijing–Tianjin–Hebei region. At the same time, the Beijing–Tianjin–Hebei Coordinated Development Plan and the Hebei Xiong'an

New District Plan are also making arrangements for coordination with Beijing (Beijing Municipal Commission of Planning and Natural Resources 2018d). The Master Plan proposes that promoting the coordinated development of the Beijing–Tianjin–Hebei region is the only way to achieve sustainable development of the capital. In terms of specific spatial strategies, it proposed that Beijing should actively participate in the construction of the regional ecological security pattern, strengthen the construction of the ecological security barrier of Yanshan–Taihang Mountains and the ecological transition zone of Beijing–Tianjin wetlands, promote regional key greening projects, build the Capital Forest Wetland Park, and build a regional greenway network linked by regional ecological corridors and water systems, thus jointly promoting cross-regional ecological environmental management projects and guiding the rational layout of ventilation corridors for urban agglomeration.

In terms of industrial layout and adjustment, the Master Plan proposes to rely on the main channels, such as Beijing–Tianjin, Beijing–Baoding–Shijiazhuang, and Beijing–Tangshan–Qinhuangdao, to promote the integration of manufacturing factors along the axis and to coordinate the construction of advantageous industrial chains such as automobiles, new energy equipment, intelligent terminals, big data, and biomedicine. Four strategic cooperation function zones, namely, the Caofeidian District, Beijing New Airport Economic Zone, Zhangjiakou/Chengde Ecological Function Zone, and Binhai New Area, are focal points, and enterprises are guided to achieve orderly transfer and accurate docking to implement the major cooperation projects.

The Beijing–Tianjin–Hebei Coordinated Development Plan and the Beijing Urban Master Plan both proposed to carry out joint prevention and control of regional environmental pollution, promote the protection and restoration of ecosystems, and strive to expand regional environmental capacity and ecological space. Focusing on the control of fine particulate matter (PM2.5) in the atmosphere, joint efforts are made to implement measures such as reducing coal combustion, controlling fuel economy, reducing pollution, and eliminating dust. Unified planning, standards, monitoring, and actions are promoted for regional air pollution prevention and control, and the cross-regional heavy pollution weather monitoring and early warning system and emergency coordination joint mechanism are improved.

In order to build a spatial pattern that is conducive to air quality improvement, strengthen the protection and construction of ecological space, and provide a higher supply of good ecological products to the people, the Beijing–Tianjin–Hebei Coordinated Development Plan and the Beijing Urban Master Plan propose to strengthen the construction of water conservation forests and promote the ecological restoration of important rivers, wetlands, and lakes, to promote the comprehensive governance of transboundary rivers, such as Yongding River, Chaobai River, Beiyun River, Juma River, and Juhe River, and to promote the construction of basin flood control and water resource control projects such as Chenjiazhuang Reservoir and Zhangfang Reservoir. These plans also propose to strengthen the monitoring of upstream water quality, clarify the responsibility of water environmental protection objectives, and lead the establishment of a horizontal ecological protection compensation mechanism for upstream and downstream regions of the basin, as well as jointly build a green ecological space, promote the construction of the Capital Forest

Wetland Park, strengthen the protection and restoration of multi-ecological elements, build a regional ecology network, construct the Yongding River—Xiaoqing River Green Wedge, Yongding River Ecological Corridor, South Central Axis Green Wedge, Chaobai River—Tonghui River Green Wedge, Pinggu—Sanhe, and Chaobai River Green Wedge and promote the organic linkage of the ecological green wedge and regional ecological pattern in Beijing.

8.6 Conclusion

Beijing as the capital of China and a megacity has a large population with rapid development. Beijing's air quality pollution is not a one-day problem. It has its own climate characteristics, large emissions and regional transport impact. Faced with air quality pollution, how to build a healthy city is a political issue of great concern to the Chinese government. After 2008, the Beijing Olympic experience transformed China's national policy for air pollution governance.

Compared with 2013, the air pollution in Beijing improved significantly in 2018. The number of haze days in Beijing has set a record of reducing by 31.4% in 4 years, benefiting from the attention and participation of governments, departments and the public at all levels, the planning and development of social linkage, regional coordinated governance, and the cooperation of all walks of life. In 2018, the institutional reform plan of the State Council was promulgated. The newly established Ministry of natural resources is to consider the land and space planning, breaking the problem of too many departments and overlapping jurisdiction. Land spatial planning from region to city to design a map is conducive to optimize the spatial pattern under the climate background. In the future, efforts will be made in urban fine governance to ensure that planning indicators are effectively implemented.

Although the air pollution problem is complex, there is a long way to go. It is believed that under the guidance of the coordinated development of the Beijing–Tianjin–Hebei region, and under the concepts of green development, ecological civilization, and resilience and security, the implementation of the new Beijing Urban Master Plan will certainly lead to excellent results in the improvement of urban environmental air quality and climate-adapted urban planning and construction, thereby providing a good solution to the Chinese model in global urban planning and construction management.

References

Beijing Municipal Commission of Planning and Natural Resources (2018a) Beijing City Master Plan (2016–2035). http://ghgtw.beijing.gov.cn/col/col5096/index.html

Beijing Municipal Commission of Planning and Natural Resources (2018b) Orderly dissolve non-capital functions and optimize the function of the capital. http://ghgtw.beijing.gov.cn/art/2018/1/9/art_5096_544312.html

Beijing Municipal Commission of Planning and Natural Resources (2018c) Improve urban governance and make cities more livable. http://ghgtw.beijing.gov.cn/art/2018/1/9/art_5096_544309.html. Accessed 9 Jan 2018

Beijing Municipal Commission of Planning and Natural Resources (2018d) Deepen the coordinated development of Beijing, Tianjin and Hebei, and build a world-class urban agglomeration with the capital as the core. http://ghgtw.beijing.gov.cn/art/2018/1/9/art_5096_544307.html. Accessed 9 Jan 2018

Beijing Municipal Environment Monitoring Center (2019) http://www.bjmemc.com.cn/index.jsp

Beijing Municipal Environmental Protection Bureau (2018) Results of the "Medium and Long-Term Planning and Research Project for Deepening Air Pollution Control in Beijing, Tianjin and Hebei and Surrounding Areas" Window of the capital. http://www.bjepb.gov.cn/wap/gzdt43/832436/index.html. Accessed 5 May 2018

Beijing Science and Technology Commission (2005) Tender Notice for Air Pollutant Conveying, Transformation and Beijing Air Quality Target Research Project in Beijing and Surrounding Areas. https://www.docin.com/p-543923619.html

Beijing Science and Technology Commission (2009) Announcement of bidding for bidding for the formation mechanism and prevention and control measures of atmospheric composite pollution in Beijing and its surrounding areas. China Government Procurement Network. http://www.mof.gov.cn/xinxi/difangbiaoxun/difangzhongbiaogonggao/200904/t20090415_133182.html

Du W, Fang X, Liu Y, He Y, He J (2016) Construction of ventilation corridors in the Beijing Central Urban Area based on meteorology and GIS technology. Urban Plan Forum 231:79–85

Fang X, Yang R, Du W (2018) Climate and urban planning: an important guarantee for the realization of ecological civilization in the city. Meteorological Press, Beijing

General Office of the State Council of China (2010) The General Office of the State Council forwarded the notice of the Ministry of Environmental Protection and other departments on promoting the joint prevention and control of air pollution and improving the regional air quality guidance. Chinese government portal. http://www.gov.cn/xxgk/pub/govpublic/mrlm/201005/t20100513_56516.html

General Office of the State Council of China (2013) The State Council issued 10 measures for the Action Plan for Air Pollution Prevention and Control. Chinese government portal. http://www.gov.cn/jrzg/2013-09/12/content_2486918.htm. Accessed 12 Oct 2013

Greenpeace (2008) Beijing Olympics Environmental Assessment Report. https://www.docin.com/p-389178589.html

Han J, Zhu H (2018) Beijing released a new round of PM2.5 source analysis results. China Environment News. http://www.cenews.com.cn/news/201805/t20180515_873987.html. Accessed 15 May 2018

Li X (2013) Beijing, Tianjin, Hebei teamed up to strictly control the production capacity of heavily polluting industries. Wen Wei Po, Hong Kong, A14. Accessed 19 Dec 2013

Ma L-M, Zhang X (2014) The spatial effect of China's haze pollution and the impact from economic change and energy structure. China Ind Econ 313(4):19–31

Mao M-J, Du R-G, Hu D-Y (2018) The influence of climatic change on air pollution in Zhejiang province. Res Environ Sci 31(2):221–230

Ministry of Environmental Protection of China (2008) Announcement on Emergency Measures for Air Pollution Control in Extreme Weather during the Olympics. Chinese government portal. http://www.gov.cn/gzdt/2008-07/31/content_1060286.htm. Accessed 31 Jul 2008

Ministry of Environmental Protection of China (2013) The Ministry of Environmental Protection issued the full text of the "Implementation Rules for the Implementation of the Air Pollution Prevention and Control Action Plan in Beijing, Tianjin and Hebei and Surrounding Areas". Polaris Power Network. http://news.bjx.com.cn/html/20130922/460851.shtml. Accessed 22 Oct 2013

Pan J-H, Shan J-J (2017) Blue book of cities in China—annual report on urban development no.10. Social Sciences Academic Press, Beijing

Shi Q-L, Guo F, Chen S (2016) "Political Blue Sky" in Fog and Haze Governance—evidence from the Local Annual "Two Sessions" in China. China Ind Econ 5:40–56
Wang W-G, Zheng G-G (2013) Green book of climate change—annual report on actions to address climate change. Social Sciences Academic Press, Beijing
Wang W, Zheng G (2014) Annual report on actions to address climate change. Social Sciences Academic Press, Beijing
Yang Q, Wang H-R, Liu H-J (2016) Does joint prevention and control of regional air pollution achieve the expected effect?—evidence from the urban agglomeration of Shandong provincial capital. Urban and Environment Studies 4:3–21
Yu S, Tie M (2015) Study on the governance of local government haze pollution under the changing environment—based on the analysis of PM2.5 data of 354 cities in 2001-2010. Jiangsu Social Science 6:86–93
Zhao Y (2013) Global Times 2013 Top Ten International News. Global Times. http://world.huanqiu.com/depth_report/2013-12/4710686.html. Accessed 31 Dec 2013
Zhou M-Z, Chen Y-J, Xu L (2017) China Integrated City Index. People's Publishing House, Beijing
Zhu T, Li X, Zhang S (2018a) Theory and practice of joint control of air pollution across provinces and cities. Peking University Press, Beijing
Zhu R, Zhang C-J, Mei M (2018b) The climate characteristics of atmospheric self-cleaning ability index and its application in China. China Environ Sci 38(10):3601–3610

Chapter 9
Singapore: An Integrated Multi-scale Urban Microclimate Model for Urban Planning in Singapore

Tian Kuay Lim, Nyuk Hien Wong, Marcel Ignatius, Miguel Martin, Hann-Ming Henry Juang, Jing Lou, and Robert Lee Kong Tiong

Abstract The urban heat island (UHI) is one of the most critical issues for dense urban environments, as high outdoor temperatures and poor wind flow in high-density built-up areas can have negative impacts on the thermal comfort and health of city dwellers by trapping air pollutants and increasing energy demand for artificial cooling. The Intergovernmental Panel on Climate Change (IPCC) Assessment Reports recognized the potential impacts of climate change on cities. With global warming, an increased frequency of heat waves and the severity will certainly amplify the current problems associated with UHI effects and heat-related illnesses (HRI) in a highly dense urban environment.

T. K. Lim
National Environment Agency, Singapore, Singapore
e-mail: tiankuay.lim@unimelb.edu.au

N. H. Wong · M. Ignatius (✉)
Department of the Built Environment, National University of Singapore, Singapore, Singapore
e-mail: bdgwnh@nus.edu.sg; bdgmarc@nus.edu.sg

M. Martin
SinBerBEST, Berkeley Education Alliance for Research in Singapore (BEARS), Singapore, Singapore
e-mail: miguel.martin@bears-berkeley.sg

H.-M. H. Juang
Environmental Modeling Center, DOC/NOAA/NWS/NCEP, Maryland, USA

J. Lou
Institute of High Performance Computing, A*STAR, Singapore, Singapore
e-mail: loujing@ihpc.a-star.edu.sg

R. L. K. Tiong
School of Civil and Environmental Engineering, Nanyang Technological University, Singapore, Singapore
e-mail: clktiong@ntu.edu.sg

C. Ren, G. McGregor (eds.), *Urban Climate Science for Planning Healthy Cities*, Biometeorology 5, https://doi.org/10.1007/978-3-030-87598-5_9

This chapter showcases how an Integrated Multi-scale Environmental Urban Model (IMEUM) can be useful for building professionals, city planners, and policymakers in their decision-making related to minimizing the environmental impact due to continuously changing dense urban environment. The concept originates by downscaling environmental models from global scale (25 km) to a site-specific parcel or building. IMEUM provides a computationally efficient method which couples multi-scale atmospheric models with statistical model and computational fluid dynamic (CFD) to estimate weather parameters. IMEUM can be applied to integrate various UHI and heat-related illnesses mitigation measures at the early stage of design process.

This chapter also showcases the IMEUM calibration by means of ground sensing. Furthermore, a series of case studies of a hypothetical office building and the microclimate simulation is provided to illustrate how the simulation output can be fully converted into a localized weather data file for cooling load simulation. These results consequently can be used for urban microclimate on heat-related illnesses impact assessment. IMEUM is part of integrated quantitative urban environment simulation tool (QUEST), a platform for analysing and testing the immediate microclimatic impact of development plans and assessing their long-term impacts under future climate change scenarios.

Keywords Multi-scale global-urban climate modelling · Urban heat island · Thermal comfort · Heat-related illnesses · Cooling load simulation · Microclimate analysis · Climate change

Key Messages

- During evening, urban areas slowly release the heat absorbed during the day. The night temperatures outside may not have cooled down enough for people whose bodies have been exposed to extreme heat all day. This also applies for children, as they are less proficient at regulating internal body temperatures, especially during physical activity, hence increasing the potential for heat-related illnesses, which may be fatal for some during heat waves.
- With global warming, an increased frequency of heat waves and severity will certainly amplify the current problems associated with urban heat island (UHI) effects and heat-related illnesses (HRI) in a highly urban environment with densely aggregated, tall buildings.
- UHI has aggravated the increase of energy demand due to achieving thermal comfort by means of heating or cooling with air-conditioner. Hence, simulation packages for building performance analysis, particularly in energy and comfort, have become a paramount process in the building, or even master plan design.

- With improved understanding of the interactions between the built environment and urban climates, many opportunities exist for those involved in urban planning and development to adopt this knowledge. Appropriate design and mitigation measures can then be incorporated upfront to minimize the risks of developing unfavourable urban climates.
- There are three different scales in terms of UHI models diversification: building, micro, and city scale. The issue pertaining to UHI studies is the analysis segregation at different spatial scales. Quantitative Urban Environment Simulation Tool (QUEST), an integrated multi-scale environmental urban model (IMEUM), couples Global-Regional Spectral Model (GRSM) to Mesoscale-Spectral Model (MSM) and uses Model Output Statistics approach (MOS), a multi-dimensional statistical model, to estimate weather variables (wind, air temperature, humidity, etc.) at precinct level.
- QUEST is an automated, operation-ready and seamless simulation tool for end-users, such as planners and architects, to refine upcoming land use and development plans to incorporate appropriate UHI and heat-related illnesses mitigation measures upfront in the design process.

9.1 Introduction

9.1.1 Urban Heat Island (UHI) in Singapore

Cities are growing towards megacities with higher density urban settings, narrower streets and more high-rise urban structures. Increasing urbanization causes the deterioration of the urban environment, as the size of housing plots decreases, thus increasing densities and crowding out greeneries. Within the built environment at micro-scale, buildings and vegetation influence the incident solar radiation received by urban surface.

Cities tend to record higher temperatures than their non-urbanized surroundings, a phenomenon known as UHI. Earlier studies show strong relation between urban morphology and increasing air temperature within city centres. Urban structures absorb solar heat during the day and release it during the night. Densely built area tends to trap heat which is released from urban structures into the urban environment, increasing urban air temperature compared to surrounding rural areas and causes UHI effect, as illustrated in Fig. 9.1. UHI affects street-level thermal comfort, health, environment quality, and may increase the urban energy demand (Oke 1973, 1987; Takahashi et al. 2004; Chen and Wong 2006; Chong et al. 2012; Jusuf and Wong 2016). UHI has the potential to exacerbate poor air quality due to higher pollutants being obstructed by buildings, roads, and other man-made structures.

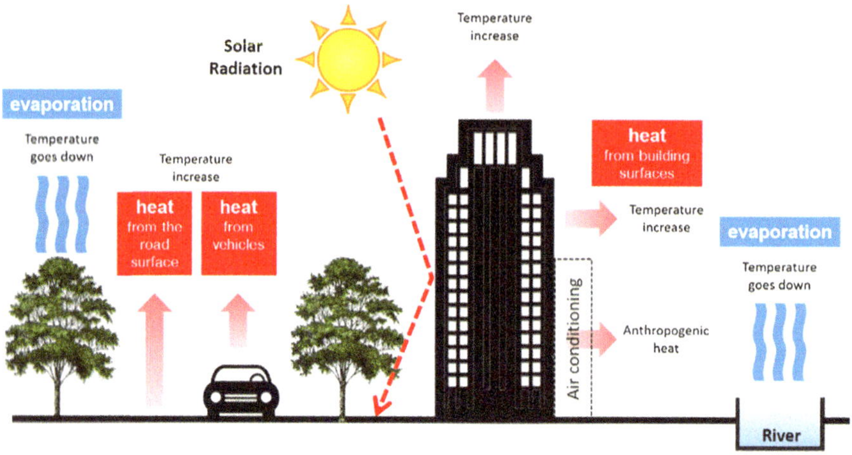

Fig. 9.1 Schematic of urban heat island phenomena (Wong and Ignatius 2016)

Current efforts to achieve desirable cooling through air-conditioning, commonly in Singapore, can paradoxically heat up cities. Air conditioners venting hot air outside can contribute to urban warming, and subsequently UHI effect. The increasing demand on air-conditioning will lead to further increasing greenhouse gasses, and subsequently buildings energy usage (Santamouris 2001; Crutzen 2004; Santamouris et al. 2015). The energy use in the cities affects the urban surface energy budget, which leads to both anthropogenic sensible heat emission and moisture release into the atmosphere (Sailor 2011). According to Santamouris (2014), the typical growth of the cooling demand of representative buildings is nearly 23%, based on 1970–2010 data. Sun and Augenbroe mentioned that the UHI considerably alters the urban climate, and without considering the UHI effect into the energy assessment, this can lead to inaccurate building energy use in hot climate zones where cooling energy use is dominant (Sun and Augenbroe 2014).

Singapore, a high-density tropical city, is similarly vulnerable to UHI negative effects. Based on both 50 years climate and spatial data, the rapid urban development in Singapore is reflected in both air and surface temperature patterns. There has been a twofold increase in the night-time UHI intensity between 1965 and 2004, in parallel with the expansion of the spatial development of new housing and industrial districts (Roth and Chow 2012). Differences in daily mean temperatures are estimated to rise 1.4–4.6 °C by 2070–2099, established on baseline period of 1980–2009. From 1972 to 2014, it has been recorded that Singapore annual average temperature has increased 1.1 °C (MEWR 2016).

Previous study done by Chen and Wong (2006) in conjunction with the Building Construction Authority (BCA), the Housing and Development Board (HDB), the National Parks Board (NParks), the Meteorological Service Singapore (MSS) and the National Environment Agency (NEA) shows that the UHI intensity in Singapore was about 4.5 °C.

9.1.2 Thermal Comfort and Heat-Related Illness (HRI)

In Singapore, increased temperature due to UHI negatively affects the thermal comfort of residents. Apart from straining city energy grids and raising air-conditioning expenses, UHI coupled with global warming can result in serious health problems or heat-related illnesses such as heat stroke, heat exhaustion, cramps and sunburn.

During evening, urban areas slowly release the heat absorbed during the day. The evening temperatures outside may not have cooled down adequately for people whose bodies have been exposed to extreme heat throughout the day. This can lead to higher exposure to heat, which may be fatal during heatwaves. A study by Pascal et al. (2006, 2018) on the 2003 heat wave in Europe which claimed more than 70,000 lives found that high night-time temperatures influenced the health impacts of heat waves in urban areas. Kodera et al. (2019) noted that in recent years, the rates of heat-related morbidity and mortality have begun to increase with the increase in global warming and the number of patients transported by ambulance in heat-related cases in Japan reached 95,137 in 2018. Recent studies also suggest that playgrounds and bounce houses create microclimates with potentially dangerous conditions elevating the risk of heat-related illnesses for children as they are less proficient at regulating internal body temperatures, especially during physical activity, increasing the potential for heat-related illnesses during periods of elevated outdoor temperatures (Labosier et al. 2019).

Thus, UHI magnifies the effects of global warming, which is expected as the number of heat waves increases. This further emphasized the importance of having better urban planning and design to address and mitigate the impacts of heat stress caused by UHI and global warming. For tropical climate, such as island-state Singapore where the humidity is high all year round, the human body feels warmer in such humid conditions as evaporative cooling of perspiration is not as effective as compared to drier climate elsewhere. Hence, cities with tropical climate could potentially face worse heat-related challenges. Better knowledge of the linkage between climate change, variability and extreme weather-related illness is needed in developing the strategies in addressing heat-related vulnerabilities for the planning of healthy cities.

With these climate challenges, alongside population growth and future development, it is paramount that the impacts are assessed early during the design stage. This is to make sure that appropriate design and mitigation measures are incorporated upfront.

9.1.3 Multi-scale UHI Studies

One of the main challenges in many UHI studies is the separation of specific analysis at different spatial scales. To address this, the solution is to link mesoscale and small-scale urban climate, which has been considered difficult due to complexity and non-linearity of urban climate system (Arnfield 2003). As highlighted by Mirzaei (2015), there are three different scales in terms of UHI models diversification: building, micro, and city scale. The idea of developing multi-scale models is to integrate these different types of models to cover the existing gap between them.

An example of this multi-scale analysis is the micro-mesoscale coupling with the building effect parameterization (BEP), building energy model (BEM), and weather research forecasting (WRF) system. Even though, various challenges remain, from initialization of the coupled model, vast number of parameterizations, and model sensitivity to uncertainty in urban canopy parameters. Furthermore, multiple nested domains as many as four were required to downscale the horizontal resolutions of the WRF domains from 24.3 km to 0.9 km (Chen et al. 2011; Li et al. 2013; Chow et al. 2014; Sun and Augenbroe 2014).

9.1.4 UHI, Urban Energy Performance, and Contextual Matter

Many have observed how UHI has exacerbated the increase of energy demand to be thermally comfortable by using heating or cooling equipment. This aspect significantly affects the building performance analysis, both in building scale and master plan. Commonly, it is assumed that weather conditions are uniformly distributed around the reference building. This results in the inaccuracy of determining the external heat gain of building envelope. Furthermore, the impact of shading, greenery, and anthropogenic heat are often neglected. This leads to oversimplification of the outdoor-indoor relationship within the energy building simulation, particularly on the thermal aspect.

Adding to the simplification issue, the available industry standard weather data for building simulation are not well suited to address the changing microclimate condition in cities (Kolokotroni et al. 2006; Jentsch et al. 2008, 2013; Chan 2011a, b; Sun and Augenbroe 2014). To assess a certain building energy performance, one requires to consider surrounding environment of a studied building. This is where the building context matters the most, which requires integration of larger-scale models with urban morphology factors at small scale. Similar studies have also investigated ways to modify the existing weather climate data, either for current study or future prediction (Li et al. 2015; Chakraborty et al. 2016; Zhu et al. 2016). Ignatius et al. (2016) illustrate the significance of using the appropriate weather data by evaluating urbanized temperature profile with the field measurement. Based on a controlled parametric study, it was observed that there was a 1–2 °C disparity for average temperature, and the difference could reach 1.2–3.5 °C over peak afternoon period. Consequently, this temperature difference can lead to an approximately 8% difference in the predicted cooling load, 20% in external conduction gain, and 17% in fresh air intake gain.

These studies highlight the importance of context in conducting building energy analysis. The coupling method previously described tries to overcome the simplification issue in building simulation analysis, by using the proper context, where the results reflect the surrounding microclimate and built environment.

9.1.5 Motivation to Develop Multi-scale Urban Microclimate Tool for Singapore

To worsen the matter, studies indicate that continuing current patterns of development without intervention would further exacerbate temperatures in urban areas (Rankin 1959; Smoyer-Tomic et al. 2003; Harlan et al. 2006). Greater greenhouse effect and global warming issue are extremely important issues to be tackled as the growing population could be further subjected to rising temperatures. Ongoing urban development mostly targets issues such as housing, transport, water and infrastructure, but it lacks comprehensive strategies addressing future climate and its impact on built environment (Coutts et al. 2010). Improving the knowledge of this interaction will benefit those involved and responsible in urban planning, as it provides many opportunities to mitigate the climate issues upfront (Arnfield 2003; Sailor 2011; Santamouris et al. 2015), to minimize the unfavourable risks (Eliasson 2000; Fehrenbach et al. 2001; Coutts et al. 2010).

9.2 Quantitative Urban Environment Simulation Tool (QUEST)

QUEST has been developed to assess the combined impacts of the UHI phenomenon and rising global temperatures due to climate change on thermal comfort and heat-related illnesses, so that planners, architects, and engineers can use it to incorporate appropriate UHI and heat-related illnesses mitigation measures upfront in the urban planning and design process.

QUEST is an automated, operationally ready and seamless simulation tool for end-users, such as planners and architects, to refine upcoming land use and development plans with the objective of maintaining acceptable outdoor temperature, urban ventilation, and consecutively thermal comfort. QUEST enables users to specify the targeted locality/region to generate a set of baselines and future scenarios based on development plans. Then, it automatically triggers a set of simulation workflow to generate the various parameters, namely temperatures, wind, and thermal comfort index for the targeted locality. QUEST comprises components and work packages as illustrated in Fig. 9.2.

The development of the simulator for QUEST is based on the integrated multi-scale environmental urban model (IMEUM) proposed by Lim et al. (2017). The

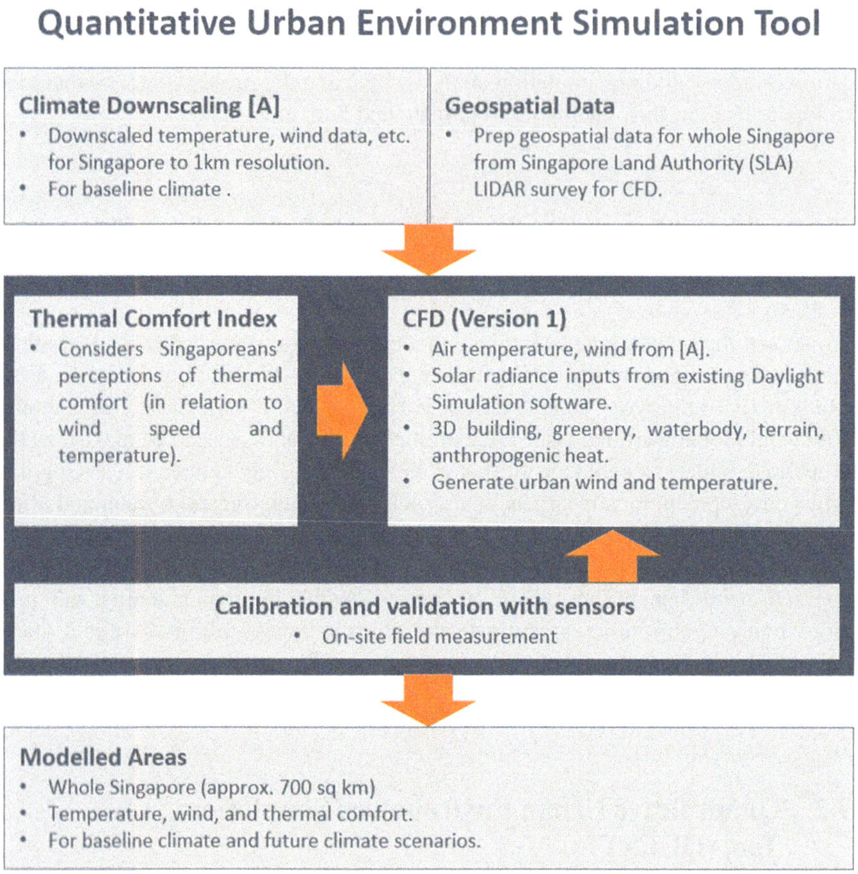

Fig. 9.2 Methodology for different components of QUEST (IHPC 2019)

IMEUM couples Global-Regional Spectral Model (GRSM) to Mesoscale Spectral Model (MSM) and uses Model Output Statistics approach (MOS), a multi-dimensional statistical model, to estimate weather variables (wind, air temperature, and humidity) at precinct level. The IMEUM couples global-regional to mesoscale atmospheric models to support urban planning and design for the management of urban heat islands and global warming, and other matters such as thermal comfort, tree failures risk, and building energy analysis.

9.2.1 IMEUM Development

Studies by Guan (2009, 2011) adopted an effective process to generate approximate future hourly weather data based on current weather condition, imposed offset, or diurnal modelling method. The methodology of scaling down the climatic outputs

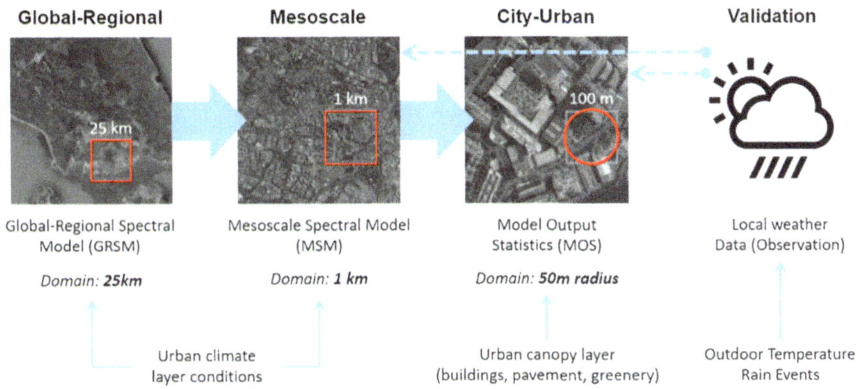

Fig. 9.3 Overall methodology workflow of IMEUM, in which the final output is the microclimate estimation at city-urban scale. This has been calibrated using on-site observation data (Lim et al. 2017)

from global climate model (GCM) to local scale forms the foundation of current study. However, the GCM outputs, typically ranging from 250 to 600 km, are too coarse to be directly used in regional climate change studies. Therefore, downscaling of global to regional climate model has been developed with the aim of a higher resolution intermediate outputs for regional and local scales (see McGregor 1997).

This downscaling process is the key method in developing IMEUM, which elaborates the downscaling method (see Fig. 9.3) from Global-Regional (25 km domain), to Mesoscale (1 km domain), and lastly City-urban scale (100 m domain). For the smallest scale, it adopts the MOS approach using the Direct Model Output (DMO) values from the MSM (dynamical approach), coupling climate with urban morphology variables. The following sections show the IMEUM development and its evaluation process.

The idea of developing IMEUM is for sustainable planning purpose and simultaneously addressing UHI and its impact on building cooling energy demand in Singapore. Largely, UHI effect is evaluated based on the surface air temperature within urban areas and built-up areas relative to the surrounding rural areas. For the surface air temperature, it is correlated to the solar heat retention in buildings, ground surfaces, emission, and re-absorption of night-time outgoing long-wave radiation by surrounding structures.

9.2.1.1 Coupled Global-Regional Spectral Model (GRSM) and Mesoscale Spectral Model (MSM)

Both GRSM and MSM derive from the operational National Centres for Environmental Prediction global spectral model (NCEP GSM) at T170L30/T384L64, NCEP regional spectral model (NCEP RSM) (Juang and Kanamitsu 1994; Juang et al. 1997) and NCEP mesoscale spectral model (NCEP MSM) (Juang

2000). The latter is the non-hydrostatic version of RSM that have been developed for high-resolution limited-area atmospheric model integration for weather forecast, seasonal prediction, and climate simulation (Chang and Hong 2011; Lee et al. 2014; Li et al. 2014; Ham et al. 2016) to support 10 km 48-h forecast over Hawaii islands, and 32 km 63-h forecast over Continental USA for short-range ensemble forecasts. These models have been modified for Singapore context, with a spatial resolution of 1 km. Using idealized tests comprising case studies of bubble convection and mountain waves at 10 m spatial resolution (Chen et al. 2011), the NCEP MSM successfully produced acceptable simulations as compared to the analytical solutions. This indicated that higher MSM resolution of sub 1 km is applicable for Singapore.

The basic equations in sigma vertical coordinates are used for both GRSM and MSM. The uniqueness of the modelling approach is in its handling of global-scale waves in global, regional, and mesoscale domains. While the large-scale motions, such as global wave numbers 1–3, are well represented in the global spectral model, these global wave numbers cannot be handled properly in the regional or mesoscale spectral model. This is due to the limited spatial domain and may lead to an erroneous simulation of the global scale in the regional or mesoscale domain. With this modelling approach, the prognostic variable for MSM is separated into two parts: (1) *Based field*, which is based on large-scale model or analysis and well represented in the global spectral model, and (2) *Perturbation field*, which is computed by dynamic and physic models for the entire model domain.

The perturbation is integrated by spectral computation method. With its high accuracy, this spectral method is capable of fast spin-up and generating high precision mesoscale features. Also, due to the base field inclusion from large-scale model, the large-scale feature for the model results would not drift away. Thus, it is not only suitable for high-resolution weather forecast, but also for long-term climate prediction, climate change impact, and downscaling GCMs simulations to smaller scales. Significantly, the GRSM and MSM coupling applies a single nested domain to downscale the horizontal resolutions from 25 km to 1 km. This method is considered to be more efficient as compared to WRF method (Li et al. 2013), which required four nested domains for downscaling process from 24.3 km to 0.9 km.

The spectral models also use a set of model physics: a package of short and long-wave radiations (with diurnal variation and cloud interaction), a soil modelling physics, land surface physics which include an option for the NCEP–Oregon State University–U.S. Air Force–National Weather Service Office of Hydrologic Development (NOAH) land surface model (LSM) (Chen and Dudhia 2001; Ek et al. 2003), a high-resolution planetary boundary layer physics with a non-local vertical diffusion (Hong and Pan 1996), gravity wave drag physics parameterization, shallow convection, cumulus convection parameterizations, and cloud physics which include cloud water, rain water, ice, snow, and graupel. Each of them can be turned on and off manually, especially convection parameterization and number of cloud constituents in cloud physics (Hong et al. 1998).

9.2.1.2 Evaluation of the IMEUM (Coupled GRSM-MSM)

The horizontal resolution/grid sizes of the Singapore MSM domain are established at 1.0 km (256 × 128 resolution). There are 42 sigma levels or vertical coordinates in the vertical direction, defined by $\overline{p} / \overline{p}_s$, where $\overline{p}$ refers to hydrostatic pressure and $\overline{p}_s$ refers to hydrostatic pressure at the surface. The time integration step was set at 10s. The initial conditions and lateral boundary conditions for the MSM were analysed from the NCEP's 6-h operational Global Forecast System (GFS) at 25 km resolution.

Local weather conditions were obtained from November 2007 to April 2008 period for the purpose of this evaluation of the IMEUM capability on simulating fair or clear weather conditions. These observations include the sensing data from weather radar (250 m resolution) and in situ rain gauges.

The evaluation refers to skill scores recommendations for the verification and inter-comparison of Quantitative Precipitation Forecasts (QPFs) and Probabilistic Quantitative Precipitation forecasts (PQPFs) from operational NWP models, World Weather Research Programme (WWRP), World Meteorological Organization (WWRP/WMO 2008). To evaluate IMEUM capability in simulating during rain events, the highly recommended skill scores, namely: Proportion correct (PC) and Probability of detection (POD), were computed, as shown in the four-cell contingency table (Tables 9.1 and 9.2) for 'Rain' and 'No Rain' events.

Table 9.1 Four-cell contingency table, where a is the number of hits, b is the number of misses, c are the false alarms and d is the number of hits for forecasting no rain correctly (Lim et al. 2017)

	Forecast	
Measured	Rain	No Rain
Rain	a	b
No rain	c	d

Table 9.2 Parameters (based on the contingency Table 9.1) applied to assess the quality of coupled Global-Regional Spectral Model-Mesoscale Spectral Model (GRSM-MSM) in simulating fair weather (no rain) and weather conditions (with rain) (Lim et al. 2017)

Description	Formula	Range	Perfect	Meaning
Percentage correct (PC)	$PC = \frac{a+d}{a+b+c+d} \times 100\%$	[0,100`++]	100%	Insensitive to rarer precipitation events
Probability of detection (POD), Hit Rate (HR) (Donaldson et al. 1975)	$POD = \frac{a}{a+b}$	[0,1]	1	Proportion of rain events, which were correctly predicted

Table 9.3 PC and POD skill scores over the study area (Lim et al. 2017)

	9 h	10 h	11 h	12 h	13 h	14 h	15 h	16 h	17 h	18 h
PC (%)	89	87	84	82	86	84	77	79	84	80
POD	0.90	0.89	0.96	0.96	0.97	0.93	0.85	0.87	0.86	0.73

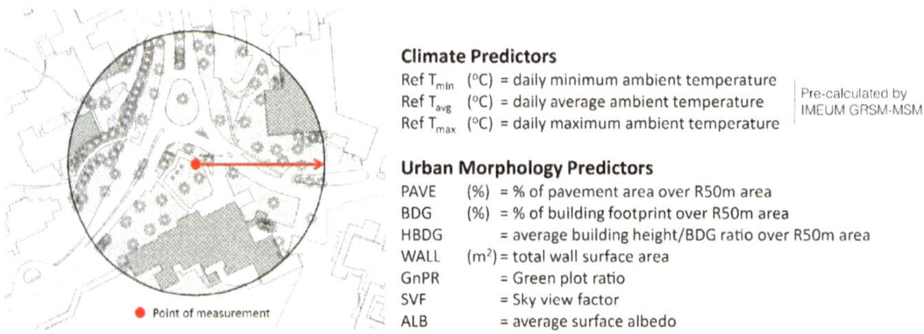

Fig. 9.4 Independent variables used in IMEUM City-urban model (CUM) (Lim et al. 2017; Ignatius et al. 2016)

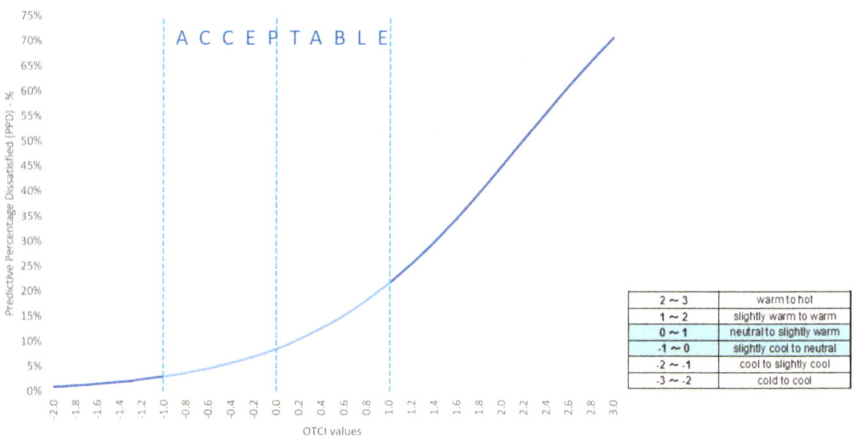

Fig. 9.5 Outdoor TCI-PD chart (IHPC 2019)

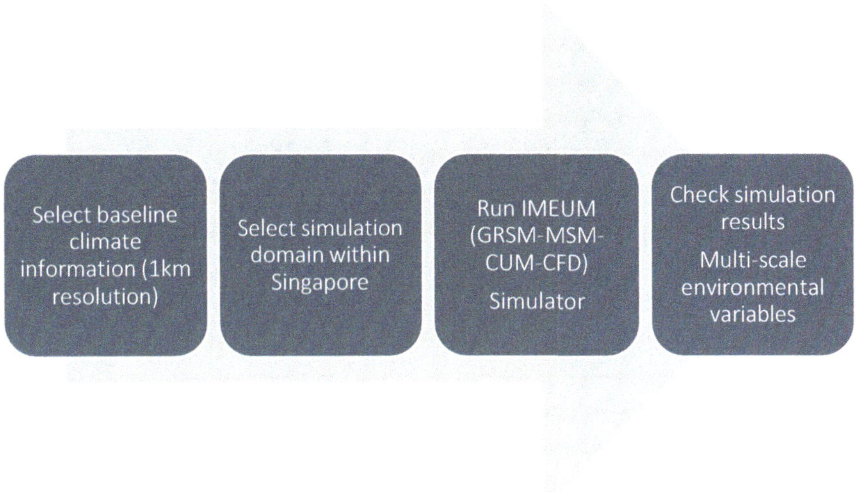

Fig. 9.6 General flowchart for the QUEST version 1.0 user PC package (IHPC 2019)

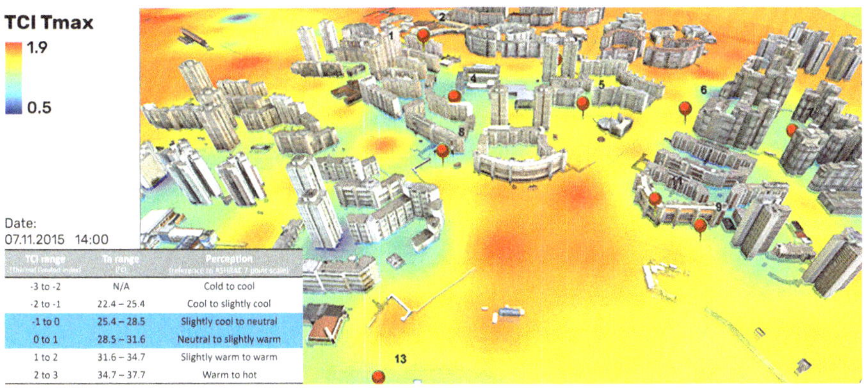

Fig. 9.7 Case study illustration of simulation result (Thermal Comfort Index) (IHPC 2019)

The skill scores for PC and POD over the study area are shown in Table 9.3, which showed that IMEUM can simulate fair or clear weather conditions. The DMO such as surface temperature, wind, and mean sea-level pressure from the were used for the City-Urban Model component as climate predictors to simulate the ambient temperature condition, combined with urban morphological variables (surrounding buildings, pavement, and proposed greenery options).

Fig. 9.8 QUEST workflow baseline climate selection interface (IHPC 2019)

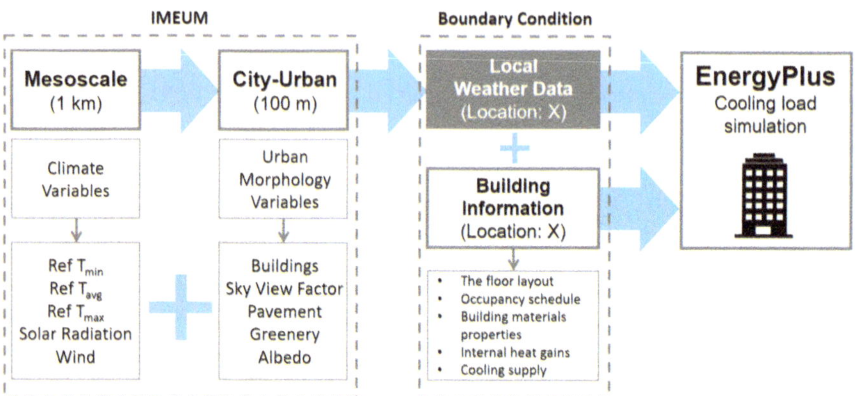

Fig. 9.9 Workflow on using QUEST to generate a local weather data for cooling load simulation (Lim et al. 2017)

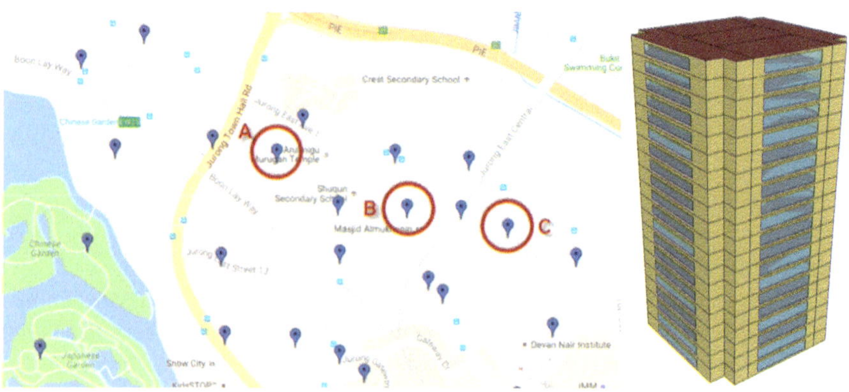

Fig. 9.10 (Left) On-site weather measurement points, where point A, B, and C were selected to be used for case study evaluation. (Right) 20-storey building used to test outputs provided by IMEUM for building energy assessment (Lim et al. 2017)

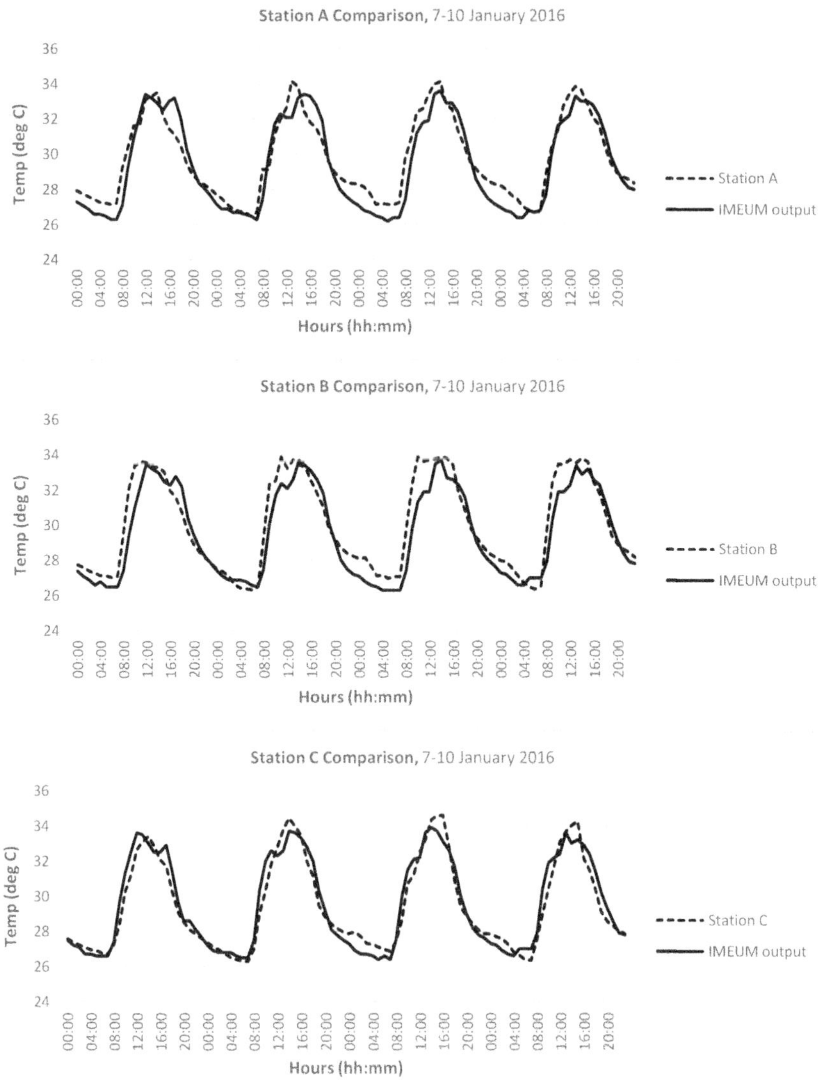

Fig. 9.11 Ambient temperature comparison between sensing data and calculated method from IMEUM across three different locations (Lim et al. 2017)

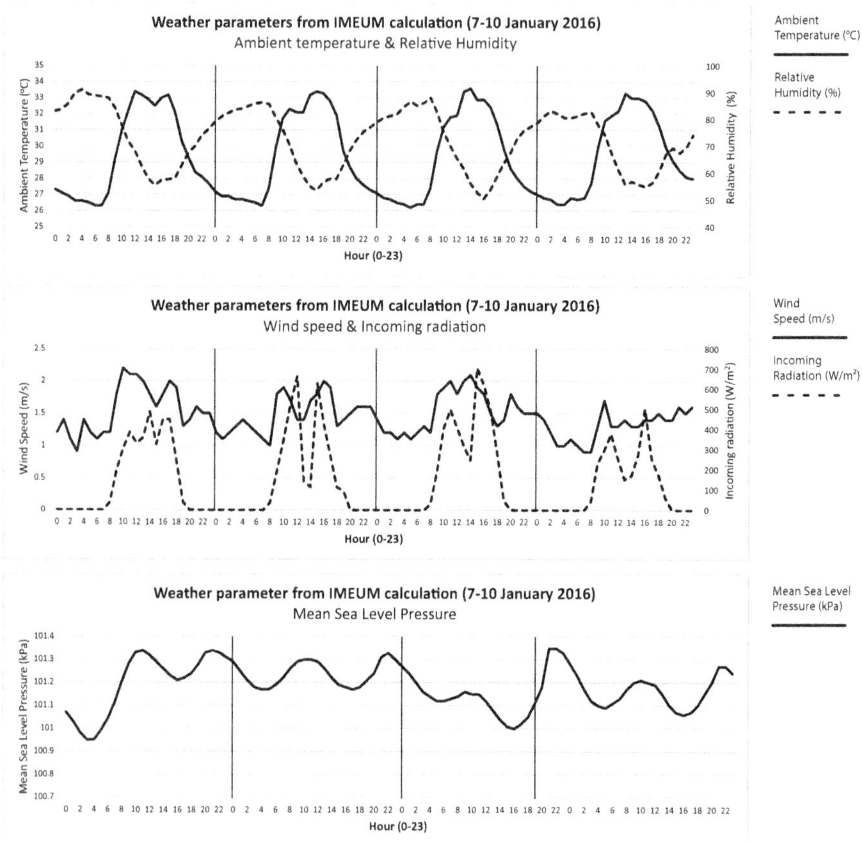

Fig. 9.12 IMEUM final output comprises weather parameters at local scale (Lim et al. 2017)

9.2.1.3 City-Urban Model (CUM), a Dynamical-Statistical Approach

A statistical downscaling approach, which makes use of the simulated high-resolution atmospheric conditions, is incorporated for the city-scale IMEUM. The purpose is to establish statistical relationships between variables simulated by the atmospheric model and local-scale observations to rectify errors. This is mainly due to the shortcomings in some of the physics parameterization in the atmospheric model.

Model Output Statistics or MOS technique, together with screening regression, has been applied to the prediction of surface wind, probability of precipitation, maximum temperature and cloud amount (Glahn and Lowry 1972; Klein and Glahn 1974). For the IMEUM CUM, the MOS technique uses the DMO values from the MSM as climate predictors coupled with urban morphology predictors (buildings, pavement, and greenery conditions) up to 100 m resolution.

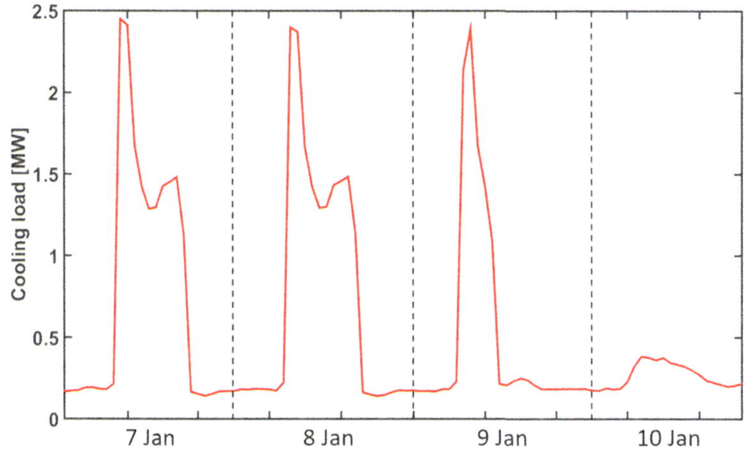

Fig. 9.13 Ideal cooling load as estimated by the EnergyPlus model using weather conditions assessed from IMEUM (Lim et al. 2017)

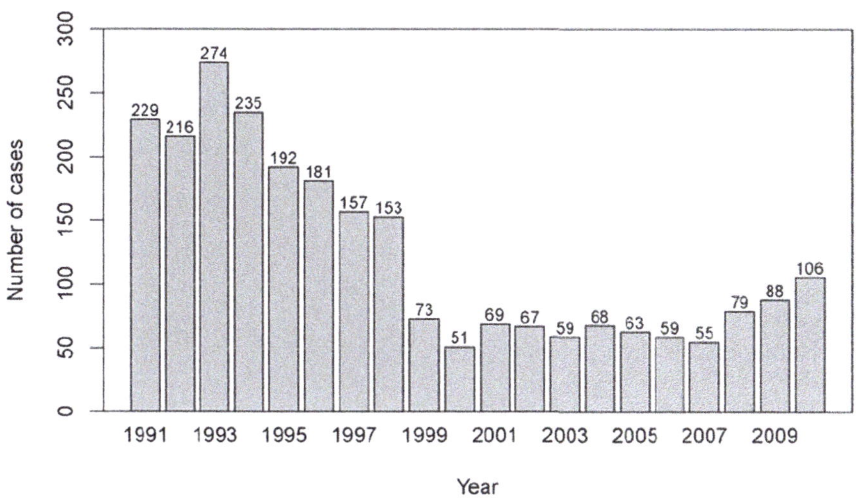

Fig. 9.14 Annual counts of hospital admissions for heat-related illness in Singapore, 1991–2010 (Xu et al. 2017)

The MOS technique refers to the already established empirical prediction models called STEVE tool, previously developed by Jusuf and Wong (Jusuf et al. 2007; Wong and Jusuf 2008; Jusuf and Wong 2009; Wong et al. 2011; Ignatius et al. 2015, 2016). STEVE tool uses environmental variables to calculate the minimum, average and maximum air temperature (T_{min}, T_{avg}, and T_{max}) at point of interest. As shown in Fig. 9.4, the predictors comprise: (1) *climate predictors* (RefT_{min}, *Ref*T_{avg}, and RefT_{max}); and (2) *urban morphology predictors* (surrounding buildings

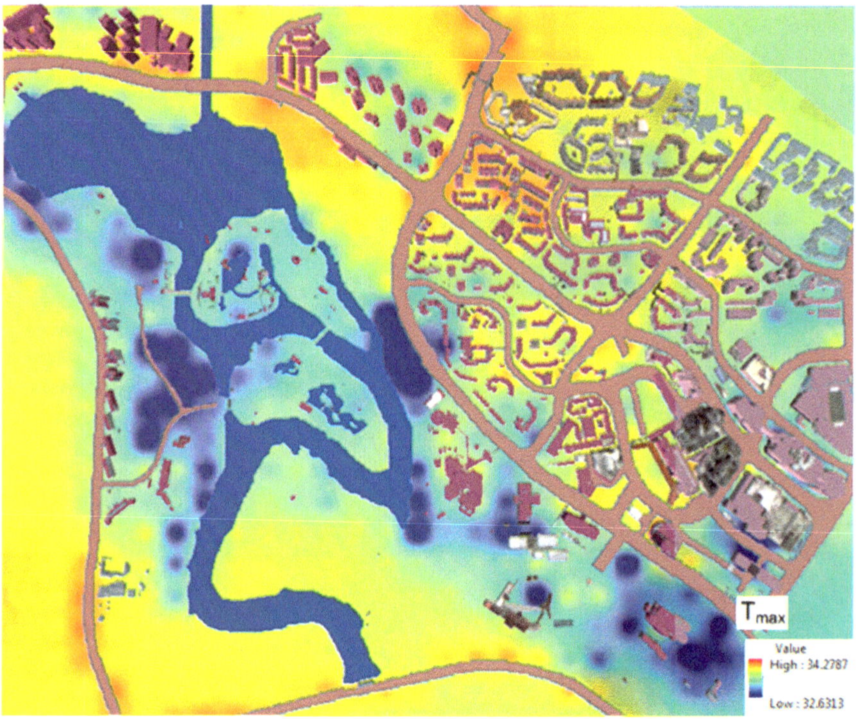

Fig. 9.15 Case study illustration of the microclimate simulation of an urban region (*T*max map) (IHPC 2019)

consideration, surface customization, and proposed greenery options), based on a 50 m radius of influence.

The air temperature prediction models are as follow:

$$T_{\max}\left({}^{\circ}\mathrm{C}\right)=4.494+0.695\,\mathrm{Ref}T_{\max}\left({}^{\circ}\mathrm{C}\right)+0.001\,\mathrm{Solar}_{\max}\left(\mathrm{W/m^2}\right)-0.39\,\mathrm{Wind}_{\max}\left(\mathrm{m/s}\right)+0.008\,\mathrm{Pave}\left(\%\right)-0.002\,\mathrm{Avg\,Height}\left(\mathrm{m}\right)+1.078\,\mathrm{SVF}+32.235\,\mathrm{ALB} \tag{9.1}$$

$$T_{\mathrm{avg}}\left({}^{\circ}\mathrm{C}\right)=2.676+0.919\,\mathrm{Ref}T_{\mathrm{avg}}\left({}^{\circ}\mathrm{C}\right)-0.201\,\mathrm{GnPR}-0.015\,\mathrm{HBDG}+1.52\mathrm{E}-5\,\mathrm{Wall}\left(\mathrm{m}^2\right)+0.24\,\mathrm{SVF} \tag{9.2}$$

$$T_{\min}\left({}^{\circ}\mathrm{C}\right)=2.116+0.932\,\mathrm{Ref}T_{\min}\left({}^{\circ}\mathrm{C}\right)-0.272\,GnPR+0.003\,\mathrm{Pave}\left(\%\right)+1.619\mathrm{E}-5\,\mathrm{Wall}\left(\mathrm{m}^2\right)-0.004\,\mathrm{HBDG}-1.33\,\mathrm{ALB} \tag{9.3}$$

To simulate the ambient temperature conditions ($T_{\min}$, T_{avg}, and $T_{\max}$) at a certain area, the models require reference weather data, which are usually obtained from weather stations. In this case, those reference data (Ref T_{min}, RefT_{avg} and Ref T_{max}) have been pre-calculated by IMEUM.

Consecutively, to develop the output into a functional weather data (e.g. for EnergyPlus simulation), the predicted temperatures need to be converted into hourly diurnal ambient temperature data. An empirical model was applied to create 24-h temperature profile for a typical day, using both reference and predicted values. This method has been used in the previous study to showcase the importance of generating local temperature condition on conducting building energy performance (Ignatius et al. 2016).

If $X_t < X_{avg}$,

$$T_t = T_{min} + (X_t - X_{min}) \frac{(X_{max} - X_{min})}{(X_{avg} - X_{min})2} \frac{(T_{avg} - T_{min})2}{(T_{max} - T_{min})} \tag{9.4}$$

If $X_t > X_{avg}$,

$$T_t = T_{max} - (X_{max} - X_t) \frac{(X_{max} - X_{min})}{(X_{max} - X_{avg})^2} \frac{(T_{max} - T_{avg})^2}{(T_{max} - T_{min})} \tag{9.5}$$

X_t is the outdoor air temperature pre-calculated by the IMEUM at day d at hour t (from 00:00 to 23:00). Subsequently, the temperatures X_{min}, X_{avg}, and X_{max} represent the climate predictors (RefT_{min}, RefT_{avg}, and RefT_{max}).

9.2.1.4 IMEUM-Computational Fluid Dynamics (CFD) modelling

The IMEUM allows for historical global climate information (e.g. wind, temperature, moisture content) to be downscaled to higher resolutions, ranging from 500 m to 50 m, for Singapore. Coupled with CFD, it provides boundary condition to simulate integrated wind and air temperature up to a spatial resolution of 1–10 m for different heights above ground and times of the day. It can conduct simulations for various urban scales, ranging from the island-wide scale down to the building scales. Details of the theory for CFD can be found in studies done by Versteeg and Malalasekera (2007).

The IMEUM-CFD modelling outputs (e.g. wind, air temperature, humidity, mean sea-level pressure) are then used for the development of various environmental applications such as thermal comfort maps for Singapore, mapping of cooling energy demand for buildings, or greenery planning in urban spaces.

9.2.1.5 Human-Environment Modelling: Outdoor Thermal Comfort Index (OTCI)

In addition to multi-variables of weather (outdoor temperature, humidity, wind, and radiant heat), QUEST has also the capability to investigate outdoor thermal comfort index or OTCI for Singapore. This validated index predicts the level of thermal comfort experienced by a typical Singaporean when exposed to different outdoor environmental conditions in Singapore. The index, which looks at multi-variables of weather (outdoor temperature, humidity, wind, and radiant heat) and thermal sensation from the surveyed people, provides an acceptable range of thermal comfort levels for most Singaporeans. This serves as basis for assessing the microclimatic performance of urban developments.

Using multiple linear regression, the OTCI model can be written as:

$$OTCI = 0.257\mathrm{T_a} - ss0.158\mathrm{WS} - 6.853 \tag{9.6}$$

Predictive Percentage Dissatisfied (PPD) equation was developed based on the relationship between thermal sensation vote and thermal acceptability. Equation (9.7) shows the PPD equation as a function of OTCI.

Figure 9.5 demonstrates the relationship between OTCI and percentage of dissatisfaction (PD). When OTCI is below −1.0, the PD is less than 3%. At OTCI of 0, the PD is 8.53%. When OTCI is 1.0, PD is 21.6%. Then the PD increases dramatically to 44.89% at OTCI of 2.0, and it ends at 70.66% when OTCI is 3.0.

ASHRAE Standard 55 specifies that acceptable thermal conditions must be acceptable at minimum 80% for typical applications. PD of 20% could be the reference level for thermal comfort design, modification and improvement, which is when OTCI value closes to 1.0 (ASHRAE 2010).

$$\mathrm{PPD} = 1 - \frac{e^{(-1.084\times\mathrm{OTCI}+2.373)}}{1 + e^{(-1.084\times\mathrm{OTCI}+2.373)}} \tag{9.7}$$

9.3 QUEST Workflow

The approach for the development of QUEST is to leverage on existing work and efforts such as the IMEUM and the coupled MSM-CFD model and daylight simulation (simplified solar irradiance model) to produce heat flux input in the urban landscape as a heat source. The general workflow of QUEST can be seen in Fig. 9.6.

A key feature of QUEST is the incorporation of large-scale weather phenomenon into local-scale environmental analysis. A very high-resolution mesoscale spectral models have been adapted and calibrated for Singapore.

Due to the huge computational resources required to conduct CFD simulation, an operational QUEST, which runs on ArcGIS, enables seamless and automatic workflow in executing the IMEUM for the simulation of temperature, thermal comfort, and wind at city-urban scale for any location in Singapore to optimize the adoption of UHI, heat-related illness and climate change mitigation strategies. Planners also have the flexibility to easily perform in-depth CFD simulation such as incorporating appropriate use of ventilation corridors combined with urban forms in improving urban ventilation (street width between buildings, distances between individual buildings, and heights of buildings) which can alleviate the urban heat island effect and lead to a more acceptable pedestrian-level wind environment and enhance the ventilation potential within inner urbanised areas.

9.3.1 QUEST Operation

QUEST comprises three components.

9.3.1.1 Integrated 3D Geospatial Platform

The priority is to:

1. Transform and integrate the Singapore data (e.g. BIM with 2D and 3D geospatial data, all CityGML format) into the integrated 3D Geospatial Platform.
2. Adapt Singapore datasets (e.g. BIM models, cadastre data, and CityGML models) and integrating the algorithm and Singapore datasets into the platform.

Catalogues and classifications of different urban objects based (buildings, landscape, water, etc.) on Singapore Land Authority (SLA) standard. User can select any historical/baseline climate data for Singapore at 1 km resolution to run the QUEST simulation within Singapore.

9.3.1.2 Integrated Suite of Simulation Modelling: IMEUM (GRSM-MSM-CUM-CFD)

In the QUEST simulation, multiple layers of objects are utilized, including roads, waterbodies, green plot ratio (GnPR), and buildings. The combined effect of the urban objects on thermal and wind conditions in urban context are considered during the simulation. The QUEST is a tool that runs as an add-on built for ArcGIS.

9.3.1.3 Multi-scale, Multi-variables Simulation for Urban Planning and Design

Figure 9.7 shows the simulation result on outdoor thermal comfort. Users can also customize the simulation parameters by editing the climate scenarios per users input (see Fig. 9.8).

The following sub-sections illustrates how QUEST simulation results can be used for further building energy simulation of a hypothetical building, within Singapore context.

9.4 QUEST: Case Studies (Singapore)

9.4.1 Microclimate Simulation for Building Energy Analysis Using QUEST

This section indicates how to use QUEST output in building simulation software, especially in cooling load simulation. Figure 9.9 shows the workflow of this process. A hypothetical office building has been prepared on a real location in the west of Singapore. In addition, this study used field sensing data from multiple measurement points located near the site to evaluate the output. As shown in Fig. 9.10, weather sensors were distributed in an area with a radius of 1.0 km. Points A, B, and C were selected as reference points to be compared with the QUEST output to assess the accuracy of the climate model up to the regional level.

Building simulation using EnergyPlus has been conducted for a 20-storey hypothetical office building near Station A. This building energy model can simulate energy performance based on the outdoor-indoor energy balance. Because EnergyPlus ignores the impact of neighbouring buildings, it lacks integration with large-scale models, especially when it comes to studying the impact of UHI on buildings. On the contrary, IMEUM can provide correct weather data according to the morphological conditions of surrounding environment.

This case study demonstrated the integration of QUEST, combining city scale and micro model, with EnergyPlus as a building energy model. The cooling load simulation considered the influence of the surrounding microclimate conditions, which are affected by the urban morphology and the presence of greenery. The purpose is to illustrate how to use the model output from QUEST (converted to hourly data) as local weather data in EnergyPlus for simulation. The simulation period has been determined to be 4 days from January 7th to 10th, 2016, with calm weather conditions. The floor layout, occupancy schedule, building material properties, internal heat gains and cooling supply of the EnergyPlus model were defined by Martin et al. (2017). The structure of the building developed with EnergyPlus is shown in Fig. 9.10.

9.4.1.1 QUEST Output: Localized Weather Data

The field measurement data was used to evaluate the ability of QUEST to simulate local hourly temperature at a resolution of 100 m for the temperature from January 7 to 10, 2016. At a resolution of 100 m, the local hourly temperature time change fitted well with the temperature observed at stations A, B, and C (see Fig. 9.11). Observed from the figure, the method provided in this article could track the rising, peak, and falling trends of sensor data. Therefore, it could simulate the ambient temperature by considering various weather factors and the influence of the built environment with sufficient accuracy.

The local hourly simulated temperature with a resolution of 100 m from IMEUM was used as the boundary condition of the EnergyPlus model to calculate the cooling load energy consumption of a hypothetical office building located near Station A.

Essential weather variables for 4 days period were compiled to be converted into an EnergyPlus weather data file (.epw format). Variables including (1) outdoor temperature, (2) wind speed, and (3) mean sea-level pressure were extracted from the QUEST output. To complete the data set, data such as relative humidity (RH) and solar radiation that are also required were obtained from the nearest weather station from the Department of Geography at National University of Singapore (NUS 2016), the nearest data available from the study location. The time series chart of the 4-day weather data is shown in Fig. 9.12.

9.4.1.2 Energy Simulation with EnergyPlus

Figure 9.13 shows the estimated ideal cooling load based on the EnergyPlus model for a 20-storey office building in Singapore. As mentioned earlier, this study used weather conditions estimated from IMEUM as input to the model. The ideal cooling load refers to the exact amount of heat (per unit time) that must be removed from the indoor air to keep the temperature and humidity within a given range. In this study, it is also assumed that the HVAC system has unlimited capacity. This result shows that cooling energy consumption is highest in the first 2 days (i.e. Thursday and Friday), lower in the second day (i.e. Saturday), and lowest in the last day (i.e. Sunday). In this study, the last day is the most interesting situation. The internal heat gain was set to zero while assuming cooling is still being provided to determine whether the weather conditions assessed from IMEUM influence the cooling load estimated by EnergyPlus. On Sunday, the cooling load peaks in the middle of the day. Therefore, the cooling load seems to follow the diurnal temperature curve on Sunday.

9.4.2 Climate Change and Heat-Related Illnesses (HRI)

Xu et al. (2017) demonstrated that 1 °C increase in mean temperature was associated with a 43% (95% confidence interval or CI: 31%, 57%) increase in the occurrence rate of HRI; similarly, 1 unit increase in maximum temperature, and relative humidity, was associated with an 11% (95% CI: 5%, 18%) and a 3% (95% CI: 1%, 5%) increase in the occurrence rate of HRI, respectively. In this study, the annual incidence of HRI cases admitted to hospital had decreased from over 200 in the early 1990s to fewer than 100 in the 2000s (see Fig. 9.14).

However, the HRI incidence has been increasing since 2008, with 55, 79, 88, and 106 reported HRI cases admitted to hospital in 2007, 2008, 2009, and 2010, respectively. This reversal in HRI trends requires a thorough understanding of the factors impacting HRI in Singapore. Factors that impact HRI may include weather, social activity, and individual health status. In our study, we focused on modelling weather impact on HRI. Other factors were also considered by categorized parameters like holidays and weekdays.

The high temperature is the key factor leading to the occurrence of HRI. Considering the concern of global climate change, we may foresee the increase of temperature in both tropical and temperature regions. In IPCC Fourth Assessment Report (EPA 2017), it is projected that the mean temperature may increase at least 1.5 °C under "all scenarios except the one representing the most aggressive mitigation of greenhouse gas emissions". With such an increase, the relative risk for the occurrence of HRI increases 72% on average in Singapore with the assumption that the current demographic, social and environmental settings remain the same. As QUEST has been developed to assess the combined impacts of the UHI and climate change on thermal comfort and heat-related illnesses, it can conduct the microclimate simulation of an urban region (Fig. 9.15), in which planners, architects and engineers can use it to incorporate appropriate UHI and heat-related illnesses mitigation measures upfront in the design process.

9.5 Conclusions

With the motivation for sustainable development and to better adaptation to climate change, this study showcases the development of IMEUM with its capability to incorporate multi-scale modelling from global to local scale in an urban environment. The method has considered multi-variables ranging from climatic parameters (such as temperature and wind) to urban morphology factors (such as buildings, pavement, and vegetation).

Comparison analysis of the model output with the in situ sensing data has shown a good agreement, in which the model is able to create the similar temperature profile of a specific urban area within 50 m radius of influence. Furthermore, the later section illustrates how the outcome of QUEST was consolidated into a typical weather data to be used in EnergyPlus. The results showed that cooling load performance of a building can be properly simulated by using the urbanized weather data. Subsequent sub-chapter also shows how the method can assess the impact of the UHI phenomenon on heat-related illnesses.

The novelty of QUEST lies in its efficient computational approach via coupled GRSM-MSM with MOS-EnergyPlus as compared to other computationally intensive approaches (e.g. the coupled WRF with energy models). This is mainly due to the coupling technique which implements dynamic statistical approach and an efficient dynamical nesting approach on scaling down the climate model into the local scale. Furthermore, it also considers the surrounding urban morphology factors at the local/precinct scale. Hence, the method facilitates straightforward simulation to optimize the adoption of UHI and climate change mitigation strategies to support urban planning and design. Planners, architects and engineers can utilize this method to evaluate upfront appropriate mitigation measures and strategies involving parameters such as greenery, building heights, footprint and wall, albedo, etc., in urban planning.

For the next 5–15 years, Singapore has planned new housing estates and other areas to meet population and economic demand by 2030. As the built environment continues to contribute heat generation and trapping, it is vital to assess the microclimatic impacts of these upcoming developments, as well as the long-term effects of climate change, in the planning process. This is to make sure that applicable and appropriate mitigation measures are laid on the table in parallel with the design.

With a better understanding of the linkage between climate and weather variability and heat-related illnesses, QUEST has been developed to function as an automated and seamless simulation tool that is operation-ready for end-users. Planners and designers in relevant both industry and agencies can modify land use and development plans for achieving acceptable thermal comfort level and providing insights related to public health policy, especially in addressing heat-related vulnerabilities for the planning of healthy cities for both the immediate and future.

Acknowledgements The authors wish to thank members of the staff at the Singapore Land Authority (SLA), the National Environment Agency (NEA) and the Urban Redevelopment Authority (URA) for the support of this work. Lim Tian Kuay also wish to thank Mr. Ronnie Tay (former CEO, NEA), Mr. Richard Hoo (URA), Dr. Raj Thampuran (Surbana Jurong) and Dr. Vijay Tallapragada (NOAA's NCEP/EMC) for their support and encouragement. We also wish to thank Dr. Chew Kian Hoe and Dr. Stefan Ma for their comments on the heat-related illnesses.

References

Arnfield AJ (2003) Two decades of urban climate research: a review of turbulence, exchanges of energy and water, and the urban heat island. Int J Climatol 23(1):1–26. https://doi.org/10.1002/joc.859

ASHRAE (2010) ANSI/ASHRAE standard 55–2010, thermal environmental conditions for human occupancy. American Society of Heating, Air-Conditioning and Refrigeration Engineers, Inc., Atlanta

Chakraborty D, Elzarka H, Bhatnagar R (2016) Generation of accurate weather files using a hybrid machine learning methodology for design and analysis of sustainable and resilient buildings. Sustain Cities Soc 24:33–41. https://doi.org/10.1016/j.scs.2016.04.009

Chan ALS (2011a) Developing a modified typical meteorological year weather file for Hong Kong taking into account the urban heat island effect. Build Environ 46(12):2434–2441. https://doi.org/10.1016/j.buildenv.2011.04.038

Chan ALS (2011b) Developing future hourly weather files for studying the impact of climate change on building energy performance in Hong Kong. Energ Buildings 43(10):2860–2868. https://doi.org/10.1016/j.enbuild.2011.07.003

Chang E-C, Hong S-Y (2011) Projected climate change scenario over East Asia by a regional spectral model. JKESS 32(7):770–783

Chen F, Dudhia J (2001) Coupling an advanced land Surface–Hydrology Model with the Penn State–NCAR MM5 modeling system. Part I: Model implementation and sensitivity. Mon Weather Rev 129(4):569–585. https://doi.org/10.1175/1520-0493(2001)129<0569:Caalsh>2.0.Co;2

Chen Y, Wong NH (2006) Thermal benefits of city parks. Energ Buildings 38(2):105–120. https://doi.org/10.1016/j.enbuild.2005.04.003

Chen F, Kusaka H, Bornstein R, Ching J, Grimmond CSB, Grossman-Clarke S, Loridan T, Manning KW, Martilli A, Miao S, Sailor D, Salamanca FP, Taha H, Tewari M, Wang X, Wyszogrodzki AA, Zhang C (2011) The integrated WRF/urban modelling system: development, evaluation, and applications to urban environmental problems. Int J Climatol 31(2):273–288. https://doi.org/10.1002/joc.2158

Chong ZMA, Ignatius M, Wong NH, Jusuf SK (2012) The effect of urban heat island on heat gain increase. Paper presented in the 8th International Conference on Urban Climates (ICUC8), Dublin, Ireland, 6–10 August 2012

Chow WTL, Salamanca F, Georgescu M, Mahalov A, Milne JM, Ruddell BL (2014) A multi-method and multi-scale approach for estimating city-wide anthropogenic heat fluxes. Atmos Environ 99:64–76. https://doi.org/10.1016/j.atmosenv.2014.09.053

Coutts A, Beringer J, Tapper N (2010) Changing urban climate and co2 emissions: implications for the development of policies for sustainable cities. Urban Policy Res 28(1):27–47. https://doi.org/10.1080/08111140903437716

Crutzen PJ (2004) New directions: the growing urban heat and pollution “island” effect—impact on chemistry and climate. Atmos Environ 38(21):3539–3540. https://doi.org/10.1016/j.atmosenv.2004.03.032

Donaldson RJ, Dyer RM, Kraus MJ (1975) An objective evaluator of techniques for predicting severe weather events. In: Norman O (ed) Preprints of the 9th conference on severe local storms. American Meteor Society, p 321–326

Ek MB, Mitchell KE, Lin Y, Rogers E, Grunmann P, Koren V, Gayno G, Tarpley JD (2003) Implementation of Noah land surface model advances in the National Centers for Environmental Prediction operational mesoscale Eta model. J Geophys Res Atmos 108(D22). https://doi.org/10.1029/2002jd003296

Eliasson I (2000) The use of climate knowledge in urban planning. Landscape Urban Plan 48(1):31–44. https://doi.org/10.1016/S0169-2046(00)00034-7

EPA (2017) Future of climate change. https://19january2017snapshot.epa.gov/climate-change-science/future-climate-change_.html. Accessed 30 Aug 2019

Fehrenbach U, Scherer D, Parlow E (2001) Automated classification of planning objectives for the consideration of climate and air quality in urban and regional planning for the example of the region of Basel/Switzerland. Atmos Environ 35(32):5605–5615. https://doi.org/10.1016/S1352-2310(01)00205-9

Glahn HR, Lowry DA (1972) The use of model output statistics (MOS) in objective weather forecasting. J Appl Meteorol 11(8):1203–1211. https://doi.org/10.1175/1520-0450(1972)011<1203:Tuomos>2.0.Co;2

Guan L (2009) Preparation of future weather data to study the impact of climate change on buildings. Build Environ 44(4):793–800. https://doi.org/10.1016/j.buildenv.2008.05.021

Guan L (2011) Sensitivity of building cooling loads to future weather predictions. Archit Sci Rev 54(3):178–191. https://doi.org/10.1080/00038628.2011.590057

Ham S, Lee J-W, Yoshimura K (2016) Assessing future climate changes in the East Asian summer and winter monsoon using regional spectral model. J Meteorol Soc Jpn Ser II 94A:69–87. https://doi.org/10.2151/jmsj.2015-051

Harlan SL, Brazel AJ, Prashad L, Stefanov WL, Larsen L (2006) Neighborhood microclimates and vulnerability to heat stress. Soc Sci Med 63(11):2847–2863. https://doi.org/10.1016/j.socscimed.2006.07.030

Hong S-Y, Pan H-L (1996) Nonlocal boundary layer vertical diffusion in a medium-range forecast model. Mon Weather Rev 124(10):2322–2339. https://doi.org/10.1175/1520-0493(1996)124<2322:Nblvdi>2.0.Co;2

Hong S-Y, Juang H-MH, Zhao Q (1998) Implementation of prognostic cloud scheme for a regional spectral model. Mon Weather Rev 126(10):2621–2639. https://doi.org/10.1175/1520-0493(1998)126<2621:Iopcsf>2.0.Co;2

Ignatius M, Wong NH, Jusuf SK (2015) Urban microclimate analysis with consideration of local ambient temperature, external heat gain, urban ventilation, and outdoor thermal comfort in the tropics. Sustain Cities Soc 19:121–135. https://doi.org/10.1016/j.scs.2015.07.016

Ignatius M, Wong NH, Jusuf SK (2016) The significance of using local predicted temperature for cooling load simulation in the tropics. Energ Buildings 118:57–69. https://doi.org/10.1016/j.enbuild.2016.02.043

IHPC (2019) Development of quantitative urban environment simulation tool (QUEST). Institute of High Performance Computing, Singapore

Jentsch MF, Bahaj AS, James PAB (2008) Climate change future proofing of buildings—generation and assessment of building simulation weather files. Energ Buildings 40(12):2148–2168. https://doi.org/10.1016/j.enbuild.2008.06.005

Jentsch MF, James PAB, Bourikas L, Bahaj AS (2013) Transforming existing weather data for worldwide locations to enable energy and building performance simulation under future climates. Renew Energy 55:514–524. https://doi.org/10.1016/j.renene.2012.12.049

Juang H-MH (2000) The NCEP mesoscale spectral model: a revised version of the non-hydrostatic regional spectral model. Mon Weather Rev 128(7):2329–2362. https://doi.org/10.1175/1520-0493(2000)128<2329:Tnmsma>2.0.Co;2

Juang H-MH, Kanamitsu M (1994) The NMC nested regional spectral model. Mon Weather Rev 122(1):3–26. https://doi.org/10.1175/1520-0493(1994)122<0003:Tnnrsm>2.0.Co;2

Juang H-MH, Hong S-Y, Kanamitsu M (1997) The NCEP regional spectral model: an update. Bull Am Meteorol Soc 78(10):2125–2144. https://doi.org/10.1175/1520-0477(1997)078<2125:Tnrsma>2.0.Co;2

Jusuf SK, Wong NH (2009) Development of empirical models for an estate level air temperature prediction in Singapore. In: Proceedings of the Second International Conference on Countermeasures to Urban Heat Islands, Berkeley

Jusuf SK, Wong NH (2016) Valuing green spaces as a heat mitigation technique. In: Santamouris M, Wong NH (eds) Urban climate mitigation techniques. Routledge, New York, pp 41–66

Jusuf SK, Wong NH, Hagen E, Anggoro R, Hong Y (2007) The influence of land use on the urban heat island in Singapore. Habitat Int 31(2):232–242. https://doi.org/10.1016/j.habitatint.2007.02.006

Klein WH, Glahn HR (1974) Forecasting local weather by means of model output statistics. Bull Am Meteorol Soc 55(10):1217–1227. https://doi.org/10.1175/1520-0477(1974)055<1217:Flwbmo>2.0.Co;2

Kodera S, Nishimura T, Rashed EA, Hasegawa K, Takeuchi I, Egawa R, Hirata A (2019) Estimation of heat-related morbidity from weather data: a computational study in three prefectures of Japan over 2013–2018. Environ Int 130:104907. https://doi.org/10.1016/j.envint.2019.104907

Kolokotroni M, Giannitsaris I, Watkins R (2006) The effect of the London urban heat island on building summer cooling demand and night ventilation strategies. Sol Energy 80(4):383–392. https://doi.org/10.1016/j.solener.2005.03.010

Labosier CF, Beckman J, Robinson T, Tennis D (2019) Preliminary findings of thermal safety in children's outdoor playhouses. Int J Biometeorol 63(9):1303–1307. https://doi.org/10.1007/s00484-019-01732-y

Lee J-W, Ham S, Hong S-Y, Yoshimura K, Joh M (2014) Future changes in surface runoff over Korea projected by a regional climate model under A1B scenario. Adv Meteorol 2014:753790. https://doi.org/10.1155/2014/753790

Li X-X, Koh T-Y, Entekhabi D, Roth M, Panda J, Norford LK (2013) A multi-resolution ensemble study of a tropical urban environment and its interactions with the background regional atmosphere. J Geophys Res Atmos 118(17):9804–9818. https://doi.org/10.1002/jgrd.50795
Li H, Kanamitsu M, Hong S-Y, Yoshimura K, Cayan DR, Misra V, Sun L (2014) Projected climate change scenario over California by a regional ocean–atmosphere coupled model system. Clim Chang 122(4):609–619. https://doi.org/10.1007/s10584-013-1025-8
Li N, Wang K, Cheng J (2015) A research on a following day load simulation method based on weather forecast parameters. Energy Convers Manag 103:691–704. https://doi.org/10.1016/j.enconman.2015.06.073
Lim TK, Ignatius M, Miguel M, Wong NH, Juang H-MH (2017) Multi-scale urban system modeling for sustainable planning and design. Energ Buildings 157:78–91. https://doi.org/10.1016/j.enbuild.2017.02.024
Martin M, Wong NH, Hii DJC, Ignatius M (2017) Comparison between simplified and detailed EnergyPlus models coupled with an urban canopy model. Energ Buildings 157:116–125. https://doi.org/10.1016/j.enbuild.2017.01.078
McGregor JL (1997) Regional climate modelling. Meteorog Atmos Phys 63(1):105–117. https://doi.org/10.1007/BF01025367
MEWR (2016) Using energy responsibly: climate change. http://www.mewr.gov.sg/topic/climate-change
Mirzaei PA (2015) Recent challenges in modeling of urban heat island. Sustain Cities Soc 19:200–206. https://doi.org/10.1016/j.scs.2015.04.001
NUS (2016) Department of Geography Weather Station. https://inetapps.nus.edu.sg/fas/geog. Accessed 10 Aug 2016
Oke TR (1973) City size and the urban heat island. Atmos Environ (1967) 7(8):769–779. https://doi.org/10.1016/0004-6981(73)90140-6
Oke TR (1987) Boundary layer climates. Routledge, London
Pascal M, Laaidi K, Ledrans M, Baffert E, Caserio-Schönemann C, Le Tertre A, Manach J, Medina S, Rudant J, Empereur-Bissonnet P (2006) France's heat health watch warning system. Int J Biometeorol 50(3):144–153. https://doi.org/10.1007/s00484-005-0003-x
Pascal M, Wagner V, Corso M, Laaidi K, Ung A, Beaudeau P (2018) Heat and cold related-mortality in 18 French cities. Environ Int 121:189–198. https://doi.org/10.1016/j.envint.2018.08.049
Rankin D (1959) Mortality associated with heat wave conditions in the Melbourne Metropolitan area, January and February 1959. Aust Meteorol Mag 26:96–98
Roth M, Chow WTL (2012) A historical review and assessment of urban heat island research in Singapore. Singap J Trop Geogr 33(3):381–397. https://doi.org/10.1111/sjtg.12003
Sailor DJ (2011) A review of methods for estimating anthropogenic heat and moisture emissions in the urban environment. Int J Climatol 31(2):189–199. https://doi.org/10.1002/joc.2106
Santamouris M (ed) (2001) Energy and climate in the urban built environment. James & James, London
Santamouris M (2014) On the energy impact of urban heat island and global warming on buildings. Energ Buildings 82:100–113. https://doi.org/10.1016/j.enbuild.2014.07.022
Santamouris M, Cartalis C, Synnefa A, Kolokotsa D (2015) On the impact of urban heat island and global warming on the power demand and electricity consumption of buildings—a review. Energ Buildings 98:119–124. https://doi.org/10.1016/j.enbuild.2014.09.052
Smoyer-Tomic KE, Kuhn R, Hudson A (2003) Heat wave hazards: an overview of heat wave impacts in Canada. Nat Hazards 28(2):465–486. https://doi.org/10.1023/A:1022946528157
Sun Y, Augenbroe G (2014) Urban heat island effect on energy application studies of office buildings. Energ Buildings 77:171–179. https://doi.org/10.1016/j.enbuild.2014.03.055
Takahashi K, Yoshida H, Tanaka Y, Aotake N, Wang F (2004) Measurement of thermal environment in Kyoto city and its prediction by CFD simulation. Energ Buildings 36(8):771–779. https://doi.org/10.1016/j.enbuild.2004.01.033
Versteeg HK, Malalasekera W (2007) An introduction to computational fluid dynamics: the finite volume method, 2nd edn. Pearson Education, Harlow

Wong NH, Ignatius M (2016) Urban heat island in Singapore: contributing factors and mitigation solutions. Innovation Magazine 15(1):33–39

Wong NH, Jusuf SK (2008) GIS-based greenery evaluation on campus master plan. Landscape Urban Plan 84(2):166–182. https://doi.org/10.1016/j.landurbplan.2007.07.005

Wong NH, Jusuf SK, Syafii NI, Chen Y, Hajadi N, Sathyanarayanan H, Manickavasagam YV (2011) Evaluation of the impact of the surrounding urban morphology on building energy consumption. Sol Energy 85(1):57–71. https://doi.org/10.1016/j.solener.2010.11.002

WWRP/WMO (2008) Recommendations for the verification and intercomparison of QPFs and PQPFs from operational NWP models. World Meteorological Organization, Switzerland

Xu H-Y, Fu X, Lim CL, Ma S, Lim TK, Tambyah PA, Habibullah MS, Lee GKK, Ng LC, Goh KT (2017) Weather impact on heat-related illness in a Tropical City State, Singapore. Atmos Clim Sci 8(1):97–110

Zhu M, Pan Y, Huang Z, Xu P (2016) An alternative method to predict future weather data for building energy demand simulation under global climate change. Energ Buildings 113:74–86. https://doi.org/10.1016/j.enbuild.2015.12.020

Chapter 10
Creating Cooler, Healthier and More Liveable Australian Cities Using Irrigated Green Infrastructure

Nigel Tapper

Abstract This chapter focuses on the health impact of heat in Australian cities and the range of strategies that can be implemented for effective urban cooling. Extreme heat is responsible for more fatalities than any other natural hazard in Australia, and projections show that without adaptation, heat-related mortality and morbidity will increase substantially in the future. In Australian cities, clear threshold temperatures exist above which adverse health outcomes escalate dramatically, meaning that even small reductions in temperature can improve health outcomes significantly. Recent research in the Australian Cooperative Research Centre for Water Sensitive Cites (CRCWSC), reviewed in this chapter, has shown the urban cooling effectiveness of irrigated green infrastructure and its potential to dramatically improve human thermal comfort and to save lives. At the microscale (backyard to street) the key role of water and green infrastructure is to reduce air and radiant temperature. Radiation loading to the body is critical for human thermal comfort, so reducing that energy load through shade and reducing surface temperature is critical for improved thermal comfort. At the local scale (neighbourhood to city) the role of water and green infrastructure is to reduce air temperature to improve thermal comfort and heat-health outcomes at that larger scale. The chapter concludes by discussing the current key impediments for implementation of effective urban cooling strategies in Australia, which are a lack of policy co-ordination across the different levels of government and a lack of clarity about the cost-benefit of cooling technologies. Some examples of current best practice for overcoming these issues are provided.

Keywords Heat-health relationships · Urban heat island (UHI) · Urban cooling · Climate change adaptation · Extreme heat events · Irrigated green infrastructure · Water sensitive urban design (WSUD)

N. Tapper (✉)
School of Earth, Atmosphere and Environment, Monash University, Clayton, VIC, Australia
e-mail: nigel.tapper@monash.edu

C. Ren, G. McGregor (eds.), *Urban Climate Science for Planning Healthy Cities*, Biometeorology 5, https://doi.org/10.1007/978-3-030-87598-5_10

Key Messages

- Extreme heat is responsible for more fatalities than any other natural hazard in Australia, with most of these fatalities occurring in urban areas.
- Clear threshold temperatures exist in Australian cities above which adverse health outcomes escalate dramatically, meaning that even small reductions in temperature can improve health outcomes significantly.
- We have shown for Australian cities that different technologies exist for application at different scales that can reduce air temperatures by up to 2.5 °C for improved human thermal comfort and to save lives.
- Key impediments to implementation of urban cooling strategies currently being addressed are a lack of policy co-ordination across all levels of government and lack of clarity about the cost-benefit of cooling technologies.
- The value of urban cooling can be shown for reduced mortality and morbidity, reduced energy costs, and increased work productivity.

10.1 Introduction

There is no doubt that extreme heat is a significant public health concern for Australia and for Australian cities. Extreme heat in Australia has been responsible for more human fatalities since 1900 than any other natural hazard, with 4555 estimated deaths from heat exposure (55% of the total) compared with 1221 from flooding (15% of the total), the next largest cause of death (Coates et al. 2014). Given that Australia with 86% of the population living in urban centres (World Bank 2020) is one of the most urbanised societies on the globe, most of these deaths have occurred in metropolitan areas. Future projections are that without adaptation, heat-related mortality and morbidity in Australia will increase substantially in the future (Bi et al. 2011; Loughnan et al. 2013; Vardoulakis et al. 2014; Guo et al. 2016). This is a critical issue for current and future city governments in Australia because cities are already considerably warmer environments than their rural surrounds (the urban heat island—UHI), and that urban warmth is superimposed upon any background global climate warming (Tapper et al. 2014). The UHI is known to impact health in various ways, by affecting rainfall patterns, worsening air pollution, increasing flood risk and decreasing water quality, but the most significant and direct impact on human health is through urban exposure to increased temperature, especially during extreme heat events (EHE) (Heaviside et al. 2017).

With urban consolidation proceeding across most Australian cities, relative urban warmth is likely to increase into the future unless urgent steps are taken to mitigate urban warming (Coutts et al. 2008). The critical nexus of urbanisation and climate change and the need to urgently develop appropriate adaptation responses has been recognised by United Nations Habitat (2011), in part because the Intergovernmental Panel on Climate Change (IPCC) Fourth Assessment Report in 2007 deemed that some climate change is now inevitable, no matter what happens with future emissions and clearly shifted focus to include adaptation as well as mitigation responses. Cities and climate adaptation remain key focus areas for the IPCC Sixth Assessment Report due in 2021.

This chapter reviews the health impact of heat in Australian cities, the strategies that are available to provide urban cooling to mitigate that impact along with the effectiveness of such strategies, with a particular focus on irrigated green infrastructure. The critical role of government, especially at the municipal level in promoting adoption of irrigated green infrastructure, is outlined. Finally, the economic value of urban heat island mitigation, including improved health outcomes, is discussed.

In Australia, provision of irrigated green infrastructure is usually within the philosophy of *Water Sensitive Urban Design (WSUD)* where stormwater is harvested, purified and conserved for a range of uses including the sustenance of urban vegetation. Similar approaches in North America and Britain were developed as *Low Impact Development (LID)* and *Sustainable Urban Drainage Systems (SuDS)*, respectively, and in China as *Sponge Cities* (Radcliff 2019). In all such approaches, returning the urban environment to a more natural vegetated and permeable landscape (nature-based solution) that promotes a range of benefits including urban cooling is the key objective. WSUD has been at the core of research undertaken by the Cooperative Research Centre for Water Sensitive Cites (CRCWSC) that was established in 2012 as a collaboration among research institutions, government and the water industry.

10.2 Current and Future Heat Vulnerabilities of Australian Cities

Issues of heat have been brought into sharp focus in recent years, with notable Australian extreme heat events in Melbourne in 2009 (47.5 °C in the western suburbs) and 2014, Sydney and Brisbane in 2017 and Adelaide in 2019. Observations show that the frequency, intensity and duration of EHE have increased in Australia and around the world (Perkins et al. 2012), and Rogers et al. (2019) have shown that for some Australian cities the UHI is accentuated during EHE. The trend towards enhanced EHE is projected to continue throughout the twenty-first century (Cowan et al. 2014; Perkins-Kirkpatrick et al. 2016), with increases of ~3.0 °C in the maximum temperatures of the hottest southern Australian heatwaves. Given recent extreme maximum temperatures for Perth, Adelaide, Melbourne and Sydney, there is potential to produce air temperatures of 50 °C or more in those cities.

Post-event analysis of EHE and mortality and morbidity has shown clear thresholds or tipping points across all climate zones of Australia, below which there was little adverse health response, but above which adverse health outcomes dramatically increase. For example, for Melbourne, Nicholls et al. (2008) were able to identify a temperature threshold above which mortality increased for older persons (65+ years). In this instance, an average 24-h (9 am to 9 am) temperature of 30 °C (e.g. a 40 °C maximum day temperature followed by a 20 °C minimum night temperature) or above resulted in a 15–17% increase in the number of deaths among the elderly. Loughnan et al. (2013) extended this approach to all Australian capital cities. That work showed that thresholds based on daily maximum, minimum or mean temperature may not be suitable for all regions, and use of a heat stress index that incorporates humidity, such as Apparent Temperature (AT), is needed in sub-tropical and tropical regions. However, in the context of this chapter it is very important to note that, because of the abrupt threshold effect shown for most metropolitan areas of Australia, *a relatively small reduction in temperature at or close to the threshold can save many lives*. In addition, it is worth noting that there is strong evidence that heat warning systems based on the above approach that provide early warning for citizens and for emergency services *reduced death rates by as much as 50% during a recent EHE in Melbourne* (Nicholls et al. 2016). The life-saving benefits of heat warning systems have also been shown for the city of Adelaide, South Australia (Williams et al. 2012).

Excess urban warmth is variable in both space and time as a result of alterations in the local surface energy balance as a function of urban density and surface characteristics (Coutts et al. 2007) and weekly and daily cycles of human activity (Earl et al. 2016). For this reason heat-health vulnerability is also variable in space and time. Spatial heat vulnerability maps derived from a range of demographic, socio-economic, health and environmental (including UHI) variables have been developed for all Australian capital cities (Loughnan et al. 2013). Figure 10.1 shows the derived spatial pattern of population heat vulnerability for Melbourne. These heat vulnerability maps and other frameworks incorporating heat vulnerability (e.g. Norton et al. 2015) are pivotal in allocating emergency services at times of EHE as well as for identifying parts of municipalities where heat mitigation should be a priority.

Although the health impacts/costs of heat dominate in any assessment of economic impacts of heat (AECOM 2012; Larsen 2015), it should be recognised that EHE also have significant impacts/costs in relation to transport delays, energy consumption, damage to built infrastructure, anti-social behaviour and tree/animal mortality. During the period 1987–2016, Australian natural disasters were responsible for 971 deaths and 4370 injuries and about nine million people were affected in total (Deloitte 2017). More than 50% of those deaths and injuries were associated with heatwaves in cities. AECOM (2012) assessed total heat costs for Melbourne for the 2012–2051 period, discounted to 2012 values, as being $1860 million of which $1603 million were human health costs. An example of partitioning of costs and benefits (health, energy, etc.) of local- (suburb) scale heat mitigation is provided later in this chapter.

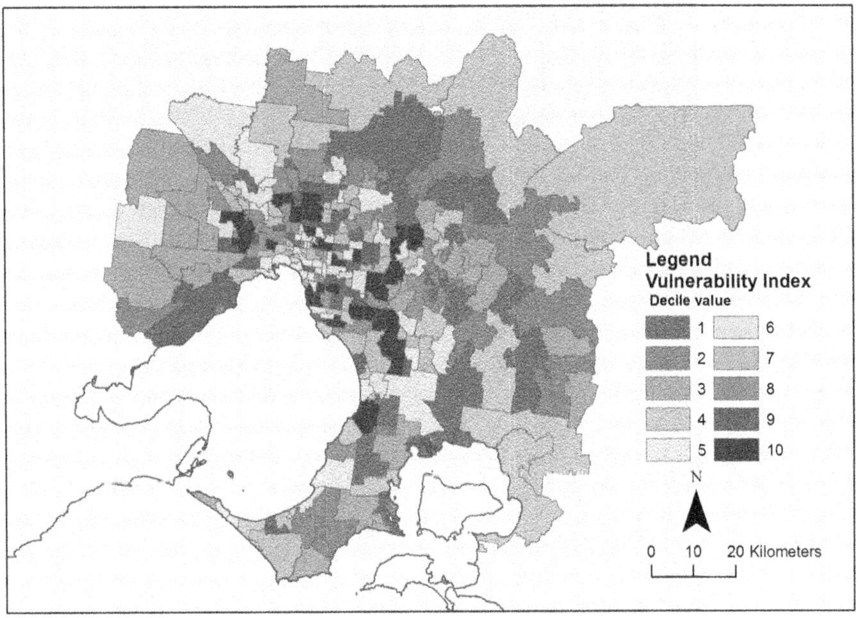

Fig. 10.1 Spatial distribution of the population heat vulnerability index for Melbourne based on statistical relationships between morbidity and a range of health, socioeconomic and environmental variables. From Loughnan et al. (2013)

10.3 Approaches to Heat Mitigation

As noted above, small reductions in temperature associated with EHE can save lives. It is fortunate that the climate processes associated with urban warming are well understood scientifically and can be mitigated. As a result of the nature of urban materials, shapes and activity, cities tend to maximise energy inputs from the sun, terrestrial environment and from artificial sources, and to minimise energy used in evaporating moisture as a result of the dominance of waterproof surfaces in cities. Consequently, there is more energy available for direct heating of the urban atmosphere and for daytime storage in urban materials, producing a much warmer urban environment, especially at night. Micrometeorological work by Coutts et al. (2007) for Melbourne shows that storage in building materials increases by ~250%, evaporation reduces by ~50%, and heating of the atmosphere increases by ~30% on moving from a fully rural to a fully urban environment. As a consequence, an urban heat island (UHI) of ~4.0 °C is typical for Melbourne at night during summer. Additionally, policies in developed countries such as Australia to densify our cities are likely to exacerbate the magnitude and extent of the UHI (Coutts et al. 2008), through promoting the climate processes that lead to the development of urban warmth. Urban consolidation is inevitable and is necessary for good sustainability reasons including reducing the carbon footprint of cities. Offsetting excess warmth of urban areas, including that produced by consolidation, through active promotion of building, urban design and planning approaches that can cool the urban landscape as much as possible is therefore imperative as urban densification proceeds.

Santamouris et al. (2017) has recently provided an assessment of the heat mitigation potential of a range of passive and active cooling technologies including evaporative, urban green, ground cooling and albedo (reflectivity) approaches. Substantial cross-over between urban green and evaporative technologies occurs, as for example in WSUD, where storm water that is otherwise waste is harvested and utilised in treatment wetlands, open water bodies and irrigated green infrastructure to provide cooling through evapotranspiration (Coutts et al. 2013a). The cooling effectiveness of such approaches depends on the budget available, the combination of approaches applied, their suitability for the climate of the city concerned and the co-benefits that might be sought. It should also be noted that the benefits not only accrue from the nature of the infrastructure (whether built or green), but also the distribution of the infrastructure in space, so considerable research is invested in optimising configurations of cooling infrastructure (Thom et al. 2016).

Direct evaporative cooling involves latent heat exchange from liquid water surfaces including lakes, ponds, fountains and misting systems and liquid droplets or films adhering to surfaces. Lakes and ponds can provide additional daytime benefit because their thermal inertia coupled with evaporation provides a relatively cool daytime surface reducing sensible heat loss to the atmosphere. Urban greening that can involve urban forests, urban parks, green walls and other WSUD features contributes to cooling through evapotranspiration and surface shading, among other processes (Coutts et al. 2013a; Santamouris et al. 2017). It is worth noting that green roofs provide only limited cooling benefit at the surface, because the site of active energy exchange is often well above the surface (Coutts et al. 2013b; de Munck et al. 2018). Cool roofing and cool paving involves the use of highly reflective materials to reduce available energy at the surface, that in turn reduces sensible heating of the atmosphere and storage of energy in the substrate, both of which provide local cooling benefits. Finally, ground cooling is an approach that uses the thermal inertia of soil/rock below the surface to cool ambient air via a circulation system and heat exchangers (the same approach can be used in winter to add warmth to buildings). The approach has only been used occasionally for outdoor cooling (Santamouris et al. 2017). Evaluation of 25 urban cooling projects (Santamouris et al. 2017) using a combination of both reflective materials and urban greening was shown to deliver a median maximum temperature reduction of ~2.0 °C. This level of cooling is similar to that shown for Melbourne by Jacobs et al. (2018, 2021) using combinations of greening, irrigation and manipulation of albedo and is in line with average air cooling across an Adelaide suburb from irrigation alone of up to 1.7 °C (Broadbent et al. 2018a). Such cooling is clearly in the range of temperature reductions that can improve human health outcomes (Loughnan et al. 2013).

10.4 Research on Urban Cooling in the CRCWSC

A fundamental objective for the CRC in a water sensitive city context was to identify the key technologies and approaches required to cool Australian cities to adapt to climate change and save lives in extreme heat. This urban climate research was undertaken at a range of spatial scales from individual tree to city scale and using a range of scientific approaches including micrometeorological observations, climate modelling and remote sensing. Figure 10.2 elaborates on the multi-scale approach and the cooling technologies investigated. At the microscale (backyard to street) the key role of water and green infrastructure is to reduce air and radiant temperature at that scale. Radiation loading to the body is critical for human thermal comfort, so reducing that energy load through shade and reducing surface temperatures (e.g. via green walls) is key to improved thermal comfort. At the local scale (neighbourhood to city) the key role of water and green infrastructure is to reduce air temperature to improve thermal comfort and heat-health outcomes at that scale.

Table 10.1 summarises findings from a number of individual pieces of research into the effectiveness of various WSUD features at a range of scales, with a subset of this research discussed in more detail below. Table 10.1 suggests that at the

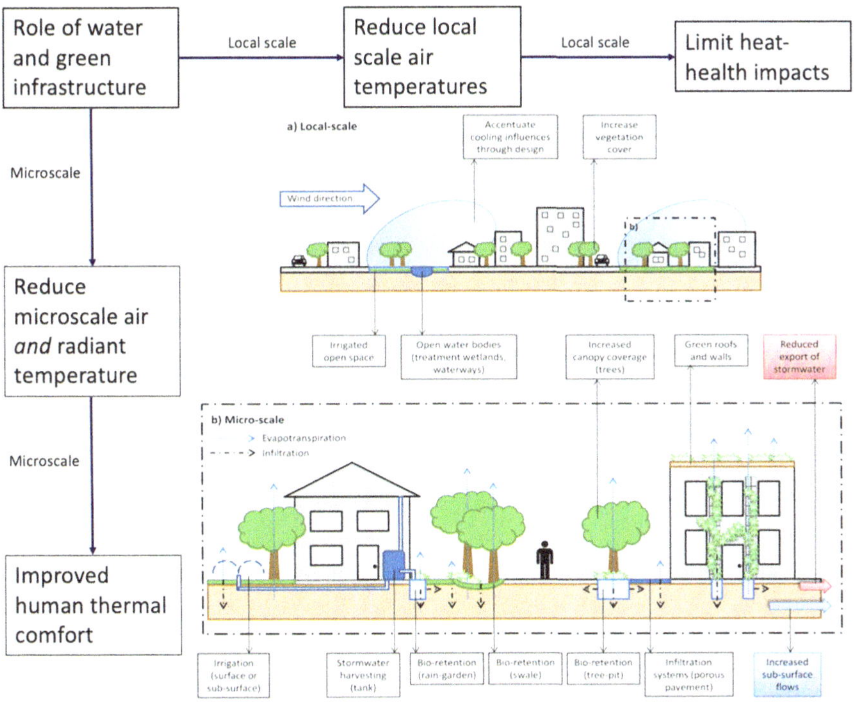

Fig. 10.2 WSUD cooling technologies employed at a range of scales, from the microscale (household to street) to the local scale (neighbourhood to city). Modified from Coutts et al. (2013a)

Table 10.1 Nature of micro- and local-scale urban cooling treatment examined by CRCWSC urban climate researchers and the level of summer cooling achieved in each case

	Nature of treatment	Summer cooling provided
Microscale	**Precinct tree cover** ***(modelling study)*** Thom et al. (2016)	Doubling tree cover (from 20 to 40%) reduces mean radiant temperature by 4 °C in summer and by up to 7 °C during EHE
	Street trees—inner city E-W oriented streets ***(observational study)*** Coutts et al. (2016)	1 °C air temperature cooling in treed v's un-treed streets; UTCI ("feels like") cooling of up to 4 °C
	Individual tree ***(observational study)*** Coutts et al. (2021)	Up to 1.0 °C air temperature cooling within and below canopy during EHE, along with up to 6 °C UTCI temperature reduction
	Green roof ***(observational study)*** Coutts et al. (2013b)	Up to 20 °C surface temperature reduction, but little impact away from the roof
	Rain garden ***(observational study)*** Miao and Tapper (2017)	Up to 1.5 °C air temperature reduction for up to 1x raingarden diameter downwind of raingarden; up to 25 °C reduction in surface temperature
Local scale	**Inner city park—lightly irrigated** ***(observational study)*** Motazedian et al. (2020)	1 °C average air temperature cooling (more in EHE); UTCI cooling of up to 10 °C in shade
	City botanical gardens—irrigated ***(observational study)*** Lam et al. (2018)	Up to 3.5 °C air temperature cooling in irrigated areas during EHE
	Large suburban lake ***(observational study)*** Broadbent et al. (2018a)	~1 °C air cooling above and up to 1x lake diameter downwind of a 1 km^2 lake
	Suburban-scale irrigation ***(modelling study)*** Broadbent et al. (2018b)	~0.5 °C air temperature cooling during EHE with light irrigation and up to 2.5 °C cooling with heavy irrigation; up to 20 °C surface temperature reduction
	City-scale tree canopy, irrigation and albedo increase ***(modelling study)*** Jacobs et al. (2021)	0.6–1 °C air temperature reduction for tree cover increase from 20 to 40%; up to 1.5 °C cooling with combined canopy, irrigation and albedo increase
	City-scale vegetation fraction increase ***(remote sensing study)*** Nury (2016)	6 °C cooling in summertime average surface temperatures with an increased vegetation fraction from 0.2 to 0.8

microscale, we can achieve air temperature reductions of up to 2 °C during EHE, and human thermal comfort improvements (determined using Universal Thermal Climate Index (UTCI; Matzarakis et al. 2010) in the order of 7–10 °C of cooling, using different WSUD approaches. Similar reductions in air temperature are achievable at the local scale. These levels of heat reduction during EHE are extremely important because they offer real opportunities to reduce heat morbidity and mortality in Australian cities.

For example, at the microscale (Fig. 10.2) isolated trees are a common urban landscape feature yet are unique compared to massed trees in a forest. They are an important feature in the urban environment where often only one large tree may be able to be accommodated in a particular location, for example, in a suburban front or back yard. Understanding how isolated trees influence the microclimate is important for designing attractive, sustainable, and thermally comfortable cities, yet observations are uncommon, mainly because of the intensive instrumentation required. Coutts et al. (2021) explored the radiation budget of an isolated evergreen urban tree (Queensland Box—*Lophostemon confertus*) and its influence on the microscale climatic environment using a north-south-oriented, cross-sectional array of 35 air temperature sensors mounted on seven masts (Fig. 10.3). Microscale temperature variations within and adjacent to the tree (Figs. 10.3 and 10.4) show air temperature below the tree to be up to 1.0 °C cooler than immediately upwind of the tree during the warmest time of the day. Neutral or weak lapse conditions characterise the tree canopy by day. Radiative differences below and beside the tree primarily drive air temperature differences. UTCI temperatures (not shown) were up to 6.0 °C cooler within the canopy, reflecting the unique ability of an individual tree to provide important cooling.

At the neighbourhood-to-suburban (local) scale landscape irrigation can provide impressive amounts of cooling during EHE, especially with larger applications of water. Using the SURFEX climate model Broadbent et al. (2018a) showed that for Mawson Lakes, an outer suburb of Adelaide, South Australia, landscape cooling averaging up to 2.5 °C across a land area of ~6 km^2 was possible using continuous (day and night) applications of up to 30 L m^2 day^{-1} (Fig. 10.5), with the local temperature reduction varying substantially with perviousness of the substrate. Much greater cooling is clearly possible when water is able to infiltrate the soil. The amount of water required to produce cooling of 2.5 °C is clearly not sustainable

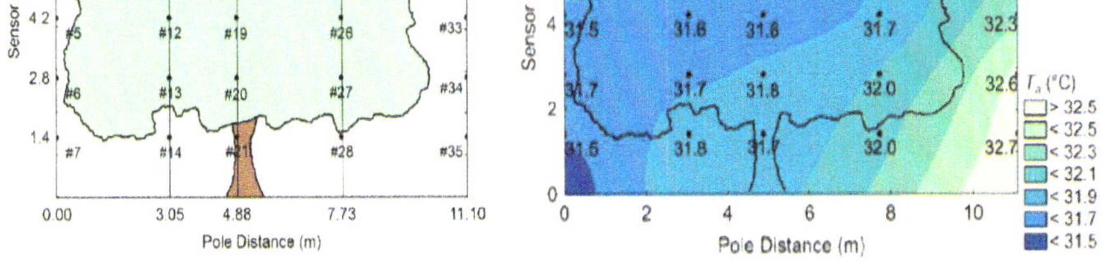

Fig. 10.3 Mast and temperature sensor placement on the isolated tree (left) and the thermal pattern (10-min average) within the tree canopy and surrounding environment (right) at 16:00 on a warm, sunny afternoon (9 March 2014) with moderate northerly winds. From Coutts et al. (2021)

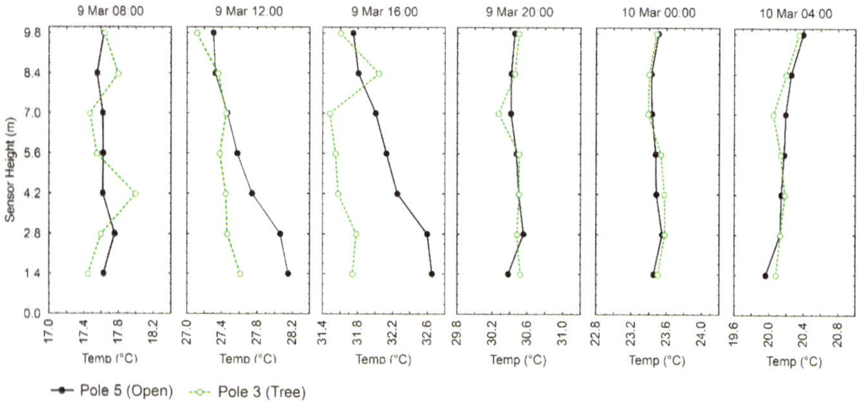

Fig. 10.4 Air temperature profiles (10-min average) from 04:00 9 March 2014 to 04:00 10 March 2014 profiles for Pole 5 (see Fig. 10.3) located on the sunlit side of the tree and Pole 3 located in the centre of the tree. From Coutts et al. (2021)

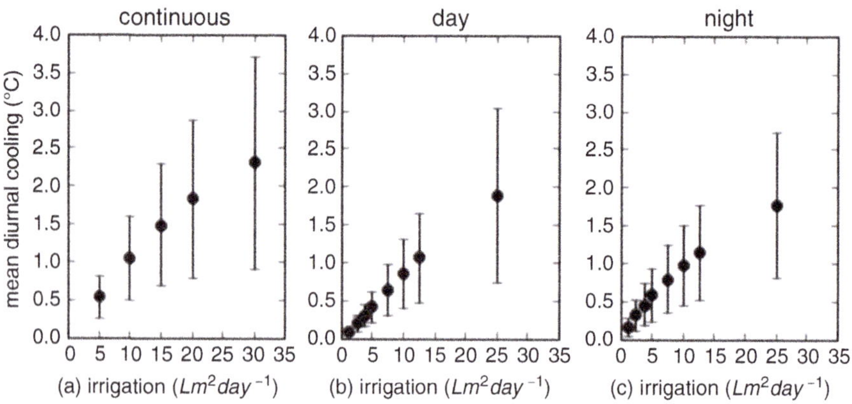

Fig. 10.5 Average diurnal cooling (with standard deviations) with different timings and quantities of water applications across the suburb of Mawson Lakes, Adelaide. From Broadbent et al. (2018a)

over long periods because of cost and availability, but such applications could be an appropriate strategy to save lives during isolated EHEs.

Further work at the city scale (larger local scale) clearly shows the synergistic benefits of multiple cooling approaches working together. Jacobs et al. (2021) used the Weather Research Forecasting Model (WRF) with a detailed urban land surface scheme to show the relative benefits of increased tree canopy cover (20 → 40%), increased rooftop albedo (0.2 → 0.7) and nocturnal irrigation for citywide cooling of Melbourne during EHE. Individual and combined simulations were undertaken for the current climate (current urban structure), mid-twenty-first century climate (current urban structure), and mid-twenty-first century climate (future urban structure). Similar absolute amounts of cooling occur for each of the three scenarios,

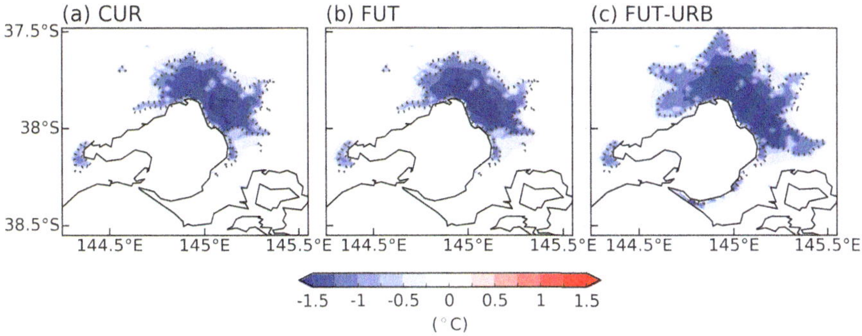

Fig. 10.6 Cooling effectiveness of increased rooftop albedo (0.2 → 0.7), increased tree canopy (20 → 40%) and nocturnal irrigation for the city of Melbourne during EHE. The graph shows combined simulations for the current climate city (current urban structure), mid-twenty-first century climate city (current urban structure), and mid-twenty-first century climate city (future urban structure). From Jacobs et al. (2021)

with cool roofs alone providing up to 0.9 °C of cooling, vegetation up to 0.6 °C of cooling and nocturnal irrigation up to 0.3 °C of cooling. Applying each of the approaches simultaneously (Fig. 10.6) produces up to 1.5 °C of cooling, confirming that a range of approaches applied together is likely to be the most effective approach.

To be effective, each of the approaches described above and in Table 10.1 requires significant amounts of available water in summer. This is an issue for the southern Australian cities with dry summer Mediterranean-like climates, but harvesting, storage and targeted use of an otherwise unused winter storm water resource, as advocated in WSUD principles, can provide an appropriate and sufficient supply. Overall it is clear that the amount of cooling delivered by WSUD approaches in southern Australian cities (i.e. up to 2.5 °C depending on the approach) is commensurate with reported cooling delivered by a range of technologies internationally (Santamouris et al. 2017).

In the following section the role of government at all levels in implementing WSUD/irrigated green infrastructure as part of urban climate adaptation is discussed. One of the most significant barriers for implementation is the perceived cost, or the lack of clarity/information about the cost-benefit of WSUD-related development. For that reason, the CRCWSC commissioned a research project dedicated to a full cost-benefit analysis of WSUD. A key part of that work was economic valuation of urban heat mitigation. Using both biophysical and climate modelling Tapper et al. (2019) assessed the degree of urban heat mitigation generated from different scales of investment in urban greening for a new suburban development. The study also identified the dollar value of the benefits of the urban cooling effect produced (Whiteoak and Saigar 2019).

The case study examined a green-field development on Melbourne's outskirts, comprising 33,000 residential dwellings, road, commercial and industrial space, and both non-irrigated and irrigated public open space. The study focused on the thermal benefits to the core residential area of three policy settings for implementation of WSUD over a 50-year period: the current level of WSUD, and two levels of enhanced WSUD (involving more green infrastructure and water use). The analysis also considered the changing climate over the period. A new urban microclimate model designed specifically for evaluating the cooling effectiveness of green infrastructure in cities (TARGET–Broadbent et al. 2018b) was used to examine the cooling benefit of the different scenarios for a cool, a mild, and a hot summer (Tapper et al. 2019). Maximum cooling was up to 2.0 °C in places at certain times, but averaged across the landscape and across all times was much less and in the order of 0.1 °C–0.5 °C. This is largely because, even at maximum implementation, only a fraction of the landscape can be utilised for WSUD because of roads, commercial/industrial space and un-irrigated public space.

In the partner economic analysis Whiteoak and Saigar (2019) calculated the value of heat reduction on mortality, morbidity, work productivity, electricity costs, as well as an assessment of willingness to pay. They drew on economic and scientific literature to identify a range of costs linking heat and economic outcomes. The economic framework used a discounted cash flow method and 50-year time period to produce benefit estimates that reflected the different scenarios. An "avoided cost" method was largely used to estimate these benefits. At the household level, the maximum greening scenario showed a heat mitigation value of over $1500 per household in present value terms over the period. This result averaged around $80 per household per year (~$2.6 m across the development), with mortality reduction and lower electricity costs being the dominant drivers (~40% each), followed by increased productivity.

10.5 Federal, State and Local Government Policy/Role in Urban Heat Mitigation

Urban livability is defined by a combination of factors related to human health and well-being, economic prosperity, social cohesion and the quality of the urban fabric (Bush et al. 2015). Concepts of livability, including addressing issues of urban heat and green space, underpin urban policy for Australian cities at all levels of government, but policy development is relatively immature, uneven, often disconnected across jurisdictions and levels of government, and until recently has suffered from a lack of evidence-based research.

Relevant policy at the federal level provides some overarching guidance, but much of the responsibility for detailed policy development (e.g. for urban design and planning, health and emergency management, climate change adaptation) is passed to state and local government where most activity is underway at the local/community level. Australian government policies aimed directly or indirectly at addressing urban heat mitigation are listed in Table 10.2. Two key Federal government instruments have been the National Water Initiative (2004) that recognised the

Table 10.2 Australian government policies directly or indirectly addressing urban heat mitigation. For brevity State/Territory and local government examples are from Victoria only (modified from Bush et al. 2015)

Policy focus	Federal	Victorian	Municipal (general)
Cities, urban design and planning	Smart Cities Plan (2016) Cooperative Research Centre Program (CRC for Low Carbon Living, 2012) National Urban Policy—Our Cities, Our Future (2011) Green Star Rating Tool	Plan Melbourne 2050 (2014) Victorian Planning Provisions	Municipal strategic statements (e.g. Inner Melbourne Action Plan) Local planning policies
Health, emergency management	National Strategy for Disaster Resilience (2011)	Victorian Public Health and Wellbeing Plan (2014) Heat Health Plan for Victoria (revised 2015) State Emergency Response Plan (revised 2016)	Municipal emergency management plans Municipal emergency management plans Municipal heatwave strategies
Water, stormwater management	Cooperative Research Centre Program (CRC for Water Sensitive Cities, 2012) Water Act (2007) National Water Initiative (2004)	State Environment Protection Policy—Water (1988–2004) Living Melbourne, Living Victoria (2012) Managing Extreme Water Shortage in Victoria—Lessons from the Millennium Drought (2016) Melbourne Water Urban Cooling Program (in development)	Integrated water management plans (e.g. CoPP Open Space Water Management Plan; WSUD Guidelines) Waterway management plans
Climate change mitigation and adaptation	Emissions Reduction Fund (2014) National Climate Change Adaptation Research Facility (NCCARF) and Program (2008)	Climate Change Act (2010) Victorian Centre for Climate Change Adaptation Research (VCCCAR) and Program (2009) Victoria's Climate Change Framework (2016) Victorian Climate Change Adaptation Plan (2017)	Municipal climate change mitigation plans (e.g. CoPP Toward Zero) Municipal climate change adaptation plans (e.g. CoPP Climate Adaptation Plan)

(continued)

Table 10.2 (continued)

Policy focus	Federal	Victorian	Municipal (general)
Open space and recreation	20 Million Trees Program (2014)	Healthy Parks, Healthy People (2015) Metropolitan Open Space Strategy (in prep)	Municipal urban forest strategies (e.g. Greening Port Phillip) Municipal open space strategies Municipal street tree strategies
Ecology and biodiversity	Environment Protection and Biodiversity Conservation Act (1999)	Environment Protection Act (1970) Environment Effects Act (1978) Flora and Fauna Guarantee Act (1988)	Urban ecology strategies
Buildings and infrastructure	National Construction Code (2011)		Municipal building regulations and permits

need for identification of environmental and other public benefit outcomes in national and state water plans and the National Urban Policy (2011) aimed at improving the productivity, sustainability and liveability of Australian cities. The Cooperative Research Centre (CRC) program established by Federal Government in 1990 to maximise benefits of research through enhancing utilisation, commercialisation and technology transfer led to the establishment of two key CRCs in the urban heat mitigation area in 2012. The CRCs for Low Carbon Living and Water Sensitive Cities, along with the National Centre for Climate Change Adaptation Research (established 2007), have been critical in providing quality research and guidance in this area to state and local government authorities, in part because of the large number of state and local government industry partners involved (>70 in the case of the CRC for Water Sensitive Cities).

At State/Territory and community level the "Millennium Drought" of 1997–2009 and associated water shortages, heat and fire led to widespread concerns about climate change and adaptation. Recent Victorian policy development reflects this concern (Table 10.2) and is representative of other States/Territories. Policy focuses on public health/emergency management, water security and climate adaptation. The contribution of urban green space to urban liveability, including heat mitigation, is present in a range of high-level assessments and strategic plans and is beginning to be supported by local research, but Bush et al. (2015) argue that the policies are still relatively immature and not yet properly integrated into urban planning and management of utilities at the local government level. However activity such as the urban cooling initiative of Melbourne Water, initiated in 2016 and driven largely by community pressure assisted by enthusiastic champions, will change this situation. Melbourne Water, a branch of State Government, manages Melbourne's major water resources on behalf of the community, including water supply and disposal as well as sewage treatment. The urban cooling program activity is directed at transforming 30 hectares of public land within metropolitan Melbourne that is managed by Melbourne Water into cool and shady refuges (Melbourne Water 2020).

Despite ambivalence and relative immaturity of policy development at higher levels of government, there is an impressive amount of "on-ground" activity relating to heat mitigation at the municipal level in Victoria. This is partly because of excellent communication between state and local government champions in areas such as water management, but also because of collaboration and information sharing at the local level where municipalities share many issues and physical conditions. Strong research links with the relevant CRCs have also fostered many local initiatives for water sensitive in-fill or green-field urban development, or for prioritising areas within the municipality for heat mitigation through WSUD.

Municipal government is taking the lead in Australia in adopting research and best practice for implementation of heat mitigation measures. The City of Port Phillip (CoPP), occupying an inner bayside area of Melbourne, provides an excellent example of leadership in this area. Relevant activity at CoPP was initiated in response to the *Inner Melbourne Action Plan* (IMAP) developed in 2005 in partnership between four inner Melbourne councils, with the active support of the Victorian State Government. An initial sustainable environment strategy *Toward Zero* launched by CoPP in 2007 led to multiple inter-related policy initiatives including the *City of Port Phillip WSUD Guidelines*, the *Climate Adaptation Plan*, the *Greening Port Phillip*, *Open Space Water Management Plan* and *Water Plan* documents. Relevant documents, including subsequent progress reports, can be found at City of Port Phillip (2020). Each of these documents promotes urban greening and storm water supported WSUD at the core of the CoPP heat mitigation strategy. This is because these approaches are considered highly effective in the climatic conditions typical of southern Australian cities (hot dry summers and cool rainy winters). Implementation of successful projects in the CoPP has also been guided by collaboration with researchers (e.g. Norton et al. 2015; Coutts et al. 2016). Using the CoPP municipality as an example, Norton et al. (2015) developed a framework for prioritisation of implementation of urban green infrastructure to mitigate heat based on heat exposure (from thermal mapping), heat vulnerability (e.g. Loughnan et al. 2013) and behavioural exposure (zones of high activity), along with relevant design guidelines (Fig. 10.7).

10.6 Conclusion

The research on heat-mortality relationships documented in this chapter has shown clear threshold temperatures above which mortality increases markedly. It follows that even relatively small reductions in temperature in the order of 1–2 °C can dramatically reduce heat mortality. This chapter has clearly shown that WSUD and irrigated green infrastructure implemented at both the micro- (household to street) and local (neighbourhood to city) scale can reduce air temperatures sufficiently to lower the heat-mortality burden. Furthermore the heat mitigation approaches

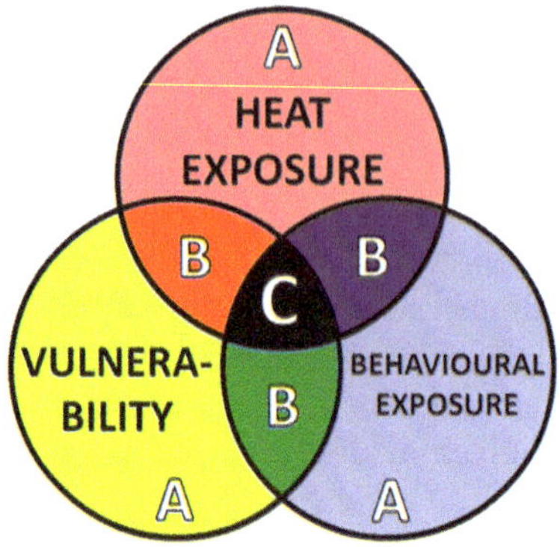

Fig. 10.7 Factors required for identification of areas of high (C), medium (B) and moderate (A) priority for implementation of heat mitigation within a municipality. The key factors are high daytime surface temperatures (heat exposure) intersecting with areas of greater population vulnerability and areas of high activity (behavioural exposure) such as train/bus stations, car parks, etc. From Norton et al. (2015)

applied at the microscale can reduce mean radiant and UTCI 'feels-like' temperatures by up to 10 °C, dramatically improving human thermal comfort.

The Australian community and various levels of government remain highly concerned about current and future climate trends and associated impacts including EHEs and wildfire. This concern has recently been heightened with 2019 being officially designated Australia's hottest year on record, with an area-averaged mean temperature 1.52 °C (2.09 °C for mean maximum temperature) above the 1961–1990 average temperature (Bureau of Meteorology 2020). It was also Australia's driest year since records began. As a consequence, interest in long-term adaptation of Australian cities to climate change has never been higher, especially where that adaptation can enhance the urban environment, such as with irrigated green infrastructure. With this interest comes a relevant opportunity to enact urban cooling initiatives. Exemplars provided by best practice at the municipal level, with researchers working hand-in-hand with developers and municipal authorities, and backed by strong evidence of the cost benefits for society of WSUD/irrigated green infrastructure, will be fundamental to the provision of more liveable, climate-adapted cities of the future.

References

AECOM (2012) Economic assessment of the urban heat island effect. Report prepared for City of Melbourne, p. 52. https://www.melbourne.vic.gov.au/SiteCollectionDocuments/eco-assessment-of-urban-heat-island-effect.pdf. Accessed 10 Mar 2017

Bi P, Williams S, Loughnan M, Lloyd G, Hansen A, Kjellstrom T, Dear K, Saniotis A (2011) The effects of extreme heat on human mortality and morbidity in Australia: implications for public health. Asia Pacific Journal of Public Health 23(2_suppl):27S–36S. https://doi.org/10.1177/1010539510391644

Broadbent AM, Coutts AM, Tapper NJ, Demuzere M, Beringer J (2018a) The microscale cooling effects of water sensitive urban design and irrigation in a suburban environment. Theor Appl Climatol 134(1):1–23. https://doi.org/10.1007/s00704-017-2241-3

Broadbent AM, Coutts AM, Tapper NJ, Demuzere M (2018b) The cooling effect of irrigation on urban microclimate during heatwave conditions. Urban Clim 23:309–329. https://doi.org/10.1016/j.uclim.2017.05.002

Bureau of Meteorology (2020) Annual Climate Statement 2019, issued 9 January 2020. http://www.bom.gov.au/climate/current/annual/aus/. Accessed 23 Jan 2020

Bush J, Aye L, Hes D (2015) Cooling cities with green space: a policy analysis framework. In: Burton P, Shearer H (eds) State of Australian Cities Conference 2015: Refereed Proceedings, Gold Coast, 9–11 December 2015. http://soacconference.com.au/soac-conference-proceedings. Accessed 10 Mar 2017

City of Port Phillip (2020) Sustainability. http://www.portphillip.vic.gov.au/sustainability.htm. Accessed 5 June 2020

Coates L, Haynes K, O'Brien J, McAneney J, de Oliveira FD (2014) Exploring 167 years of vulnerability: an examination of extreme heat events in Australia 1844–2010. Environ Sci Pol 42:33–44. https://doi.org/10.1016/j.envsci.2014.05.003

Coutts AM, Beringer J, Tapper NJ (2007) Impact of increasing urban density on local climate: spatial and temporal variations in the surface energy balance in Melbourne, Australia. J Appl Meteorol Clim 46(4):477–493. https://doi.org/10.1175/jam2462.1

Coutts A, Beringer J, Tapper N (2008) Investigating the climatic impact of urban planning strategies through the use of regional climate modelling: a case study for Melbourne, Australia. Int J Climatol 28(14):1943–1957. https://doi.org/10.1002/joc.1680

Coutts AM, Tapper NJ, Beringer J, Loughnan M, Demuzere M (2013a) Watering our cities: the capacity for water sensitive urban design to support urban cooling and improve human thermal comfort in the Australian context. Prog Phy Geog 37(1):2–28. https://doi.org/10.1177/0309133312461032

Coutts AM, Daly E, Beringer J, Tapper NJ (2013b) Assessing practical measures to reduce urban heat: green and cool roofs. Build Environ 70:266–276. https://doi.org/10.1016/j.buildenv.2013.08.021

Coutts AM, White EC, Tapper NJ, Beringer J, Livesley SJ (2016) Temperature and human thermal comfort effects of street trees across three contrasting street canyon environments. Theor Appl Climatol 124(1):55–68. https://doi.org/10.1007/s00704-015-1409-y

Coutts A, Moore C, Thom J, Tapper N, White E (2021) Radiative and advective influences on cooling from an isolated tree in inner Melbourne. Submitted to Geographical Research

Cowan T, Purich A, Perkins S, Pezza A, Boschat G, Sadler K (2014) More frequent, longer, and hotter heat waves for Australia in the twenty-first century. J Clim 27(15):5851–5871. https://doi.org/10.1175/jcli-d-14-00092.1

de Munck C, Lemonsu A, Masson V, Le Bras J, Bonhomme M (2018) Evaluating the impacts of greening scenarios on thermal comfort and energy and water consumptions for adapting Paris city to climate change. Urban Clim 23:260–286. https://doi.org/10.1016/j.uclim.2017.01.003

Deloitte (2017) Building resilience to natural disasters in our states and territories. Deloitte Access Economics. 112 pp. http://australianbusinessroundtable.com.au/assets/documents/ABR_building-resilience-in-our-states-and-territories.pdf. Accessed 5 June 2020

Earl N, Simmonds I, Tapper N (2016) Weekly cycles in peak time temperatures and urban heat island intensity. Environ Res Lett 11(7):074003. https://doi.org/10.1088/1748-9326/11/7/074003

Guo Y, Li S, Liu DL, Chen D, Williams G, Tong S (2016) Projecting future temperature-related mortality in three largest Australian cities. Environ Pollut 208:66–73. https://doi.org/10.1016/j.envpol.2015.09.041

Heaviside C, Macintyre H, Vardoulakis S (2017) The urban heat island: implications for health in a changing environment. Curr Environ Health Rep 4(3):296–305. https://doi.org/10.1007/s40572-017-0150-3

Jacobs SJ, Gallant AJE, Tapper NJ, Li D (2018) Use of cool roofs and vegetation to mitigate urban heat and improve human thermal stress in Melbourne, Australia. J Appl Meteorol Clim 57(8):1747–1764. https://doi.org/10.1175/jamc-d-17-0243.1

Jacobs S, Gallant A, Tapper N (2021) Cooling effectiveness of cool roofs, tree canopy and irrigation for Melbourne, Australia under current and mid-century climatic conditions. Submitted to Geographical Research

Lam CKC, Loughnan M, Tapper N (2018) Visitors' perception of thermal comfort during extreme heat events at the Royal Botanic Garden Melbourne. Int J Biometeorol 62(1):97–112. https://doi.org/10.1007/s00484-015-1125-4

Larsen L (2015) Urban climate and adaptation strategies. Front Ecol Environ 13(9):486–492. https://doi.org/10.1890/150103

Loughnan M, Tapper N, Phan T, Lynch K, McInnes J (2013) A spatial vulnerability analysis of urban populations during extreme heat events in Australian capital cities. National Climate Change Adaptation Research Facility, Gold Coast, p 128

Matzarakis A, Rutz F, Mayer H (2010) Modelling radiation fluxes in simple and complex environments: basics of the RayMan model. Int J Biometeorol 54(2):131–139. https://doi.org/10.1007/s00484-009-0261-0

Melbourne Water (2020) Why we're giving Melbourne shade. https://www.melbournewater.com.au/community-and-education/about-our-water/liveability-and-environment/urban-cooling-program. Accessed 5 June 2020

Miao S, Tapper N (2017) Land surface temperature and soil moisture distribution characteristics for a raingarden in Fitzroy Gardens, Melbourne. Journal of Southeast University (English Edition) 33(3):355–361. https://doi.org/10.3969/j.issn.1003-7985.2017.03.016

Motazedian A, Coutts AM, Tapper NJ (2020) The microclimatic interaction of a small urban park in central Melbourne with its surrounding urban environment during heat events. Urban For Urban Green 52:126688. https://doi.org/10.1016/j.ufug.2020.126688

Nicholls N, Skinner C, Loughnan M, Tapper N (2008) A simple heat alert system for Melbourne, Australia. Int J Biometeorol 52(5):375–384. https://doi.org/10.1007/s00484-007-0132-5

Nicholls N, Loughnan M, Tapper N (2016) Building evidence that effective heat alert systems save lives in Southeast Australia. In: Climate services for health. WMO and WHO Joint Office for Climate and Health, Geneva, pp 190–191. http://public.wmo.int/en/resources/library/climate-services-health-case-studies. Accessed 10 March 2017

Norton BA, Coutts AM, Livesley SJ, Harris RJ, Hunter AM, Williams NSG (2015) Planning for cooler cities: a framework to prioritise green infrastructure to mitigate high temperatures in urban landscapes. Landsc Urban Plan 134:127–138. https://doi.org/10.1016/j.landurbplan.2014.10.018

Nury S (2016) Informed implementation of greening as a heat mitigation measure in Melbourne, Australia: a remote sensing study. Unpub. PhD thesis, Monash University, Australia, 364pp

Perkins SE, Alexander LV, Nairn JR (2012) Increasing frequency, intensity and duration of observed global heatwaves and warm spells. Geophy Res Lett 39(20):L20714. https://doi.org/10.1029/2012gl053361

Perkins-Kirkpatrick SE, White CJ, Alexander LV, Argüeso D, Boschat G, Cowan T, Evans JP, Ekström M, Oliver ECJ, Phatak A, Purich A (2016) Natural hazards in Australia: heatwaves. Clim Chang 139(1):101–114. https://doi.org/10.1007/s10584-016-1650-0

Radcliff J (2019) History of water sensitive urban design/low impact development adoption in Australia and internationally. In: Gardner, Begbie (eds) Approaches to water sensitive urban design potential, design, ecological health, urban greening, economics, policies, and community perceptions. Elsevier, Sharma, pp 1–28

Rogers CDW, Gallant AJE, Tapper NJ (2019) Is the urban heat island exacerbated during heatwaves in southern Australian cities? Theor Appl Climatol 137(1):441–457. https://doi.org/10.1007/s00704-018-2599-x

Santamouris M, Ding L, Fiorito F, Oldfield P, Osmond P, Paolini R, Prasad D, Synnefa A (2017) Passive and active cooling for the outdoor built environment—analysis and assessment of the cooling potential of mitigation technologies using performance data from 220 large scale projects. Sol Energy 154:14–33. https://doi.org/10.1016/j.solener.2016.12.006

Tapper N, Coutts A, Loughnan M, Pankhania D (2014) Urban population vulnerability to climate extremes: mitigating urban heat through technology and water-sensitive urban design. In: Lehmann S (ed) Low carbon cities: transforming urban systems. Earthscan book series on sustainable design, vol 3. Routledge, London, pp 361–375

Tapper N, Lloyd S, McArthur J, Nice K, Jacobs S (2019) Estimating the economic benefits of Urban Heat Island mitigation—biophysical Aspects. Cooperative Research Centre for Water Sensitive Cities, Melbourne, 26pp

Thom JK, Coutts AM, Broadbent AM, Tapper NJ (2016) The influence of increasing tree cover on mean radiant temperature across a mixed development suburb in Adelaide, Australia. Urban For Urban Green 20:233–242. https://doi.org/10.1016/j.ufug.2016.08.016

United Nations Habitat (2011) Cities and climate change: global report on human settlements 2011. United Nations Human Settlements Programme, Earthscan, London, 279pp. https://unhabitat.org/sites/default/files/download-manager-files/Cities%20and%20Climate%20Change%20Global%20Report%20on%20Human%20Settlements%202011.pdf

Vardoulakis S, Dear K, Hajat S, Heaviside C, Eggen B, McMichael AJ (2014) Comparative assessment of the effects of climate change on heat- and cold-related mortality in the United Kingdom and Australia. Environ Health Perspect 122(12):1285–1292. https://doi.org/10.1289/ehp.1307524

Whiteoak K, Saigar J (2019) Estimating the economic benefits of urban heat island mitigation—economic analysis. Cooperative Research Centre for Water Sensitive Cities, Melbourne, 33pp

Williams S, Nitschke M, Sullivan T, Tucker GR, Weinstein P, Pisaniello DL, Parton KA, Bi P (2012) Heat and health in Adelaide, South Australia: assessment of heat thresholds and temperature relationships. Sci Total Environ 414:126–133. https://doi.org/10.1016/j.scitotenv.2011.11.038

World Bank (2020) Urban population. https://data.worldbank.org/indicator/SP.URB.TOTL.IN.ZS?view=chart. Accessed 5 June 2020

Chapter 11
What Alternatives Does Paris Have to Adapt to Future Heat Waves?

Aude Lemonsu, Cécile de Munck, Raphaëlle Kounkou-Arnaud, Valery Masson, and Vincent Viguié

Abstract In 2003, the city of Paris faced a particularly severe heat wave that resulted in an exceptional excess mortality rate for the urban population. Subsequently, several research projects in partnership with public stakeholders investigated adaptation strategies that could be implemented to address the combined effects of climate change (leading to an increase in occurrence of heat waves) and the urban heat island phenomenon. Action levers have been tested at different scales to mitigate air temperature in the city under heat wave conditions: urban design strategies from the neighbourhood to the city scale based on implementation of reflective materials on buildings, greening, or pavement watering, but also large-scale territorial planning policies related to urban expansion and landscape planning around the city. All action levers tested for local-scale urban design provide a substantial cooling effect, but their efficiency varies spatially according to the urban typology. Greening is relevant in medium-density urban areas where vegetation coverage is enough to bring noticeable cooling through evapotranspiration. Reflective materials and pavement watering are most effective in central Paris, the first acting on the radiative balance of buildings and the second on surface evaporation at pavement level. All these levers can therefore be combined for an increased and additive efficiency. Larger-scale planning policies are another interesting way to adapt and regulate local climate, while bringing multiple other ecosystemic services. These two adaptation pathways finally lead to a comparable urban heat island attenuation of about 1.3–1.5 °C. Nonetheless, the studies highlighted the predominant effect of residents' practices, particularly with regard to energy use, on the effectiveness of

A. Lemonsu (✉) · C. de Munck · V. Masson
CNRM, Université de Toulouse, Météo-France/CNRS, Toulouse Cedex, France
e-mail: aude.lemonsu@meteo.fr; cecile.demunck@meteo.fr; valery.masson@meteo.fr

R. Kounkou-Arnaud
Meteorological Services Direction, Météo-France, Saint-Mandé Cedex, France
e-mail: raphaelle.kounkou@meteo.fr

V. Viguié
CIRED, École des Ponts ParisTech/CNRS/EHESS/AgroParisTech/CIRAD, Nogent-sur-Marne, France
e-mail: viguie@centre-cired.fr

C. Ren, G. McGregor (eds.), *Urban Climate Science for Planning Healthy Cities*, Biometeorology 5, https://doi.org/10.1007/978-3-030-87598-5_11

the strategies envisaged. In particular, massive use of air conditioning tends to increase street-level air temperature and degrade outdoor comfort conditions, and can consequently counterbalance all benefits of the other adaptation levers.

Keywords Heat waves · Climate change · Urban heat island · Adaptation strategies · Urban planning and urban design · Individual practices

Key Messages

- Adaptation strategies can be implemented at different scales in time and space.
- Greening is relevant in medium-density urban areas, while reflective materials and pavement watering are most effective in central districts.
- The combination of these action levers ensures an increased and additive efficiency.
- Local-scale urban design and larger-scale planning policies lead to a comparable UHI attenuation of about 1.3–1.5 °C.
- Massive use of air conditioning increases street-level air temperature and degrades outdoor comfort conditions, that can counterbalance all benefits of the other adaptation levers.

11.1 Introduction

In the study of climate change and its impacts, cities are high-stakes places (IPCC 2007, 2014). Climate risk, which results from the interaction between climate hazard, exposure, and vulnerability, is particularly enhanced in cities since they concentrate populations, capital goods, and economic activities, all of which are part of the exposure factor. Locally, cities will be subject in the coming decades to the regional impacts of climate change and to those of the urban climate, especially to the urban heat island (UHI). This phenomenon already inherent to all cities (even those of medium size) is highly dependent on meteorological conditions (Oke et al. 2017). It is favoured by hot and sunny anticyclonic conditions. Especially, it can reach very high intensities during heat waves and therefore worsen the already extreme conditions to which inhabitants are subjected. These situations can lead to a significant increase in mortality and morbidity (Pascal et al. 2018). Today, climate projections agree that the frequency of heat waves will increase in the future, which will further increase the vulnerability of cities and urban populations if no action is taken (IPCC 2014).

Nonetheless in response to global climate challenges, several national programmes have been launched in France since 2000 to combat climate change and control greenhouse gas emissions (Colombert 2008; Lambert-Habib et al. 2013). These programmes are the result of more local actions, in particular aimed at planning regional and urban development by integrating stronger environmental considerations. Issues of urban adaptation to the effects of climate change and UHI mitigation are now emerging in some urban planning documents and tools. More and more French cities are integrating these considerations into their local urban planning schemes or implementing actions through their environmental charters or territorial climate-energy plans. It is often a question of establishing an initial diagnosis of the environmental problems linked to urban sprawl, and proposing local solutions such as regulating the sealing of natural soils, reintroducing vegetation and water surfaces, and improving buildings.

Motivated by these increasing concerns from the institutional world and society and the collective awareness due to dramatic effects experienced in France during the 2003 heat wave, several research projects and actions have been conducted in the last 15 years regarding adaptation and/or mitigation strategies for French cities. Many of these studies focused on the city of Paris (and its region), which is the most populated French city with 2.2 million inhabitants and 21,000 inhabitants per m^2. Paris also suffered the full force of the 2003 heat wave, resulting in an exceptional excess mortality rate that was higher than that recorded for other cities (Canouï-Poitrine et al. 2006).

11.2 Future Climate and Heat Waves Over Paris Region

Heat waves are the subject of numerous studies for evaluating adaptation strategies of cities because they have the particularity of reinforcing the UHI phenomenon and potentially aggravating health impacts (Basara et al. 2010; Tan et al. 2010). Case studies rely most of the time on past events by considering them as representative of what summers could become in the future climate. Studying such events that left their mark on people's minds is also an effective way to raise awareness to issues of impacts and adaptation. As such, the heat waves of 2003 and 2006 have received attention as case studies (Dousset et al. 2011; Laaidi et al. 2012; de Munck et al. 2013a; Masson et al. 2013). It is nevertheless necessary to understand the evolution of these events to face future climate hazards at best.

11.2.1 Future Heat Waves

For the Paris region an analysis of the evolution of heat waves by the middle and the end of the twenty-first century has been performed by Lemonsu et al. (2014). A method for identifying heat waves was developed to retrieve these events from long time series of daily minimum and maximum temperature indices, calculated as a 3-day moving average of daily temperature time series. The method was first built and evaluated using observations, and then applied to nine regional climate models' projections produced for the European project ENSEMBLES (Christensen et al. 2008; van der Linden and Mitchell 2009). These projections were established at 25 km spatial resolution, all based on the A1B greenhouse gas emission scenario

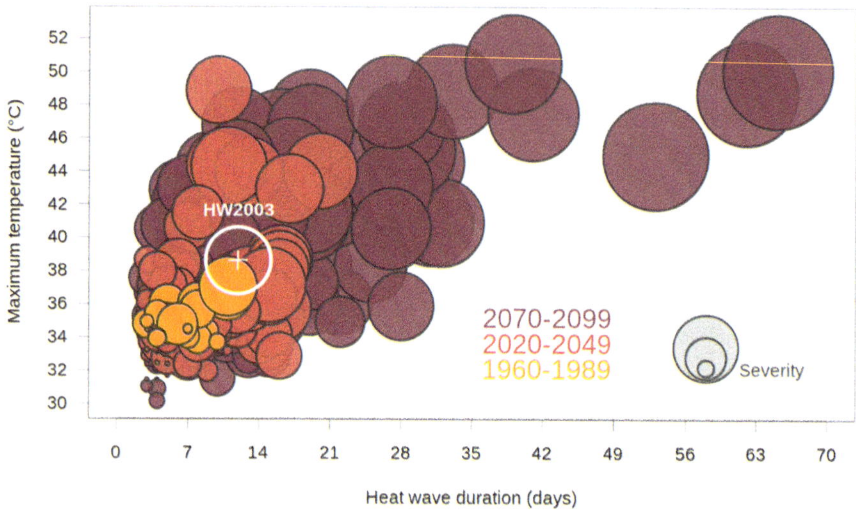

Fig. 11.1 Comparison of the 2003 heat wave (HW2003) with heat waves extracted from climate projections in A1B scenario over the time periods 1960–1989, 2020–2049, and 2070–2099. Each disk represents a heat wave characterized by a duration (indicated on the abscissa) and a maximum intensity (on the ordinate). The size of the disk is proportional to the severity of the heat wave

(IPCC 2007), but by using different couples of regional climate models and general circulation models as drivers.

Heat waves were extracted over three time periods, i.e. 1960–1989, 2020–2049, and 2070–2099, to investigate the evolution of the probability of occurrence of these events, as well as their duration and intensity. Based on this study, all projections anticipated a systematic increase in the frequency of occurrence of heat waves over time: one heat wave every 2 years on average for the period 2020–2049 and at least one heat wave per year and up to two heat waves per year on average over the period 2070–2099 (Fig. 11.1). These heat waves would also become longer, with average durations of 6–12 days by the end of the twenty-first century, with some exceptionally long events of 5–9 weeks.

11.2.2 The 2003 Heat Wave Compared to Future Events

The representativeness of the 2003 heat wave as an exceptional past event, which could approach standard summer conditions in the future, can be investigated in the light of the statistical distribution of future events by 2050 and 2100. Its characteristics were compared with those of events extracted from climate projections (Fig. 11.1) using a severity index, calculated as the product of the average of maximum daily temperatures (36.5 °C) and of the duration of the event (11 days). This made it possible to estimate that the 2003 heat wave would have a return period of

27 and 6 years over the periods 2020–2049 and 2070–2099, respectively. This analysis displayed the marked increase in the frequency of occurrence of such a heat wave, while confirming the exceptional nature of this event, both in terms of its duration and intensity.

11.2.3 Urban Impact on Climatic Trends

The regional impacts of climate change locally interact with the urban climate, and especially with the UHI. Nonetheless, the temperature differences between urban and rural areas do not necessarily increase with time. The work of Lemonsu et al. (2013), based on long-term surface simulations (in stand-alone mode forced by climatic projections according to emission scenarios A1B and A2), showed that the nocturnal UHI remains at a maximum over the most compact and built-up areas, mainly driven by urban processes. However, rural and suburban areas warm up more than the city centre area during the day. This trend results from the substantial drying of natural soils in response to a decrease in precipitation rate, leading to a 30% decrease in soil evaporation and plant transpiration. Consequently, the exposure of the city and its population to high temperatures is highly dependent on regional-scale evolutions in soil water status. A good knowledge of the evolution of climatic conditions in future decades, in interactions with local urban climate, is then necessary to propose adaptation solutions with the aim of preventing or mitigating impacts.

11.3 Multi-scale Solutions and Multi-criteria Evaluation of Paris's Adaptation

11.3.1 Modelling Framework for Multi-scale and Multi-criteria Analyses

To counter UHI effects and mitigate climate change impacts, urban planning initiatives can be defined at different spatial scales. They obviously imply varying implementation times and different financial and political resources. The long-term urban expansion influences the evolution of city's shape and size, and consequently the evolution of the UHI (Stone and Rodgers 2001; Adachi et al. 2012, 2014; Lemonsu et al. 2015). This level of intervention requires the urban planning policies to be anticipated over time (Masson et al. 2013): e.g. the creation of a greenbelt or blue corridor, or the implementation of an urban densification policy to prevent uncontrolled urban sprawl. More locally, through construction or urban renewal projects at the district level, the actions on urban forms and urban developments, or on materials and coatings, can play a significant role in the regulation of local climate

(House-Peters and Chang 2011; Santamouris 2013; Dimoudi et al. 2014; Perini and Magliocco 2014). Such measures can be implemented more quickly, due to the less ambitious level of intervention.

The evolution of outdoor air temperature and UHI intensity under the influence of climate change is only one of many dimensions to be considered. The expected impacts are diverse, multi-sectoral, and often interconnected, hence adaptation and mitigation strategies must be designed and assessed based on an ensemble of considerations and criteria. To meet the needs for a systemic and interdisciplinary analysis of the city, a relatively sophisticated modelling methodology has been developed for adaptation strategy studies, particularly in the Paris region (Masson et al. 2014; Lemonsu et al. 2015).

The urban canopy model TEB (Masson 2000; Lemonsu et al. 2012; de Munck et al. 2013b; Pigeon et al. 2014) simulates urban climate from city to neighbourhood scales, with the ability to represent different urban typologies and different vegetation rates (up to hectometric horizontal resolution). It not only models the evolution of outdoor microclimatic variables but also that in buildings, as well as the universal thermal climate index (UTCI) (UTCI, Bröde et al. 2012; Fiala et al. 2012). It also models the energy balance inside buildings, the operation of air conditioning systems and their energy consumption. This model is forced by atmospheric conditions provided either by a meteorological model to simulate realistic episodes (such as the 2003 heat wave) or by climate projections. It is also supported by maps describing land cover and land use (Bonhomme et al. 2013; Viguié et al. 2014), as well as building and vegetation characteristics. These data can be real data or scenario-based for adaptation studies.

11.3.2 Different Research Initiatives for Studying Paris's Adaptation

The studies conducted in Paris area addressed and evaluated different levels of adaptation of the city and region to the effects of climate change and UHI mitigation. The results of five main projects, Grand Pari(s) (Descartes 2009; Masson et al. 2013), EPICEA (EPICEA 2012; Kounkou-Arnaud et al. 2014), CLIM2 (Trémeac et al. 2012; de Munck et al. 2013a), VURCA (VURCA 2013; Lemonsu et al. 2015; Daniel et al. 2018; Viguié et al. 2020), and MUSCADE (MUSCADE 2014; de Munck et al. 2018), are presented here. These were research initiatives (described in Table 11.1) motivated by related scientific issues and based on somewhat different but comparable approaches.

Although the modelling methodologies applied in these projects differed somewhat, they nevertheless relied on a common framework (notably the use of the same model TEB) that makes it possible to place the different results in context and to analyse them jointly. The VURCA and MUSCADE projects were fundamental and interdisciplinary research projects, also involving institutional partners. They

Table 11.1 Description of research projects related to Paris area's adaptation to climate change

Project	Instigators/Context	Objectives	Methodology
EPICEA	Collaboration between researchers in urban climate and in building techniques, Paris City council, and the Paris urban planning agency, launched within the framework of the Climate Plan of the City of Paris (adopted in October 2007)	Evaluation of realistic scenarios for the urban design of the Paris "intra-muros", in view of existing technologies and urban constraints, and with a time frame of 2020 in accordance with the objectives of the Climate Plan of the City	Real-case meteorological simulations of 2003 heat wave with the Meso-NH non-hydrostatic atmospheric model coupled to the urban canopy model TEB 250-m spatial resolution
Grand Pari(s)	Consortium of urban climate researchers, urban planners, and architects (Descartes group) to answer the consultation instigated in 2008 by the French Ministry of Culture	Proposition of development projections for the Paris conurbation by 2030, regarding the large territory and not the existing inner Paris city	Idealized meteorological simulations of 2003 heat wave with the Meso-NH non-hydrostatic atmospheric model coupled to the urban canopy model TEB 250-m spatial resolution
CLIM2	Research project initiated by the CLIMESPACE Company, which operates and develops a large district cooling system for Paris	Quantification of the impact of heat discharges from air-conditioners (for different equipment scenarios) on outdoor micrometeorological conditions	Real-case meteorological simulations of 2003 wave with the Meso-NH non-hydrostatic atmospheric model coupled to the urban canopy model TEB 250-m spatial resolution
VURCA	Interdisciplinary research project involving experts in climate, urban meteorology, building techniques, and economy.	Evaluation of the vulnerability of Paris and its population to future heat waves (with a time frame of 2100). Comparison of several integrated city evolution scenarios, combining different hypotheses on urban expansion, buildings, air conditioning use, etc.	Urban climate simulations for "future" heat waves of variable intensity and duration. The ranges of variations were estimated from analysis of climate scenarios over the Paris region by 2100 1-km spatial resolution
MUSCADE	Interdisciplinary research project involving experts in climate, urban meteorology, building techniques, geography, economy, as well as urban planning agencies	Evaluation of integrated city evolution scenarios combining different hypotheses on urban expansion, buildings, energy use, greening strategies, etc.	10-years continuous urban climate simulations for allowing analyses of impacts at the scale of the 2003 heat wave but also at seasonal and annual scales 1-km spatial resolution

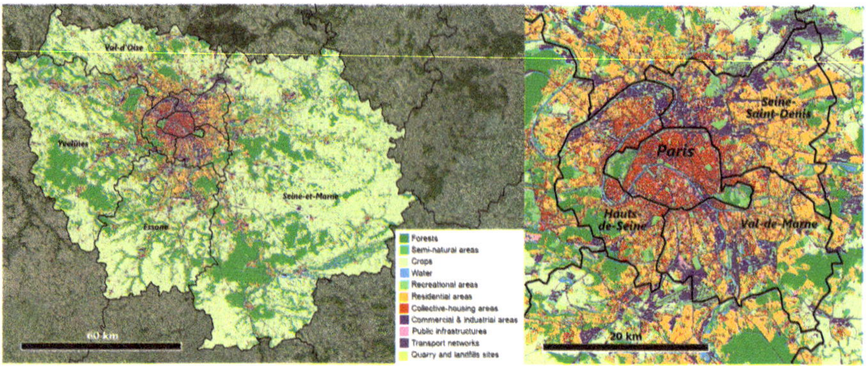

Fig. 11.2 Land use map of the Ile-de-France administrative region (left) based on the simplified MOS 2017 database produced by the Institute of Planning and Urban Development of the Parisian region (IAU). The thick black lines delimit the departments. A zoom on the area of Paris "intra-muros" and its first suburbs is presented to the right. Source: http://data.iau-idf.fr/datasets/75ab6d1bb58043729a8b05172937cfb0_40

addressed urban planning strategies up to long time horizons (end of twenty-first century), in coherence with the scale of analysis for climatic trends. The other three projects were specific in that public stakeholders and professionals in charge of urban planning, who involved urban climate researchers in their approach, initiated these. EPICEA was launched by the City of Paris following its Climate Plan in 2007 while Grand Pari(s) responded to a consultation instigated in 2008 by the French Ministry of Culture. The CLIM2 impact study was initiated by the CLIMESPACE Company, which operates and develops a large district cooling system for Paris. The scenarios studied in these projects were developed with time horizons of 10–30 years.

Finally, the scales of the territory involved in adaptation differ according to the projects. CLIM2 and EPICEA focused on Paris "intra-muros" (Fig. 11.2, right), a term that refers to the city of Paris excluding its suburbs. MUSCADE and VURCA studied the Paris urban unit, composed of Paris and close municipalities in the first suburbs and part of the second suburbs (Fig. 11.2, right), that is more than ten million inhabitants. Grand Pari(s) investigated adaptation of natural areas of the Ile-de-France region (Fig. 11.2, left).

11.4 Urban Design Strategies from Neighbourhood to City-Scale

Many of the urban studies conducted in Paris focused on urban development and sensitive-climate design strategies dedicated to mitigating UHI. However, indirectly, countering the UHI effect is also a way to adapt locally to climate change, particularly to the increase in heat waves with time, heat waves having the effect of intensifying the UHI phenomenon. Several action levers were evaluated based on realistic urban scenarios, taking into account current technical and architectural constraints. The aim here was to address relatively close time horizons (10–20 years) that are often those considered by institutional stakeholders.

11.4.1 Reflective Materials

Many studies have mentioned the benefits of reflective materials on buildings, especially cool roofs. Scenarios were tested for Paris intra-muros by modifying the reflective properties of both the walls and facades of all buildings. Albedos and emissivities were prescribed for this purpose at 0.9 (Kounkou-Arnaud et al. 2014).

Compared to a reference simulation for the existing city, a substantial cooling effect on 2-m air temperatures was noted for all Paris intra-muros, both at daytime and night-time. The efficiency of this measure however evolves spatially, depending on urban morphology and time of day. The high urban density of the central districts of Paris intra-muros, ranging from 50 to 70%, supports the relevance of this type of strategy. On average, a 1 °C drop in street-level air temperature was obtained in Paris intra-muros during daytime, but a localized cooling up to 2 °C was noted in the most densely built districts of western Paris' centre. It was in the order of 0.5–0.8 °C during the night, with still a maximum effect (−1 °C) in western central districts (Fig. 11.3, bottom left panel).

Whatever the neighbourhood, the efficiency of this action lever is systematically greater for daytime than night-time, which suggests that the albedo of artificial surfaces has a more substantial effect than emissivity on UHI mitigation (Colombert et al. 2011). However we should recall that the roofs of Paris' city core historic Haussmannian buildings are made of zinc, a material whose emissivity is already very low (0.6) compared to other coatings. Hence, the gain in reflectivity at night, with higher emissivity materials, is necessarily lower for these roofs. On the other hand, the effect of modifying the emissivity of walls predominates over that of roofs at night: the radiative trapping of infrared emissions inside the canyon due to multiple reflections and energy absorption (that is known as the key process limiting the night cooling of the air) is significantly reduced.

Another example of building modification tested for the Paris area is the installation of solar panels on roofs. Besides their function of renewable energy production, these panels modify the radiative characteristics of surfaces and therefore potentially the urban climate. Masson et al. (2014) investigated these effects by simulating a scenario of massive solar panel implementation in the Paris conurbation. A large part of buildings was equipped depending on roof orientation, favouring thermal panels for housing buildings and a mix of thermal and photo-voltaic panels for others. Panels darken the roofs albedo, which increases the absorption of incident radiation. However, this effect is counterbalanced by the conversion of the radiation received into another form of energy (electrical or thermal) that is not involved in the surface energy balance of roofs. Consequently, a slight decrease in the UHI of

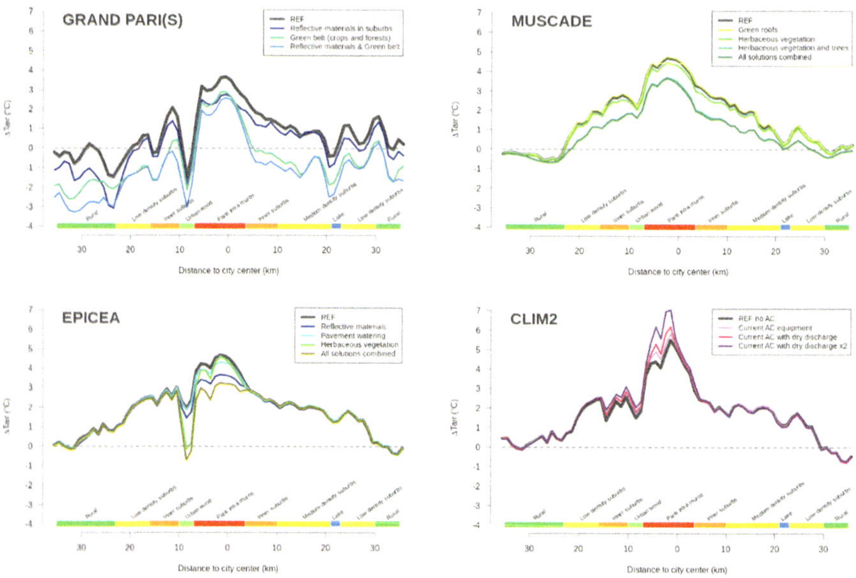

Fig. 11.3 For each research project, comparison of temperature anomaly profiles at night (0000–0200 LT) simulated for the reference case (black line) and the adaptation scenarios (colour lines) along a west-to-east cross section passing through Paris intra-muros. The temperature anomaly is calculated for each project as the difference between the temperature profile of each simulation and an average rural temperature estimated from the reference case

0.2 °C by day and 0.3 °C at night was noted, especially in densely built first suburbs. These results therefore support the relevance of this strategy, which constitutes a mitigation lever to climate change, without degrading urban climate conditions.

11.4.2 Greening

In cities, the surfaces potentially free for greening are bare ground areas, impervious ground-based surfaces such as pavements, roundabouts, car parks, and flat roofs. The hypotheses for Paris greening scenarios varied noticeably from one project to another. For MUSCADE especially, different greening rates (from 25 to 75%) of available ground-based surfaces were tested, by implementing for each of them either herbaceous vegetation only or a mix of herbaceous and tree vegetation (de Munck et al. 2018). Green roofs on adapted buildings were implemented, i.e. individual and dense collective buildings and warehouses, but not on the Haussmannian buildings of the historical city centre.

The results obtained for the 2003 heat wave indicated that the ground-based vegetation might cool the air during the day, especially when the greening rate is high: a 2 °C maximum mitigation of Tmax was obtained. Nonetheless, this is only possible because urban vegetation is watered. Indeed, Kounkou-Arnaud et al. (2014) previously showed that comparable greening scenarios, but without watering, did not bring any microclimate regulation benefits. In this case, plants are water-stressed and close their stomata to prevent water loss, which limits or stops evapotranspiration. At night, the herbaceous ground-based vegetation slightly cools the ambient air. The effect was more pronounced for mixed vegetation (up to 1.3 °C in Paris intra-muros) because the water supply for watering is partly intercepted by plant foliage. Its direct evaporation causes rapid local cooling (Fig. 11.3, top right panel). The cooling effect induced by vegetation varies spatially according to urban typologies and their morphological characteristics. The same applies to thermal comfort conditions in the outdoor environment. The gain in thermal comfort reaches a maximum in collective housing areas where urban density is moderate enough to make substantial greening possible: up to 1 h less per day under very high heat stress conditions, i.e. for the universal thermal climate index (UTCI) greater than 38 °C.

Finally, de Munck et al. (2018) found that green roofs had no effect on air temperature at street level if not irrigated. Even irrigated, their cooling potential remains very low (less than 0.5 °C) compared to other greening strategies (Fig. 11.3, top right panel). Under heat-wave conditions, their main interest is to maintain longer thermal comfort inside buildings, or to reduce energy consumption for buildings equipped with air conditioning systems.

11.4.3 Pavement Watering

Largely inspired by the experimental works of Takahashi et al. (2010) an original scenario for watering pavements was also tested for Paris. The protocol for water sprinkling was calibrated by a sensitivity study to optimize the amount of water consumed and avoid any water loss through surface runoff to sewer systems. A sprinkling rate of 0.2 L m^{-2} during 3 min at the beginning of each hour was prescribed from 7 am to 9 pm Local Time (LT).

Simulation results suggested that this type of strategy was not suitable for low built-density districts, where the evaporative capacity of wet pavements remained lower than the transpiration capacity of vegetation, except in the case of severe drought. Nonetheless, this is an interesting alternative in very dense urban centres, where urban greening is technically complicated (Kounkou-Arnaud et al. 2014; Daniel et al. 2018). The cooling effect due to surface water evaporation on pavements is highly localized and instantaneous. As a result, it was mostly occurred during daytime and reached up to −0.9 °C in Paris intra-muros, where the fraction of pavement is highest. A slight cooling of 0.3 °C persisted at night in the same areas (Fig. 11.3, bottom left panel).

Convinced by these results based on modelling, the Paris City council launched an experiment in the summers of 2013 and 2014 to confirm the efficiency of this technique and to assess its feasibility. A street was instrumented with a weather station and sensors measuring solar irradiance and ground heat flux. Water was sprayed on pavements by public watering trucks at an optimal frequency determined according to solar exposure. During the day, measurements showed a cooling of 0.8 °C for the street-level air temperature, and of 1.5 °C for the UTCI comfort index (Hendel et al. 2015, 2016).

11.4.4 Combination of Action Levers

The modelling framework implemented in these studies made it possible to quantify the efficiency of different action levers independently and then combined in supposedly optimal scenarios. Kounkou-Arnaud et al. (2014) showed that combining the adaptation of buildings with reflective materials, the greening of available ground-based surfaces, and the watering of pavements results in a maximum mitigation of the UHI of 3 °C during the day, and of 1.6 °C at night when the UHI is the highest (Fig. 11.3, bottom left panel). The effect of these action levers is cumulative due to each acting through different physical processes. They contribute to the cooling of the air with different efficiencies, depending on the time of day and the typo-morphological characteristics of the places where implemented.

Similarly, de Munck et al. (2018) showed that energy savings for air conditioning reach a maximum for the scenario combining greening of ground-based surfaces which cools the outside air and reduces the need for air conditioning, and implementation of green roofs, which insulates and prevents building warming (Fig. 11.3, top right panel).

11.5 Large-Scale Territorial Planning Policies

Soil-sealing is the main dynamic behind the UHI effect. With a continuous world population growth, combined with a systematic increase in urban population, urban sprawl is intensifying at the expense of natural areas. The choices in long-term policies for urban planning and the large territory development raise various issues related to land use and occupation, transport networks and modes, energy consumption, quality of life, and of course urban climate.

11.5.1 Impact of Urban Expansion Trends

In particular, it is difficult to find a clear consensus on the issue of the compact city or the extended city, as the associated effects are multiple and variable. Lemonsu et al. (2015) have addressed this issue in the VURCA project, by investigating under heat-wave conditions how the Paris' climate might respond to different urban expansion trends. Socio-economic simulations (Viguié et al. 2014) led to five contrasting projections for the Paris urban unit by 2100. These included a compact city resulting from the control of urban sprawl and densification of the existing city, a spread-out city derived from non-binding urbanization trends, and three Green-city alternatives for which a minimum greening rate of 10, 30 and 50% was imposed on all urbanized areas.

The city's vulnerability to heat waves was estimated by calculating the proportion of urbanized area affected by different UHI's intensities. At night, this proportion was shown to increase with the degree of compactness of the whole urban unit, so that the Compact city is the most affected. For events equivalent to the 2003 heat wave, about 8.0% of the Compact city is subject to a UHI of 2 °C, compared to 5.1 and 2.7% for the Spread-out city and the 30% Green city, respectively (Fig. 11.4, left panel). For daytime on the contrary, the Compact city is generally the most efficient since less than 1% of the urban unit experiences a UHI of 2 °C. The predominant compact urban forms of this scenario favour the shading effects at daytime, preventing warming by direct sunlight. Conversely, the cooling capacity by evapotranspiration expected for the 10 and 30% Green city scenarios is counterbalanced and even cancelled out over the entire residential fabric by radiative processes.

From a health study perspective, rather than a climatological one, it is also relevant to relate indicators to population density instead of urban density (Lemonsu et al. 2015). By doing this, the results showed that the percentage of inhabitants affected by the UHI effect (Fig. 11.4 middle panel) was comparable to the fraction of urban unit during the day. However, at night, the fractions expressed relative to population are generally higher than that expressed relative to urbanization due to greater spatial disparities in population density, varying between 7.1 and 27.8%. The level of population exposure to the UHI is highest for the Compact city because population is largely concentrated in Paris intra-muros and first suburbs that are also the areas most affected by the UHI. These results were confirmed by the comparison of the average time during which inhabitants experience strong heat stress conditions (i.e. UTCI>32 °C) during heat-wave days (Fig. 11.4 right panel). This is highest for the Compact city and decreases with an increase in greening rate of the city.

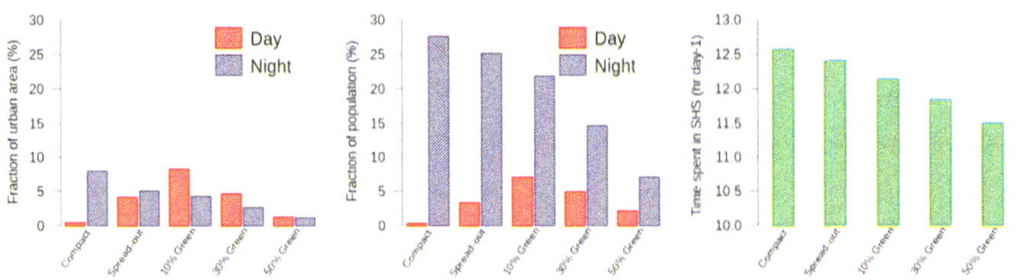

Fig. 11.4 Left panel: Fraction of the Paris urban unit affected by an UHI of 2 °C or more during the night (blue) and during the day (red), according to the different urban expansion scenarios. Middle panel: Same than previously but for the fraction of urban population. Right panel: Number of hours per day spent by population in strong heat stress (SHS) conditions

11.5.2 Landscape Planning and Land Use Modification Around the City

Modifying the existing urban centres such as Paris intra-muros poses obvious difficulties due to technical constraints as well as heritage and architectural preservation. However, the control and regulation of urban climate can also result from a reflection on the development and management of territories on a larger scale, and of the natural spaces surrounding cities. It can be part of a broader approach for the valuation of ecosystem services. This involves promoting both biodiversity and preservation of natural habitats, as well as local resource production (agriculture, wood) and short distribution circuits, and access to leisure areas for inhabitants.

In this sense, the works of the Descartes group (2009) for the Grand Pari(s) initiative focused on development projects for the Paris conurbation by 2030, with the main objective of "rethinking the relationship between the city and the countryside". Scenarios involved major modifications of the land use and land cover around the city, combining (1) the creation of lakes, (2) the extension of forest areas, and (3) the conversion of part of the cereal crops into summer market gardening, but also an adaptation of buildings in suburbs areas by (4) increasing the reflective properties of roofs, walls and roads.

From a climatological point of view, such large-scale scenarios can have a significant impact on local or regional meteorology. The modification of land surface properties alters the energy exchange between surfaces and the atmosphere, as well as the characteristics of the atmospheric boundary layer. This can result in horizontal air temperature gradients and therefore advection of air masses over long distances. Masson et al. (2013) investigated the impact of the Descartes group's scenarios on the local climate of Paris under the 2003 heat wave. They showed that Paris intra-muros benefit from the cool air transported from forested and market gardening areas implemented within a distance of 60 km, and from residential areas equipped with reflective materials. This leads to a maximum attenuation of air temperature of up to 2.5 °C in the first suburbs and 1.3 °C in the Paris intra-muros at night (Fig. 11.3 top left panel).

11.6 Influence of Individual Practices

11.6.1 Impact of Air Conditioning on Microclimate

Climate projections indicate an increase in heat-wave occurrence in the Paris region during the twenty-first century, with events that may become longer and more intense (Lemonsu et al. 2014). The results of the VURCA project showed that it will be difficult, or even impossible, to avoid a systematic use of air conditioning (AC) in summer in the future to ensure acceptable comfort conditions for inhabitants

(Viguié et al. 2020). However, we must be vigilant about the consequences of intensive air conditioning use.

The impact of heat release by air-conditioners on the urban microclimate of Paris has been investigated by de Munck et al. (2013a). Four alternatives were compared by modifying the spatialized inventories of heat and humidity emissions related to the operation of air-conditioners. These included: (1) a reference case without air conditioning; (2) a case taking into account the current density and distribution of air conditioning equipment in Paris, as well as their operation, which results in both dry and wet heat releases outside buildings; (3) a case similar to (2) but with only dry heat release; and (4) a prospective case planning a doubling in air conditioning equipment in Paris in the near future, which would generate twice as much emissions of hot and dry air in the streets.

Comparison of simulated temperatures under these different scenarios showed that hot air emissions from air conditioning have a significant effect on street air temperature, even if emission release is at roof level. Compared to a city without air conditioning, the prospective scenario leads to an amplification of the night-time urban heat island by 2 °C. This effect is very localized in Paris intra-muros (Fig. 11.3 bottom right panel). This not only degrades the thermal comfort conditions in outdoor spaces, but also indirectly in the non air-conditioned dwellings. In addition, heating the exterior environment of buildings results in a positive feedback loop on air conditioning needs and associated energy consumption.

11.6.2 Promoting Moderate Use of Air Conditioning

Contrary to what might have been expected, the adaptation scenarios evaluated by the VURCA project for future heat waves clearly highlighted the predominant role of inhabitants' practices on the efficiency of adaptation measures. Indeed, promoting moderate use of air conditioning, by regulating set-point temperatures to 26 °C in offices and 28 °C in housing instead of 23 °C, made it possible to save 43% of the energy consumption for AC at the large scale of Paris conurbation (Viguié et al. 2020). This gain was much higher than that generated by a policy of massive improvement in the energy performance of buildings (estimated to 17%).

Closing the shutters during the day to limit solar gain inside homes was another individual action that had a strong impact on air conditioning consumption. This practice, which is widespread in cities of southern France on hot summer days (at least for domestic buildings), is much less common in the north of France. However, this was found to reduce summer energy consumption for AC use by 45% (MUSCADE 2014; Viguié et al. 2020).

11.6.3 Water Consumption for Watering Gardens

It is still difficult to conclude on the issue of water resources solely based on present studies, but important messages and lines of research are emerging. Our work has shown that urban vegetation could no longer play its role as a microclimatic regulator by evapotranspiration in case of water stress. The scenarios studied for the MUSCADE and VURCA projects therefore assumed systematic watering of green areas during summer months. This consequently implied considerable volumes of water on the scale of the entire Paris urban unit, i.e. up to 5 Mm^3 day^{-1} of water in summer (Daniel et al. 2018; de Munck et al. 2018). For comparison, this volume represents 17% of the current mean flow of the Seine River but 87% under low flow conditions (especially during heat waves). Not to mention that a 30% decrease in the flow of the Seine river is expected by 2100 due to climate change and increased droughts (Ducharne et al. 2009).

11.7 Conclusions

Heat-wave conditions, such as those experienced during the 2003 event, combined with the UHI effect, induce high night temperatures that can reach 30 °C in Paris. The studies presented here have shown that different kinds of action levers could be implemented to mitigate these effects. They could be applied at city and neighbourhood levels by adapting buildings through technical innovations (i.e. vacuum insulation, cool roofs, etc.) and by greening ground and rooftops where existing urban design allows. Each lever provides a cooling effect but with an efficiency that varies spatially according to the urban typology. Greening is relevant in medium-density urban areas, while reflective materials and pavement watering are most effective in central Paris. It is therefore interesting to note that the combination of levers allows for an increased and additive efficiency. Studies carried out on Paris intra-muros, such as EPICEA, have initiated real interest and commitment from the City of Paris, which remains actively involved in research and experimentation activities. Its orientations for urban planning and environment set out in recent documents for adaptation strategy (2015) and resilience (2016), as well as the new Climate Plan of the city (2018), have been informed by these research projects. They highlight the importance of preserving and reintroducing nature and water into public spaces to create cool islands. In addition, larger-scale planning policies—governing the rules of urban expansion and the development of natural areas surrounding cities—are another interesting way to adapt and regulate local climate.

These two adaptation pathways that refer to different time and space scales lead to a comparable UHI attenuation of about 1.3–1.5 °C. On the other hand, heat discharges associated with significant use of air conditioning degrade outdoor comfort by increasing air temperature by 2 °C as well, which could negate any benefits of other adaptation methods. In this sense, the VURCA and MUSCADE projects both highlighted the indisputable role of the inhabitants' energy behaviour on adaptation strategies. It seems unrealistic to go against a more frequent use of air conditioning in the coming decades in view of climate projections for the Paris region. Such an increase in AC use however could be moderated by improving technologies for building construction and renovation, and above all by raising public awareness of the need to adopt more sustainable behaviours. Encouragingly, the MUSCADE project has given rise to initial thinking on compensation strategies. Masson et al. (2014) showed that, on an annual basis and at the scale of the whole Paris urban unit, local energy production through the installation of solar panels could balance energy consumption for heating and cooling at the end of the century. In addition, de Munck et al. (2018) estimated that the total annual volume of water lost by runoff on impervious surfaces was comparable to the amount of water needed for watering vegetation during summer. However, in both cases, the question of the storage and redistribution of energy and water remains.

Adaptation strategies should be assessed based on multi-sectoral indicators taking into account, as far as possible, the interactions between the multiple mechanisms that occur in urban areas. The present studies already include multiple aspects related to the specificities of Paris. However, the issue of water resources still needs to be explored further using more appropriate modelling tools, which will be possible with the next version of the TEB model (Stavropulos-Laffaille et al. 2018). We can also question the interactions between urban development strategies, urban climate, air pollution (strongly driven by local emissions of buildings, transport and vegetation), and health. Note that a preliminary study of a pollution episode over the Paris region showed that the dynamics of the urban boundary layer could significantly influence the distribution of pollutants above and around the city (Sarrat et al. 2006). Finally, the quantification of CO_2 emissions in cities is crucial in the evaluation of adaptation and mitigation strategies. This would require completing the present studies using a better representation of transport networks and modes, and their evolution over time.

Faced with increasingly high institutional and societal expectations about what the City of tomorrow should be, scientists must address this subject by acknowledging its complexity, and the fact that physical, economic and social phenomena collide on multiple scales of time and space. Anticipating what the urban climate of cities could be in the coming decades (but also associated thermal comfort conditions and energy demand), or even recommending adaptation and/or mitigation strategies, not only requires understanding current city system dynamics but also projecting these for distant time horizons. This raises important methodological questions and challenges that constitute a field of research that is still very exploratory.

References

Adachi SA, Kimura F, Kusaka H, Inoue T, Ueda H (2012) Comparison of the impact of global climate changes and urbanization on summertime future climate in the Tokyo Metropolitan Area. J Appl Meteorol Climatol 51(8):1441–1454. https://doi.org/10.1175/jamc-d-11-0137.1

Adachi SA, Kimura F, Kusaka H, Duda MG, Yamagata Y, Seya H, Nakamichi K, Aoyagi T (2014) Moderation of summertime heat island phenomena via modification of the urban form in the Tokyo Metropolitan Area. J Appl Meteorol Climatol 53(8):1886–1900. https://doi.org/10.1175/jamc-d-13-0194.1

Basara JB, Basara HG, Illston BG, Crawford KC (2010) The impact of the urban heat island during an intense heat wave in Oklahoma City. Adv Meteorol 2010:230365. https://doi.org/10.1155/2010/230365

Bonhomme M, Masson V, Adolphe L, Faraut S (2013) GENIUS: a tool for multi-disciplinary and multi-scalar databases. Paper presented at the AGU Fall Meeting Abstracts, GC53A-1018, 01 December 2013

Bröde P, Fiala D, Błażejczyk K, Holmér I, Jendritzky G, Kampmann B, Tinz B, Havenith G (2012) Deriving the operational procedure for the Universal Thermal Climate Index (UTCI). Int J Biometeorol 56(3):481–494. https://doi.org/10.1007/s00484-011-0454-1

Canouï-Poitrine F, Cadot E, Spira A (2006) Excess deaths during the August 2003 heat wave in Paris, France. Rev Epidemiol Sante Publique 54(2):127–135. https://doi.org/10.1016/S0398-7620(06)76706-2

Christensen JH, Boberg F, Christensen OB, Lucas-Picher P (2008) On the need for bias correction of regional climate change projections of temperature and precipitation. Geophys Res Lett 35(20). https://doi.org/10.1029/2008gl035694

Colombert M (2008) Contribution to the analysis of various means to take into account urban climate in urban planning. PhD thesis, Engineering Science, University of Paris-Est, France

Colombert M, Diab Y, Salagnac J-L, Morand D (2011) Sensitivity study of the energy balance to urban characteristics. Sustain Cities Soc 1(3):125–134. https://doi.org/10.1016/j.scs.2011.05.003

Daniel M, Lemonsu A, Viguié V (2018) Role of watering practices in large-scale urban planning strategies to face the heat-wave risk in future climate. Urban Clim 23:287–308. https://doi.org/10.1016/j.uclim.2016.11.001

de Munck C, Pigeon G, Masson V, Meunier F, Bousquet P, Tréméac B, Merchat M, Poeuf P, Marchadier C (2013a) How much can air conditioning increase air temperatures for a city like Paris, France? Int J Climatol 33(1):210–227. https://doi.org/10.1002/joc.3415

de Munck C, Lemonsu A, Bouzouidja R, Masson V, Claverie R (2013b) The GREENROOF module (v7.3) for modelling green roof hydrological and energetic performances within TEB. Geosci Model Dev 6(6):1941–1960. https://doi.org/10.5194/gmd-6-1941-2013

de Munck C, Lemonsu A, Masson V, Le Bras J, Bonhomme M (2018) Evaluating the impacts of greening scenarios on thermal comfort and energy and water consumptions for adapting Paris city to climate change. Urban Clim 23:260–286. https://doi.org/10.1016/j.uclim.2017.01.003

Descartes (2009) (F. Leclercq, Ateliers Lion, M. Mimram, S. Nivet for the Groupe Descartes). Propositions du Groupe Descartes pour une «Initiative Paris Métropole 2021. Synthesis report 19p. https://www.apur.org/sites/default/files/documents/08_plaquette_Descartes.pdf (in French). Accessed 7 Apr 2020

Dimoudi A, Zoras S, Kantzioura A, Stogiannou X, Kosmopoulos P, Pallas C (2014) Use of cool materials and other bioclimatic interventions in outdoor places in order to mitigate the urban heat island in a medium size city in Greece. Sustain Cities Soc 13:89–96. https://doi.org/10.1016/j.scs.2014.04.003

Dousset B, Gourmelon F, Laaidi K, Zeghnoun A, Giraudet E, Bretin P, Mauri E, Vandentorren S (2011) Satellite monitoring of summer heat waves in the Paris metropolitan area. Int J Climatol 31(2):313–323. https://doi.org/10.1002/joc.2222

Ducharne A, Habets F, Pagé C, Sauquet E, Viennot P, Déqué M, Gascoin S, Hachour A, Martin E, Oudin L, Pagé C, Ribstein P, Sauquet E, Thiéry D, Terray L, Viennot P, Boé J, Bourqui M, Crespi O, Gascoin S, Rieu J (2009) Project RExHySS: climate change impact on the Water Resources and Hydrological Extremes of the Seine and Somme river basins. Final report (in French). http://www.cerfacs.fr/~page/publications/rapport_final_rexhyss.pdf. Accessed 7 Apr 2020

EPICEA (2012) Etude pluridisciplinaire des impacts du changement climatique à l'échelle de l'agglomération parisienne, research project "Paris 2030". Final report (in French). https://www.umr-cnrm.fr/IMG/pdf/epicea-rapport-final_v4.pdf. Accessed 7 Apr 2020

Fiala D, Havenith G, Bröde P, Kampmann B, Jendritzky G (2012) UTCI-Fiala multi-node model of human heat transfer and temperature regulation. Int J Biometeorol 56(3):429–441. https://doi.org/10.1007/s00484-011-0424-7

Hendel M, Colombert M, Diab Y, Royon L (2015) An analysis of pavement heat flux to optimize the water efficiency of a pavement-watering method. Appl Therm Eng 78:658–669. https://doi.org/10.1016/j.applthermaleng.2014.11.060

Hendel M, Gutierrez P, Colombert M, Diab Y, Royon L (2016) Measuring the effects of urban heat island mitigation techniques in the field: application to the case of pavement-watering in Paris. Urban Clim 16:43–58. https://doi.org/10.1016/j.uclim.2016.02.003

House-Peters LA, Chang H (2011) Modeling the impact of land use and climate change on neighborhood-scale evaporation and nighttime cooling: a surface energy balance approach. Landsc Urban Plan 103(2):139–155. https://doi.org/10.1016/j.landurbplan.2011.07.005

IPCC (Intergovernmental Panel on Climate Change) (2007) Climate change 2007: the physical science basis. In: Contribution of Working Group I to the 4th assessment report of the Intergovernmental Panel on Climate Change. Cambridge University Press, Cambridge

IPCC (Intergovernmental Panel on Climate Change) (2014) Climate change 2014: impacts, adaptation and vulnerability: regional aspects. Cambridge University Press, Cambridge

Kounkou-Arnaud R, Desplat J, Lemonsu A, Salagnac J-L (2014) Epicea: étude des impacts du changement climatique à Paris. La Météorologie 42(84):42–48. (in French)

Laaidi K, Zeghnoun A, Dousset B, Bretin P, Vandentorren S, Giraudet E, Beaudeau P (2012) The impact of heat islands on mortality in Paris during the August 2003 heat wave. Environ Health Perspect 120(2):254–259. https://doi.org/10.1289/ehp.1103532

Lambert-Habib ML, Hidalgo J, Fedele C, Lemonsu A, Bernard C (2013) How is climatic adaptation taken into account by legal tools? Introduction of water and vegetation by French town planning documents. Urban Clim 4:16–34. https://doi.org/10.1016/j.uclim.2013.04.004

Lemonsu A, Masson V, Shashua-Bar L, Erell E, Pearlmutter D (2012) Inclusion of vegetation in the Town Energy Balance model for modelling urban green areas. Geosci Model Dev 5(6):1377–1393. https://doi.org/10.5194/gmd-5-1377-2012

Lemonsu A, Kounkou-Arnaud R, Desplat J, Salagnac J-L, Masson V (2013) Evolution of the Parisian urban climate under a global changing climate. Clim Chang 116(3):679–692. https://doi.org/10.1007/s10584-012-0521-6

Lemonsu A, Beaulant AL, Somot S, Masson V (2014) Evolution of heat wave occurrence over the Paris basin (France) in the 21st century. Clim Res 61(1):75–91

Lemonsu A, Viguié V, Daniel M, Masson V (2015) Vulnerability to heat waves: impact of urban expansion scenarios on urban heat island and heat stress in Paris (France). Urban Clim 14:586–605. https://doi.org/10.1016/j.uclim.2015.10.007

Masson V (2000) A physically-based scheme for the urban energy budget in atmospheric models. Bound-Layer Meteorol 94(3):357–397. https://doi.org/10.1023/A:1002463829265

Masson V, Lion Y, Peter A, Pigeon G, Buyck J, Brun E (2013) "Grand Paris": regional landscape change to adapt city to climate warming. Clim Chang 117(4):769–782. https://doi.org/10.1007/s10584-012-0579-1

Masson V, Marchadier C, Adolphe L, Aguejdad R, Avner P, Bonhomme M, Bretagne G, Briottet X, Bueno B, de Munck C, Doukari O, Hallegatte S, Hidalgo J, Houet T, Le Bras J, Lemonsu A, Long N, Moine MP, Morel T, Nolorgues L, Pigeon G, Salagnac JL, Viguié V, Zibouche

K (2014) Adapting cities to climate change: a systemic modelling approach. Urban Clim 10:407–429. https://doi.org/10.1016/j.uclim.2014.03.004
MUSCADE (2014) Modélisation Urbaine et Stratégies d'adaptation au Changement climatique pour Anticiper la Demande et la production Energétique. Research project ANR-09-VILL-0003, final report (in French). http://www.umr-cnrm.fr/vurca/IMG/pdf/muscade_rapportfinal.pdf. Accessed 7 Apr 2020
Oke TR, Mills G, Christen A, Voogt JA (2017) Urban climates. Cambridge University Press, Cambridge
Pascal M, Wagner V, Corso M, Laaidi K, Ung A, Beaudeau P (2018) Heat and cold related-mortality in 18 French cities. Environ Int 121:189–198. https://doi.org/10.1016/j.envint.2018.08.049
Perini K, Magliocco A (2014) Effects of vegetation, urban density, building height, and atmospheric conditions on local temperatures and thermal comfort. Urban For Urban Gree 13(3):495–506. https://doi.org/10.1016/j.ufug.2014.03.003
Pigeon G, Zibouche K, Bueno B, Le Bras J, Masson V (2014) Evaluation of building energy simulations with the TEB model against EnergyPlus for a set of representative buildings in Paris. Build Environ 76:1–14
Santamouris M (2013) Using cool pavements as a mitigation strategy to fight urban heat island—a review of the actual developments. Renew Sust Energ Rev 26:224–240. https://doi.org/10.1016/j.rser.2013.05.047
Sarrat C, Lemonsu A, Masson V, Guedalia D (2006) Impact of urban heat island on regional atmospheric pollution. Atmos Environ 40(10):1743–1758. https://doi.org/10.1016/j.atmosenv.2005.11.037
Stavropulos-Laffaille X, Chancibault K, J-m BRUN, Lemonsu A, Masson V, Boone A, Anthony AH (2018) Improvements to the hydrological processes of the Town Energy Balance model (TEB-Veg, SURFEX v7.3) for urban modelling and impact assessment. Geosci Model Dev 11(10):4175–4194. https://doi.org/10.5194/gmd-11-4175-2018
Stone B, Rodgers MO (2001) Urban form and thermal efficiency: how the design of cities influences the urban heat island effect. J Am Plan Assoc 67(2):186–198. https://doi.org/10.1080/01944360108976228
Takahashi R, Asakura A, Koike K, Himeno S, Fujita S (2010) Using snow melting pipes to verify water sprinkling's effect over a wide area. In: NOVATECH, 7th international conference on sustainable techniques and strategies in urban water management, Lyon, France, 27 June–1 July
Tan J, Zheng Y, Tang X, Guo C, Li L, Song G, Zhen X, Yuan D, Kalkstein AJ, Li F, Chen H (2010) The urban heat island and its impact on heat waves and human health in Shanghai. Int J Biometeorol 54(1):75–84. https://doi.org/10.1007/s00484-009-0256-x
Trémeac B, Bousquet P, de Munck C, Pigeon G, Masson V, Marchadier C, Merchat M, Poeuf P, Meunier F (2012) Influence of air conditioning management on heat island in Paris air street temperatures. Appl Energy 95:102–110. https://doi.org/10.1016/j.apenergy.2012.02.015
van der Linden P, Mitchell JF (2009) ENSEMBLES: climate change and its impacts: summary of research and results from the ENSEMBLES project. Met Office Hadley Centre, Exeter
Viguié V, Hallegatte S, Rozenberg J (2014) Downscaling long term socio-economic scenarios at city scale: a case study on Paris. Technol Forecast Soc Change 87:305–324. https://doi.org/10.1016/j.techfore.2013.12.028
Viguié V, Lemonsu A, Hallegatte S, Beaulant A-L, Marchadier C, Masson V, Pigeon G, Salagnac J-L, Somot S (2020) Early adaptation to heat waves and future reduction of air-conditioning energy use in cities. Environ Res Lett 15:075006
VURCA (2013) Cities vulnerability to future heat waves & adaptation strategies. Research project ANR-08-VULN-013, final Report (in French). http://www.umr-cnrm.fr/vurca/IMG/pdf/vurca_final_report_1_1-1.pdf. Accessed 7 Apr 2020

Chapter 12
Urban Climate in Urban Planning: The Experience from Stuttgart

Ulrich Reuter and Rainer Kapp

Abstract Because of the special topographical and orographically situation of the City of Stuttgart lying in a basin surrounded by hills, urban climate in urban planning has a long tradition since 1938. The global climate change is an additional challenge. In this chapter first there is a summary of the instruments and basics that are used to take into account urban climate in urban planning. Besides of measurements and numeric calculations an important point is the Urban Climate Atlas with a lot of thematic maps with climate information and with planning hints, which allow the city planners to take care of climate in planning and microclimate modelling. Subsequently some best practice examples of climate in urban planning are demonstrated with the aspects of green in the city, ventilation of the city, climate-optimised inner city development and climate-optimised construction of public spaces. Concrete measures shown are green belts, roadside greenery, green rails, green roofs and fresh air corridors. At the same time successful methods and procedures are revealed implementing these issues. It is described how urban climate is established in the Stuttgart city administration affecting or controlling administration activities. Finally the chapter deals with the question of important climate legislation possibilities, success factors and obstacles to take into account climate in urban planning.

Keywords Urban climate · Urban planning · Climate atlas · Green city · Ventilated city · Hillside development plan · Inner city development · Climate legislation · City governance

---

U. Reuter (✉)
The University for Applied Sciences, Stuttgart, Germany

R. Kapp
Department of Urban Climatology, Office for Environmental Protection, Stuttgart, Germany
e-mail: Rainer.Kapp@stuttgart.de

C. Ren, G. McGregor (eds.), *Urban Climate Science for Planning Healthy Cities*, Biometeorology 5, https://doi.org/10.1007/978-3-030-87598-5_12

Key Messages

- It is an important issue to take into account urban climate in urban planning. The global climate change is an additional challenge.
- Taking into account urban climate in urban planning is an ongoing procedure and needs legislation and the implementation in the city governance, e.g. with an own department for climatology.
- Detailed climatic information are necessary to take into account climate in planning. The development and working out of a climate atlas is recommended.
- Best practice measures are green belt policy, green space planning, a good ventilation in the city and a climate-optimised inner city development.
- The experience from Stuttgart shows that this strategy works.

12.1 Introduction

The total area of the city covers 207 km^2 with 50% settlement and 24% forest. The remainder are traffic, water, recreation and agricultural area. In some districts the population density is more than 11,000 inhabitants per km^2.

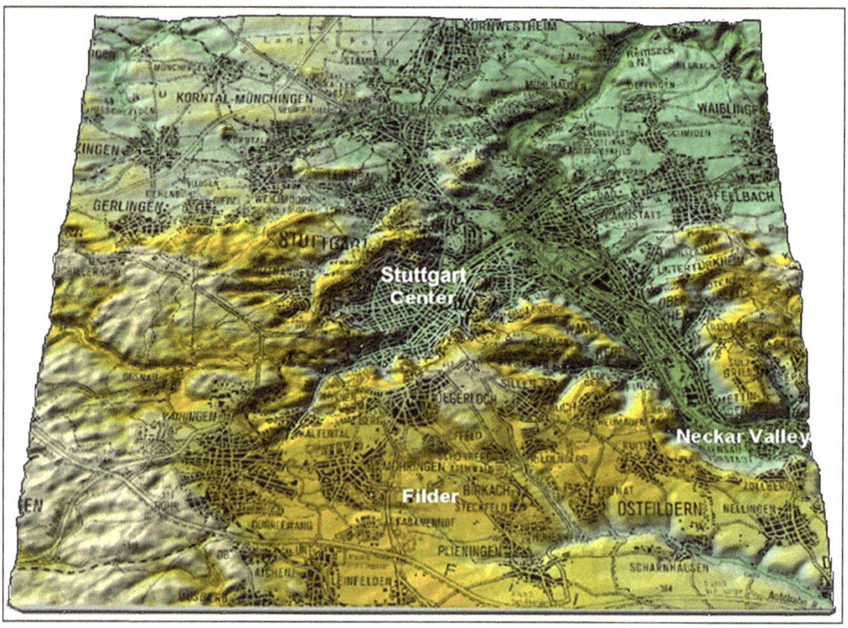

Fig. 12.1 Orographic situation of Stuttgart city centre. (Source: Department of Urban Climatology, Municipality of Stuttgart)

The local climate of the city of Stuttgart is influenced by its site in the valley of the river Neckar and by the lee situation with the Black Forest in the West and the Swabian Alb in the South. In addition, the city centre is situated in a basin with hills around (Fig. 12.1). This influences the local climate.

The climate of Stuttgart is characterised by mild temperatures and a strong urban heat island effect in the city. The annual mean temperature in the city centre is about 10 °C and about 8.5 °C on the hills around the centre. In summer this leads to a markable number of days with heat stress. The main wind direction is Southwest. This is why the hills around the city shield the centre with a mean wind velocity of only 1.5 m/s with an insufficient natural air ventilation. The elevated outskirts have a mean wind velocity of about 2.5 m/s, which is still low. This is because Stuttgart is situated in the southwestern part of Germany with in general low mean wind speed.

The special situation of climate and air quality have long been of interest in Stuttgart. Already in the year 1938 the city council decided to employ a meteorologist to care about urban climate in relation to urban development. This was the beginning of the Department of Urban Climatology with more than ten employees nowadays.

Climate change in addition causes a strong challenge. This means the need to reduce carbon emissions (mitigation) and the adaptation to climate change, for example, by taking into account the urban climate in urban planning (Landeshauptstadt Stuttgart 2010).

The efforts invested over decades by the Section of Urban Climatology of the Office for Environmental Protection in Stuttgart to protect the local climate are also effective to adjust to climate change, especially concerning the increasing urban heat island effects by the global warming.

The main basis of legislation that takes into account climate in urban planning is the Federal German Building Code (Bundesrepublik Deutschland 2004), article 1, which sets out the objectives and principles of construction management planning and includes the consideration of environmental concerns. According to this legislation the "air" and the "climate" are raised to the status of planning factors.

In the Federal German Building Code also a list of possible regulations is defined, which can be used in development plans in order to implement climate-friendly planning principles. On this basis an urban planning directive considering the global climate change has been published (Verein Deutscher Ingenieure e.V. 2020).

Given the long history of awareness of the interaction between urban development and climate in Stuttgart the purpose of this chapter is to describe how urban climate science (urban climatology) has been used to inform and to affect urban design and planning in this city. The rest of this chapter is organised as follows. Section 12.2 summarises the instruments and the basic climate information used and taken into account in urban planning in Stuttgart followed by a consideration of practice examples in Sect. 12.3. In Sect. 12.4 the main principles of climate and planning legislation are outlined along with a discussion of the success factors and obstacles related to incorporation of climate information in urban planning. The application of knowledge about urban climatology in local urban planning meanwhile is subject of many projects implemented in evermore cities national (e.g.

Stadt Freiburg 2019) and international (e.g. Mahlkow et al. 2016; Webb 2017; Hidalgo et al. 2019).

12.2 Basic Materials for Climatic Information

Before taking into account urban climate in urban planning basic facts concerning urban climate and air pollution are necessary. The city of Stuttgart did all this of its own and with external partners such as the German Weather Forecasting Service or the State Environmental Protection Agency and with university institutes (e.g. Ketterer and Matzarakis 2016) and private special bureaus. Vehicle-based measurements of climate and air pollution with mobile instrumentation and measurements at fixed stations were done. In addition, infrared-thermographic measurements from aircraft were done, as well as detailed numerical simulations concerning the wind field, cold air flows and the general climatic situation in Stuttgart. On the basis of all these information in 1992, the first Climate Atlas as a basis of the common land use plan for Stuttgart and the associated municipalities was worked out (Nachbarschaftsverband Stuttgart 1992). In 2008, an actualised digital climate atlas was developed (Verband Region Stuttgart 2008).

The methods and results in the context of drawing up the Climate Atlas have influenced the VDI (Association of German Engineers) Guideline 3787, section 1 (Verein Deutscher Ingenieure e.V. 2015). Parts of the Climate Atlas are available via the internet as "Climate Viewer" (Landeshauptstadt Stuttgart 2011). The Viewer covers urban climate, air pollution and noise providing detailed information for the citizens of Stuttgart with more than 40 different data. The data are visualised in an easy way in detailed maps for the city of Stuttgart. This helps citizens to have detailed information to take part in municipal planning processes.

The State of Baden-Württemberg has also published knowledge about urban climate and urban planning in the "Climate Booklet for Urban Development-References for Urban Planning" (Ministerium für Verkehr und Infrastruktur Baden-Württemberg 2015). This booklet up to now serves as a good example in the field of urban climate and urban planning in Germany and other countries.

12.2.1 Climate Atlas Region Stuttgart 2008

With the background of the needs of new residential and business areas and large railway planning considerations (Stuttgart 21) in the Stuttgart region a lot of fundamental studies on climate and air quality were done as basis for the land use planning. Climatic aspects such as cold air production and cold air flow, which are important for human health, are regional aspects. So the Climate Atlas from 1992 had to be updated, because data were only available for parts of the region. The "Verband Region Stuttgart" (a political organisation for the regional development of

Greater Stuttgart) developed the Climate Atlas for the Region of Stuttgart (Verband Region Stuttgart 2008). The work was done by the climatologists of the Department of Urban Climatology in the Office for Environmental Protection of Stuttgart. The new Climate Atlas has become a very good basis for all planners and planning processes in the Stuttgart Region. Planners normally work with maps. So it was important also to have all climatic and air quality information in cartographic representations. A broad worldwide overview about urban climatic maps for a sustainable urban planning is given by Ng and Ren (2015).

The necessity for addressing climatic concerns in the context of updating the regional plan is not least the compelling result of the new legal demands of the Strategic Environmental Impact Assessment *EIA* (European Union 2017). A qualified analysis of the implications of the regional plan for the protective good Climate was a major requirement for the legal security of the planning.

The Climate Atlas for the Region of Stuttgart was constructed on a digital basis. It is GIS based and covers the whole area of the region (3654 km^2), including all 179 towns and municipalities. This is a significant functional improvement compared to the 1992 Climate Atlas. New tools were integrated, such as the possibility to calculate wind fields or cold air flow simulations at any location of the region.

The Climate Atlas is published in printed form. A CD-ROM with a comprehensive set of maps as a PDF file is added to the printed version. Texts and maps are made available as PDF files on the homepage of the Verband Region Stuttgart (Verband Region Stuttgart, Körperschaft des öffentlichen Rechts, Stuttgart 2008) and the homepage of the Department for Urban Climate (Abteilung Stadtklimatologie, Amt für Umweltschutz der Landeshauptstadt Stuttgart 2008; http://www.stadtklima-stuttgart.de).

In addition to the general accessibility of the information, the data are also available for towns and municipalities for the use in their local planning and zoning plans. The importance and the need of improvement of such information and data are indicated by Grimmond et al. (2010).

After a general description of the urban climatic situation of the Stuttgart Region, the climatic and air quality concerns for planning are presented in text and cards. Some of the resources associated with the Climate Atlas are described below.

12.2.1.1 Thermal Maps

Thermal maps show in a snapshot the surface temperatures of the city on a fine day in two situations, first in the evening after sunset and second in the next morning before sunrise. The temperature differences between the two measurements done by aircraft allow to study the nocturnal cooling of the city structure and they give hints on the urban heat island effects. In addition, maps concerning the heat stress in the city on the basis of numerical calculations for the situation nowadays and as scenario for the future with the effects of global change also give further information on the urban heat island.

12.2.1.2 Production and Drainage of Cold Air, Wind Field Calculations

Besides the thermal situation in the city, the wind in the city is very important. The Climate Atlas illustrates in different maps the mean wind velocity in the urban structure, the prevailing wind directions, especially influenced by local nocturnal cold air flows. Maps also show the thickness and intensity of this cold air drainage, calculated numerically based on digital height and land use models. The nocturnal cold air flows are important for the cooling down of the built-up areas and for the transport of air pollution. So these maps are valuable information for planning and zoning concepts.

In addition, the Climate Atlas shows all the measured wind fields in the region graphically and allows the synthetic calculation of wind roses at any location in the whole Stuttgart Region.

12.2.1.3 Climate Analysis Maps

The topographic maps, maps of climate elements, land use maps, aerial photography and more are the basis for the climate analysis maps, which integrate all available information in a set of maps, which is a substantial part of the Climate Atlas (Fig. 12.2). This set of maps is a compact summary of all climatic and air quality information, which is especially important for changes in land use. It is a set of maps, because it is constructed in the scale of the land use plan. The whole Stuttgart Region could not be illustrated in one map.

One main feature of the maps is "Climatopes". Climatopes designate spatial units, in which the meteorologically most important factors are relatively

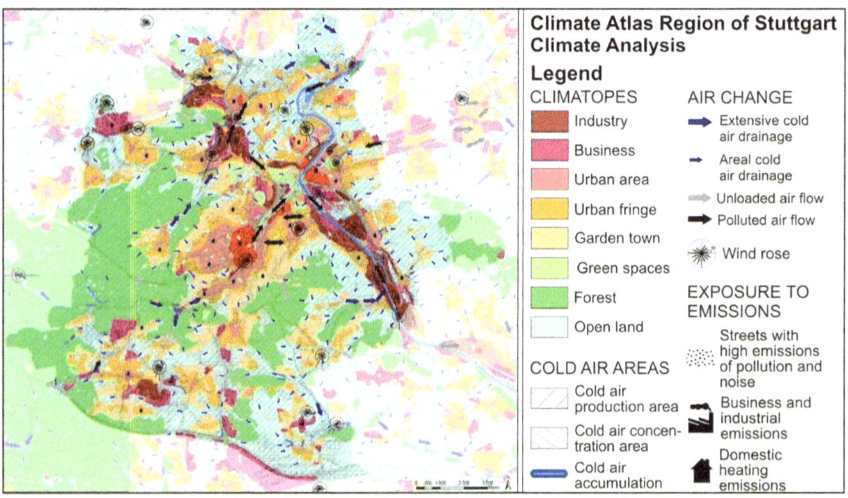

Fig. 12.2 Climate Analysis Map of Stuttgart (Verband Region Stuttgart 2008)

homogeneous and the microclimatic characteristics do not differ considerably. The following factors have a significant climate impact: Land use, density of development, sealing level, surface structure, roughness, relief, vegetation structure and type, and location in the urban area (Verein Deutscher Ingenieure e.V. 2015).

About ten different climatopes are defined: water body climate, open land climate, forest climate, urban green space climate, garden city climate, suburban climate, city climate, inner city climate, commercial/industry climate and climate of railway lines.

The classification of climatope and cold-air collection areas is not parcel-specific. The scale allows ranges up to 100 m. Technically detailed appraisals are necessary for more precise results.

Further on the climate analysis maps illustrate areas with cold air production and cold air drainage as well as obstacles to cold air drainage and areas with cold air stagnation, for example, in valleys or in lowland levels. Also important ventilation lanes caused by regional winds and wind field changes by building structures are marked.

There also exist signatures for air pollution by traffic, industry and domestic heating.

12.2.1.4 Maps with Recommendations for Planning

In addition to the numerous climatic maps, the Climate Atlas includes maps with recommendations for planning in different area classifications. These maps could give support to the city planners, who are not so much interested in climatic details but primarily in the question, what are from the point of view of urban climate the planning requires for a specific area, including the challenges and effects of global climate change. The indications refer to the effects of structural uses and changes of use on the climatic situation (Fig. 12.3).

The recommendations are a planning related summary of all the different climatic aspects in the climate and climate analysis maps, giving hints to the planners how sensitive land use changes are to urban climate. So the planners are able to develop planning measures, which can help to optimise the climatic conditions in the planning and zoning process.

Planning recommendations are not developed for single parcels at the small scale but more to a range of about 100 m. In addition to the recommendations of the Climate Atlas in some cases detailed climatic questions have to be investigated in special expertise, especially in climatic and air quality very sensitive areas.

The following principles form the basis for the planning recommendations in the Climate Atlas (Verband Region Stuttgart 2008):

1. Areas of vegetation have an important effect on the local climate, since on the one hand they cause the nightly fresh/cold-air production, and on the other they exert a balancing thermal effect when present as a high proportion of the land coverage. Green spaces in the city and nearby areas exert a positive influence on

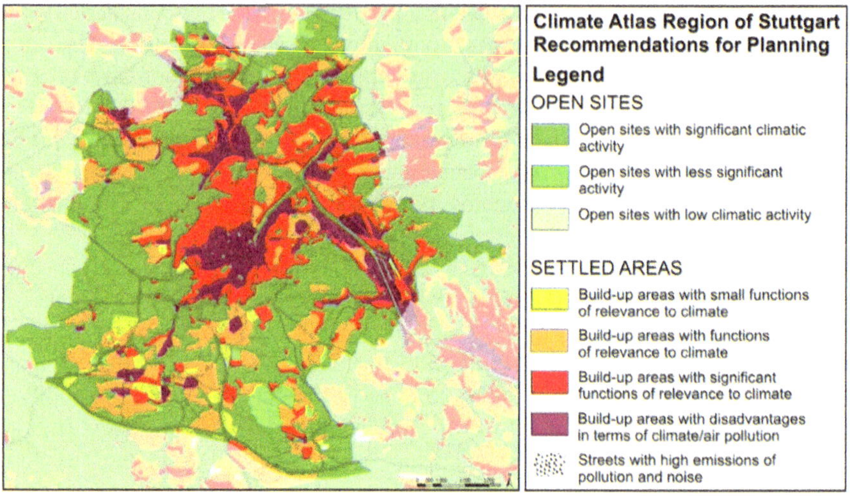

Fig. 12.3 Map with recommendation for Planning in Stuttgart (Verband Region Stuttgart 2008)

their immediate vicinity in a microclimatic sense; vegetation on the border of developments also contributes to air exchange. Larger, connected green spaces represent the climatic and air quality potential for regeneration. Particularly in the present spatial context of the built-up area, such green spaces are very important for air exchange. Consequently as far as possible, open spaces should not be converted to development from a climatic perspective.

2. Development in valleys can also be judged as negative in general, since the movement of cold, fresh air takes place in valleys under weak wind conditions. Valleys also serve as air delivery corridors for stronger regional winds.
3. Hillsides in extended built-up areas should remain undeveloped, especially when development exists in valleys, since intensive cold- and fresh-air transport occurs here. The same is valid for gullies and ridges along these hillsides.
4. Saddle-like topography on the backside of mountains serve as air induction corridors and should not be developed.
5. From a climatic and air quality perspective encircling development with as much green space as possible as well as crisscrossing it with green corridors oriented to topographic features (e.g. ventilation passages; air induction corridors) in order to support air exchange is recommended.
6. Urban sprawl from numerous developments strewn across the landscape, as well as the emergence of disruptive belts of built-up areas (e.g. through the convergence of neighbouring communities), are to be avoided. Urban development must be accompanied by close, large, fresh- and cold-air production areas and ventilation corridors.

12.3 Application Experience in Stuttgart

In this section best practice examples are presented of how to take into account urban climate in urban planning.

12.3.1 Green in the City

Green space planning is the most likely municipal measure achieving the aims of urban climatology and climate protection. This fact necessitates an appropriate green belt policy. The provision of green areas and open spaces benefits the appeal of the city in equal measure.

Green spaces serve as place-keepers, and in this way eliminate other uses which could negatively impact climate protection due to potential emissions of relevance for the climate. Where low and less dense vegetation exists, green spaces also promote ventilation within built-up areas. This in turn reduces exposure to pollutants and dissipates the buildup of heat and thermal stress. The significance of vegetated areas as a heat-sink through evapotranspiration is important.

According to the Climate Atlas, the city of Stuttgart is defined as a city core climatope with a heat island effect and significant influences on the wind field and thus ventilation. In terms of planning this hints as a need to characterise the city core as a built-up area with disadvantages in terms of climate and air pollution with the need for renewal from an urban climate perspective.

Besides forests, agricultural and other green spaces, the following possibilities for greening in the city exist.

12.3.1.1 Greened-over Urban Railway Tracks (Grass Tracks)

Stuttgart forwards the construction of tram lines with grassed-over tracks (Fig. 12.4). The total rail length of the tram-train is over 260 km, of which 60 km are green rails. As many rails tracks are underground these cannot be greened.

Grass track constructing is a comparatively complex process and increases the laying costs for new track sections considerably. In the long term, certain savings may be expected in the process of track maintenance, as the cleaning and exchanging of the rail ballast and tamping are no longer required. Mowing of the track lawn is still performed manually but is relatively simple, and in the future this process is certain to be further rationalised through the use of mechanised alternatives.

In individual cases it has succeeded to plant greenery over existing tracks without reconstructing the base.

Fig. 12.4 Green rails in Stuttgart. (Source: Department of Urban Climatology, Municipality of Stuttgart)

12.3.1.2 Roadside Greenery

There are over 35,000 trees planted alongside Stuttgart's streets and all traffic islands have already been greened over. By maintaining and expanding this significant existing plant stock, major urban climatological air quality objectives and also municipal climate protection targets can be pursued alongside the process of implementing attractive urban planning proposals. The enhanced attraction of planted areas for pedestrians and cyclists (Fig. 12.5) and also the associated reduction in the traffic speed and volume of private road users where renaturation projects are implemented (reduced lane widths for cars) deserve particular mention.

Extensive greening brings about bioclimatic benefits, and avenue-style tree planting not only provides a CO_2 reducing effect, which is beneficial as a means of climate protection, but also offers a way of adapting to climate change (providing shade, cooling, a more even moisture balance). Also importantly, roadside trees with dense foliage help absorb air pollution and filter particulates from the ambient air. The uncapped soil area accommodating the roadside planting is able to absorb and store rainwater, helping to relieve the stress on the sewage system, which is likely to occur with the expected increasing frequency of heavy rainfall.

Fig. 12.5 Roadside trees in Stuttgart. (Source: Department of Urban Climatology, Municipality of Stuttgart)

12.3.1.3 Roof Greening

Roof greening can also be set out as a legal requirement in the development plan. The benefits of roof greening in terms of the ecology, building structure and architectural/urban planning, have been extensively analysed in the past and have been scientifically proven (Fig. 12.6).

To increase the extent of greening in Stuttgart, houses with flat roofs normally have to be built with green roofs. Since many years ago there has been a municipal grant programme for citizens for the renewal of roofs with roof greening and for the greening of courtyards and facades.

Roof greening helps to alleviate the surface temperature extremes both over the diurnal and seasonal cycles. This brings the annual roof surface temperature fluctuation of around 100 °C on non-greened roofs down to only 30 °C where planting is

Fig. 12.6 Green roof in Stuttgart. (Source: Department of Urban Climatology, Municipality of Stuttgart)

present. The reduced thermal load means lower material wear of the roof construction, which serves to reduce costs.

In summer, a large proportion of the solar energy that radiates a greened roof is used to evaporate water, which helps to provide protection against summer heat. In winter, the vegetation layer and the roof substrate create an additional component layer to reduce the passage of heat to the atmosphere. As a result, roof greening makes a modest contribution towards saving energy used for heating in winter, and for cooling in the summer. This in turn reduces the release of damaging greenhouse gases and helps protect the climate as one of the natural bases for life. Other advantages of green roofs are the retention of rainwater, dust fixation and new living spaces for animals.

In order to maximise the positive environmental aspects of using solar energy, alongside the potential for greening, flat roofs also offer ideal conditions for the installation of a solar thermal or photovoltaic system.

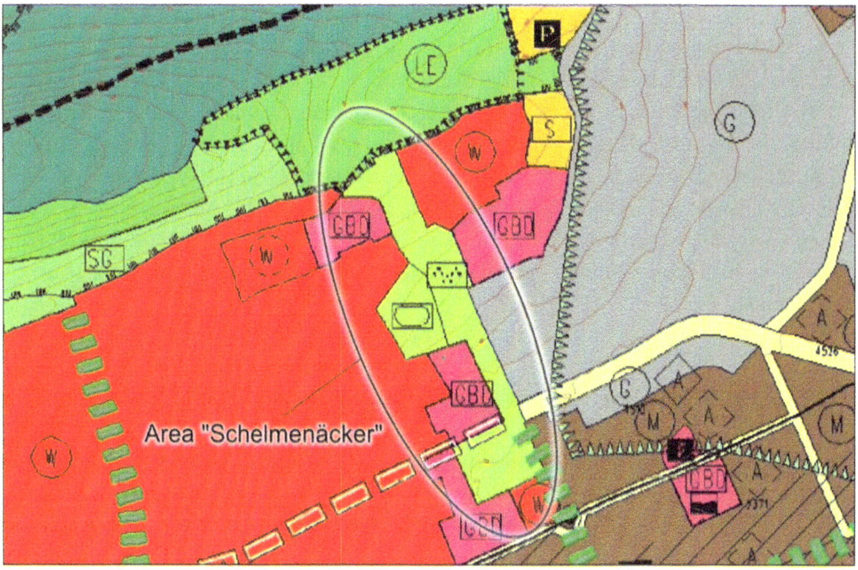

Fig. 12.7 Urban land-use plan "Schelmenäcker" with green corridor (Landeshauptstadt Stuttgart, Amt für Stadtplanung und Stadterneuerung 2010)

12.3.2 Ventilation in the City

Due to their beneficial orographic situation valleys constitute natural green belts and at the same time preferred pathways for ventilation. These must be implicitly kept free, given that aspects of landscape and nature conservation also support the urban climatology arguments.

The planning ethos is set in favour of retaining undeveloped fresh air corridors by the Land Use Plan, in which the structures to be developed should be already set out in the overriding regional plan by the depicted green belts.

12.3.2.1 Case Example: Structural Expansion in a Hillside Area

A first planning provides the extension of an existing residential building area (Schelmenäcker) near a hillside zone extending down from a green space to the dense town centre envisaging development was intended through the whole corridor. By means of the information and presentations in the Climate Atlas a green divide of around 100 m in width could be prevailed. In according to that the conditions were created in which the climatic and ecological balancing function of the area could be able to effectively impact the town centre. In particular, the possibility of air exchange as the result of a cold air downslope flow close to the ground has to be highlighted. Figure 12.7 shows the urban land-use plan Schelmenäcker with a green corridor.

Figures 12.8 and 12.9 show the climatic situation and the recommended planning options as presented in the Climate Atlas for the region of Stuttgart (Verband Region Stuttgart 2008). Figure 12.10 shows the green corridor.

The result was a completely changed development concept which was implemented in the form of two separate development islands, each developed outward from their outer boundaries instead of centrally across the hillside zone and separated by a wide green divide. The illustrations show also the use made of the green space created by the changed plan. This not only upgrades the surrounding residential areas, but also assumes a climatic function which has a positive impact on the town centre.

12.3.2.2 Hillside Development Plan

In the form of the Hillside Development Plan (Rahmenplan Halbhöhenlagen) (Landeshauptstadt Stuttgart 2008), what has been termed a "different planning model" exists for the districts of Stuttgart's inner city basin. Performed prior to the process of urban Land Use Planning, this model has a relevant role to play in urban planning and has helped to ensure the creation and maintenance of green and open spaces in the sensitive system existing in the hillside areas.

The hillside areas, which are not be built on help support thermally induced air exchange close to the ground. They contribute to improved air quality conditions and nightly cooling in summer hot spells.

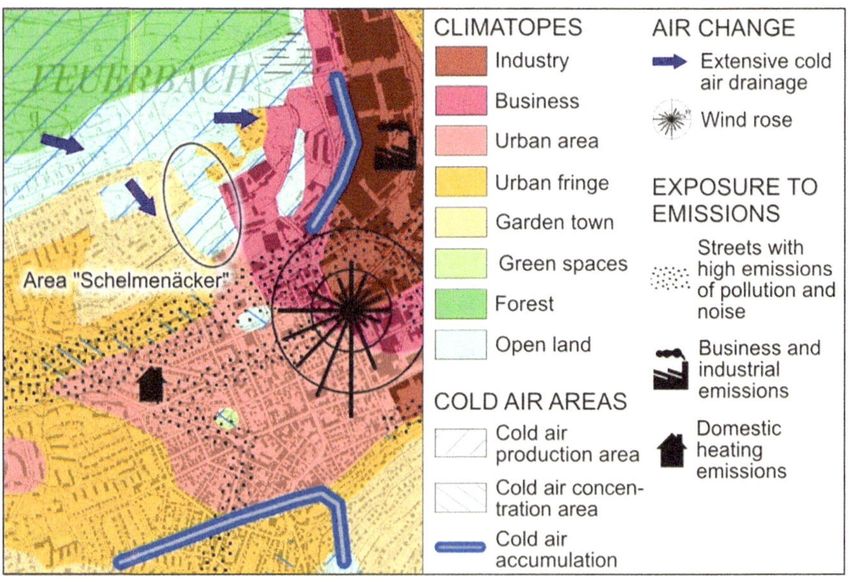

Fig. 12.8 Climate analysis map Schelmenäcker district (Verband Region Stuttgart 2008)

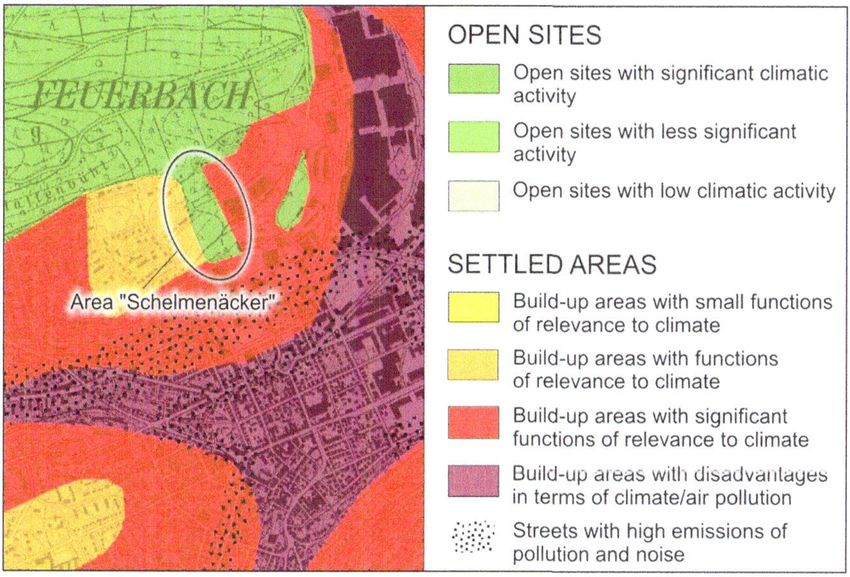

Fig. 12.9 Map with recommendations for Planning "Schelmenäcker" district (Verband Region Stuttgart 2008)

Fig. 12.10 Green corridor "Schelmenäcker" district. (Source: Department of Urban Climatology, Municipality of Stuttgart)

The nightly downslope flows of cold air are crimped by extended hillside development in a mechanical and also in a thermical way.

The quality areas summarising the climatological arguments are assigned in relation to distinctive requirements and measures outlined by the plan (Fig. 12.11).

At the same time, the outline plan also takes consideration of the fact that isolated urban climatological assessments of minor individual building projects encounter problems of scale if the changes anticipated as a result of the building project have been described in quantitative terms.

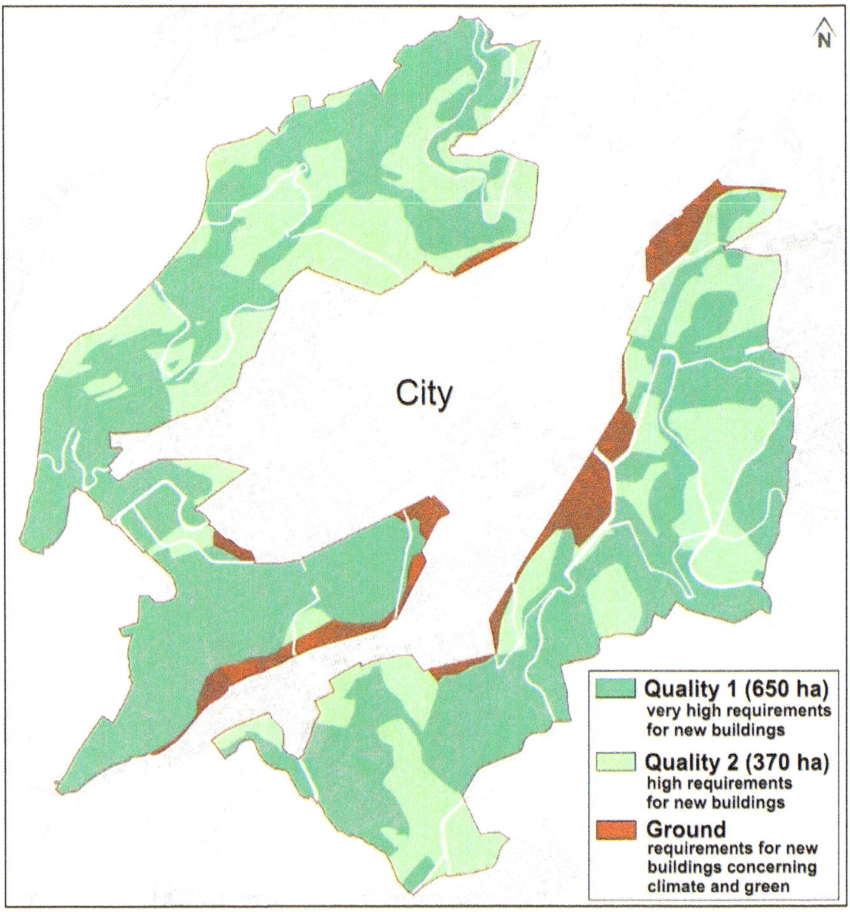

Fig. 12.11 Hillside Development Plan Stuttgart (Landeshauptstadt Stuttgart 2008)

12.3.3 Inner Development Before Outer Development

The Municipality of Stuttgart works towards the sustained development of the city based on the principle of "inner development before outer development".

A structural concentration can contribute towards reduced energy consumption and so help to protect the climate and to preserve cold air generating areas. At the same time it serves to intensify the heat island effect and so may be considered counterproductive thereby generating a new challenge for planning; the prioritisation of the inner city development induces a densification and an associated limit in quality with respect to urban climate. In order to counteract the negative effects of this densification, functionally networked free spaces and their careful design are needed, as well as minimal soil capping, the thermal insulation of houses, and also greening and shading effects using deciduous trees.

Additionally, there is an urgent need for adaptation to climate change and its local consequences. In this context measures like unsealing and revegetating existing courtyards, particularly in redevelopment areas, is required (Fig. 12.12). Furthermore it is important to pay attention to the local nocturnal cold air flows from the surrounding country into the city centre and to improve ventilation. The integration of such factors contributes to achieving an effective limitation of heating of the urban/residential environment and a reduction of thermal stress.

Fig. 12.12 Example of climate-optimised inner development (Area "Rossbollengässle" in Stuttgart-West). (Source: Department of Urban Climatology, Municipality of Stuttgart)

12.3.4 Climate-Optimised Inner City Development

The consequences of a high degree of soil capping and a greater building density are known to be overheating to the detriment of the bio/urban climate and the frequent occurrence of a sweltering and oppressive atmosphere.

Fundamentally the inner city development in favour of larger cohesive free spaces and cold air generating areas can be viewed positively. Despite this—particularly also against the background of the legally regulated simplification of planning projects for inner development—it is important to work towards a *climate-optimised* concentration of urban structures.

Green corridors and spaces that are well designed have been shown to ensure ventilation throughout the area, bringing numerous climate benefits such as cooling down the air temperature which buffers climate stress during heat waves and reduces air pollution, while bringing many other benefits such as biodiversity conservation, carbon sequestration and recreational benefits.

The *Valley Bottom Development Outline Plan* (Landeshauptstadt Stuttgart 2018) is a planning instrument which on the basis of statements relating to climate adaptation evinces measures counteracting heat stress and improving the quality of residential environments. This Outline Plan is successfully engineered using the tools provided by urban climatology hand in hand with the legal instruments of spatial planning. As will be shown in the following, the urban heat island effect can be directly and indirectly counteracted by the implementation of a sensible green belt policy which seeks to maintain fresh air lanes and air exchange corridors. For those subscribing to the adaptation agenda, this type of urban development measure means that the city's population can be shielded from the more unpleasant aspects of climate change such as rising air temperature.

Green axes and networks are essential for obtaining climate objectives while small isolated green spaces (<1 ha) have only a minimal remote impact to the surrounding. However, this consideration does not in any way detract from the temperature-sinking importance of all vegetation areas.

As a consequence the Valley Bottom Development Plan pursues two main objectives related to urban climate issues, namely:

1. The elaboration and depictive representation of a green network including roadside greenery and building greening (see Sect. 12.3.1, Figs. 12.13 and 12.14).
2. For the purpose of reducing thermal stress in residential environments a suitable urban design strategy has been proposed, developed and applied as a long-term prevention method, based on the former urban development plan.

It includes as portrayed in Fig. 12.15:

- Gaps between the buildings to facilitate the ventilation of the inner area.
- Greened courtyards.
- Buildings inside the block with green roofs and limited height.
- Buildings around the block should be equipped with green roofs, but it is not compulsory.

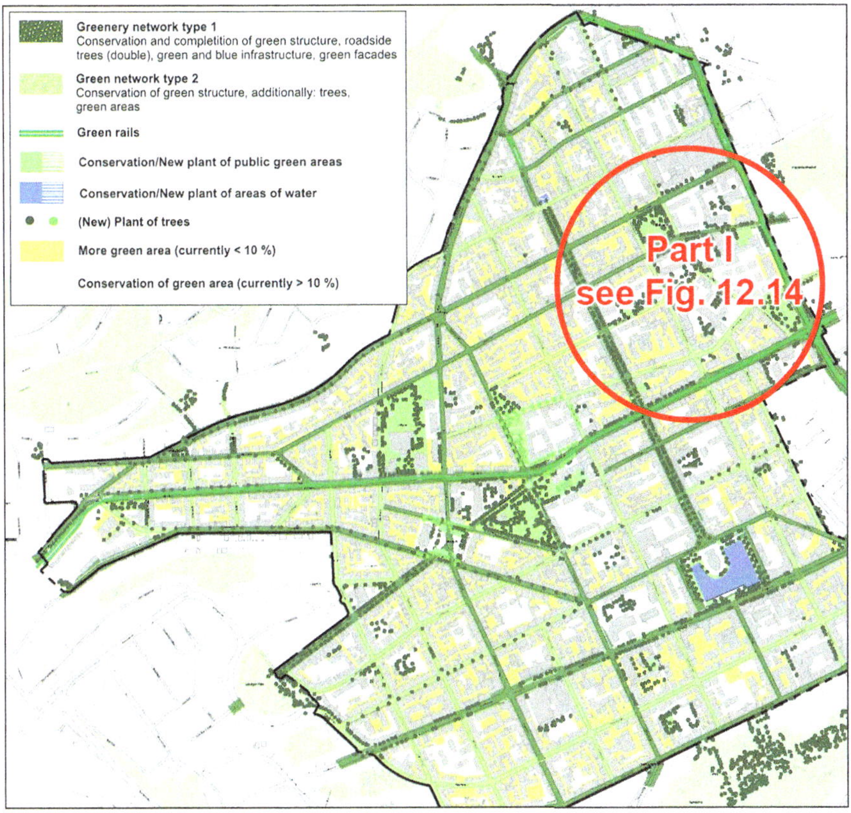

Fig. 12.13 Greenery network map, Valley Bottom Development Outline Plan (Landeshauptstadt Stuttgart 2018)

- Facades which are orientated to the south have to be greened or designed with cool materials.

The greenery measures facilitate a significant reduction of thermal stress in the inner area as the mean total green effect (inner area, 10–16 ha, trees, grassland, green roofs/facades) provides a 3.7 °C reduction of PET (Physiologically Equivalent Temperature as one of the significant assessment indices to quantify the level of human heat stress).

12.3.5 *Climate-Optimised Construction of Public Spaces*

Urban development planning mainly focusses on how to adapt to heat in practice. To reduce the increasing number of days with heat stress and thermal discomfort in Stuttgart green areas, green roofs, green facades, trees in the streets and greening

Fig. 12.14 Detailed Part I, "Berliner Platz/Diakonissenplatz" (Landeshauptstadt Stuttgart 2018)

public transport rail lines are significant. An important factor is also the construction of public spaces according to climate needs.

Figure 12.16 shows an example ("Neckarpark") in the Neckar valley, an area with a marked urban heat island effect. Logically the development plan defines and locates a central green space as a recreation area for adjoining residents, but it is also made available to the public.

The green space serves as a retention area including an additional retention ditch (Fig. 12.17). The water collected here can be used for temporarily running fountains. The central green space integrates existing free spaces and green areas. Areas with a southerly aspect and associated walking areas are required to plant shade-spreading trees. Planned buildings receive green roofs and facades (30% minimum).

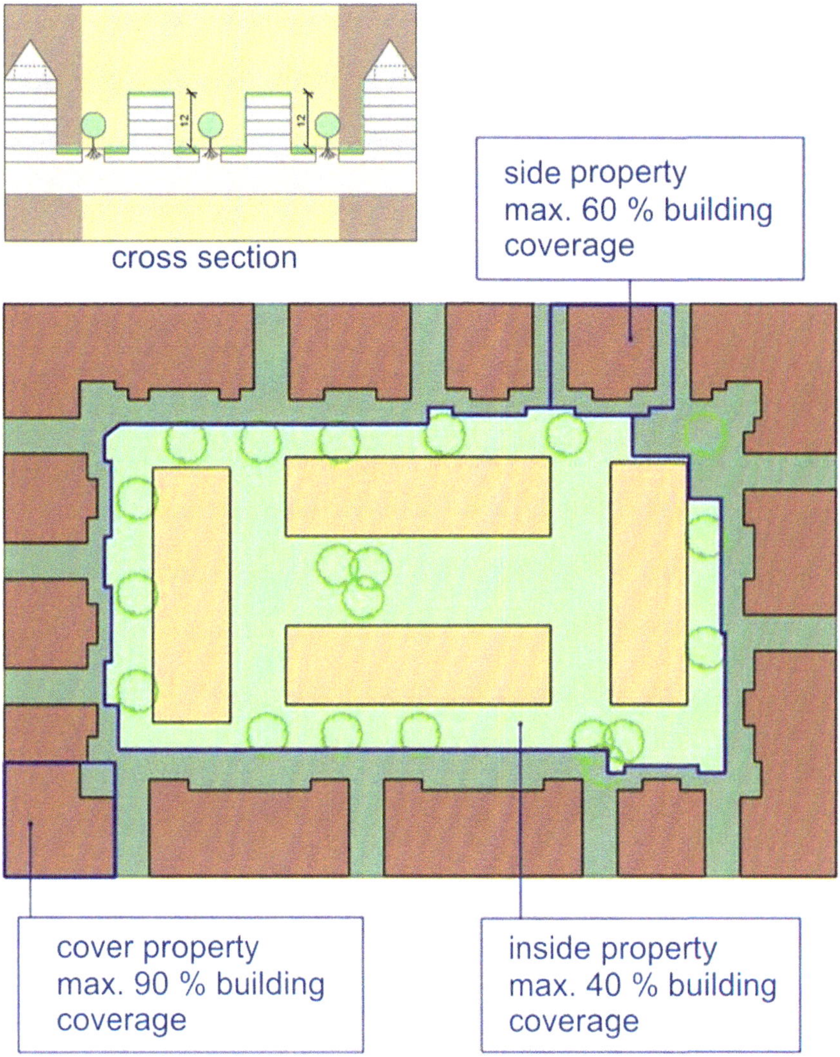

Fig. 12.15 Valley Bottom Development Outline Plan, Model of optimised building structure (Landeshauptstadt Stuttgart 2018)

Pedestrian walkways with a water permeable surface have been generously incorporated in the design. A further enhancement is that the whole area is a car-free zone; only in exceptional cases it is permitted to use the walkways for vehicular access (emergencies, delivery of goods, etc.).

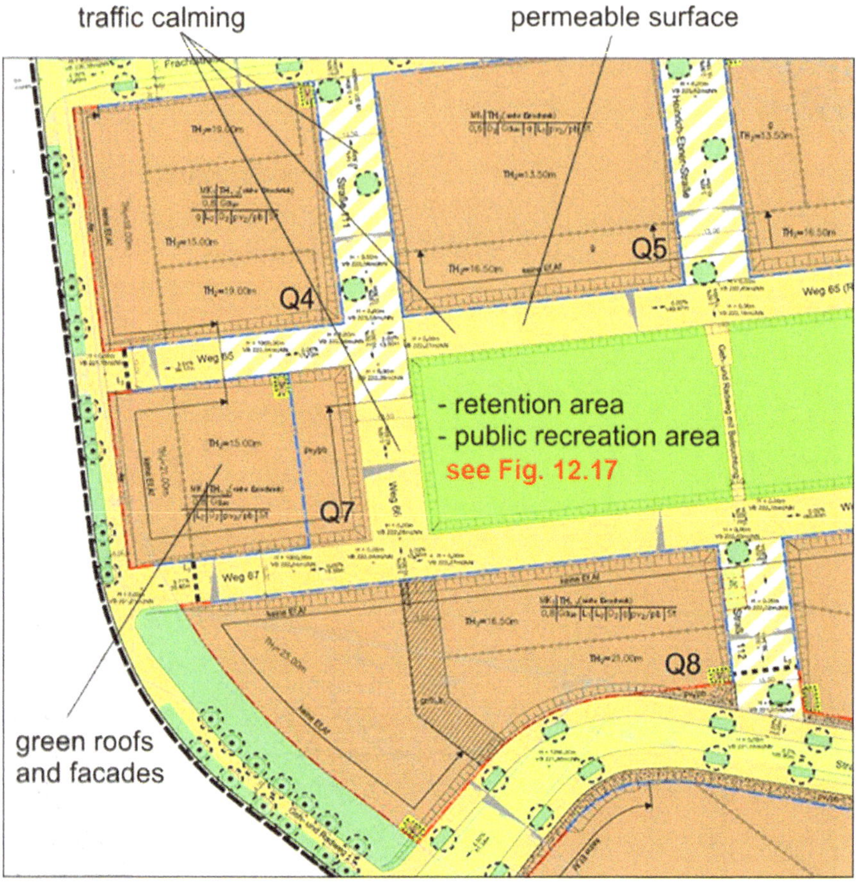

Fig. 12.16 Part of the urban development plan "Neckarpark" with legally binding "Green" and Central green space

12.4 Urban Climate Challenges: City Governance Dimensions

A mounting imperative is that cities have to take into account climate in urban planning. It is evident that the efforts invested over decades by the Section of Urban Climatology of the Office for Environmental Protection in Stuttgart to protect the local climate serve, at the same time, as an effective approach to the process of adjusting to climate change. In Stuttgart, this applies in particular to heat in the city centre. Measures undertaken have been outlined in the foregoing sections of this chapter.

The main policy basis that incorporates climate in urban planning is the Federal German Building Code (Baugesetzbuch), Article 1 (Bundesrepublik Deutschland 2004) which sets out the objectives and principles of construction management

Fig. 12.17 "Neckarpark": Implemented Central green space, Opening 2017. (Source: Municipality of Stuttgart)

planning, including consideration of environmental concerns, directing that the natural basis for human existence, in particular the soil, climate and air, be maintained and safeguarded. This raises "air and climate" to the status of planning factors.

Since 2011 the Act for the Promotion of Climate Protection in Town and Municipal Development (Gesetz zur Förderung des Klimaschutzes bei der Entwicklung in den Städten und Gemeinden) requires measures in urban planning to work against climate change as well as serving adaptation to climate change. Especially the added § 1a (5) of the German Federal Building Code determines this double objective of a municipal climate protection policy and its relevance in the weighting process.

Additionally, it is the purpose of landscape and open space plans to realise the goals of nature conservation and landscape management within the field of urban land use planning. They cover an inventory of the natural conditions and the use requirements of the particular area. The natural conditions to be analysed and taken into consideration are, for example, climate and air quality including an ecological evaluation of the identified conditions and use conflicts. To this effect the Federal Nature Conservation Act (§ 1 (3) no. 4) (Bundesrepublik Deutschland 2009) demands the protection of the air and the climate. This shall apply especially with regard to areas with favourable air quality or climatic effects, such as areas in which fresh or cold air develop or pathways for air exchange.

Depictions in the landscape plan or open-space control plan are as far as necessary and suitable, transferred to urban development plans and made legally binding through corresponding regulations (designations).

The statutory category of designations that may be admissibly listed in development plans opens up a range of effective possibilities for the implementation of principles developed from the aspect of urban climatology and of climate protection. There is, however, no single designation that can bring about on its own, a healthy urban climate.

The designation of the type and degree of building and land use (§ 9 (1) und (1a)) determines, for instance, whether the land is utilised with due consideration to climate protection.

Also of significance are the following possible designations declared by the Code in relation to the creation of green spaces in urban conurbations, providing scope for the planning of green areas and open spaces with due consideration given to areas for cold air generation and fresh air corridors:

For example, there is the possibility setting spaces to be kept free from built development, public and private green spaces, such as parks, agricultural land and woodland. Additionally measures for the protection, conservation and development of the natural environment could be assigned and even the planting of trees, shrubs and greenery of any other kind including the preservation in a mandatory way.

A further legal instrument is the "urban planning contract" (§ 11 of the German Building Code) and the "project completion plan" (§ 12). These accommodate considerations of urban climate interests in several regards.

The representation of the concrete project in the project completion plan allows for the examination and assessment of all climatic effects on the surrounding area resulting from the forms of built structures. This has relevance for the aspects of sunlight, lighting, possibilities for solar energy use and bioclimatic conditions.

The range of activities associated with urban development sanitation and reconstruction measures (§ 171a (3) of the German Federal Building Code) opens up to cover climate-friendly development functions as well. What comes more and more to the fore is the preservation and development of unemployed and cleared areas, which have a big potential for climate-friendly urban development, especially that of green spaces playing a major role for a healthy urban climate (e.g. with regard to the production of cold air).

Taking into account urban climate in urban planning engenders at the same time a combination of some obstacles and successes. Climate experts and city planners must always work together with all development plans worked out or co-produced from the beginning. In the case of climate experts they need to provide recommendations, if possible in the context of the instruments of the planning legislation. Urban planners always use the instrument of maps. Thus climate experts should present all thematic results as maps at the scale which is normally used by planning authorities. While best efforts will be made by climate experts and urban planners to produce recommendations that resonate with the aspirations of urban climate-based planning, it is important to recognise that the final decisions concerning planning projects are not done by the experts but by municipal politicians. Given this awareness raising amongst politicians is very important. Having said that, the information provided by climate experts to politicians needs to be articulated in such a way that it is easy to understand and quantitatively very concrete, or in other words non-abstract. If the climate imperative in urban planning is not able to convince the politicians, other aspects, especially the economic advantages associated with a planned development, will influence and determine the final political decision.

Thus the success for climate-friendly planning means good basic climate facts, good collaboration between climatologists and planners and an excellent awareness raising strategy for local politicians. Also the need of an effective knowledge transfer to city practitioners is considered in Hidalgo et al. (2019).

12.5 Conclusions

Taking into account urban climate in urban planning has a long tradition in Stuttgart. It is an important topic, which is enshrined in the German Federal Building Code. Stuttgart's practices can very well serve as a model for other cities in the world. In facing global climate change, urban climatology is growing more and more important especially with regard to the applications of urban climate science principles in the development of urban climate related adaptation to climate change.

Measures that have been undertaken in the city of Stuttgart and could be applied in other cities include the maintenance and expansion of green spaces such as woodland and parks, safeguarding vital fresh air corridors and air exchange channels as well as roadside, railway track and roof greening. Furthermore for optimal climate outcomes planning projects for inner city development need to respect a *climate-optimised* concentration of urban structures and public spaces incorporating additional features like retention areas. Lastly there is no escaping the fact that the success of climate-friendly planning depends on a detailed meteorological data base comprising urban climate/planning relevant variables, good collaboration between climatologists and urban planners and a well-formed and intensive awareness raising strategy for the local politicians, who command the final say regarding the planning decisions.

References

Abteilung Stadtklimatologie, Amt für Umweltschutz der Landeshauptstadt Stuttgart (2008) Klimaatlas Region Stuttgart. https://www.stadtklima-stuttgart.de/index.php?klima_klimaatlas_region. Accessed 2008

Bundesrepublik Deutschland (2004) Baugesetzbuch. Federal German Building Code 2004. BGBl. I S. 2414. Last update: 20.07.2017 (BGBl. I S. 2808)

Bundesrepublik Deutschland (2009) Bundesnaturschutzgesetz. Federal Nature Conservation Act, 29.07.2009 (BGBl. I S. 2542). Last update: 15. September 2017 (BGBl. I S. 3434). https://www.gesetze-im-internet.de/bnatschg_2009

European Union (ed) (2017) Environmental Impact Assessment *EIA*. http://ec.europa.eu/environment/eia/pdf/EIA_rulings_web.pdf
Grimmond CSB, Roth M, Oke TR, Au YC, Best M, Betts R, Carmichael G, Cleugh H, Dabberdt W, Emmanuel R, Freitas E, Fortuniak K, Hanna S, Klein P, Kalkstein LS, Liu CH, Nickson A, Pearlmutter D, Sailor D, Voogt J (2010) Climate and more sustainable cities: climate information for improved planning and management of cities (producers/capabilities perspective). Procedia Environ Sci 1:247–274. https://doi.org/10.1016/j.proenv.2010.09.016
Hidalgo J, Lemonsu A, Masson V (2019) Between progress and obstacles on urban climate interdisciplinary studies and knowledge transfer to society. Ann N Y Acad Sci 1436(1):5–18. https://doi.org/10.1111/nyas.13986
Ketterer C, Matzarakis A (2016) Mapping the physiologically equivalent temperature in urban areas using artificial neural network. Landsc Urban Plan 150:1–9. https://doi.org/10.1016/j.landurbplan.2016.02.010
Landeshauptstadt Stuttgart (ed) (2008) Rahmenplan Halbhöhenlagen. Hillside Development Outline Plan, Stuttgart
Landeshauptstadt Stuttgart (ed) (2010) Global climate change—adaptation and mitigation: the new challenge facing urban climatology, Authors: Reuter U., Hoffmann U., Kapp R.. Stuttgart: Publication series No. 3/2010. https://www.stadtklima-stuttgart.de/index.php?klima_klimawandel_heft-3-2010
Landeshauptstadt Stuttgart (2011) Climate Viewer. http://www.stadtklima-stuttgart.de/index.php?climate_urban_climate_viewer
Landeshauptstadt Stuttgart (ed) (2018) Rahmenplan Talgrund-West. Valley bottom development outline plan. https://www.stuttgart.de/img/mdb/item/644334/131783.pdf
Landeshauptstadt Stuttgart, Amt für Stadtplanung und Stadterneuerung (ed) (2010) Flächennutzungsplan 2010. Urban land-use plan. https://www.stuttgart.de/aktueller-fnp
Mahlkow N, Lakes T, Donner J, Köppel J, Schreurs M (2016) Developing storylines for urban climate governance by using Constellation Analysis—insights from a case study in Berlin, Germany. Urban Clim 17:266–283. https://doi.org/10.1016/j.uclim.2016.02.006
Ministerium für Verkehr und Infrastruktur Baden-Württemberg (ed) (2015) Climate Booklet for Urban Development-References for urban planning. Ministerium für Verkehr und Infrastruktur Baden-Württemberg, Stuttgart. http://www.staedtebauliche-klimafibel.de
Nachbarschaftsverband Stuttgart (ed) (1992) Klimaatlas. Climate study for the area of the Stuttgart Regional Federation and bordering areas of the Stuttgart region. Municipality of Stuttgart, Stuttgart
Ng E, Ren C (eds) (2015) The urban climatic map: a methodology for sustainable urban planning. Routledge, New York
Stadt Freiburg (ed) (2019) Klimaanpassungskonzept (Climate Adaption Conception)—Ein Entwicklungskonzept für das Handlungsfeld "Hitze". Freiburg. https://www.freiburg.de/pb/site/Freiburg/get/params_E-416449108/1342674/Klimaanpassungskonzept.pdf
Verband Region Stuttgart (ed) (2008) Klimaatlas Region Stuttgart. Stuttgart: Schriftenreihe Verband Region Stuttgart No. 26
Verband Region Stuttgart, Körperschaft des öffentlichen Rechts, Stuttgart (2008) Klimaatlas Region Stuttgart. Regionales Rauminformationssystem Stuttgart—RegioRISS. https://www.region-stuttgart.org/klimaatlas/ and https://www.region-stuttgart.org/geoinformation/. Accessed 2008
Verein Deutscher Ingenieure e.V. (Association of German Engineers) (ed) (2015) VDI Guideline 3787, section 1. Environmental meteorology—climate and air pollution maps for cities and regions. Beuth, Berlin
Verein Deutscher Ingenieure e.V. (Association of German Engineers) (ed) (2020) VDI Guideline 3787, section 8. Environmental meteorology—urban development and global climate change. Beuth, Berlin
Webb B (2017) The use of urban climatology in local climate change strategies: a comparative perspective. Int Plan Stud 22(2):68–84. https://doi.org/10.1080/13563475.2016.1169916

Chapter 13
India: Heat Wave and Action Plan Implementation in Indian Cities

Mahaveer Golechha, Dileep Mavalankar, and Subhash Chander Bhan

Abstract Climate change is causing an increase in severity and frequency of extreme weather events and disasters. Heat waves are among the most dangerous of natural hazards, but rarely receive adequate attention. Extreme heat can lead to dangerous, even deadly, health consequences, including heat stress and heatstroke, it can exacerbate chronic cardiovascular and respiratory diseases.

India is vulnerable extremely to the impact of climate change and the casualties due to heat waves have increased over the past decades. There have been 25,716 deaths recorded from 1992 to 2016 in various parts of the country. Climate adaptability and resilience is the way forward to tackle with the effects of climate change on health. The increase in heat wave events and other climate change events calls for urgent action for preventing and controlling the impact on health at state and national level.

The Indian Institute of Public Health-Gandhinagar (IIPHG) is a pioneer in developing threshold-based heat action plans with other partners, including Ahmedabad city government, NRDC, USA, IMD, Government of India and University of Washington. Heat Action Plans aim to provide a framework for the early warning, and implementation of effective action to reduce the negative health impacts of extreme heat and prevention and control of heat-related illnesses. The Ahmedabad Heat Action plan has proven highly impactful in preventing and controlling morbidity and mortality due to heat-related illnesses in the city. Looking at this, Maharashtra Public Health Department requested IIPHG to help develop similar plans for seven cities in Nagpur region (Vidharbha) of the state in 2015. Following this guidance by NDMA and technical inputs by IIPHG, 17 states and more than 130 cities have developed and implemented heat action plans.

Heat Action Plans are important contributors to public health for reducing climate sensitive diseases and also to make Indian cities more climate resilient.

M. Golechha (✉) · D. Mavalankar
Center of Excellence for Climate Change (Heat Health), Indian Institute of Public Health-Gandhinagar, Gandhinagar, Gujarat, India
e-mail: dmavalankar@iiphg.org

S. C. Bhan
India Meteorological Department, Ministry of Earth Sciences, Government of India, New Delhi, India

C. Ren, G. McGregor (eds.), *Urban Climate Science for Planning Healthy Cities*, Biometeorology 5, https://doi.org/10.1007/978-3-030-87598-5_13

Keywords Climate change · Heat waves · Heat action plan · Heat-related illnesses · Ahmedabad heat action plan · IIPH-G · Climate resilient

Key Messages

- Heat Action Plan is successful cost-effective environmental public health intervention for developing countries.
- Community Awareness and public outreach is essential for prevention of heat-related illnesses.
- High Exposure Vulnerable Groups mapping and focused attention.
- Evaluated health outcomes (mortality and morbidity studies).
- Coalition between academics and implementers (Coalition Building and Outreach).
- Special attention to death registration and cause of death.
- All-cause mortality and temperature correlation is required for threshold determination.

13.1 Heat Waves in India

Significant increasing trends have been observed in global and regional surface air temperatures in last century (Jones and Moberg 2003; IPCC 2014). According to IPCC, a linear warming trend of 0.85 °C over the period 1880–2012 has been observed for globally averaged combined land and ocean surface temperature (IPCC 2014). The IMD annual climate report shows that during the recent 117 years (1901–2017), the country averaged annual mean land surface air temperature have increased at a rate of 0.66 °C per 100 years with significant increasing trend of 1.06 °C/100 years in the maximum temperature (Ministry of Earth Sciences 2018). Also, 11 out of the 15 warmest years were from the past 15 years (2003–2017), and the past decade (2008–2017) was the warmest decade on record with anomaly of 0.54 °C above the 1971–2000 average. Temperatures in India start increasing from the month of March with northwards movement of the Sun. With commencement of southwest monsoon rains, the southern and eastern parts of the country cool down significantly. However, northwestern parts are still very warm till end of June, and sometimes into July also when onset of monsoon rains is delayed.

It is a challenging task to determine the impacts of urbanization and land use changes on mean surface temperatures. For Indian scenario, Sinha Ray et al. (1997) found that the phenomenon is due in part to the increase in urbanization-related minimum temperature. Sati and Mohan (2018) conducted study on impact of urbanization in five representative years of each decade—1972, 1981, 1993, 2003, and 2014 and reported an increase of 3–5 °C in land surface temperature and 2–3°C in air temperature. Further, the urbanized and built-up area went up by 17% in five decades. Urban land cover changes (ULCC) due to urbanization are mainly caused by removal of vegetation cover, which affects the surface climate (Oke 1987).

India generally experiences heat wave during the period between April and June. However, occasional episodes during March and July have also been experienced in the past. According to the World Meteorological Organization (WMO 2015) there is no universally accepted definition for heat wave. They are understood to be periods of unusually hot and dry or hot and humid weather that have a subtle onset and cessation, a duration of at least 2–3 days, usually with a discernible impact on human and natural systems. Because there is no absolute universal value, such as a given temperature that defines what is extreme heat, heat waves are relative to a location's climate: the same meteorological conditions can constitute a heat wave in one place but not in another. In India, heat wave is defined when the actual maximum temperature of a station reaches a certain threshold and is above the normal value of temperature by a certain magnitude. Heat wave is considered if maximum temperature of a station reaches at least 40 °C, 37 °C, 30 °C for plains, coastal areas, and hilly regions, respectively. Once these thresholds are reached, it is defined as Heat Wave (HW) if departure of maximum temperature from normal is 4.5–6.4 °C, and Severe Heat Wave (SHW) if departure is >6.4 °C. Also, it is called HW when actual maximum temperature ≥45 °C, and SHW when actual maximum temperature ≥47 °C, irrespective of deviation from normal. A good number of past studies on the HWs over India provided the climatological characteristics of HWs over India (Raghavan 1966; Subbaramayya and Surya Rao 1976; Chaudhury et al. 2000; De et al. 2005; Pai et al. 2004, 2013). As incidences of heat waves during March and July are rather sporadic, most of these studies are confined to the period April to June only. Pai et al. 2013 (Fig. 13.1) have reported that, on an average, most parts of northern India, eastern India, parts of central India, and east coast experience up to 8 days of HW whereas the number of SHW are one to two only. The duration of individual heat wave spells, however, could be quite longer. Pai et al. (2017) (Fig. 13.2) have reported continuous spells of HW of ≥10 days over most parts of northern India, eastern India, parts of central India, and east coast. Continuous

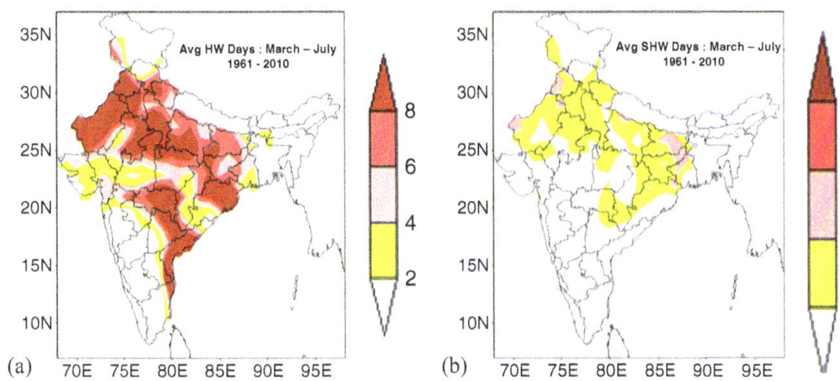

Fig. 13.1 Average number of (**a**) HW days and (**b**) SHW days over India during the hot weather season (March–July) for the period 1961–2010 (Pai et al. 2013). (Courtesy: Editor MAUSAM—India Meteorological Department)

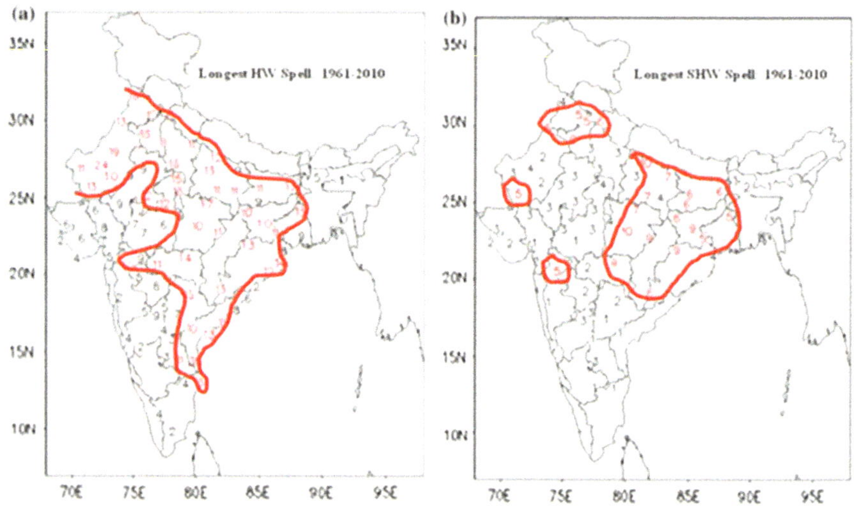

Fig. 13.2 Duration of the longest HW spells (**a**) and longest SHW spells (**b**) during 1961–2010. The areas bounded by red color lines represent HW spells of ≥10 days duration and SHW spells of ≥5 days duration (Pai et al. 2017). (Courtesy: Editor MAUSAM—India Meteorological Department)

spells of SHW of ≥5 days have been reported over pockets of northwest India and eastern India. In 1994, Phalodi, a small town of northwest India, experienced the longest continuous HW spell of 24 days (17 May–9 June, 1994).

With the changing climate, considerable increase in incidence of HW has been experienced in India. Pai et al. (2013) using data of 103 stations in the country have reported that number of HW days per year totaled for all these stations have increased from about 510 to 680, and the SHW days from 80 to 100 (Fig. 13.3). Spatial extent of incidence of HW has also increased considerably from the decade of 1960–1970 to 2001–2010 (Fig. 13.4).

13.2 Review of Heat-Related Mortality in India

In the last century, 16 of the 18 hottest summers have been in the past 20 years, and predictions indicate that this upward trend is likely to continue, with an increase in heat waves and mortality due to heat. However, compared to Europe and North America, there is still a notable gap in research on heat-related negative health events and mortality in low- and middle-income countries, especially in India. According to the 2011 census, nearly two million people in India were found to be homeless and dependent on streets for shelter. Even for those with access to dwellings of some type, the unreliable supply of electricity implies that fans and air-conditioners are often unable to provide the intended indoor temperature regulation.

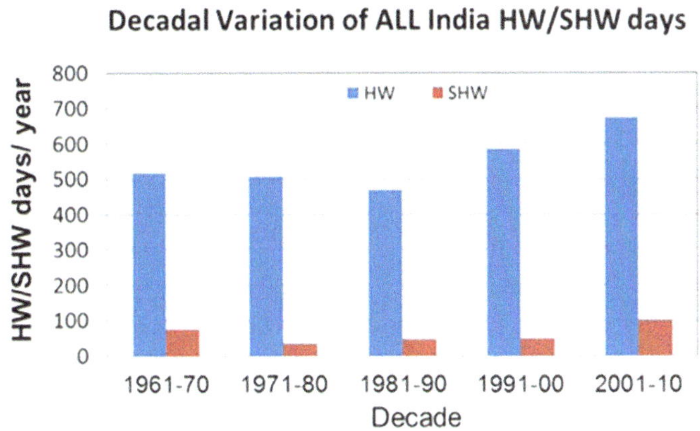

Fig. 13.3 Decadal variation of the all India 6+. HW/SHW days (days/year) during the hot weather season (March–July) for the period 1961–2010. (Pai et al. 2013). (Courtesy: Editor MAUSAM—India Meteorological Department)

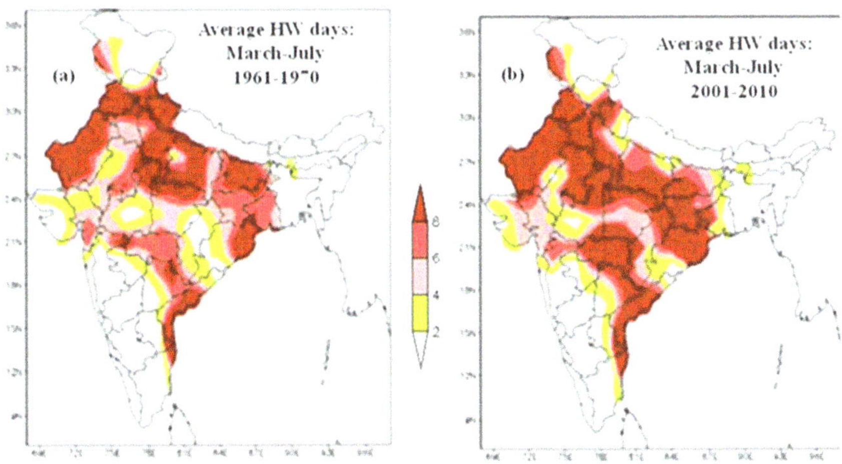

Fig. 13.4 Average number of HW days during the hot weather season (March–July) over India during (**a**) 1961–1970 and (**b**) 2001–2010. The climatology was computed by averaging the number of HW days during the respective decades. (Courtesy: Editor MAUSAM—India Meteorological Department)

Combined with a lack of clean and accessible drinking water sources, and a weak social and public health infrastructure, these conditions expose large numbers of people across India to negative health consequences, and even death, from extreme hot and cold temperatures.

Due to lack of proper death records and cause of death analysis in India, many deaths due to heat may not be recorded as such. It is concerning that while the all-cause mortality in India is nearly 25 million, deaths attributed to heat number only

Table 13.1 Year wise mortality in India (NDMA 2017)

Year	No. of deaths due to heat waves
2009	1071
2010	1274
2011	793
2012	1247
2013	1216
2014	1677
2015	2442
2016	1111
2017	375
2018	25

from 200 to 2000, annually (Deccan Herald 2018). In the 24-year period between 1992 and 2016, only 25,716 heat-related deaths were reported, with 1111 and 2040 deaths reported in 2015 and 2016, respectively, as indicated in Table 13.1 (NDMA 2017). This is a consequence of the inadequacies in the national death reporting system. It is therefore highly likely that a significant number of heat-related deaths are unreported. Specifically, this loss of data can be attributed to the lack of granularity in classifying deaths, including but not limited to the absence of cause, and place of residence data for each deceased. This is further complicated when reporting heat wave deaths among older adults, since this population is more likely to be multimorbid compared to their younger counterparts. Therefore, we estimate that there is a 50–100-fold gap in the number of reported and actual deaths attributable to heat.

In Indian climatic patterns, there is a temperature increase of almost 2 °C from normal maximum temperature, in the summer months, starting from March, to the pre-monsoon period in June. This stretches even up to July when disruptions in the south westerly wind patterns cause a delayed onset of the monsoon. From 2010 to 2018, abnormally high temperatures during this time period were observed, most notably in the central and western regions of the country. Such increases in temperature not only contribute to death and ill-health in humans, but also have devastating effects on the ecosystem by way of wildlife deaths.

Although there are several factors which affect the heat-related mortality and data accuracy of reporting. The sudden rise from 2010 to 2015 is typically because the awareness and capacity building of different stakeholders and involvement of political will, leading to improved data capture.

Public health advocates feel this has been a triumph of advocacy on heat-related health issues. A combination of public awareness campaigns, training for medical staff and simple policy changes and multiple actions such as provision of water, and cooling places in cities, etc. may have averted thousands of deaths for little cost in recent years.

13.3 Experience of City of Ahmedabad During Heat Wave of 2010 and Its Impact on Mortality

Ahmedabad has a population of 7.2 million. Ahmedabad is among the top ten rapidly growing cities in India and it is important hub for automotive and pharmaceutical industries. It is poised to become one of India's leading metropolises by 2030.

The extreme heat events are one of the important significant public health challenges for the city of Ahmedabad, especially when daily maximum temperatures reach 45 °C (113 °F). Heat causes negative impact on population health through dehydration, heat exhaustion, heat stroke, and exacerbations of chronic cardiovascular and respiratory diseases. Environmental factors such as poor air quality can worsen during extreme heat and further lead to respiratory and other illnesses (Knowlton et al. 2014).

In May 2010, Ahmedabad experienced severe heat wave events. When the temperature reached as high of 46.8 °C (116 °F) on 21st May, this caused a spike in reported heat-related illness and death. The May 2010 heat wave was associated with significant excess in all-cause mortality: 4462 all-cause deaths occurred, comprising an excess of 1344 all-cause deaths, an estimated 43.1% increase when compared to the reference period (3118 deaths) (Azhar et al. 2014). The heat waves stimulated Ahmedabad's public health community. It helped identify heat as a public health hazard. It also led to development of public health adaptation to climate change and provided additional motivation for pursuing adaptive action at city government level.

To understand the relationship between extreme heat events and all-cause mortality in the city of Ahmedabad, an ecological analysis was conducted. The May 2010 has been chosen as study period because of unprecedented heat waves in the month. The team adopted methodology of Anderson and Bell (Anderson and Bell 2012).

The Ahmedabad Municipal Corporation (AMC) is the local government body responsible for registration of birth and deaths in the city. With support from AMC the team got the day wise mortality data for the years 2009–2011. Temperature data was obtained from the METAR system located at Ahmedabad Airport.

To establish the relationship between all-cause mortality with temperature, deaths on each day in 2010 was considered and compared to mean of all-cause death counts from corresponding days in May from 2009 and May 2011. The average daily mortality during the heat wave days was estimated at 143.9 ± 48.13, significantly higher than the average of 100.6 ± 10.34 for the reference period (Fig. 13.5) (Azhar et al. 2014).

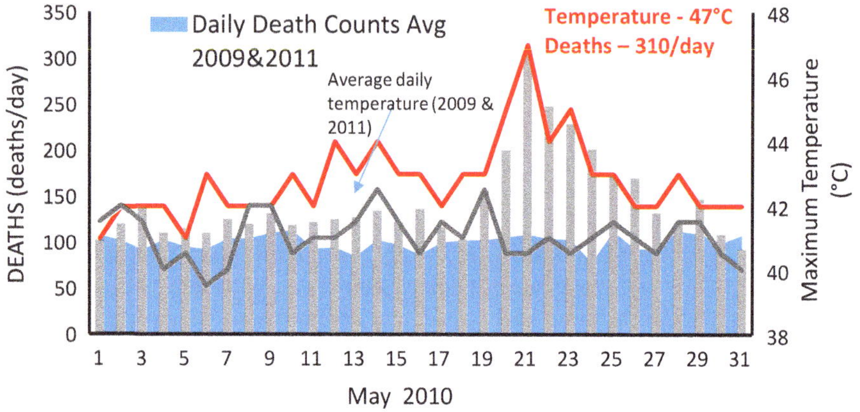

Fig. 13.5 Daily Temperature and all-cause mortality correlation graph

13.3.1 Ahmedabad Heat Action Plan (HAP) Development Process

A coalition was formed to prepare for the rising heat-related health threats in the city of Ahmedabad as part of their climate adaptation strategy, even without the national policy on extreme heat in 2010. The main partners are public health, environmental and academic groups, led by the Ahmedabad Municipal Corporation (AMC), IIPH-Gandhinagar (IIPH-G), the Public Health Foundation of India (PHFI), and the Natural Resources Defense Council (NRDC), in partnership with Emory University's Rollins School of Public Health, Icahn School of Medicine at Mount Sinai and Georgia Institute of Technology. The coalition aimed to develop an early warning system and heat preparedness plan for the city of Ahmedabad, which is the first plan in South Asia. Later, this led to the development of Ahmedabad Heat Action Plan (HAP).

The key goals of the coalition are:

1. To assess heat-health impacts of the 2010 heat wave, with a focus on vulnerable populations.
2. To use the best available evidence to develop a suite of protective measures for the city.
3. To use the 2010 heat wave to generate estimates of temperature thresholds at which management interventions should be initiated.
4. To develop an early warning system that could provide highly reliable temperature forecasts earlier than those that are currently available from the India Meteorological Department (IMD).
5. To evaluate the program's impacts and outcomes, particularly on vulnerable populations.

13.3.2 *Process of Development and Implementation of HAP in the City of Ahmedabad*

Strategies undertaken for development of Ahmedabad Heat Action Plan (HAP):

1. Planning and Conceptual Model Development
2. Needs Assessment
3. Baseline Data Collection
4. Coalition Building and Outreach
5. Intervention Development
   (a) Heat Action Plan:
   - Community Outreach
   - Initiating Early Warning System
   - Capacity building and training of Health Professional
6. Intervention Implementation:
   (a) Early Warning System for Extreme Heat Events
   (b) Communications
7. Project Evaluation

   Components of HAP are given in Fig. 13.6.

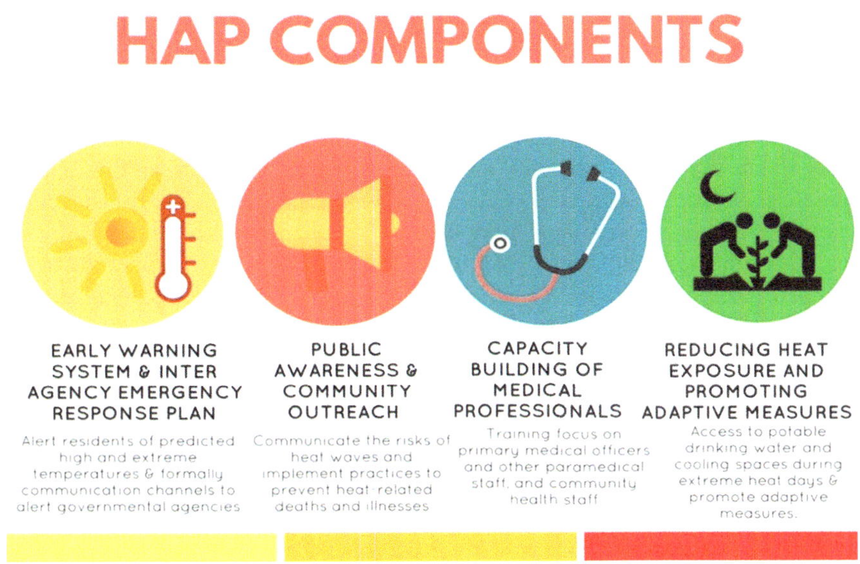

Fig. 13.6 Heat action plan components

As Ahmedabad HAP was the first such plan in Indian and south Asia it became the model HAP in the country and Southeast Asia region which other cities followed to safeguard their citizens from this increasing health danger.

The five reasons why HAP Ahmedabad works and save lives (NRDC 2020):

1. Simplicity: The Ahmedabad HAP uses a simple, local threshold-based color-coded, early warning "heat alert" system that alerts residents and city offices of predicted high and extreme temperatures.
2. Partnership: The partnership between HAP development agencies, implementing agencies (city administration and health system), and IMD for heat alerts is the base of the plan. This coordination led AMC to deploy Nodal officer. The interagency coordination framework and formal communication plan has been institutionalized to alert all line departments, community groups, hospitals, schools about forecasted extreme temperatures by IMD.
3. Communication: Behavior Change Communication is at core to create awareness and save lives from extreme heat events. The communication strategy for generating public awareness is vital to implementing practices to prevention and management of heat-related illnesses. Ahmedabad HAP used print, electric and social media to release heat alerts.
4. Innovation: A threshold-based scientific early warning system for alerting residents about extreme heat events.
5. Leadership: Strong local political and administrative will enables the highly effective implementation of HAP.

13.3.3 Impact of Ahmedabad Heat Action Plan on Mortality

The Heat Action Plan implementation led to significant reduction in all-cause mortality by 25% in the month of May (Hess et al. 2018). The study suggested that Ahmedabad heat action plan saves hundreds of lives every summer. Since the Heat Action Plan was implemented, fewer deaths have been reported during Ahmedabad's hottest months than before the plan was not in place. The community awareness, capacity building of municipal corporation officials, training of medical officers, and providing necessary preventive care to vulnerable communities for the health threats of extreme heat, in addition to providing early warnings and interventions led to significant reduction in mortality and morbidity due to heat waves (Hess et al. 2018).

13.4 Scaling-Up of Heat Action Plan (HAP) in Other Indian Cities and States

The Ahmedabad Heat Action Plan (HAP) was implemented in 2013. Based on successful and enthusiastic implementation by the city government and IIPH-Gandhinagar, the heat action coalition convened a workshop of health department top officers from western India in April of 2015. Learning from Ahmedabad efforts, the Secretary, Public Health Department of Maharashtra, requested IIPH-Gandhinagar to help develop similar plans in seven cities near Nagpur (Vidharbha region) in central India. IIPH-Gandhinagar and NRDC team also had several meetings and presentations to National Disaster Management Authority (NDMA) and Indian Meteorological Department (IMD) head office in New Delhi regarding impact of heat wave on mortality in Ahmedabad and other parts of the country. The IMD has started forecasting maximum temperature for many Indian cities in 2016. In this year NDMA also convened meetings of experts and regional- and state-level officers to create awareness about heat waves and its impact as well as help start development of Heat Action Plans at state and city levels throughout India. Such meetings by NDMA were repeated in 2017 and 2018 with very good effect. By 2020, based on guidance provided by the central government, 17 states and over 130 cities have adopted and developed heat action plans (NRDC 2020).

13.4.1 Sample City Activities

The states of Gujarat, Maharashtra, and Jharkhand have been the pioneers in establishing Heat Action Plans in the cities of Ahmedabad, Surat, Nagpur, Chandrapur, and Hazaribagh. Similar initiatives were also launched in Jhansi, Uttar Pradesh, Sagar, Madhya Pradesh, Rajkot in Gujarat, and Bhubaneswar in Odisha, with the support of the Department of Science and Technology of the Government of India, in 2018. Some work has also stated in one of the municipal governments of New Delhi in 2019.

Ahmedabad city, Gujarat—Starting with 2013 each year HAP was launched in end of March or starting of April in Ahmedabad, by the Mayor or the municipal commissioner in presence of media. Each year the plan was modified a bit with new elements added based on previous years learnings. Sixth version of HAP was launched in 2018, with partners NRDC and IIPH-G, Ahmedabad Municipal Corporation. This plan expanded included expanding the "Cool Roof" initiative where the tops of buildings were given a coating of reflective paint to avoid heat absorption and ensure lower indoor temperatures (AMC 2018). Innovative health promotion measures were also taken by incorporating the support of religious organizations to enhance population awareness about extreme heat and the related adverse health effects. The plan also aimed at strengthening Information, Education and Communication (IEC) activities to raise awareness of taking precautions during heat season and enhancing capacity of medical professionals and frontline health workers to help protect community in times of extreme heat.

Surat city, Gujarat—Studies carried out by the Urban Health and Climate Resilience Centre in Surat found strong correlations between mortality and high temperatures. This research also revealed that high Heat Index was related to a larger number of deaths (18% per day), rather than maximum temperature (11% per day) (Desai et al. 2015).

Chandrapur city—This city is a small city in central India in the state of Maharashtra and has been working on HAP since 2016. In 2018, as part of the heat action plan Chandrapur city in central India took up community outreach programs that range from awareness marches to "WhatsApp" alerts to reach people based on earlier efforts (The Hitavada 2018). Laudable intersectoral coordination was also observed, especially in the area of prevention and health promotion. Under the Intensive Pulse Polio Immunization program, community health workers visiting house-to-house distributed pamphlets to increase awareness about extreme heat and its impacts. Public parks and gardens were also instructed to remain open by the city authorities during the heat wave period, and several kiosks were set up to increase public access to clean potable water.

It also ensured active participation of medical emergency services like 108 Ambulance Services. City also developed a policy to change working hours of workers doing vigorous physical labor. The municipal administration also conducted capacity building program for civic medical officers and frontline health workers to build their capacity in management and prevention of heat-related illnesses. Sociocultural activities such as religious processions and festivals like the *Mahakali Devi Yatra* were provided with support including cool water, shades at temple entrances, and overhead sprinklers inside temples to prevent suffocation due to heat and crowding. Large displays such as hoardings were also put up around religious places to utilize the large waves of people in order to spread heat awareness.

Nagpur city—A major city in central India in the state of Maharashtra, has also developed HAP similar to Ahmedabad. The municipal administration is using Ahmedabad threshold system based on yellow, orange and red alerts for informing population about extreme heat events. They appointed a dedicated nodal officer who took lot of interest and developed and implemented the heat action plan in an ideal manner. Nagpur is even hotter in summer than Ahmedabad, but is similarly having dry heat.

The NRDC and IIPHG developed several resources to support other cities in the development of HAPs. One of these, the City Resilience Toolkit, was developed as a response to heat waves and to help city authorities prepare for irregular increases in temperature. This toolkit included among other things a "How-to" Manual containing important learnings from the Ahmedabad HAP, presented in an easily adaptable format, as an aid for cities to develop their own HAPs in accordance with local peculiarities and population needs.

13.4.2 State-Level Efforts

Droughts are very common in India and hence there has been long tradition of developing drought plans. After the ill famous Bhopal gas leak disaster and the implementation of National Disaster Management Act, National and state disaster planning started in India. But Heat waves were not considered disaster under the act as there is not much property damage in heat waves. States had the liberty of adding it to list of disasters based on local situation.

Odisha was the first state to develop heat action plan at state level as it suffered major heat waves in 1998 when more than 2000 deaths were recorded in the state in that year (Down to Earth 2016). After 2010 following development of Ahmedabad heat action plan and leadership of NDMA in heat wave preparedness, many states started to prepare heat action plans. Andhra Pradesh, Bihar, Delhi, Gujarat, Haryana, Jharkhand, Karnataka, Maharashtra, Rajasthan, Tamil Nadu, Telangana, and Uttar Pradesh had developed locally adapted heat action plans taking guidance from NDMA. The innovative activities undertaken by states that bore extreme impacts of the heat wave are highlighted below.

13.4.3 State Heat Wave Prevention Action Plan-Related Activities

13.4.3.1 Rajasthan

Urban areas often have the capacity as well as the resources to cope with detrimental effects of the climate. The public health system is stronger in the urban context. In a country like India which has a very high rural population the possibility is highly likely to have vulnerability to climate change and other climate sensitive diseases. Rajasthan falls in the area of greatest climatic sensitivity, maximum vulnerability and lowest adaptive capacity with scarcity of water resources, geography and largest desert area. The Government of Rajasthan has partnered with UNICEF, Rajasthan, and IIPH-Gandhinagar for development and implementation of country's first heat action plan for rural settings at sub-district level (block) of Rajasthan. In the first phase the plan was implemented in Rishabhdeo block of Udaipur district, Sanchore and Chitalwana blocks of Jalore district (IIPHG, UNICEF 2017).

This effort led to better preparedness of rural community, strengthening of health system to tackle climate sensitive diseases, and capacity building of healthcare professionals, teachers, and community members. The heat action plan is comprehensive and involves early warning system used during heat waves, health education, and awareness promotion among vulnerable community about prevention from heat-related illnesses, awareness programs in schools, etc. This was supplemented with the development of vernacular IEC resources, and in-depth training initiatives for healthcare professionals. Another example of intersectoral coordination was observed in sending out citizen alerts for high temperature predictions, and to map areas and communities at highest risk of heat-related health events. This Heat Action Plan is also unique in that it was the first such plan created for rural areas in India. The activities undertaken in order to build resilience against extreme heat and diseases sensitive to climate change are listed below (NRDC 2020):

Communication need assessment carried out for developing a comprehensive communication strategy for prevention and management of climate sensitive diseases.

- Community vulnerability assessment to analyze the effect of climate change and extreme heat events on rural population
- Vernacular IEC resources focusing on health education and health promotion for community health workers, children, teachers, and the community at large
- Nearly 200 community health workers and healthcare practitioners, and 60 teachers trained in the prevention and management of heatstroke and other climate-related health issues
- Cooperative and intersectoral workshops for line department officials at the block and district levels to build capacity and sensitize

13.4.3.2 Andhra Pradesh

The State Disaster Management Authority (APSDMA) coordinated with voluntary organizations to set up 66,453 booths for drinking water, and 70,134 summer shelters for the homeless and other street-dwellers. In addition, over 200,000 awareness pamphlets highlighting heatstroke prevention and close to 1.24 million oral-rehydration packets were also distributed throughout the state (The Hans India 2018). 1168 automatic weather stations (AWS) were set up to enhance temperature monitoring efforts (Jaiswal 2018), which made day-to-day heat forecasts possible for 670 sub-district administrative zones. A bilingual mobile phone application named *Varuna* was launched to deliver these forecasts in real time. In addition, the APSDMA also released a Heat Wave Atlas, which identified heat wave hotspots based on an analysis of heat data from 2010 onwards (APSDMA 2018).

13.4.3.3 Maharashtra

Maharashtra, a western Indian state, has put in place a comprehensive state-wide planning and also implemented heat action plan in several cities. The state has also implemented a regional approach for combating extreme heat events. A steering committee has also been formed, including the Maharashtra State Disaster Management Authority, the departments of urban and rural development, education, police, labor, publicity, the medical education councils and the National Health Mission. The chair of this committee is the secretary of the department of public health which ensures that the focus is maintained on reducing negative health outcomes and mortality from heat for the population. The state plan calls for the development and publication of IEC plans, training, and awareness materials by the appropriate departmental officers, who are also tasked with delivering workshops and training for ASHA and *Anganwadi* workers to ensure that heat health awareness reaches the grassroots level. The state health department also issued a pragmatic list of Do's and Don'ts to all public healthcare facilities in the state including 34 district hospitals, 1811 PHCs, and 365 rural hospitals. Cold wards were also set up in healthcare facilities, equipped with icepacks and medication necessary to treat heat-related illnesses (Times of India 2018). The MSDMA, similar to its counterparts in other states, enhanced access to drinking water in cities, and also supported the police by creating shelters for traffic police personnel and restructuring the duty hours of personnel above the age of 45 years.

13.4.3.4 Odisha

As stated earlier, Odisha was one of the first states to develop state-level heat action plan as it faced many deaths due to heat waves in 1998 (Down to Earth 2016). The Odisha State Disaster Management Authority (OSDMA) is the designated nodal authority in the state, led by the special relief commissioner. This organization coordinates with the local meteorological center, other line departments, and even civil society organizations to ensure the successful implementation of the multifaceted HAP including the early warning system, community awareness drives, and mitigation strategies when heat waves strike.

One of the striking features of the Odisha HAP is the early heat warning thresholds which are created based on analyses of temperature and humidity data. The state has focused on increasing public access to both natural and man-made cooling spaces such as parks, lakes, pools, and shopping malls, green urban planning strategies such as cool and green roofs, and tree planting. This is supported by the state government's focus on energy conservation and clean energy technology (OSDMA 2018).

Scientific efforts have also been undertaken by researchers for identifying trigger points related to heat stress threshold in the capital city Bhubaneswar of Odisha.

These trigger points are limits beyond which atmospheric heat would contribute to an increased risk of mortality in the city (Dutta et al. 2020). Upper and lower thresholds were identified at 36.2 °C and 40.5 °C, respectively. Additionally, the Odisha HAP draws greatly on a study of the Urban Heat Island (UHI) effect. UHIs are built areas in urban spaces that register higher levels of heat compared to their surrounding rural areas, as a result of constructions that absorb and consequently emit large amounts of heat. As a result, the effects of heat waves in these areas are much higher and pose considerable risks to vulnerable population groups such as children and older adults (US EPA 2012). The HAP also highlights the need to reduce the urban heat island (UHI) effect through longer term adaptive measures.

13.4.3.5 Telangana

Due to its landlocked geography and proximity to central India, Telangana also contains some very hot regions. The HAP in the state is unique in that it addresses heat-related health issues not only among humans, but also livestock, thereby preventing the loss of livelihood for poor agriculturalists. Similar to the HAPs in other states, the Telangana HAP, prepared by the Revenue and Disaster Management Department with support from the NDMA, IMD, Telangana State Development Planning Society and UNICEF, emphasizes intersectoral coordination and provides actors with strategies to specifically identify the risks to vulnerable population groups and implement appropriate strategies.

One of the salient measures taken by the state includes the 2018 HAP which enhanced the availability of drinking water at large traffic junctions, ORS distribution drives, capacity building of public healthcare facilities, and summertime awareness campaigns across the state (The Hindu 2018). Food and water were also arranged by the state for livestock and wildlife, and families of heatstroke victims were financially compensated, under the *Apath Bandhu* scheme. At the district level, thorough death investigations and verification were carried out by 3-member teams in order to accurately report on heat-related deaths (Revenue Disaster Management Department 2018).

13.5 National Efforts on HAP by National Disaster Management Authority (NDMA) and Other Government of India Departments

The NDMA has taken stewardship in the development of HAPs. The body identified 17 states prone to disastrous heat effects and in 2016 created the first National Guidelines on Heat Waves (NDMA 2017). It is responsible for convening yearly workshops at the national level to address and strengthen heat preparedness. The NDMA also brings together subject matter experts to share information, design strategies for heat monitoring, and lends its support to state- and city-level authorities in the development of locally relevant HAPs.

NDMA also supports and regularly monitors activities implemented by state revenue departments and disaster management authorities. These guidelines were since revised in 2017 and 2019 by incorporating the lessons learnt and feedback received from the states.

In 2018, NDMA has started the campaign "Beat the Heat India" through television and social media for raising community awareness on prevention against extreme heat events. On its website, under the tab "Citizen's Corner," NDMA has dedicated a section that provides a range of resources on combating extreme heat (NDMA 2018).

The Indian Meteorological Department (IMD), which is a part of Ministry of Earth Sciences of Government of India, provides seasonal and daily forecasts to over 650 cities. The IMD forecasting system is very crucial for implementation of threshold-based early warning system.

These systems are critical in providing advance notice to communities and sufficient time for authorities to prepare for extreme heat events. For instance, the early warning system allows the public welfare and health departments to set up drinking water stations, and stock ORS and icepacks in healthcare facilities. Regional offices of the IMD are responsible for providing daily maximum and 5-day temperature forecasts to over 650 cities across India. Their daily bulletins to city governments start at 0730 h and emphasize intersectoral coordination. This forecast information also allows citizens to plan their activities so as to protect themselves from the heat. Since 2016, the IMD has been issuing the Seasonal Outlook for Temperature for the Hot Weather Season (April–June). Lastly, the IMD and the Indian Meteorological Society (IMS) often carry out collaborative activities with the aim of bringing together the perspectives of city authorities, health functionaries, and civil society groups.

The research arm of Ministry of Earth Sciences called Indian Institute of Tropical Meteorology (IITM) situated in Pune city has been doing research on heat waves and has published several papers on it. They are also involved in long-term climate change modeling and prediction. IITM also issues forecasts for temperatures in the extended range scale (valid for next 4 weeks) every week to help in season preparedness in case heat waves are expected over an area during next 4 weeks.

13.5.1 Ministry of Health and Family Welfare, Government of India

Preparedness measures against extreme heat such as ORS stocking, setting up health posts in crowded areas, reconfiguring health facilities for heat-related health events, and appropriate training for healthcare providers are helmed by the Ministry of Health and Family Welfare. The ministry is also responsible for collecting illness and mortality data to improve monitoring efforts and single national data repository of all heat-health events, and deploying rapid response teams to affected communities.

Recently, Climate Change and Health division has also established within Ministry of Health and Family Welfare to address issues related to climate change on human health, including heat-related illnesses. The Director General of Health Services (DGHS), Government of India, has also developed national guidelines for diagnosis and treatment of heat-related illnesses (Emergency Medical Relief 2015).

13.5.2 Department of Science and Technology, Government of India (DST 2010, 2017)

Department of Science and Technology, Government of India, has initiated National Network Programme on Climate Change and Human Health as part of National Mission on Strategic Knowledge for Climate Change (NMSKCC). Under this initiative several grants are given to various research organizations to work on climate change and human health. Under NMSKCC, Heat Action Plans are developed and implemented in three Indian cities. More scientific approach will be incorporated in developing heat action plan of Nagpur (Maharashtra), Rajkot (Gujarat) and Jhansi (Uttar Pradesh). A Heat Stress Index will be created through the correlation of climatic data from the past two decades with morbidity and mortality data collected by city authorities. This will then be used to generate thresholds for each city in order to inform future protective efforts. In addition, satellite imaging and GIS mapping will be used to zero in on vulnerable communities and consequently create tailor-made responses. Heat Action Plans will also be assessed for their effect on reducing illness and deaths caused by heat.

13.6 Efforts in the South Asian Association for Regional Cooperation (SAARC) Region

The health risks of extreme heat waves are an emerging priority in the world especially in South Asia Region. Consecutive extreme heat waves in 2015 and 2016 resulted in thousands of deaths, a range of heat stress and dehydration-related illnesses, losses in productivity, and overburdening of health systems, particularly in India and Pakistan (WHO-WMO Joint Office for Climate and Health 2016). The larger part of Asia has observed increase in number and frequency of extreme heat events since middle of the twentieth century. These extreme weather events have catalyzed the public health community to improve the scientific for development of threshold-based early warning system and heat action plans for prevention and management of heat-related illnesses in South Asian region.

13.6.1 The South Asian Climate Outlook Forum (SASCOF)

The SASCOF was established in 2010 and is coordinated by the India Meteorological Department (IMD). It covers India, Afghanistan, Bhutan, Bangladesh, Maldives, Myanmar, Nepal, Pakistan, and Sri Lanka. Under this forum Indian Meteorological Department (IMD) has taken responsibility for the preparation of annual regional forecast outlook for the SW Monsoon Season rainfall as a Regional Climate Centre of the World Meteorological Organization. In recent SASCOFs special sessions are organized on heat wave action planning (SASCOF 2010).

13.6.2 Improving Health Preparedness for Extreme Heat Events in South Asia: Colombo 2016 (WHO-WMO Joint Office for Climate and Health 2016)

This meeting was held along with SAFCOF and was supported by WMO and WHO. It is focused on heat health and brought together over 25 experts in public health, climate and meteorology from various south Asian countries to discuss the needs and opportunities for multi-sectoral collaboration to better understand and address these health risks in the region.

This meeting concluded with the following recommendations ((WHO-WMO Joint Office for Climate and Health 2016; WMO and WHO 2015):

1. To communicate, advocate, and share resources on heat health in South Asia by disseminating meeting outcomes and taking steps to reach a broader regional public health community which can help identify further research and public health practice partners.
2. To advance product research and development on the IITM experimental 3-week Regional Heat Outlook by proposing steps that can be taken in the short term to improve product communication and suitability, as well as a longer term plan for forecast validation and the development of appropriate extreme heat monitoring and forecasting products.
3. To develop national action profiles to scope national activities and propose pilot projects and technical activities that respond to identified needs of the health sector in South Asia.
4. To develop a South Asia action plan to establish a Regional Heat Health Network based on the inputs from the first South Asia Climate Services Forum for Health and the National Integrated Heat Health Information System (NIHHIS) Framework that can unite partners around common directions in research, capacity building, partnerships, and data and information to accelerate heat health action.

13.6.3 SAARC Training Programme on Heat Wave Preparedness and Response in SAARC region: Gandhinagar

SAARC Disaster Management Centre, located at Gujarat Institute of Disaster Management (GIDM), Gandhinagar, in collaboration with the Indian Institute of Public Health, Gandhinagar, had conducted a 3-day regional workshop on Heat Wave Preparedness and Response in GIDM campus in March 2018. The objectives of this was to: (1) understand Heat Wave trends and its related Health impacts in the region; (2) development and implementation of early warning-based Heat Action Plan (HAP); and (3) Discuss the Capacity Building needs for Heat Wave

Preparedness. Around 40 representatives from all the SAARC countries participated in this event.

Thus there are systematic efforts to build heat wave action planning in many countries of South Asia.

13.7 Conclusions, Challenges, and Way Forward

We are already seeing warmer summers because of climate change. This century (post-2000) has recorded 16 of the 18 hottest summers of the past 100 years. It is expected that each consecutive year will be warmer than last year. Although the heat-related mortality and reporting precision are affected by several variables. The sudden increase from 2010 to 2015 is typically due to the fact that distinct stakeholders' involvement and capacity building and involvement of political will, lead to data accuracy. Public health advocates say this has been a triumph of common sense. A combination of public awareness campaigns, training for medical staff and simple and easy policy changes have prevented thousands of mortality at low price.

The heat action plan is comprehensive mechanism involving early warning system used during heat waves, health education and awareness of vulnerable community about prevention from heat-related illnesses, awareness programs in schools, inter-departmental coordination to alert citizens of predicted high temperatures, training for healthcare professionals to prevent and treat heat-related illnesses and death, developing innovative IEC material in local language and mapping of high-risk areas and community. Ahmedabad heat action plan was first of its kind in India which has since expanded to a large number of cities and states. A study by Hess et al. (2018) showed decreased summertime all-cause mortality rates, with largest declines at highest temperatures after implementation of HAP. Ahmedabad's plan become a model template for other cities attempting to increase resilience against extreme heat events.

Ahmedabad Heat Action plan has proven highly impactful in preventing and controlling morbidity and mortality due to heat-related illnesses in the city.

Numerous challenges were faced during development and implementation of HAP and these challenges were quite similar in every part of India.

Few of them are listed below.

Challenges	Description	Response	Way Forward
Lack of recognition of Heat as a disaster and growing health threat	Heat Waves were not initially considered disasters requiring high level preparation as no studies were done of impact on human health	IIPHG study in Ahmedabad and Nagpur and other studies in Bhubaneshwar and Surat have now provided needed data on health impact Research agencies need to be funded more to do research in this area of climate change impact	City, state and national technical and administrative and political leaders should understand the impact of heat waves on health and develop and implement comprehensive early warning and response plans to extreme heat. More funds should be allocated to studies
Interagency coordination and communication	Limited communication between municipal and government agencies, many agencies except health are still not serious about heat waves impacts. Hospitals and the general public still do not take heat waves as serious health threat	Data on health impact needs to be assess in each major city and district and results disseminated to the various departments to make them aware of the health impact. An Interagency coordination framework should be developed and implemented. Increase interagency research and advocacy	Enhancing interagency coordination and formal communication. Designating a nodal officer for managing heat action plan. The chief executive officer or the highest civil servant should take review meetings during heat season to monitor the health impacts
Health data collection and analysis of health impact very weak	Collecting health data-hospitalization, mortality data is challenging. Many cities do not even keep data on gross all-cause mortality. No city or town keeps data on cause of death	Health department must strengthen the public health function of death registration and cause of death recording and analysis. Special focus should be given to recording heat-related causes especially in summer season	The local, state and national registrars of birth and death should pay special attention to death registration and cause of death. The country chief statistician and statistical office should also help improve the health data
Lack of budgetary provisions for Heat Action Plans and political will	Very little budget given by state and national government for heat-related planning and implementation. Climate change department and health departments are also not paying adequate attention	Policy planers should identify and prioritize policies and programs that would have the biggest impact on climate change and health. Social and administrative advocacy will be needed	Need to reach the top political and administrative levels to endure adequate prioritization in the budget and other resources

References

Ahmedabad Municipal Corporation (AMC) (2018) Ahmedabad heat action plan 2018: guide to extreme heat planning in Ahmedabad, India. https://www.nrdc.org/sites/default/files/ahmedabad-heat-action-plan-2018.pdf. Accessed 14 Apr 2020

Anderson GB, Bell ML (2012) Lights out: impact of the August 2003 power outage on mortality in New York, NY. Epidemiology 23(2):189–193. https://doi.org/10.1097/EDE.0b013e318245c61c

Andhra Pradesh State Disaster Management Authority (APSDMA) (2018) Atlas on heat wave conditions over Andhra Pradesh 2010–2017. http://www.apsdma.ap.gov.in/latestupdate_pdfs/heatwave/heatwave2018/Heat%20wave%20atlas%202018.pdf

Azhar GS, Mavalankar D, Nori-Sarma A, Rajiva A, Dutta P, Jaiswal A, Sheffield P, Knowlton K, Hess JJ (2014) Heat-related mortality in India: excess all-cause mortality associated with the 2010 Ahmedabad heat wave. PLoS One 9(3):e91831. https://doi.org/10.1371/journal.pone.0091831

Chaudhury SK, Gore JM, Ray KCS (2000) Impact of heat waves over India. Curr Sci 79(2):153–155

De US, Dube RK, Prakasa Rao GS (2005) Extreme weather events over India in the last 100 years. J Ind Geophys Union 9(3):173–187

Deccan Herald (2018) Does your city have a heat action plan?. https://www.deccanherald.com/amp/content/670619/does-your-city-have-heat.html. Accessed 1 Apr 2021

Department of Science and Technology (DST) (2010) Mission Document, National Mission on Strategic Knowledge for Climate Change (NMSKCC). Department of Science and Technology, Government of India. http://www.dst.gov.in/sites/default/files/NMSKCC_mission%20document%201.pdf. Accessed 14 Apr 2020

Department of Science and Technology (DST) (2017) Year wise Projects CCP-SPLICE Division 2017–18, National Mission on Strategic Knowledge for Climate Change (NMSKCC). Department of Science and Technology, Government of India. http://www.dst.gov.in/sites/default/files/Year%20wise%20Projects%20CCP_2017-18.pdf. Accessed 14 Apr 2020

Desai VK, Wagle S, Rathi SK, Patel U, Desai HS, Khatri K (2015) Effect of ambient heat on all-cause mortality in the coastal city of Surat, India. Curr Sci:1680–1686

Down to Earth (2016) How to fight heat wave, the Odisha way. https://www.downtoearth.org.in/blog/how-to-fight-heat-wave-the-odisha-way-49950. Accessed 14 Apr 2020

Dutta A, Bhattacharya S, Ak K, Pati S, Swain S, Nanda L (2020) At which temperature do the deleterious effects of ambient heat "kick-in" to affect all-cause mortality? An exploration of this threshold from an eastern Indian city. Int J Environ Health Res 30(2):187–197. https://doi.org/10.1080/09603123.2019.1587389

Emergency Medical Relief (2015) Guidelines on Prevention and Management of Heat Related Illnesses. Emergency Medical Relief, Directorate General of Health Services, Ministry of Health and Family Welfare, Nirman Bhawan, New Delhi. http://health.bih.nic.in/Operational%20guideline/EMR%20Mohfw%20heat%20illnesses%20-2015.pdf. Accessed 14 Apr 2020

Hess JJ, Lm S, Knowlton K, Saha S, Dutta P, Ganguly P, Tiwari A, Jaiswal A, Sheffield P, Sarkar J, Bhan SC, Begda A, Shah T, Solanki B, Mavalankar D (2018) Building resilience to climate change: pilot evaluation of the impact of India's first heat action plan on all-cause mortality. J Environ Public Health 2018:7973519

IIPHG, UNICEF (2017) Rajasthan climate change project: heat action plan. http://iiphg.edu.in/images/pdfs/2019/RCCPHAP.pdf. Accessed 14 Apr 2020

IPCC (2014) Climate change 2014: the scientific basis, contribution of working group-I to the fifth assessment report of intergovernmental panel on climate change (IPCC). Cambridge University Press, Cambridge

Jaiswal A (2018) Preparing for the heat: Indian cities. https://www.nrdc.org/experts/anjali-jaiswal/preparing-heat-indian-cities. Accessed 14 Apr 2020

Jones PD, Moberg A (2003) Hemispheric and large-scale surface air temperature variations: an extensive revision and an update to 2001. J Clim 16(2):206–223. https://doi.org/10.1175/1520-0442(2003)016<0206:Halssa>2.0.Co;2

Knowlton K, Kulkarni SP, Azhar GS, Mavalankar D, Jaiswal A, Connolly M, Nori-Sarma A, Rajiva A, Dutta P, Deol B, Sanchez L, Khosla R, Webster PJ, Toma VE, Sheffield P, Hess JJ, Ahmedabad Heat and Climate Study Group (2014) Development and implementation of South Asia's first heat-health action plan in Ahmedabad (Gujarat, India). Int J Environ Res Public Health 11(4):3473–3492

Ministry of Earth Sciences (2018) Seasonal Outlook for the Temperatures during the Hot Weather Season (April to June). Ministry of Earth Science, Press Information Bureau, Government of India. https://pib.gov.in/newsite/PrintRelease.aspx?relid=178240. Accessed 14 Apr 2020

National Disaster Management Authority (NDMA) (2017) Guidelines for preparation of action plan—prevention and management of heat-wave. https://ndma.gov.in/images/guidelines/heatwaveguidelines2017.pdf

National Disaster Management Authority (NDMA) (2018) Resources on heat wave. https://ndma.gov.in/en/media-publicawareness/disaster/natural-disaster/heat-wave.html

Natural Resources Defense Council (2020) Expanding heat resilience across India: heat action plan highlights. https://www.nrdc.org/sites/default/files/india-heat-resilient-cities-ib.pdf

Odisha State Disaster Management Authority (OSDMA) (2018) Heat action plan for Odisha. http://www.osdma.org/Download/heat-wave-action-plan.pdf

Oke TR (1987) Boundary layer climates. Taylor & Francis Group, London

Pai DS, Thapliyal V, Kokate PD (2004) Decadal variation in the heat and cold waves over India during 1971-2000. Mausam 55(2):281–292

Pai DS, Nair SA, Ramanathan AN (2013) Long term climatology and trends of heat waves over India during the recent 50 years (1961–2010). Mausam 64(4):585–604

Pai DS, Srivastava AK, Nair SA (2017) Heat and cold waves over India. In: Rajeevan MN, Shailesh N (eds) Observed climate variability and change over the Indian Region. Springer Geology. Springer, pp 51–71

Raghavan K (1966) A climatological study of severe heat waves in India. Indian J Met Geophys 17(4):581–588

Revenue Disaster Management Department (2018) Telangana heat wave action plan. Revenue Disaster Management Department, Government of Telangana. https://www.telangana.gov.in/PDFDocuments/Telangana-HeatWave-Action-Plan.pdf. Accessed 14 Apr 2020

Sati AP, Mohan M (2018) The impact of urbanization during half a century on surface meteorology based on WRF model simulations over National Capital Region, India. Theor Appl Climatol 134(1):309–323. https://doi.org/10.1007/s00704-017-2275-6

Sinha Ray KC, Mukhopadhya RK, Chowdhary SK (1997) Trend in maximum and minimum temperature and sea level pressure over India. INTROPMET-97, IIT New Delhi:2–5

Subbaramayya I, Surya Rao DA (1976) Heat wave and cold wave days in different states of India. Indian J Meteorol Hydrol Geophys 27:436–440

The Hans India (2018) Heatwave deaths come down to single digit Number in AP. http://www.thehansindia.com/posts/index/Andhra-Pradesh/2018-05-26/Heatwave-deaths-come-down-to-single-digit-Number-in-AP/384182. Accessed 14 Apr 2020

The Hindu (2018) Government buckles up for summer. https://www.thehindu.com/news/cities/Hyderabad/govt-buckles-up-for-summer/article23344584.ece

The Hitavada (2018) With summer round the corner, CMC set to implement heat action plan. http://www.ehitavada.com/encyc/6/2018/03/16/VLpage_6.pdf

The South Asian Climate Outlook Forum (SASCOF) (2010) The South Asian Climate Outlook Forum (SASCOF). http://www.wmo.int/pages/prog/wcp/wcasp/rcofs/webpage/SASCOF.html

Times of India (2018) Action plan on the cards to shield citizens from heat. https://timesofindia.indiatimes.com/city/pune/action-plan-on-the-cards-to-shield-citizens-from-heat/articleshow/64278977.cms

United States Environment Protection Agency (US EPA) (2012) Reducing urban heat islands: compendium of strategies. https://www.epa.gov/sites/production/files/2014-06/documents/basicscompendium.pdf

WHO-WMO Joint Office for Climate and Health (2016) Improving health preparedness for extreme heat events in South Asia, 2016. In: 1st South Asia Climate Services Forum for Health (CSF-Health), Meeting Report, Colombo, Sri Lanka. 26–28 Apr 2016. http://www.wmo.int/pages/prog/wcp/wcsp/rcofs/sascof/health/documents/1st-South-Asia-Climate-Services-Forum-for-Health.pdf. Accessed 14 Apr 2020

WMO (2015) Heatwaves and health: guidance on warning-system development. WMO-No. 1142. p 114

WMO and WHO (2015) Heatwaves and health: guidance on warning-system development. Geneva. Accessed 14 June 2021

Part III
Future Challenges and the Way Forward

Chapter 14
Guidance on Integrated Urban Hydro-meteorological, Climate and Environmental Services: Challenges and the Way Forward

Paul Joe, Alexander Baklanov, Sue Grimmond, Veronique Bouchet, Luisa T. Molina, K. Heinke Schluenzen, Gerald Mills, Jianguo Tan, Brian Golding, Valery Masson, Chao Ren, James Voogt, Shiguang Miao, Humphrey Lean, Bert Heusinkveld, Anahit Hovespyan, Giacomo Terrugi, Patrick Parrish, Jens H. Christensen, Joachim Fallmann, Ranjeet Sokhi, Reinhard Spengler, and Junichi Yoshitani

Abstract This chapter describes a new World Meteorological Organisation (WMO) approach to the provision of science-based services to assist the planning of safe, healthy, resilient and climate-friendly cities. The approach is outlined in full in the WMO *Guidance on Integrated Urban Hydrometeorological, Climate and Environmental Services*. Volume I, Concepts and Methodology publication. WMO Guidance for Urban Integrated Hydrometeorological, Climate and Environmental Services. Volume I: Concept and Methodology. WMO-No: 1234, https://library.wmo.int/doc_num.php?explnum_id=9903, 2019a, which has been approved for promotion and communication to WMO Members. Highlighted here are the underlying concepts and methods that underpin the services to help manage cities on a

P. Joe (✉)
WMO Consultant, Toronto, Canada

A. Baklanov · A. Hovespyan · G. Terrugi · P. Parrish
WMO Secretariat, Geneva, Switzerland
e-mail: abaklanov@wmo.int; ahovsepyan@wmo.int; gteruggi@wmo.int; pparrish@wmo.int

S. Grimmond
University of Reading, Reading, UK
e-mail: c.s.grimmond@reading.ac.uk

V. Bouchet
Environment and Climate Change Canada, Montreal, Canada
e-mail: Veronique.bouchet@canada.ca

L. T. Molina
Molina Center for Energy and the Environment (MCE2), La Jolla, USA
e-mail: ltmolina@mit.edu

C. Ren, G. McGregor (eds.), *Urban Climate Science for Planning Healthy Cities*, Biometeorology 5, https://doi.org/10.1007/978-3-030-87598-5_14

day-to-day basis, responding to short-term hazards but also to prepare for climate change are described. Heterogeneous dense observation networks, high-resolution forecasts, multi-hazard early warning systems, long-term climate prediction and service applications all working in combination are at the heart of such services, the implementation of which will support mitigation and adaptation strategies for achieving thriving climate resilient cities and promotion of the United Nations

K. H. Schluenzen
University of Hamburg, Hamburg, Germany
e-mail: heinke.schluenzen@uni-hamburg.de

G. Mills
School of Geography, University College Dublin, Dublin, Ireland
e-mail: gerald.mills@ucd.ie

J. Tan
Shanghai Meteorological Bureau, China Meteorological Administration, Beijing, China

B. Golding · H. Lean
Met Office, Reading, UK
e-mail: brian.golding@metoffice.gov.uk; humphrey.lean@metoffice.gov.uk

V. Masson
CNRM, Université de Toulouse, Météo-France/CNRS, Toulouse Cedex, France
e-mail: Valery.masson@meteo.fr

C. Ren
Faculty of Architecture, The University of Hong Kong, Pok Fu Lam, Hong Kong
e-mail: renchao@hku.hk

J. Voogt
Western University, London, ON, Canada
e-mail: javoogt@uwo.ca

S. Miao
Institute of Urban Meteorology, China Meteorological Administration, Beijing, China
e-mail: sgmiao@ium.cn

B. Heusinkveld
Wageningen University, Wageningen, the Netherlands
e-mail: Bert.heusinkveld@wur.nl

J. H. Christensen
Danish Meteorological Institute, København, Denmark
e-mail: hesselbjerg@nbi.ku.dk

J. Fallmann
KIT, Karlsruhe, Germany

R. Sokhi
University of Hertfordshire, Hatfield, UK
e-mail: r.s.sokhi@herts.ac.uk

R. Spengler
German Weather Service, Freiburg im Breisgau, Germany
e-mail: Reinhard.Spengler@dwd.de

J. Yoshitani
Shinshu University, Matsumoto, Japan
e-mail: yoshitani@shinshu-u.ac.jp

Sustainable Development Goals. To meet the needs of urban areas, the concept of **I**ntegrated **U**rban Hydrometeorological, Climate and Environmental **S**ervices (IUS) is embedded in a multidisciplinary approach that identifies the requisite partnerships, including research agencies, city governments, international organizations, and private sector stakeholders, to establish and sustain urban services. As this is an emerging and diverse service, the requirements are broad with considerable research, development, capacity building and cross-service integrations yet to be done. In addition to providing an overview of the concept, we set out the challenges and recommended path forward to develop and implement IUS.

Keywords Integrated urban services · Multi-hazard early warning systems · Weather · Climate · Environment · Air quality · Hydrology · World Meteorological Organization · Challenges and the way forward · Research and development · Seamless earth system prediction · Scales

Key Messages

- WMO's participation in the UN New Urban Agenda includes developing both the concept and guidance for the provision of multi-hazard early warning to long-term urban planning services for cities.
- Integration of weather, hydrology, environment and climate urban services (IUS) will lead to accurate, efficient and consistent services.
- National Hydrometeorological Services are positioned to lead the promotion and development of Integrated Urban Services (IUS).
- Understanding capabilities, capacities and roles and responsibilities amongst different levels of government, universities, agencies and decision-makers is the first challenge followed closely by data access and the co-design of the IUS.
- Research and science will play a significant role to develop multidisciplinary scientists to bridge the scientific gaps and to demonstrate the effectiveness of IUS.

14.1 Introduction

The United Nations' (UN) New Urban Agenda (UN 2016) and the UN Sustainable Development Goal SDG11: Sustainable Cities and Communities (UN 2019) amongst others focus attention on cities and human settlements as places where equitable opportunities for human development are needed. The novel concept and approach of Integrated Urban Hydrometeorological, Climate and Environmental Services (or Integrated Urban Services, IUS) serves this agenda through promoting the idea of services for: (i) sustainable development and (ii) multi-hazard early-warning systems for cities. Allied with these international agendas is the Sendai Framework for Disaster Reduction 2015–2030, a core aim of which is to substantially reduce the mortality, economic losses and damages and disruption of basic services arising from extreme climate events as well as the mitigation of technological and security risks. Given this, the ideal is for services to be impact/risk based (WMO 2015a) and include hazard prediction as well as hazard exposure, vulnerability and other effects, to inform governments, business sectors and the public about how hydro-meteorological hazards may impact lives, livelihoods, property and economic activity so that appropriate actions and plans can be formulated. Because defining disaster risk and forecasting hydro-meteorological impacts is generally beyond the remit of meteorologists and hydrologists (WMO 2015a), engaging with disaster management officials and other relevant experts will yield benefits for disaster risk reduction (DRR). As argued by Rogers and Tsirkunov (2013), National Hydrological and Meteorological Services (NMHS) may be well placed to meet such challenges given their capacity to: understand the dynamic relations between extreme weather events and impacts, their record of engagement with the DRR community and burgeoning expertise in integrated risk communication across a range of stakeholders and affected public and private services. Moreover, atmospheric processes and associated hydrometeorological and climate extremes know no national boundaries and are interrelated. Accordingly, global-scale collaboration and integration is essential for the development and beneficial application of cognate services for reducing the impacts arising from meteorological, climate, environmental and hydrological extremes. WMO Members, often represented by NMHS, provide a framework for such international cooperation. Here, we articulate the IUS concept for which numerous types of agencies (e.g. government institutes, universities and private companies) may be providers of the requisite components for creating integrated services at the urban scale (WMO 2019a).

To move towards development of IUS, WMO formed an interdisciplinary Urban Expert Team, with scientists representing the different WMO research and service development programmes. From within the Global Atmospheric Watch (GAW) programme, the Urban Meteorology Science Advisory Group (GURME SAG) led the development of the Guidance for Integrated Urban Hydro-Meteorological, Climate and Environmental Services. Eventually the guidance comprises two volumes. Here we discuss Volume I, which addresses the concept, methods, gaps, challenges and the path forward for the development of an operational IUS (WMO 2019a). The second volume contains demonstration city examples and case studies (WMO 2021). The new Study Group for Integrated Urban Services (SG-URB) under the WMO Service Commission will address implementation guidelines. As with other WMO documentation, these will be updated as needed. While Volume I addresses lessons learned from examples of IUS, these are discussed in detail elsewhere (Grimmond et al. 2020). In this chapter we, as the original authors of the WMO publication, provide a summary of the *Guidance on Integrated Urban Hydrometeorological, Climate and Environmental Services. Volume I, Concepts and Methodology* (WMO 2019a), describe the need for IUS (Sect. 14.2), present the IUS concepts, methods and challenges (Sect. 14.3), discuss implementation of IUS (Sect. 14.4), briefly summarize gaps and the path forward (Sect. 14.5) and draw conclusions (Sect. 14.6).

14.2 The Need

14.2.1 Growth of Cities

A driving change in people's daily lives has been the accelerating growth of cities. Urban populations, especially in developing countries, have created crowded cities that are centres of creativity and economic transformation. However, air pollution, extreme weather conditions, flooding and other hazards create substantial challenges in urban environments (Bulkeley 2021; Grimmond et al. 2020; Pelling 2003; World Bank 2010). As a response to the challenges posed by rapid city growth and the advent of megacities, the United Nations (UN) HABITAT-III conference in October 2016 adopted the New UN Urban Agenda (UN 2016) which makes urban resilience, climate and environment sustainability as well as disaster risk management key foci (Grimmond et al. 2020).

Because of their dense, complex and interdependent urban infrastructure and the reliability of urban populations on these, cities are becoming increasingly vulnerable to the impacts of extreme hydrometeorological events. Through cascading downstream or "domino" effects, spatially widespread and far-reaching breakdown of critical city infrastructure can arise from a single extreme event (Grimmond et al. 2020; WMO 2021), wreaking havoc on services and people's lives. As noted by Grimmond et al. (2020), as the components of urban systems are tightly intertwined, the availability of good predictions, tailored and refreshed at appropriate spatial and temporal scales, will facilitate effective and efficient urban operations, an imperative made even more critical when extreme events occur. As what constitutes an extreme varies for different elements of urban infrastructure (across and between cities), understanding what "all" the critical thresholds are, is essential. Figure 14.1 illustrates the cascading effects from a typhoon event. There may be immediate impacts requiring multiple services followed by potential structural changes to the city fabric leading to health impacts for example caused by the proliferation of vector-borne disease catalysed by standing water that offers mosquito breeding sites.

If exposure units (e.g. vulnerable communities/neighbourhoods, sectors, critical assets) are to optimize their response, small area forecasts are needed that identify which parts of the city are most likely at risk to any given hazard. Combining forecasts with detailed community and infrastructure information facilitates resources to be used most appropriately. Services drawing on new communication methods and technologies will not only ensure rapid assimilation of data by short-term prediction systems to enhance forecasts but also facilitate rapid communication of tailored products to targeted end users. Given this, IUS should include impact predictions that consider the interplay between hazard, exposure and vulnerability in the generation of risks. Further, IUS can provide tools to support appropriate long-term planning of cities, as inferred in a number of chapters comprising Part II of this book. Importantly ensuring sustainable approaches to the development of cities will benefit not only the urban populace but also the global environment, as cities are the largest sources of greenhouse gases (WMO-GFCS 2019).

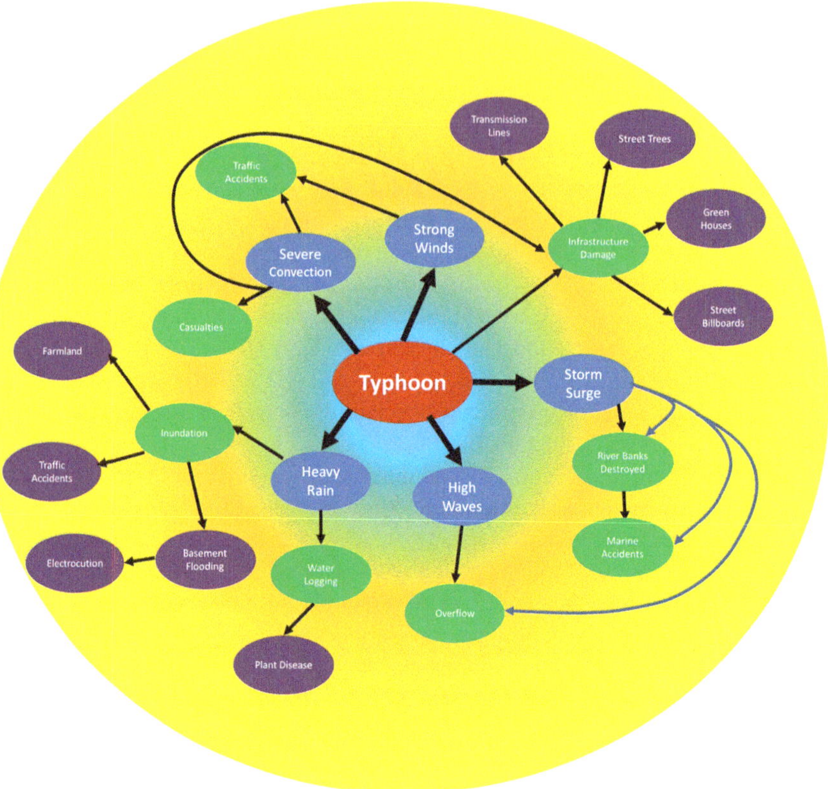

Fig. 14.1 Schematic showing the domino effect using a typhoon (or hurricane) event as an example. The event (red) creates several hydro-meteorological hazards (blue) that generate immediate impacts (green) with related effects (purple) requiring different services. (Source: WMO (2019a) Guidance for Urban Integrated Hydrometeorological, Climate and Environmental Services. Volume I: Concept and Methodology. WMO-No: 1234, ISBN 978-92-63-11234-7. https://library.wmo.int/doc_num.php?explnum_id=9903)

14.2.2 Resilient Cities

In response to climate change and sustainable development concerns over the last two decades, a range of cities have formed networks for developing strategies for achieving climate resilient cities (e.g. Local Governments for Sustainability (ICLEI 2020); C40 Cities Climate Leadership Group (C40 2020), 100ResilientCities (100ResilientCites 2019) and its replacement Resilient Cities Catalyst—https://www.rcc.city/). The resolutions and statements arising from these networks provide additional motivation for the development of the IUS, including:

- "*A resilient city is prepared to absorb and recover from any shock or stress while maintaining its essential functions, structures, and identity, adapting and thriving in the face of continual change*" 100ResilientCites 2019)

- "*A smart city has embedded 'smartness' into its operations, […]. It analyzes, monitors and optimizes its urban systems, be they physical (e.g. energy, water, waste, transportation and polluting emissions) or social (e.g. social and economic inclusion, governance, citizen participation).*" (ICLEI 2020)
- "*An EcoMobile city fulfills its objective of creating a more livable and accessible city by utilizing Sustainable Urban Mobility principles to achieve significant reductions in greenhouse gas emissions and energy consumption, improvements to air quality, and increased mobility opportunities for all citizens.*" (ICLEI 2020)
- "*A view of resilience that includes not just the shocks—earthquakes, fires, floods, etc.—but also the stresses that weaken the fabric of a city on a day to day or cyclical basis. Building urban resilience requires looking at a city holistically: understanding the systems that make up the city and the interdependencies and risks they may face.*" (100ResilientCites 2019)

A large number of cities subscribe to these objectives irrespective of size or level of socio-economic development. Ensuring population safety and resilience to extreme weather and climate events are central reasons for many cities' calls for action, with societal expectations of good living standards, sustainable economic development, cities as places with a competitive edge to invest in and do business and politically supported to do so increasingly important.

14.2.3 Urban Service Requirements

The underlying aspiration of IUS is to provide support for policymaking (at all time scales) from regulations to emergency services to the public, businesses, institutions and responsible authorities on a day-to-day and long-term planning basis at the city level (WMO 2019a). For urban management services to evolve and be fit for a rapidly changing set of climate and broader environmental circumstances, new knowledge is required. Added to this is a capability to deliver and cut across many entities, scientific disciplines and time scales (WMO 2019a; Baklanov et al. 2018). A compound event such as a heatwave embedded in a moderate yet prolonged dry spell serves to demonstrate the type of approach required. For example, an affected city may need to address concurrently, enhanced energy demands, increased pressure on healthcare systems, water quality and supply issues and wildfires at the urban-rural interface, while a global pandemic occurs where building ventilation regulations are critical. Critical in relation to the potential magnitude of the impact will be understanding the nature of the pre-existing drought conditions, preparatory lead-time and the ability to anticipate how the heatwave may evolve and affect services and the integrity of urban systems at large. Being able to integrate all the requisite information would enhance coping capacity, assist identification of vulnerable infrastructure and foretell cascading effects arising from infrastructure failure (WMO 2019a).

The nature of first-order meteorological/climatological information required by cities for managing climate-related risk is very much determined by a city's climate, geographical and geophysical settings as well as its form (three-dimensional structure) and function (social, cultural, economic and other activities). First-order needs include all year round monitoring and prediction of conditions so extremes (e.g. severe weather and climate events) and typical conditions that might pose a threat to the functioning of a city can be assessed. These might include: typhoons/hurricanes, extra-tropical storms, convective storms, tornadoes, droughts, extreme heat and cold stress, typical winds, temperatures and humidity by season,[1] sand and dust storms, wild fires, flash floods, landslides, river and lake flooding, storm surges or swell, long-term sea level rise, coastal inundation, air and water pollution, chemical and other harmful matter dispersion events, harmful UV radiation, pollen and other aerobiological allergens, factors related to food-, water- and vector-borne disease, and changes in soils[2] (WMO 2019a). Longer-term needs encompass high-resolution climate services for building codes, zoning, planning and design. A central tenet of IUS is working across systems. For example, an eco-system's urban design and planning approach, such as "Blue-Green Solutions", which incorporates green (vegetation) and blue (water) elements is very dependent on the micro-climate and water resource availability and timing. If the design scale is the city-block, then urban processes need to be linked with meteorological, climatological, hydrological and air pollution processes (e.g. understanding the science for model predictions) to develop a good design that makes an area more resilient rather than a design that increases vulnerability.[3] To achieve what is necessary, existing solutions and scientific tools are fundamental as well as academic multidiscipline curricula and capacity building, which will provide a starting point for bringing together different professionals (e.g. architects, engineers, urbanists, policy makers, meteorologists) concerned about the resilience of cities. Shared knowledge of data repositories, models, examples of application needs and how to provide direct access between the components of information systems, all need to be established. Cities need to be furnished with warnings, including climate watch advisories, to ensure appropriate urban system design and operations and the appropriate location of critical assets such as hospitals, schools, power stations, commercial centres and other key land use types.

---

[1] Critical for building design, landscape planning, outdoor workers, etc.

[2] Note extremely dry soils compared to typical conditions can result in the collapse of buildings or cracking of external facades making them vulnerable to even light rainfall or wind events.

[3] Water needs to be available to sustain vegetation or maintain water features. For example, if energy is needed to supply the water (e.g. to roof top gardens) the environmental costs may outweigh the benefits. Some vegetation can have detrimental influence son air quality through impacts to chemistry and/or flow.

14.2.4 Benefits of Integration

Integration (weather, climate, hydrology and the environment) is an efficient and effective evidenced based practice in seamless prediction (from urban and nowcasting scales, short- and medium-term forecasting, sub-seasonal and seasonal spatio-temporal scales) of the Earth System (Brunet et al. 2015; Grimmond et al. 2015) and for the provision of multi-hazard early warning services (WMO 2015a). The challenges and problems are co-dependent and hence multidisciplinary science approaches are required to resolve gaps and inconsistencies (WMO 2019a).

IUS are developing within the WMO concept of the seamless Global Data-Processing and Forecasting System (GDPFS) that include issues related to heterogeneous observations, their processing and management, high-resolution prediction systems, product generation, services and their verification (WMO 2019b). As stated by the WMO, the GDPFS aims to enable Member states and partners to support decision-makers and risk-based warnings (WMO 2004). The seamless GDPFS builds upon existing systems. With the additional goal of greater agility and adaptability to support local, regional or national needs, different application programmes including non-weather related predictions with co-designed products for impact-based and risk-based forecast and warning services to support decision-making, taking into account vulnerability and exposure information (e.g. for urban emergency preparedness and sustainable development) (WMO 2004).

14.3 The Integrated Urban Services (IUS) Concept and Challenges

14.3.1 The Concept

IUS can be provided at a range of spatial scales from the large metropolitan area to smaller neighbourhood to city block scales. However, the nature of IUS is highly dependent on the application, requirements, and local and regional factors. The urban domain, often defined by local governments, can include contiguous cities, the area and roads in-between, rural watersheds (or catchments) and the location of industries as all are critical to city functioning (WMO 2021).

The generic components of IUS are set out in Fig. 14.2. This provides a framework for discussing the concept as expanded on below. It is worth noting that Integration of Urban Services can be delivered at the technical, product or service levels, or in all permutations and combinations.

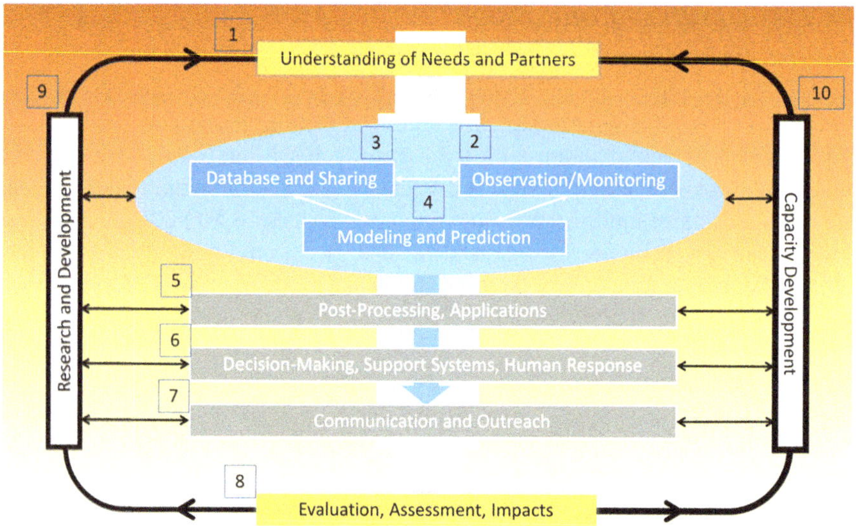

Fig. 14.2 Schematic of the components of an Integrated Urban Service (IUS) System. The numbers (#) are referred to in the text (Source: WMO 2019a Guidance for Urban Integrated Hydrometeorological, Climate and Environmental Services. Volume I: Concept and Methodology. WMO-No: 1234, ISBN 978-92-63-11234-7. https://library.wmo.int/doc_num.php?explnum_id=9903)

14.3.2 City Requirements/Local Tailoring

Cities can amplify or mitigate hazards. For example, impervious surfaces can amplify flooding at many scales up to the regional scale, flooded subways or underpasses create microscale problems, dense tall buildings can trap air pollution, and greenspace can mitigate large-scale heatwaves for some wards (WMO 2019a). Accordingly, the needs and objectives of IUS will be city-specific, necessarily diverse at the intra-urban scale and need to be "fit for purpose" (Fig. 14.2, #1). Thus, well-designed IUS will be sensitive to local needs even if developed at the national scale. Meaningful dialogue on requirements and initial steps may be the greatest challenge in developing an IUS. A mutual understanding between service providers and city authorities of capability, vision, roles and responsibilities is a prerequisite for success. Communication of science to city authorities who have vastly different expertise, cultures, skills and knowledge may pose a potential barrier to the process to form partnerships. Therefore, a common trust and language must be developed with leadership, experience and vision crucial.

14.3.3 Observations

Generally, urban areas are observation sparse, a result of challenges associated with the appropriate siting of meteorological instruments (e.g. rain gauges, wind, air quality sensors) for representative measures needed across the IUS domains. Typically, IUS observation networks are not dense enough to detect intra-urban variations in IUS fields. Further, urban observations are not always utilized in current operational numerical weather prediction (NWP), air quality modelling and/or assimilation systems, as urban atmospheric and other key processes change rapidly (vertically and horizontally), so are either not represented or are coarsely parameterized. In contrast, urban hydrological networks are better developed, with water level and flow gauges located in sewers, natural streams and engineered urban watercourses (WMO 2008, 2010a). Resource limitations can also be a significant limiting factor, especially in low- and middle-income countries.

For urban observations, four ideal requirements exist: (1) real-time monitoring and scientific understanding; (2) data assimilation for numerical weather forecasting; (3) climatology for model evaluation and verification; and (4) evaluation and verification of services (including information on human perception and response) (WMO 2019a).

Although urban siting requirements are clearly prescribed by the WMO (2006), they may not serve the requirements of the current/next generation of urban models capable of resolving processes at the street canyon and building scales. Perhaps the best way forward is to apply guiding principles rather than rules, and to retain a flexible approach for siting urban stations. Without doubt, a comprehensive review is required as part of any discussions related to the siting of urban meteorological instruments to serve the purposes of model evaluation and verification. For example, the type of observation required may be numerical model dependent and thus determine such things as instrument type, its siting and sampling frequency (Grimmond and Ward 2021). One can imagine in the case of a numerical model that only assimilates vertical profiles of above surface layer wind versus surface layer (i.e. inertial sub layer—roughness sub layer—canopy layer) winds, the configuration of observations will be quite different.

The complex three-dimensional nature of urban environments places heavy demands on measurements and sensor locations, as these need to be representative of the area under consideration. This will vary between high and low building density areas having many variations including land cover (e.g. green space, roads, buildings) and emissions (e.g. heat, water, chemical) between differing land uses (e.g. central business district, and residential/suburban). Added to this is determining the appropriate temporal and spatial sampling that includes the vertical dimension, both within and above the urban canopy, and through the boundary layer. Complete watersheds, upwind areas, and a range of urban densities may need enhanced observations. The appropriate mix of observations, scale and frequency of sampling is a challenge and will evolve over time.

Non-homogeneity of data also poses a challenge as urban development and regeneration regularly alters the environment around observation sites. Like all observation sites, instruments undergo regular evaluation, re-calibration and replacement. Consequently station metadata, with frequent updates (especially because of the changing urban environment), is as important as the data, because it provides the foundations for homogeneity testing and establishing the link between sensors, measurements and the surrounding area and for use in applications (WMO 2018).

Technological developments have opened up a range of exciting possibilities for real-time monitoring of IUS processes (e.g. surface exchanges, air quality, water flow) via atmospheric profiling using ground-based measurement networks. Instrumentation such as ceilometers, Doppler lidars, microwave radiometers, small radars, wind profilers, and Doppler sodars (latter is usually restricted to airports because of noise constraints) can provide information on the nature of urban boundary layer processes (e.g. thermal structure, momentum flux) as well as cloud, thermal, moisture and precipitation fields (Grimmond and Ward 2021). Complementing ground-based measures are those made from satellite platforms that provide a bird's eye view of the urban surface. Encouragingly the accuracy of ultra-high resolution (scale of metres) satellite thermal and other monitoring of the urban surface (including building density and urban fabric roughness) has improved over the recent years because of the emergence of high-resolution, multi-sensor systems, and new processing techniques, facilitating the assimilation of associated measurement into integrated urban prediction models. The development of the field of satellite-based urban meteorology and climatology continues at pace, with space-borne thermal radiometers and hyperspectral infrared imagers either in operation or planned for the near future (Guanter et al. 2015), with advances in vertical profiling measurements also expected (e.g. Meteosat Third Generation). Understanding and managing the uncertainty of measurements from new observation platforms and exercising scientific judgement and flexibility will be an ongoing priority as the research and applications community strive to use such novel and innovative techniques to improve short-term forecasts and high-impact predictions. Another potential source of urban data is that from small mobile sensors (e.g. phones, cars) and citizen science networks. However, the quality and uncertainties associated with this type of data need to be fully tested and quantified.

In addition to knowing about the nature of IUS fields, information on the physical and socio-economic make-up of urban environments is required (Grimmond and Souch 1994; Ching et al. 2018). This is because the urban surface in its widest sense, including both indoor and outdoor built environments, interacts with the overlying atmosphere. Accordingly, information on urban form, comprised of the three-dimensional structure of cities, the nature of the urban fabric (natural and built materials) and land cover type, and anthropogenic heat emissions are critical as input into urban meteorological models. Allied data such as environmental, geomorphological, socio-economic, population and economic activity are also essential for providing important contextual information to support analyses, for model applications and for reflection upon how the outcomes of measurement, monitoring, modelling and evaluation campaigns are relevant to policy development, urban planning, emergency response and public health.

14.3.4 Databases and Data Sharing

The data required to manage IUS risks is typified by its diversity and magnitude. Accordingly, knowing about existing databases and data sharing and getting people on board are critical to the notion of IUS. Different institutions within a city play key roles in information gathering and processing. It is essential to the success of IUS that the appropriate people in city authorities, related agencies, companies, research institutes, are identified, a common language established so that work can be undertaken collaboratively to build integrated urban information systems, that have meaningful communication of what will be "big data". To facilitate this, legislation or memoranda of understanding may engender co-operative partnerships. Data exchange is not only an important enabler of IUS, but also creates a vehicle and opportunity to communicate, interact and exchange knowledge and to understand the needs as highlighted in Fig. 14.2, #3. Indeed, data exchange can serve as a catalyst for the creation of IUS (WMO 2019a, 2021). Notwithstanding the challenges associated with making observations open for operational use, it cannot be emphasized enough how data sharing is critical to fostering IUS developments and collaboration.

It is also critical to make data (observation, model outputs, etc.) accessible in formats that are usable amongst partners. The goal should be for open data and easy access to information (WMO n.d.). Essential is agreement amongst different network-members about metadata definitions, data storage protocols and clear usage policies. For example, Muller et al. (2013) and Creutzig et al. (2019) provide useful guidance materials for urban observing systems and metadata.

14.3.5 Modelling and Prediction

Models (Fig. 14.2, #4), whether numerical (e.g. hydrological, chemistry-atmosphere forecast model), physical (e.g. hardware models within wind tunnels or climate chambers), statistical (e.g. probabilities, empirical relations between components, machine learning/artificial intelligence (AI)), and/or conceptual (e.g. schematics showing relations between components), are an integral part of IUS (WMO 2019a). In the IUS context, the overall purpose of models is to shed light on the response of urban surface and atmospheric processes, to urban form (structure, fabric and land cover) and urban function (human activities) and changing weather, climate, hydrological and environmental conditions, at a range of time scales. The corollary of this is the modelling chain should serve the purposes of urban services and impact prediction, assessment and risk quantification, to assist decision-makers and related stakeholders, with a programme of sustained research in place focused on model capability, consistency and improvement.

Amongst a range of modelling possibilities are Regional/Seamless Earth modelling systems. These attempt to model land-atmosphere, atmosphere-hydrological and ocean-atmosphere feedbacks and processes at very high resolutions—hundreds of metres or less—in order to describe urban to intra-urban scale processes. Such modelling systems are valuable tools and crucial for some IUS requirements. As noted by Brunet et al. (2015), integration at the appropriate level requires collaboration amongst diverse and highly technical disciplines. That said, although the full integration of earth or urban system processes may be a logical step and a long-term strategic goal, challenges exist (Grimmond et al. 2015). Some of these derive from the fact that the requisite disciplines "speak" different languages, work with contrasting prediction and modelling paradigms, set objectives that are not wholly compatible and link with a variety of end users in different ways. To address such challenges, the formation of multi- and inter-disciplinary partnerships with a commonality of purpose and protocols is critically important (WMO 2019a). For effective operational deployment of IUS models, management is necessary. For example, extending to high resolution NWP or to coupled Numerical Environmental Prediction models is complex and no management rules are generally in place (WMO 2019a). As outlined by the US National Research Council, the normal practice in meteorology and air quality modelling is to move to a model upgrade when there are compelling scientific grounds, with clear modelling system performance gains and no major degradation of key variables, hence making model intercomparisons an integral part of the governance process (NRC 2012).

In the case of the application of Earth-system models to cities, as full a set of physical, biogeochemical, and socio-economic processes as possible need to be represented (Masson et al. 2020), with their relative importance in many ways determined by the nature of the intended modelling applications. For example, for heatwave alerts or urban planning directed to the mitigation of urban thermal effects, the focus will be external and indoor temperature and atmospheric and surface heating processes. This is because as external temperatures rise so will those inside buildings. Where air conditioning (AC) is ubiquitous, as in some high-density high-income cities such as Osaka or New York, this will result in an elevation of energy demand for building cooling, resulting in an increase in the emission of waste (anthropogenic) heat into the environment and possible further elevation of urban air temperatures (Takane et al. 2019). Nearby districts without AC may become exposed to this "extra" heat via advection processes (Huang et al. 2019). Accordingly, knowing, via mapping, the heat vulnerability characteristics of "down-wind" urban populations as part of the information contained in heat action plans may assist with managing the unintended consequences of AC-related heat emissions (Takane et al. 2020). For the situation of modelling environmental hazards for coastal cities to determine exposure and vulnerability, it will be essential to account for ocean-atmosphere-wave interactions, ocean-wave-riverine interactions, and their interplay with land cover (WMO 2019a). For urban flooding, changes in coastal or river flood plains, surface changes due to building or infrastructure construction will be crucial.

Urban green space, whether in the form of tree-lined streets, gardens, parks and green roofs, can modify urban temperatures at a range of scales depending on its scale of application as a heat mitigation strategy. Green space also provides benefits for flood management. Despite its potential attractiveness as an urban planning approach to IUS-related risk management, the conscious large-scale implementation of green infrastructure for climate-specific sensitive planning in cities remains confined to only a few examples, as described elsewhere in this volume (e.g. Reuter and Kapp 2021; Tapper 2021). Integration of green infrastructure-related atmospheric, environment and water processes into urban systems models is an area of active research requiring coupling models across scales that simulate soil-vegetation-atmosphere interactions (WMO 2019a). Integration of the assessment of the benefits and costs of these actions on a broader range of urban activities is also possible within an IUS community, hopefully preventing unintended consequences. For example, to understand the impact of climate risk management adaptation and mitigation strategies at the intra-urban to urban scales, urban planners and designers require fields of relevant IUS variables at high spatial resolutions for current and future city configurations and climate scenarios. Developers of IUS therefore need to be cognizant of this requirement.

14.3.6 Applications

IUS have the potential to contribute to the long-term sustainability and resilience aspirations of cities via a wide range of potential applications (Fig. 14.2, #5). Although not an exhaustive list, these might include:

- Urban-specific disaster risk reduction as might be required for managing the outcomes of severe storms or periods of extreme heat
- CBNE response—from industrial accidents or terrorism that release of chemical, biological, radioactive nuclear and explosive
- Better city operations through forewarnings of IUS-related risk for critical infrastructure
- Acute and chronic human health problems
- Sustainable city economic development and planning at scales from months to seasons
- Managing weather/climate risks in business, commercial and outdoor recreation sectors
- Managing urban flooding
- Managing environment and health
- Climate change adaptation
- Climate sensitive design and planning at the urban (~10–15 km) to neighbourhood (<1 km) scales
- Short-term renewable energy forecasts and infrastructure management
- Fire risk and responses

- Snow-clearance and road-icing

Applications should not be developed independently of potential end users; the so-called "loading dock" approach whereby a product is developed without end user engagement then delivered with little consultation (figuratively dropped on the loading dock) must be avoided (Sect. 14.3.8). Accordingly, application developers and IUS governance models need to engage actively and meaningfully with the target sectors or stakeholders in a specific IUS risk problem such as the energy, water supply and sanitation, environmental, agricultural, transport and tourism sectors (WMO 2019a).

14.3.7 Decision Support

Arising from the applications component of IUS is information for decision-making in a wide variety of services (Fig. 14.2, #6). Over the past couple of decades, there has been a move away from single hazard focused systems to Multi-Hazard Decision Support Systems. These are designed to present consistent (integrated and seamless), relevant and often uncertain and conflicting information, to both technical experts (e.g. severe weather or flood forecaster) and users (e.g. emergency managers) for assessment, to support alert- and warning-related decision-making. Considerations often include societal impacts, action statements (e.g. issuance of severe weather warnings for typhoon tracks with uncertainty information, river level predictions nearing flood thresholds, heat health alert) and human behaviour especially how recent events may have influenced population response to warnings and advice (WMO 2019a). In order to assess the quality of the impact-based products, predictions are presented alongside observations and their historical performance taken into account. Assessments of whether appropriate messages (including probabilistic information about uncertainty) for the location and whether these landed with the target audience are also acted upon. In making informed decisions based on application products, the history of false positives and false negatives is important as over/under-warning of previous events may engender a lack of trust in forecasts and predictions and thus influence end user or stakeholder response to warnings. Over time, an iterative process will operate such that through developer and end user interaction, collaboration and critical reflection, feedbacks to the Decision Support System component or even beyond to the more fundamental observation, analysis and modelling components of the IUS system will occur so that IUS efficacy is enhanced (WMO 2019a).

14.3.8 Service Delivery, Communications and Outreach

The path from raw data to product and services development and eventual delivery to end users and stakeholders is complex and requires an understanding of and engagement with partners (Fig. 14.2, #7). Because of the wide range of potential end users and stakeholders (e.g. emergency management authorities, urban planners, health workers) the mode and form of information delivery is likely to be diverse. For example, emergencies require the information/warnings to be disseminated with as long a lead-time as possible to enable quick and timely implementation of preventive measures. In contrast, conventional climatological statistics and prediction at longer time scales for climate change projections, rather than real-time weather forecasts, are needed for longer-term urban design and planning. Independent of delivery time scales, vital for any information presented to end users is ensuing it is comprehensible and tailored to their level of expertise with detailed and voluminous information often counter-productive. Crucial is the way information on uncertainty is presented and addressed, and how IUS stakeholders might respond and subsequently use the products. "Uncertainty" training will be required. Thus, capacity and capability development will be important for all involved in IUS to understand the various constraints (WMO-GFCS 2019).

To ensure effective dissemination of information, a well-honed communication strategy, with supporting mechanisms, sensitive to end user and stakeholder needs is imperative (WMO 2014, 2015a). For wide and effective information dissemination, an integrated products/services delivery platform, providing access to all identified users as determined by their needs, represents the ideal with a programme of awareness raising and training in the application of IUS products complementing this. A range of possible dissemination technologies exist including dynamic web portals, Digital Display Board Systems, Integrated Voice Response Services, as well as communication through traditional avenues such as like TV, radio, E-mail alerts, press releases, media conferencing and short message service alerts (WMO 2019a). Leaflets, publications, workshops, conferences and symposiums form effective ways for awareness raising and wider public education. To ensure quality control, the avoidance of misinformation and the creation of confusion and panic amongst the wider public, the responsibility of dissemination of information via social media (e.g. Facebook, Twitter, WeChat) best lies with those trained in risk communication and the expert use of IUS products. Developing the "trusted/authoritative voice" for warnings will be critical. Over time, and assuming IUS community members develop a "community of practice" (COP) philosophy, ways of working and communication will develop and mature to a level that ensure seamless operation of IUS.

14.3.9 Evaluation and Methodologies

The true test of an IUS system is the extent it delivers its goals (e.g. reduced loss of life, improved livelihood, reduced societal and economic costs for regular city operations, improved preparedness) (Travis and Bates 2014). Accordingly, evaluation and assessment of performance is a critical aspect of the IUS framework (Fig. 14.2, #8). "Before-After" event comparisons are possible for heat-health action plans as these can evaluate individual events or seasons (e.g. Benmarhnia et al. 2016). Tailoring evaluations of IUS delivery to the intended user-oriented outcomes is crucial. For example, Amorim et al. (2018) describe how collaboration with the health sector can lead to the development of indicators (e.g. heat-induced mortality, long-term mortality from combined exposure to NO_2 and $PM_{2.5}$) for evaluation and assessment of where service improvements are required.

As overall system performance is an outcome of all the parts, evaluation of individual IUS components (Fig. 14.2) is crucial to understand their effect on overall results. To achieve this, diagnostic/process-oriented causal assessments as well as statistical evaluations complete with uncertainty analysis should be undertaken. Such an approach will inform the iterative co-design of the IUS. Further, dynamic evaluation techniques using the same IUS for very different urban services scenarios can be applied to impacts evaluation (WMO 2019a). As new partners, stakeholders and users broaden the IUS scope new evaluation approaches will need to be developed. For model development and robust diagnostic evaluation and quantitative user evaluation, "Test Beds" with data from all fields relevant for the target parameters and variables (e.g. flooded area, soil water content, population response) will be needed along with independent data sets for initialization and assimilation (WMO 2019a).

14.3.10 Research and Development

Given the novelty of IUS as a multidisciplinary approach to climate risk management, an iterative and continuous enhancement process will make a sustained programme of research and development (R&D) necessary (Fig. 14.2, #9), focusing not only on all IUS components but also on specific city systems, socio-economic impacts, population response and benefit accrual. Although practitioners are not responsible for R&D, their involvement as co-producers will have mutual co-benefits with participation R&D programmes. The potential array of players in R&D is vast with partnerships and Test Beds between research and operational bodies key, in order to foster the rapid conversion of research results to operational use and to train the next generation of multidisciplinary scientists (WMO 2019a).

14.3.11 Capacity Development

Once initiated, IUS will evolve rapidly, requiring capacity development efforts to be resilient (Fig. 14.2, #10) with the identification of strategies to facilitate a community of practice (COP, working together) approach to service development, evaluation and improvement key. Players are likely to include universities, public and private research institutions and agencies, NMHSs, international organizations, professional organizations, governmental and non-governmental organizations and community-based programmes (WMO 2019a). Capacity development will necessarily encompass multidisciplinary training in urban Earth system science, socio-behavioural and economic studies, urban planning and architecture, urban observations and network design, data sciences, user workshops for high resolution products, terminology and knowledge gaps, success metrics and impacts, decision support system use, risk communication and knowledge management (WMO 2019a).

14.4 Implementation

14.4.1 General Comments

Central to the success of an IUS will be seamless provision of services across a wide range of time scales including consistency and efficiency. As individual IUS will not be static, IUS developers, operators and managers need to be attuned to the necessity of an iterative dynamic process for continual IUS evolution and the provision of the requisite scientific knowledge and human resources for long-term sustainability. The ethos of IUS embodies a multidisciplinary philosophy from the outset in order to facilitate an inclusive approach to partnership building. For example, urban planning legal documents addressing air quality and vegetation feedbacks require innovative actions and partnerships involving social science and law research put into practice. A catalyst for instigation may be an extreme IUS event that engenders a government level response and call to action, or when a significant international event (e.g. Olympics, Expositions) is planned and staged requiring high resolution point warnings (Joe et al. 2018). Although usually funded independently of normal operating budgets, such events can be leveraged for the initiation of the development of IUS with international and national demonstration projects a proven vehicle for knowledge sharing, triggering inspiration and initial IUS implementation.

14.4.2 Role of the National Meteorological and Hydrological Services (NMHS)

As noted earlier, IUS processes in general know no national bounds. Accordingly, cooperation at global, regional and local scales is essential for managing IUS-related risks. The WMO, a United Nations Specialized Agency, is an intergovernmental organization with 193 national states and territories (called Members). Through its relevant programmes and technical commissions, it coordinates and guides its Members. Members of WMO are states and represent the disciplines of weather, climate, hydrology and the environment whether it is at the national, regional or local level or whether government, academia, research institutes or private sector (WMO 2015b). This is an important distinction as most IUS cannot be delivered by a single organization. This highlights the important coordination role that WMO can play in IUS. Consequently, one key aspiration of IUS is to establish collaborations through either legislation or co-operative agreements. Generally, the focus of NMHS is at the national scale. At the urban scale, city authorities will play a significant role in the overall specification, scope, definition, implementation, delivery and utilization of IUS with the requisite services coming from a variety of different providers. In this regard and given the increasing recognition of cities as "at risk" spaces, now is an opportune time for WMO members to consider generalizing and specializing their infrastructure (e.g. observations, models, products) for managing IUS-related risk (WMO 2021). Notwithstanding the largely national focus of national agencies and a mandate to support but not necessarily provide services at the municipal level, NMHS possess specialized capabilities, products and warning communications that extend to more complex multi-hazard early warning systems delivering integrated services within urban settings (WMO 2019a). In summary, NMHS are well-placed to play a central role in this because:

- *"NMHS operational systems and service delivery mechanisms are able to connect data, predictions, products and warnings with the type of applications related to city needs. However, important developments are needed to support these applications with an adequate scientific basis (e.g. urban areas need to be better represented in observations and in models).*
- *NMHS collect observations that are critical to IUS, but they need to be complemented by a wide range of observation types from other organisations. This requires a broadening of methods to integrate these data and data providers.*
- *NMHS can best harness the potential and results of specialized research now and in the future.*
- *NMHS already recognize themselves as key players in bringing that potential to fruition/operation in the provision of IUS"* (WMO 2019a, p. 19).

As outlined in WMO (2019a) first steps to achieving the pivotal role of NMHS in IUS delivery will necessarily include the following:

- Capacity building to introduce concepts but also to develop self-awareness, to educate and train meteorologists and potential partners (WMO 2010b, 2017a, b).
- Critical reflection upon the suitability of existing predictions systems for urban settings and associated institutional, governance and partnership arrangements and capacities and capabilities.
- Identification of the needs, capabilities and capacity of local governments, the private sector and other city-level stakeholders to better define IUS requirements (e.g. Grimmond et al. 2010; Mills et al. 2010).
- A willingness to shift to a service paradigm more suited to the notion of IUS, by going beyond traditional services and understanding the subtlety of urban system stressors, as well as cumulative/compounded impacts over different timescales.
- Build understandings of individual end user needs through a process of mutual engagement, tailor products to fit the defined risk management problem and consider the utility of case studies for illustrating the links between hydrometeorological processes and impacts.
- Consider how existing legal frameworks may facilitate partnership development and where possible move to a National Framework for Climate Services (NFCS[4]) approach as espoused by the WMO.
- Build narratives about the societal benefits of IUS that may arise from implementation related to specific events and the co-benefits (win-wins) for other urban services (WMO 2019a).

14.5 The Path Forward

14.5.1 Resources for Implementation of Integrated Urban Services (IUS)

Fundamental to IUS are partnerships that span the producers of IUS products, stakeholders and end users. These are essential to mobilizing and optimizing resources (Fig. 14.2, #1). Significant efforts are required to begin the dialogue process, to understand roles and responsibilities, to appreciate mutual constraints, capabilities and capacities, and to develop mutual trust. Engagement is necessary to settle on a common strategic vision and avoid barriers to IUS development and implementation. Ideally, all relevant service institutes should collaborate in undertaking the achievement of IUS. Some fundamental requirements for IUS implementation include:

(a) Programme resources

Observations, whether provided by the scientific or end use community, are fundamental to IUS and underpin service development (e.g. forecasts) and eval-

[4] Step-by-step Guidelines for Establishing a National Framework for Climate Services, https://library.wmo.int/opac/doc_num.php?explnum_id=4335

uation. This makes access to observations and provision of metadata for quality assurance purposes by all partner agencies essential, given the likely diversity of measurement methods and sites. Given the much smaller scales for IUS products (cf. traditional NMHS operational scales) high (spatial) resolution modelling systems that simulate key urban processes are required. To enable smooth IUS creation and operation, early definition of partner mandates, interactions, roles and responsibilities is ideal, with this best facilitated via a well-designed partner governance/legal framework and a communication office so the multi-agency joint action can be coordinated (WMO 2019a).

(b) Technical expertise

The pool of technical expertise available for IUS instigation and development is potentially vast. For example, WMO members have access to a wide range of WMO programmes and activities, such as:

- World Weather Research Programme—https://community.wmo.int/activity-areas/wwrp
- World Climate Programme (WCP)—https://public.wmo.int/en/programmes/world-climate-programme
- World Climate Research Programme (WCRP)—https://www.wcrp--climate.org/
- Global Framework for Climate Services (GFCS)—https://gfcs.wmo.int/
- Hydrology and Water Resources Programme—https://public.wmo.int/en/programmes/hydrology-and-water-resources-programme
- Global Atmospheric Watch (GAW)—https://public.wmo.int/en/programmes/global-atmosphere-watch-programme
- Global Cryosphere Watch (GCW)—https://globalcryospherewatch.org/
- Public Weather Service (PWS)—https://public.wmo.int/en/programmes/public-weather-services-programme
- Global Data Processing and Forecast Systems (GDPFS)—https://community.wmo.int/activity-areas/global-data-processing-and-forecasting-system-gdpfs
- Disaster Risk Reduction Programme—https://public.wmo.int/en/programmes/disaster-risk-reduction-programme
- WMO Integrated Global Observing Systems (WIGOS)—https://public.wmo.int/en/about-us/vision-and-mission/wmo-integrated-global-observing-system
- WMO Information Systems (WIS)—https://public.wmo.int/en/about-us/vision-mission-strategic-priorities/wmo-information-system-wis
- Global Atmospheric Watch (GAW) Urban Research Meteorology and Environment (GURME)—https://community.wmo.int/activity-areas/gaw/science/gurme
- HiWeather Research Projects—https://public.wmo.int/en/resources/bulletin/hiweather-10-year-research-project
- WMO Technical Commissions—https://community.wmo.int/governance/commission-membership
- WMO Regional Climate Centres (RCCs)—https://public.wmo.int/en/our-mandate/climate/regional-climate-centres

City authorities or countries contemplating IUS can turn to WMO Members already experienced in IUS, the development of Multi-Hazard Early Warning Systems (MHEWS) projects, and/or cities engaged in IUS demonstration projects (WMO 2021). A vast range of knowledge related to urban processes and their interactions of relevance to IUS developers also resides in other international and national agencies, universities, and public and private research institutes.

(c) Financial resources

Long-term commitment to urban safety and management of cities given their dynamic nature requires appropriate and sustained investment mechanisms to support all IUS components. In addition to national and local government funding, pay services for special users, especially businesses, may provide effective avenues for revenue generating opportunities to sustain IUS operations and improvements (UNDP 2016; WMO 2019a).

(d) Service resources

Key to the success of IUS will be its ability to deliver on its goals in an efficient and effective manner. Accordingly, rolling reviews of the efficacy of the underlying traditional and new services in the areas of meteorology, hydrology and air quality will be required.

(e) Cost-benefit/social impact

The potential for IUS to produce substantial cost savings via assisting with the development and operation of more efficient, sustainable and resilient cities is immense. However, IUS value and its ability to make a difference will need to be proven by the conduct of systematic cost-benefit and social impact analyses similar to those that have been conducted for a range of NHMS (e.g. Barrett et al. 2021; Freebairn and Zillman 2002; Gray 2015). The demonstration of clear value will assist with raising the political appetite for IUS development.

14.5.2 Science/Knowledge Gaps

The nature of IUS will be unique to each individual city and reflect a city's distinct IUS risk profile as an outcome of the interaction between exposure, hazard type and vulnerability. With the implementation of IUS in a number of locations, it has become apparent that a range of scientific, technical and knowledge gaps exist; examples include (see also WMO 2019a; Grimmond et al. 2020; Baklanov et al. 2020):

- Making and utilizing representative high resolution observations.
- Representation of urban form (three-dimensional structure, fabric, land cover) and metabolism in models.
- Data assimilation schemes suited to three-dimensional urban observations including representativeness assessment and uncertainty analysis.

- Identification of modelling systems (coupled) and approaches (e.g. ensemble) that capture urban environmental processes are fit for probabilistic hazard prediction and facilitate the understanding of the origins of prediction uncertainty.
- Balancing service/user relevant scales for prediction and generation of tailored products with what is possible via downscaling techniques.
- Understanding and projecting the impact of a changing climate on cities for a range of possible IUS futures as portrayed by Representative Concentration Pathways (RCPs) and Shared Socio-Economic Pathways (SSP).
- Physical and socio-economic impact of cities on weather, climate, water and the environment now and in the future.
- Physical and socio-economic consequences of non-hydrometeorological hazards (e.g. earthquakes/volcanic eruptions/space weather) and their interactions with IUS processes and drivers.
- Understanding user and wider public response to IUS Decision Support Systems.
- Consensus on IUS-related common language.
- Finding effective ways to narrate, communicate and manage risk.
- Establishing IUS value for money.
- Identifying critical threshold (meteorological, hydrological) values associated with step changes in human and infrastructure consequences arising from extreme hydrometeorological events.
- Development of user relevant and friendly service delivery platforms that utilize advances in communication technologies (WMO 2019a).

Although not exhaustive, this list indicates areas of research needed in the coming years—in many ways a next steps agenda—and serves as a pointer to the matters that will require serious consideration for those contemplating IUS development (WMO 2019a).

14.6 Summary

Cities are densely populated and experience domino effects from extreme weather, flooding and air pollution hazards which can be extremely disruptive to societal and economic activities (e.g. industry, transportation) that adversely affect the environment and quality of life. The United Nations adoption of a "New Urban Agenda" and with that the WMO's concept of IUS has been developed and articulated to meet the multi-hazard early warning weather, climate, air quality and hydrology warning needs for cities including long-term planning and design. Integration of services requires multidisciplinary science and multi-sector (levels of government and application sectors from, diverse decision-making) engagement and a basis for science-based consistent and efficient decision-making.

As weather, climate, water and environment pay no respect to political boundaries, the WMO is responsible for coordinating and guiding its Members and the scientific community. It has developed Guidance for IUS and identified National

Hydrological and Meteorological Services (NHMS) as being in the best position to lead IUS development and delivery. However, an imperative for IUS development is deep engagement of key political players in cities and practitioners including IUS research providers, civil protection agencies, transportation managers, urban planners, legislators and regulators.

Cities have specific requirements arising from their diverse geography, climate, weather, economies, governance, amongst many other factors, making it necessary for services and products to be tailored for, and effectively communicated to, decision-makers and users. For example, IUS could help cities build resilience to climate-related hazards. It is openly acknowledged that considerable gaps and challenges, including mutual understanding of capabilities, capacities, roles and responsibilities, exist. These need attention, if IUS are to deliver on their aspirations.

Beyond describing the concept and components of IUS, some of the imperatives for taking IUS forward have been outlined in this chapter. We conclude with some key starting points to creating IUS (WMO 2019a):

- Waiting for a disaster to strike is not the reason to take action, a proactive approach to IUS development based on a vision for achieving resilient cities is best.
- For cities contemplating development of IUS, well-functioning IUS in other cities with different risk profiles (WMO 2021) already exist to provide some initial guidance.
- At the outset—"the conceptual design phase"—engagement with potential end users and relevant stakeholders will assist with socializing the idea of IUS, raise awareness and garner important feedback to inform development.
- NHMS are well positioned to play a central role in the lead, promotion, development and coordination of IUS in addition to facilitating wider accessibility to data via influencing ownership issues and providing technical support (WMO 2019a).
- Sustained multidisciplinary cross-cutting research programmes are fundamental for developing IUS as are well-honed protocols for evaluation.
- Governance structures and legal frameworks that clearly articulate partner roles and responsibilities will benefit IUS development, implementation and operation.

Acknowledgements Many people contributed to the development of the Guidelines for Integrated Urban Hydro-meteorological Climate and Environment Services, Volume I and include: Zaidi B. Zainal Abidin, Elena Akentyeva, Jorge H. Amorim, Paolo Capizzi, Jason Ching, Felix Creutzig, Bénédicte Dousset, Patricia Ee, Karolin Eichler, Lais Fajersztajn, Tobias Fuchs, Christopher Gan, Teodoro Georgiadis, Lars Gidhagen, Ricardo Prieto González, Pablo Hernandez, Julie Hildago, Jasmin Hoefgaertner, Cyrille Honore, Kangning Huang, Karro Ilmar, Leena Järvi, Erland Kallen, Ranier Kapp, Ari Karppinen, Eigil Kass, Said El Khatri, Paul Kirshen, John Labadie, Tsz-Cheung Lee, Sylvie Leroyer, Tsz-Cheung Lee, Kenneth Kai Ming Leung, Deal Lockett, Ulrich Loehnert, Marina Makarova, Tanya Müller, Jae-Cheol Nam, Hannah Nissan, Kenji Oshio, Ttuukka Petaja, Brenda Phillips, Kobus Pienaar, Raizan Rahmat, Rayk Rinke, Matthias Roth, Pablo Saide, Rachid Sebbari, Evgenia Semutnikova, Karen Seto, Ayub Shaka, Knud-Jacob Simonsen, A. [Sena] Sopaheluwakan, Gert-Jan Steeneveld, Oksana Tarasova, Mariana Matera Veras, Roman Vilfand, Felix Vogel, Damian Wilson, Chin Ling Wong and Chui Wah Yap. The Guidelines could not have been produced without this assistance from the community.

References

100ResilientCites (2019) 100 Resilient Cities. http://www.100resilientcities.org/resources. Accessed 8 Feb 2020

Amorim JH, Asker C, Belusic D, Carvalho AC, Engardt M, Gidhagen L, Hundecha Y, Körnich H, Lind P, Olsson E, Olsson J, Segersson D, Strömbäck L, Joe P, Baklanov A (2018) Integrated Urban Services for European cities: the Stockholm case. WMO Bull 67(2):33–40

Baklanov A, Grimmond CSB, Carlson D, Terblanche D, Tang X, Bouchet V, Lee B, Langendijk G, Kolli RK, Hovsepyan A (2018) From urban meteorology, climate and environment research to integrated city services. Urban Clim 23:330–341. https://doi.org/10.1016/j.uclim.2017.05.004

Baklanov A, Cárdenas B, Lee T, Leroyer S, Masson V, Molina LT, Müller T, Ren C, Vogel FR, Voogt J (2020) Integrated urban services: experience from four cities on different continents. Urban Clim 32:100610. https://doi.org/10.1016/j.uclim.2020.100610

Barrett S, Ndegwa W, Maggio G (2021) The value of local climate and weather information: an economic valuation of the decentralised meteorological provision in Kenya. Clim Dev 13(2):173–188. https://doi.org/10.1080/17565529.2020.1745739

Benmarhnia T, Bailey Z, Kaiser D, Auger N, King N, Kaufman JS (2016) A difference-in-differences approach to assess the effect of a heat action plan on heat-related mortality, and differences in effectiveness according to sex, age, and socioeconomic status (Montreal, Quebec). Environ Health Perspect 124(11):1694–1699. https://doi.org/10.1289/ehp203

Brunet G, Jones S, Ruti PM (eds) (2015) Seamless prediction of the earth system: from minutes to months WMO No. 1156. https://library.wmo.int/doc_num.php?explnum_id=3546. Accessed 9 Feb 2020

Bulkeley H (2021) Climate changed urban futures: environmental politics in the Anthropocene city. Environmental Politics 30(1–2):266–284. https://doi.org/10.1080/09644016.2021.1880713

C40 (2020) C40 Cites Climate Leadership Group. http://www.c40.org. Accessed 8 Feb 2020

Ching J, Mills G, Bechtel B, See L, Feddema J, Wang X, Ren C, Brousse O, Martilli A, Neophytou M, Mouzourides P, Stewart I, Hanna A, Ng E, Foley M, Alexander P, Aliaga D, Niyogi D, Shreevastava A, Bhalachandran P, Masson V, Hidalgo J, Fung J, Andrade M, Baklanov A, Dai W, Milcinski G, Demuzere M, Brunsell N, Pesaresi M, Miao S, Mu Q, Chen F, Theeuwes N (2018) World Urban Database and Access Portal (WUDAPT): an urban weather, climate, and environmental modeling infrastructure for the Anthropocene. Bull Am Meteorol Soc 99(9):1907–1924. https://doi.org/10.1175/bams-d-16-0236.1

Creutzig F, Lohrey S, Bai X, Baklanov A, Dawson R, Dhakal S et al (2019) Upscaling urban data science for global climate solutions. Global Sustainability 2:E2. https://doi.org/10.1017/sus.2018.16

Freebairn J, Zillman J (2002) Economic benefits of meteorological services. Meteorol Appl 9:33–44. https://doi.org/10.1017/S1350482702001044

Gray M (2015) Public weather service value for money review https://www.metoffice.gov.uk/binaries/content/assets/metofficegovuk/pdf/about-us/what-we-do/public-weather-service/pws_value_for_money_review_-_march_2015.pdf. Accessed 18 July 2021

Grimmond CSB, Souch C (1994) Surface description for urban climate studies: a GIS based methodology. Geocarto Int 9(1):47–59. https://doi.org/10.1080/10106049409354439

Grimmond S, Ward HC (2021) Urban measurements and their interpretation. In: Foken T (ed) Handbook of Atmospheric Measurements. Springer Nature, Switzerland, pp 1393–1425. https://doi.org/10.1007/978-3-030-52171-4_52

Grimmond CSB, Roth M, Oke TR, Au YC, Best M, Betts R, Carmichael G, Cleugh H, Dabberdt W, Emmanuel R, Freitas E, Fortuniak K, Hanna S, Klein P, Kalkstein LS, Liu CH, Nickson A, Pearlmutter D, Sailor D, Voogt J (2010) Climate and more sustainable cities: climate information for improved planning and management of cities (producers/capabilities perspective). Procedia Environ Sci 1:247–274. https://doi.org/10.1016/j.proenv.2010.09.016

Grimmond C, Carmichael G, Lean H, Baklanov A, Leroyer S, Masson V, Schluenzen K, Golding B (2015) Urban-scale environmental prediction systems. In: Brunet G, Jones S. Ruti P (eds)

(Chap 18) Seamless prediction of the earth-system: from minutes to months, WMO-No 1156, pp 347–370

Grimmond S, Bouchet V, Molina LT, Baklanov A, Tan J, Schluenzen KH, Mills G, Golding B, Masson V, Ren C, Voogt J, Miao S, Lean H, Heusinkveld B, Hovespyan A, Teruggi G, Parrish P, Joe P (2020) Integrated urban hydrometeorological, climate and environmental services: concept, methodology and key messages. Urban Clim 33:100623. https://doi.org/10.1016/j.uclim.2020.100623

Guanter L, Kaufmann H, Segl K, Foerster S, Rogass C, Chabrillat S, Kuester T, Hollstein A, Rossner G, Chlebek C, Straif C, Fischer S, Schrader S, Storch T, Heiden U, Mueller A, Bachmann M, Mühle H, Müller R, Habermeyer M, Ohndorf A, Hill J, Buddenbaum H, Hostert P, Van der Linden S, Leitão PJ, Rabe A, Doerffer R, Krasemann H, Xi H, Mauser W, Hank T, Locherer M, Rast M, Staenz K, Sang B (2015) The EnMAP Spaceborne imaging spectroscopy mission for earth observation. Remote Sens 7(7):8830–8857. https://doi.org/10.3390/rs70708830

Huang B, Ni G-H, Grimmond S (2019) Impacts of upwind urban expansion on regional climate during heat waves. Atmos 10(7):364. https://doi.org/10.3390/atmos10070364

ICLEI (2020) ICLEI Corporate Report 2018–2019. https://worldcongress2018.iclei.org/wp-content/uploads/Corporate%20Report%202018-2019.pdf. Accessed 8 Feb 2020

Joe P, Belair S, Bernier NB, Bouchet V, Brook JR, Brunet D, Burrows W, Charland JP, Dehghan A, Driedger N, Duhaime C, Evans G, Filion AB, Frenette R, de Grandpré J, Gultepe I, Henderson D, Herdt A, Hilker N, Huang L, Hung E, Isaac G, Jeong CH, Johnston D, Klaassen J, Leroyer S, Lin H, MacDonald M, MacPhee J, Mariani Z, Munoz T, Reid J, Robichaud A, Rochon Y, Shairsingh K, Sills D, Spacek L, Stroud C, Su Y, Taylor N, Vanos J, Voogt J, Wang JM, Wiechers T, Wren S, Yang H, Yip T (2018) The environment Canada Pan and Parapan American Science Showcase Project. Bull Am Meteorol Soc 99(5):921–953. https://doi.org/10.1175/bams-d-16-0162.1

Masson V, Lemonsu A, Hidalgo J, Voogt J (2020) Urban climates and climate change. Annu Rev Environ Resour 45:411–444. https://doi.org/10.1146/annurev-environ-012320-083623

Mills G, Cleugh H, Emmanuel R, Endlicher W, Erell E, McGranahan G, Ng E, Nickson A, Rosenthal J, Steemer K (2010) Climate information for improved planning and management of mega cities (needs perspective). Procedia Environ Sci 1:228–246. https://doi.org/10.1016/j.proenv.2010.09.015

Muller CL, Chapman L, Grimmond CSB, Young DT, Cai X-M (2013) Toward a Standardized Metadata Protocol for Urban Meteorological Networks. Bull Am Meteorol Soc 94(8):1161–1185. https://doi.org/10.1175/BAMS-D-12-00096.1

National Research Council (2012) Urban meteorology: forecasting, monitoring, and meeting users' needs. The National Academies Press, Washington. https://doi.org/10.17226/13328. Accessed 10 Feb 2020

Pelling M. (2003) The vulnerability of cities: natural disasters and social resilience. London; Sterling, VA, Earthscan Publications, 212p

Reuter U, Kapp R (2021) Urban climate in urban planning—the experience from Stuttgart. In: Ren C, McGregor G (eds) Urban climate science for planning healthy cities. Springer, London

Rogers DP, Tsirkunov VV (2013) National meteorological and hydrological services. In: Rogers DP, Tsirkunov VV (eds) Weather and climate resilience: effective preparedness through national meteorological and hydrological services. World Bank. https://doi.org/10.1596/9781464800269_Ch03. Accessed 10 Feb 2020

Takane Y, Kikegawa Y, Hara M, Grimmond S (2019) Urban warming and air-conditioning use in a future climate: evidence and importance of a positive feedback process. NPJ Climate and Atmospheric Science 2:39. https://doi.org/10.1038/s41612-019-0096-2

Takane Y, Ohashi Y, Grimmond S, Hara M, Kikegawa Y (2020) Asian megacity heat stress under future climate scenarios: impact of air-conditioning feedback. Environmental Research Communications 2:015004. https://doi.org/10.1088/2515-7620/ab6933

Tapper N (2021) Creating cooler, healthier and more liveable australian cities using irrigated green infrastructure. In: Ren C, McGregor G (eds) Urban climate science for planning healthy cities. Springer, London

Travis WR, Bates BC (2014) What is climate risk management? Clim Risk Manag 1:1–4
UN (2016) New Urban Agenda—HABITAT III. http://habitat3.org/the-new-urban-agenda. Accessed 2 March 2020
UN (2019) UN sustainable development goals. https://www.un.org/sustainabledevelopment/sustainable-development-goals. Accessed 8 Feb 2020
UNDP (2016) Revenue-generating opportunities through tailored weather information products. United Nations Development Programme Report. https://www.adaptation-undp.org/sites/default/files/resources/revenue-generating-opportunities-for-tailored-weather-information-productions-undp-june-2-2016_0.pdf. Accessed 18 July 2021
WMO (2004) Guidelines on bio-meteorological and air quality forecasts, WMO/TD-No. 1184; PWS-No. 10. https://library.wmo.int/doc_num.php?explnum_id=9250. Accessed 9 Feb 2020
WMO (2006) Initial guidance to obtain representative meteorological observations at urban sites, WMO/TD-No. 1250; IOM Report-No.81. https://library.wmo.int/doc_num.php?explnum_id=9286. Accessed 9 Feb 2020
WMO (2008) The guide to hydrological practices (WMO-168). https://library.wmo.int/?lvl=notice_display&id=21815#.YP-b_o4zY2w (Vol. I) and https://library.wmo.int/index.php?lvl=notice_display&id=543#.YP-cJI4zY2w (Vol. II). Accessed 27 July 2021
WMO (2010a) Manual on stream gauging WMO-1044. https://library.wmo.int/index.php?lvl=notice_display&id=540#.YP-cZ7ris2x (Vol. I) and https://library.wmo.int/index.php?lvl=notice_display&id=538 (Vol. II). Accessed 27 July 2021
WMO (2010b) WMO inter-regional CLIPS training workshop on urban climatology. http://www.kerenvis.nic.in/Viewevents.aspx?Id=914. Accessed 26 Jul 2021
WMO (2014) The WMO strategy for service delivery and its implementation plan. https://library.wmo.int/doc_num.php?explnum_id=7854. Accessed 26 Jul 2021
WMO (2015a) Guidelines on multi-hazard impact-based forecast and warning services, WMO No. 1150. https://library.wmo.int/doc_num.php?explnum_id=7901. Accessed 26 Jul 2021
WMO (2015b) Role and operation of National Meteorological and Hydrological Services: a statement by the World Meteorological Organization for Directors of NMHSs. https://library.wmo.int/doc_num.php?explnum_id=6403. Accessed 26 Jul 2021
WMO (2017a) The hydrometeorological enterprise: the benefits of partnerships. https://public.wmo.int/en/resources/bulletin/hydrometeorological-enterprise-benefits-of-partnerships-0. Accessed 26 Jul 2021
WMO (2017b) Urban meteorology course 28 Aug–8 Sep 2017 Malaysia. https://www.urban--climate.org/2017/07/urban-meteorology-course-28-aug-8-sep-2017-malaysia. Accessed 26 Jul 2021
WMO (2018) Guide to instruments and methods of observation. WMO No. 8. https://library.wmo.int/index.php?id=12407&lvl=notice_display#.YP7exugzYuU. Accessed 26 Jul 2021
WMO (2019a) Guidance for urban integrated hydrometeorological, climate and environmental services. Volume I: Concept and methodology. WMO-No: 1234, https://library.wmo.int/doc_num.php?explnum_id=9903. Accessed 26 Jul 2021
WMO (2019b) Manual on the global data-processing and forecasting system, WMO No. 485. https://library.wmo.int/doc_num.php?explnum_id=10164. Accessed 26 Jul 2021
WMO (2021) Guidance for Urban Integrated Hydrometeorological, Climate and Environmental Services. Volume II: Demonstration cities. WMO-No: 1234, https://library.wmo.int/doc_num.php?explnum_id=105547. Accessed 22 Jul 2021
WMO (n.d.) Data access and exchange. https://community.wmo.int/data-access-and-exchange. Accessed 26 Jul 2021
WMO-GFCS (2019) Global framework for climate services. https://gfcs.wmo.int/. Accessed 9 Feb 2020
World Bank (2010) Cities and climate change: an urgent agenda. Urban development series; knowledge papers no. 10. Washington, DC. © World Bank. https://openknowledge.worldbank.org/handle/10986/17381. License: CC BY 3.0 IGO

Chapter 15
Characterising Urban Morphology for Urban Climate Modelling

Gerald Mills, Jason Ching, and Benjamin Bechtel

Abstract Cities represent a challenge to conventional climate modelling. Globally and even regionally, they are relatively small in extent and characterised by an extremely complex and heterogeneous landscape. At the same time, cities are of particular significance due to the exposure of the population to a variety of hazards. Over the past two decades improvements in the capacity of climate models to process large quantities of data at finer spatial resolutions have increased the demand for information on urban landscapes. At the same time, developments in urban climate science, based on observational programmes coupled with theoretical understanding of near-surface processes, now allow the integration of cities into regional scale models. Whereas, previously such models treated the urban surface as simply warm, dry and rough, newer models incorporate the variations in building dimensions and layout that describe numerically the urbanised landscape. These descriptions are termed urban canopy parameters (UCPs) that relate to aspects of urban form and functions, including impervious surface cover, the nature of construction materials, the morphology of the landscape, and the occupation patterns (transport, housing, etc.). This chapter focuses on morphology, defined here as the 3D dimensions and placement of urban elements (that is, mainly buildings and trees). Morphology alters radiation exchanges at the surface as a result of shadowing, multiple reflections and horizon screening. Moreover, it alters near surface airflow, creating a deep layer of turbulence that affects the exchange of heat, materials and momentum between the surface and the overlying atmosphere. Validated urban climate models can contribute to urban environmental management that includes supporting planning and design, hazard prediction and disaster response. However, such models need morphological data to perform simulations; unfortunately for many cities in the world, these data do not exist or are not available. This chapter identifies sources of these data and proposes methods to generate such data for any city.

G. Mills (✉)
School of Geography, University College Dublin, Dublin, Ireland
e-mail: gerald.mills@ucd.ie

J. Ching
Institute for the Environment, The University of North Carolina at Chapel Hill, Chapel Hill, NC, USA

B. Bechtel
Department of Geography, Ruhr-University Bochum, Bochum, Germany
e-mail: benjamin.bechtel@rub.de

C. Ren, G. McGregor (eds.), *Urban Climate Science for Planning Healthy Cities*, Biometeorology 5, https://doi.org/10.1007/978-3-030-87598-5_15

Keywords Urban morphology · Climate models · Urban canopy parameters · Data sources

Key Messages

- Urban climate models have become progressively more sophisticated as the meteorological processes over city surfaces are better understood.
- These models require information on the urban surface (known as urban canopy parameters or UCPs) to simulate climate.
- Information on the morphology of urban surfaces is needed to support model applications to individual cities.
- These data are often unavailable and need to be generated.
- Validated urban climate models can contribute to urban environmental management that includes supporting planning and design, hazard prediction and disaster response.

15.1 Introduction

Most people on the planet live in cities and experience an environment that is profoundly modified by the landscape changes that accompany urbanisation. The UN estimates that there are more than 1000 cities on the planet with populations of over 500,000 and half of these has populations over one million (UN 2016). Taken together, the urbanised part of the planet is very small; Liu et al. (2014) estimates the global built-up area in 2010 was about 0.65% of the world's ice-free land area. However, these landscapes are major drivers of local, regional and global climate changes. These climate effects are linked to aspects of the urban form and functions. The former describes the physical composition and layout of the city that affect the natural exchanges of energy, mass and momentum at the Earth's surface. The latter refers to the occupation of the city and the throughput of materials, energy, food and water that sustains it. This resource exchange results in wastes that are emitted into the atmosphere, water and soil degrading the quality of ecosystems inside and outside the city. From a global health perspective, urban areas are critical owing to the density of population and the intensity of the landscape changes that result in poor air quality, elevated air and surface temperatures, as well as loss of sunlight and of natural ventilation. Urban climate models (UCMs) can predict many of these consequences but need information on urban form and functions to perform simulations and meteorological data to evaluate the results. Ideally, UCMs would be a key component in an integrated weather service that would link scientific understanding of undesirable impacts to the actions needed to mitigate the risks to human health and wellbeing (Baklanov et al. 2018).

Here, urban form is defined by three properties: surface cover, construction fabric and morphology. Cover defines the extent of the urbanised landscape, which is often captured as the contiguous extent of paved and impermeable surface including the gardens, parks and water bodies that are interspersed. Fabric refers to the natural and artificial materials that are used to construct the city—these have distinct radiative, thermal and permeability properties. Morphology here is used to describe the three-dimensional character of cities, that is, the height, width and depth of buildings (and trees) and their layout. Together, these form properties are spatially correlated and vary systematically across the urbanised landscape. As an aside, it should be acknowledged that other fields use the term "morphology" to refer to all three properties of form; here we retain the distinction as each property has different climate effects.

This chapter introduces the urban climate effect (Sect. 15.2) in order to discuss UCMs and their data needs (Sect. 15.3). It then summarises the available sources of this information (Sect. 15.4) and how properly evaluated UCMs assess climate-related health risks in cities (Sect. 15.5).

15.2 The Urban Climate Effect

The effects of cities on the overlying atmosphere are expressed in the formation of an urban boundary layer (UBL), which has distinctive aerodynamic, thermal and chemical properties that are linked to the underlying urban surface (Fig. 15.1). The depth of the UBL grows from the upwind urban edge where it encounters a surface that is relatively rough and warm. The turbulence generated carries the influence of the city vertically where it mixes with air influenced by upwind "natural" surfaces (e.g. pasture, forests, water). The UBL grows in depth as ambient air progresses across the urbanised landscape; if the city is sufficiently extensive, rough and warm, the UBL will grow to 1–2 km depth by late afternoon on a clear day. At the leeward urban edge, a new boundary layer starts to form over the non-urban surface as the UBL is advected downwind as an elevated plume. Over the past three decades considerable progress has been made in understanding and characterising processes within the UBL through a combination of modelling (both physical and numerical) and real-world observations. This research work has shown that the heterogeneous urban surface can be decomposed into local-scale (>1 km^2) differences; this local scale corresponds with distinct types of urban "neighbourhood" that are associated with different urban functions, such as industrial, commercial and residential. Each of these neighbourhoods makes a distinctive contribution to the evolving UBL.

Atmospheric boundary layers are comprised of a Surface Layer (SL), which occupies the lowest 10% and an overlying Mixed Layer (ML), which is influenced by exchanges with the SL and contributions from upwind surfaces (Fig. 15.1). The SL consists of a Roughness Sub-Layer (RSL) where turbulence is generated by roughness elements (e.g. trees and crops) and microscale processes dominate. Above this layer lies the Inertial Sub-Layer (ISL) where the microscale contributions are blended to create a signal that represents the local-scale landscape. Boundary layer theory, which was developed over extensive homogenous surface,

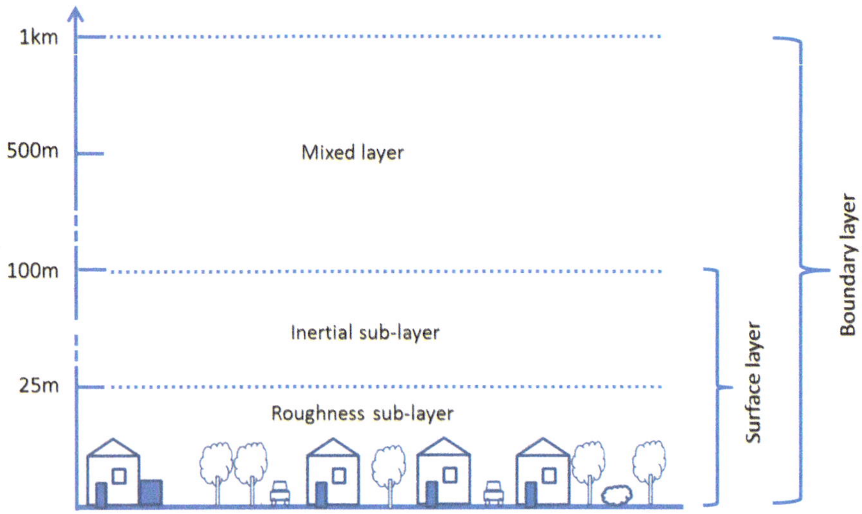

Fig. 15.1 The vertical structure of the urban atmosphere above a low-density, well-vegetated urban landscape. The surface layer (SL) occupies the lower 10% of the boundary layer, which can extend from 1 to 2 km depth during the daytime. The SL over urban landscapes consists of a deep roughness sub-layer that extends from two to four times the heights of the roughness elements (that is, the buildings and trees). The inertial sub-layer is where the Monin-Obukov Similarity Theory (MOST) can be applied

links the character of the surface to the overlying atmosphere via a set of relationships that predicts the exchanges between the SL and ML based on parameters that describe the roughness elements and vertical gradients of associated properties, such as horizontal velocity and air temperature (this is the Monin-Obukhov Similarity Theory or MOST). Similarly, mesoscale models can predict turbulent characteristics in the SL based on simulations at the base of the ML.

One of the challenges in urban climate studies has been to show that MOST can be applied to the urban boundary layer (UBL) despite the spatially heterogeneous morphology present. The evidence indicates that MOST can be applied with the caveat that the RSL is especially deep (at least two to four times the height (H) of buildings) over cities (Fig. 15.1). This means that, for example, the mean airflow at the top of the RSL can be related to flow at a reference level and two surface parameters, the displacement height (d) plus the roughness length (z_0); z_0 is a measure of the efficiency of roughness elements in extracting momentum from the overlying air. For a short grass surface, d $\approx$ 0 m and $z_0 \approx 0.01$ m but in cities, typical values of z_0 range from 0.5 m to >1 m and d $\approx$ 0.66 H. These parameters are used to define the effective aerodynamic "surface", where the mean airflow is estimated to be zero; below (z_0+d), that is between the roughness elements, there is no mean airflow. In cities, the separation between the ground and the aerodynamic surface is relatively deep as a consequence. These relations are needed to apply concepts used in macro- and mesoscale models to urban landscapes.

For the study of the climate-related health consequences of urbanisation, the focus shifts to the microscale environments to which humans are exposed. This corresponds to the urban canopy layer (UCL), which is immersed in the RSL but is located below the mean building height (Fig. 15.1). The UCL is the zone of human occupation and the source of the majority anthropogenic emissions from vehicles and buildings. The outdoor climate at this level is controlled by microscale processes that are regulated by the immediate physical environment (the dimensions and placement of buildings, the building and street fabrics, the presence of traffic, etc.) and exchanges within the UCL and with the overlying air. The outdoor environment is characterised by extreme variation over short distances associated with areas of shelter/exposure and shade/sunlight that is largely controlled by the morphology. At this scale, climate modelling and observations have relied on simplified archetypes of common urban elements to understand the relationship between processes and climate responses; these archetypes include arrays of cube-shaped buildings in different alignments and of parallel rows of buildings (urban canyons) to represent streets. For specific urban design evaluations, detailed descriptions of buildings and the outdoor spaces are needed for modelling. A comprehensive survey of urban climates and their study can be found in Oke et al. (2017).

15.3 Urban Climate Models and Their Evolution

Large-scale atmospheric modelling started in the post-war period and progress has been closely coupled to both theoretical insights and advances in computer technology. While the former have come about through observation programmes and the application of atmospheric physics, progress has always been **constrained** by the capacity of computers. All models must make compromises on the range of processes that can be included in conjunction with the appropriate spatial and temporal resolution of simulations. These decisions are necessary if the model is designed to provide forecasts, in which case the simulation runs must be completed within a limited time frame. A numerical model of the atmosphere has a core set of governing equations (expressing the conservation of momentum, mass and energy), which are applied to a set of grid cells that comprise a spatial domain. At the edges of this domain, boundary conditions are stipulated to handle the inputs and outputs at these limits; the lower boundary corresponds to the Earth's surface and captures the heterogeneous topography at appropriate scales. The time step employed in simulations is constrained by grid cell size such that exchanges must be transferred through neighbouring cells; hence, smaller grid sizes use more precise time increments. For many models, the cell size varies within the domain and is smallest where exchanges are most intense, close to surfaces.

Climate models can broadly be classified into global (macro), regional (meso) and local/micro scales to deal with domains of varying sizes. Although the basic physics of all models are the same, their application depends on the available computational resources and the most important processes relevant to that scale and application.

Urban climate models that can simulate the mean and turbulent aspects of the weather and predict wind, temperature, humidity, air quality, comfort and heat stress, etc. require relevant descriptions of the form and functions of the urbanised landscape. These descriptions take the form of data on impervious surface cover, material composition and morphology and on human occupation patterns that regulate emissions of waste heat, gases and materials. The modelling challenge has been to identify the relevant parameters and specify how these affect surface-air exchange processes. Garuma (2018) provides a review on urban surface parameterisations for Urban Canopy Models (UCMs); examples of urban canopy parameters (UCPs) include gridded values on roughness (z_0) and displacement heights (d), building spacing, frontal (λ_f) and plan (λ_b) area, and the Sky View Factor (see Fig. 15.2). Of course, each modelling scale also corresponds with different levels of detail on the underlying landscape and to connect the processes at each modelling scale, it is common to "nest" models of a smaller domain within a larger one (e.g. an urban mesoscale model within a regional model). In these circumstances, larger-scale models provide the atmospheric boundary conditions for the smaller domain, while

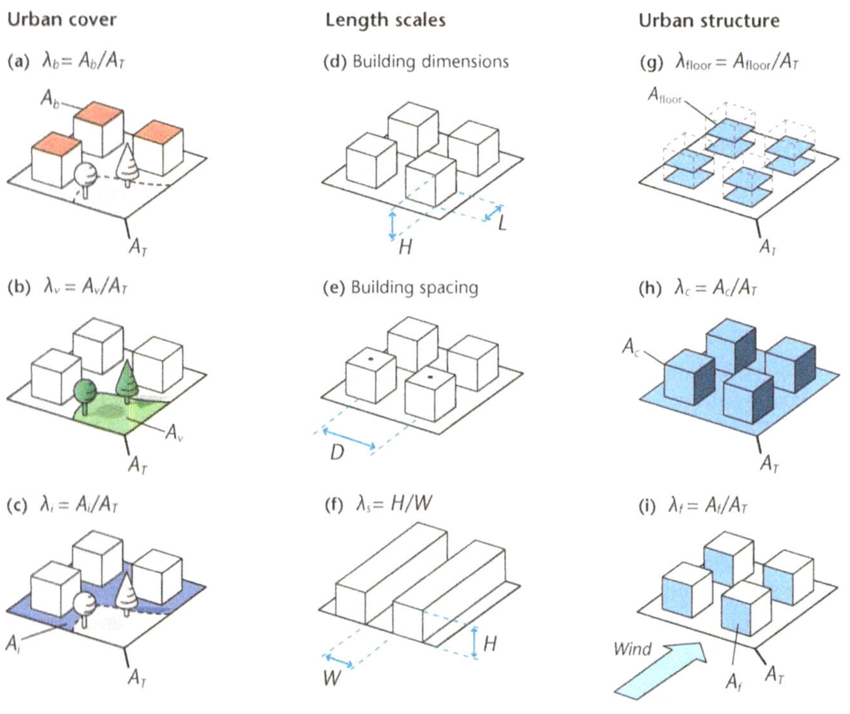

Fig. 15.2 Urban Canopy Parameters (UCPs) describing urban cover, length scales and urban structure (Oke et al. 2017). Cambridge University Press, with permission

the smaller-scale effects captured at this scale become inputs for the larger-scale model.

For urban climate studies, mesoscale models employ MOST to represent the roughness, moisture and thermal effects of the underlying surfaces, including urban areas. This approach is satisfactory for most cities when the spatial resolution of grid cells is ≥ 1 km^2 and the underlying surface can be described by UCPs (Fig. 15.2). For example, the Weather, Research and Forecast (WRF) modelling system permits options to include UCPs that describe the structure of the urban canopy layer and attendant processes (e.g. Kusaka et al. 2001; Martilli et al. 2002; Chen et al. 2011). Moreover, building energy model (BEM) schemes that address the indoor-outdoor links associated with anthropogenic heating/cooling in buildings can also be included (e.g. Salamanca et al. 2009). These advances allow researchers to examine complex interactions between urban climate, building form and human decision-making. Urban microscale models, by comparison, have focussed on the 3D structure of the canopy layer and on exchanges within the UCL. In general, these models have paid particular attention to the radiation exchanges between buildings, the energy budget of urban surfaces and airflow circulation patterns that can develop in city streets (Arnfield 2003). There are a number of models that have been developed that simulate the UCL climate in response to generic parameters (such as street height to width ratio (λ_S), Fig. 15.2) and ambient climate conditions.

Other models focus on specific urban settings and are used to evaluate the impacts of design decisions (e.g. building placement and outdoor landscaping) on the outdoor weather and climate. Some have been designed to simulate the exposure of individuals to environmental stresses and account for the biometeorological processes that govern human heat gain and loss (e.g. RAYMAN in Matzarakis et al. 2010). Some models employ Computation Fluid Dynamics (CFD), which applies the governing equations over limited spatial domains; these models can generate simulations that are extraordinarily precise in space and time but require extremely detailed building data and considerable computer resources (e.g. Hanna et al. 2006). Simpler CFD models are becoming available that can evaluate the climate consequences of different urban designs (see for example, ENVI-Met in Ng et al. 2012). Model nesting can aggregate the effects of local-scale urban processes into larger-scale weather and climate simulations while providing inputs for local-scale climate impacts on the indoor and outdoor environment (e.g. Santiago and Martilli 2010). Finally, new model architectures that harness the potential of parallel computational power of systems have begun to integrate urban processes directly into larger-scale models (see Maronga et al. 2020).

All urban climate models need input data that can describe the urbanised landscape in sufficient detail to meet the needs of the model (based on its atmospheric physics) and capture the appropriate the scale of variations in the underlying surfaces. Although information on both form and function is needed, morphological data is key to simulating mean airflow and turbulence properties over cities, which transfers other properties such as heat and pollutants horizontally and vertically. Figure 15.2 shows a range of common UCPs that are based on landcover fractions and the distribution of 3D building elements.

15.4 Urban Morphological Data

A "complete" database suited for all urban climate models would include information on all the physical elements that comprise the urbanised landscape (buildings, trees, parks, gardens, roads, etc.) and all of the functions that generate anthropogenic emissions. This information would account for spatial and temporal variations, such as the seasonal responses of vegetation and the diurnal patterns of energy use. Such primitive data could be mined at different spatiotemporal resolutions to derive the information suited to different models and applications. However, data at this level of detail is not available (nor is it likely to be) at urban scales. Practically, creating an urban database suited for climate modelling must avail of any information appropriate to its application. Here, we consider a number of criteria for evaluating sources of urban data, namely: accuracy, consistency, actuality, completeness, availability and accessibility, interoperability, cost (ideally free to use), documentation, and readiness to use. We restrict our attention here to morphology, mostly building dimensions and relative placement. Table 15.1 summarises the types of urban data that may be available and indicates its consistency, completeness, availability, accessibility and cost, level of documentation and readiness for models.

A useful framework for the organisation of morphological data is provided by the Open Geospatial Consortium (OGC), which classifies 3-D city models by a Level of Detail (LoD) that is encoded in its City Geography Markup Language (CityGML). Each LoD corresponds the completeness of information available. For a building object: LoD0 defines a building footprint; LoD1 includes its height; LoD2 provides additional semantic and geometric information but is still simplified; LoD3 provides details on the envelope including exact geometry and fabric distinctions (e.g. windows and doors) and; LoD4 includes interior structures such as walls, floors and furniture. The same principles can be applied to other urban objects, like trees and bridges. Moreover, CityGML permits generalisation from higher to lower LoD (Löwner et al. 2013). For most urban climate purposes, LoD1 and LoD2 are relevant for neighbourhood scale (1 km^2) applications while LoD3 and LoD4 are ideally suited to microscale applications.

For individual cities, official municipal sources and national mapping agencies often maintain databases on building dimensions, including footprints and heights (Fig. 15.3). In particular circumstances, these sources may also include information on construction materials, roof configuration and building use but this information may not be available at the same scale or be part of the same database. Similarly, data on urban trees may be incomplete (e.g. species, canopy area, dimensions may be absent) and fractured into multiple independent databases. Municipalities often use field surveys and manual delimitation from stereoscopic imagery and cadastral data to update their building inventory. More recently alternative methods to derive 3D building models from spaceborne high-resolution optical and Synthetic-Aperture Radar (SAR) images (Xu et al. 2015; Wang et al. 2017) have generated impressive results but since they are not applied routinely they are beyond the scope of this chapter.

Table 15.1 Example of sources of urban morphology data and associated characteristics

Source/ type	Accuracy	Global consistency	Global completeness	Availability, accessibility, and cost	Documentation	Readiness for models
Municipal 3D models	Very high	Low	Low	Depends on local data policy, often difficult	Medium-high	Medium
LIDAR point clouds	Very high	High	Low	Very high cost		Large processing required
IFSAR	High	Very high	High	Very high cost		Extensive processing required
OSM building	Medium	High	Low-medium	Free	Very good	Mostly footprints, 3D information sparse
WUDAPT L0	Low	Medium-high	Low, but can be achieved for any city	Free	Good	High, tools to connect with different models
WUDAPT L1/2	High	High	Very low	Free		In preparation

In many cities, the basic data on buildings is acquired using Light Detection and Ranging (LIDAR) systems (Fig. 15.4). These can generate LoD2–3 data using a laser to detect the distance between the airborne platform and underlying surface (Haala and Brenner 1999). Rapid laser pulses produces a dense sample of very precise height measurements, which is termed a point cloud. Building detection is conducted by creating three grids from these data: a digital surface model (DSM), a digital terrain model (DTM) and a normalised digital surface model (nDSM). The latter is the difference between the DSM and DTM and differentiates open and elevated areas. To distinguish trees from buildings, either shape or roughness parameters are applied or additional information from optical data (colour, texture) is included. These data are used then to create objects. Building reconstruction optimises the digital representation of each building and creates the geometry of the building footprint and the roof. Typically, a number of assumptions are made, such as walls are vertical and roofs are composed of planar surfaces. Model-driven techniques assume a main orientation of the building with parallel and perpendicular walls and fit the point cloud to a predefined library of known roof shapes. By comparison, a data-driven technique does not assume a specific shape and reconstructs the building by segmenting the point cloud into the planar surfaces that constitute the object (Li et al. 2011).

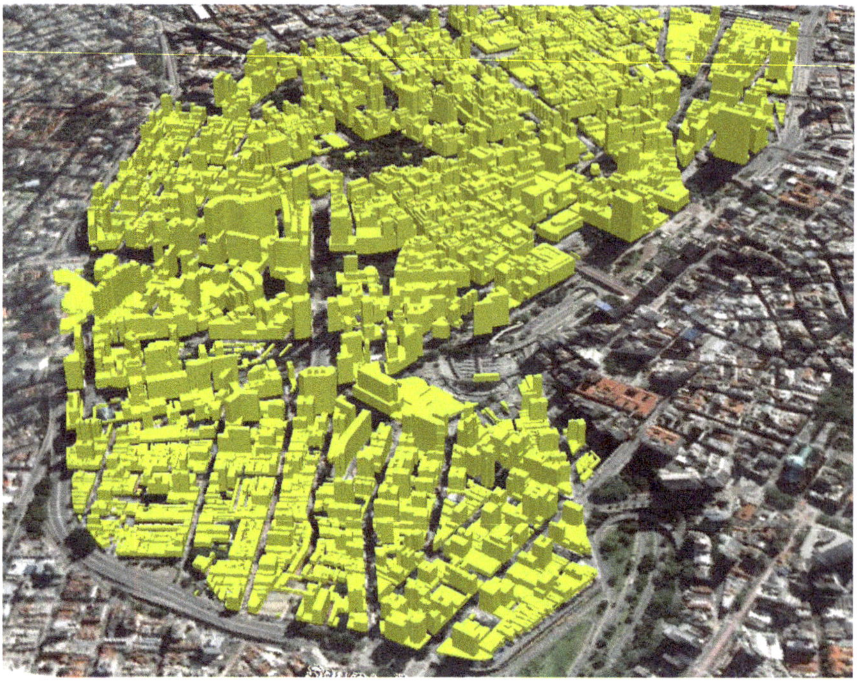

Fig. 15.3 Vector building file for the Republica neighbourhood in Sao Paulo available at https://geosampa.prefeitura.sp.gov.br/. The information consists of building footprints and heights. Here, this information is used to generate a 3D image that is superimposed on a satellite image

An alternative source of urban morphological data is Volunteered Geographic Information (VGI), which includes data collected by untrained citizens (See et al. 2016). With more than a million contributors OpenStreetMap (OSM) is the most relevant VGI source for urban data worldwide. In addition to information on streets and points of interest, OSM also includes a large number of building footprints (LoD0); recently, Microsoft added 124 million building footprints in the United States. Although the spatial coverage may be "patchy" and the sources variable, the accuracy in terms of position, completeness, semantics, and shape of these footprints has been shown to be excellent (see Fan et al. (2014a) for the case of Munich, Germany). Also, while the semantic information of the footprints is typically sparse and height information is often absent, there are approaches to estimate the functional building type (such as residential, industrial, or public) from the footprints (Fan et al. 2014b). Moreover, there are a number of ongoing projects (e.g. OSM-3D and OSM2World) that seek to visualise 3D buildings from OSM data. Goetz and Zipf (2012) have presented a methodology for an automatic VGI-based creation of 3D building models encoded in CityGML; these are mainly LoD1 and LoD2 models or LoD2 with simple roof geometries. Although these approaches are very promising, the sparse spatial coverage in poorer parts of the world is a big limitation to wider development.

Fig. 15.4 Shadow patterns for noon on 21 December generated by the Solweig model (Lindberg et al. 2008) using Lidar data for London at 1 m spatial resolution. The data are available from https://data.gov.uk/dataset/80c522cc-e0bf-4466-8409-57a04c456197/lidar-composite-dsm-1m

There are a few initiatives that are assembling global databases using consistent methodologies (e.g. satellite and population data) to derive urban data globally. For example, the Global Human Settlement project derives built-up density using Landsat satellite data at a nominal resolution of 38 m at the Equator (Pesaresi et al. 2016). Taking a different approach, the World Urban Database and Portal Tools (WUDAPT) project aims to derive, store and disseminate universally consistent, climate-relevant information for medium to large urban areas in the world (Ching et al. 2018); it is modelled on NUDAPT (National Urban Database and Access Portal Tool) which produced extremely detailed information for a limited number of North American cities (Ching et al. 2009). A primary focus of WUDAPT is the generation of UCPs for a range of macro- to mesoscale modelling studies using a data hierarchy that corresponds to levels of detail (and model needs). The lowest level uses the Local Climate Zone (LCZ) scheme, which links a lookup table scheme of (ranges of values of) UCPs keyed to each LCZ types (Stewart and Oke 2012). Maps of LCZ are derived using expert knowledge, available remote sensing data and machine learning techniques (Bechtel et al. 2015). The WUDAPT technique for generating LCZ data can be applied to any city quickly and the gridded UCP output has been shown to enhance mesoscale and energy balance model simulations (Ching et al. 2018). However, higher levels of data are needed to produce more precise UCP values that can capture the spatial heterogeneity of city surfaces (Fig. 15.3). WUDAPT is currently evaluating a Digital Synthetic City (DSC) tool to simulate the morphology of cities. The methodology applies a Procedural Model approach based on a set of rules that uses geographic information on road networks and population coupled with detailed satellite data. Initial comparison of the UCPs derived for simulated and actual urban morphology for selected cities have been positive and permit the derivation of UCPs at scales relevant to model purposes (Ching et al. 2019).

15.5 Urban Climate Models and Health

The need for UCMs to support decision-making across a range of urban based applications is clear, especially in the context of population and climate changes, and concomitant risks (e.g. Baklanov et al. 2018). These include, but are not limited to, guidance for urban planning, risk assessment, evaluating adaptation and mitigation strategies and mapping paths to improved urban resilience and sustainability. The models that can support these types of decisions are becoming increasingly sophisticated but the lack of relevant urban data remains an obstacle. To illustrate the value of this science we outline the use of these models for the management of heat stress and air quality.

The public health impact of heatwave events in cities is increased for a number of reasons, including both population exposure and urban climate effect. The latter reduces wind-speed near the ground and raises surface and air temperatures, also known as the urban heat island (UHI). The magnitude of the UHI is generally strongest in centrally located densely built parts of cities where there is little vegetation, energy use is concentrated and the urban morphology both impedes air movement and slows heat loss. UCMs can be employed to evaluate the urban contribution to heat stress if the relevant physics and urban canopy parameters are incorporated. For example, Ching et al. (2018) showed the potential for the urbanised WRF model to simulate heat stress in New Delhi, India. In this experiment the impact of urban growth was examined by deriving the urban canopy parameters for 1975 and 2015 and simulating a heatwave event from 2015 for both; the results showed that urbanisation over this period increased the magnitude and extent of the warming effect. Other UCMs can be employed to explore the links between urban design and climate impact. Lau et al. (2016) employed the SOLWEIG model (see Fig. 15.4), which requires detailed information on urban morphology to estimate spatial and temporal variations in mean radiant temperature (T_{MRT}) in Hong Kong; T_{MRT} is a key measure of the thermal environment experienced by humans and can be used to evaluate heat stress. SOLWEIG allowed the researchers to assess the potential for urban design to moderate heat stress in this sub-tropical, high density urban environment.

Poor air quality is a global environmental risk but is focussed in cities where emissions and population are concentrated. Predicting air quality in cities is especially complex owing to the distribution of local pollutant emissions from roadways, industry, dwellings, etc., which are added to pollutant that are transported regionally. Knowledge of the geography of emissions and of the character of the urban landscape is needed to model the composition of the atmosphere at a given point and time in cities. In particular, morphological properties are essential to simulate both the turbulent processes that link the UCL (where near-surface emissions occur) to the overlying air and the street-level advection that transports air pollutants within the UCL. There are well-established models that are community-based and free to use. The CMAQ modelling system (Byun and Schere 2006) simulates atmospheric chemistry and pollutant transport and concentration. As an example, Streets et al. (2007) used CMAQ to distinguish between local and regional sources of air pollution in Beijing prior to the Olympic Games; the results indicated that about 34% of PM2.5 on average and 35–60% of Ozone during high pollution episodes at the Olympic Stadium site could be attributed to sources outside Beijing. It follows that emission controls that focussed on the Beijing area alone would have a limited effect. At a microscale, there are several schemes that examine the role of street geometry in regulating air quality within the UCL; for example, the Street in Grid (SinG) scheme links a 3D Eulerian chemical-transport model (Mallet et al. 2007) with a street-network model (Soulhac et al. 2011).

15.6 Conclusions

Over the past two decades our understanding of the urban processes that modify the climate at a hierarchy of time and space scales has greatly improved. Much of this knowledge is embedded in urban climate models (UCMs) of varying sophistication that are designed for different purposes, which include applications that have direct and indirect implications for public health. Many of these models have been evaluated by comparison with independently derived observations in different urban settings and are available for use generally. However, to apply these UCMs requires urban data that describes how the city surface affects exchanges of heat, moisture, momentum, gases and materials. The relevant properties describe aspects of form (surface cover, materials and 3-D morphology) and function that are distilled into urban canopy parameters (UCPs) that are used in models to simulate these exchanges. Morphology is especially relevant as the dimensions of buildings (and trees) and their layout exert a strong control on airflow and turbulence close to the ground.

The need for urban-scale modelling will grow exponentially in the coming decades driven by continued urban population growth and the expected significant temporal vicissitudes of urban changes (even within the next one or two generations) and the need to mitigate and adapt to environmental changes at a hierarchy of scales. While our physical understanding of the climate system has improved immensely over the past few decades a major impediment to the application of this knowledge and its transfer to rapidly growing cities in poorer countries is severely hampered by the absence of suitable urban data. This chapter has highlighted the nature of the challenge facing those engaged in modelling urban environments, indicated where existing urban data can be obtained, and identified likely future sources.

References

Arnfield AJ (2003) Two decades of urban climate research: a review of turbulence, exchanges of energy and water, and the urban heat island. Int J Climatol 23(1):1–26. https://doi.org/10.1002/joc.859

Baklanov A, Grimmond CSB, Carlson D, Terblanche D, Tang X, Bouchet V, Lee B, Langendijk G, Kolli RK, Hovsepyan A (2018) From urban meteorology, climate and environment research to integrated city services. Urban Clim 23:330–341. https://doi.org/10.1016/j.uclim.2017.05.004

Bechtel B, Alexander PJ, Böhner J, Ching J, Conrad O, Feddema J, Mills G, See L, Stewart I (2015) Mapping local climate zones for a worldwide database of the form and function of cities. ISPRS Int J Geo Inf 4(1):199–219. https://doi.org/10.3390/ijgi4010199

Byun D, Schere KL (2006) Review of the governing equations, computational algorithms, and other components of the models-3 community multiscale air quality (CMAQ) modeling system. Appl Mech Rev 59(2):51–77. https://doi.org/10.1115/1.2128636

Chen F, Kusaka H, Bornstein R, Ching J, Grimmond CSB, Grossman-Clarke S, Loridan T, Manning KW, Martilli A, Miao S, Sailor D, Salamanca FP, Taha H, Tewari M, Wang X, Wyszogrodzki AA, Zhang C (2011) The integrated WRF/urban modelling system: development, evaluation, and applications to urban environmental problems. Int J Climatol 31(2):273–288. https://doi.org/10.1002/joc.2158

Ching J, Brown M, Burian S, Chen F, Cionco R, Hanna A, Hultgren T, McPherson T, Sailor D, Taha H, Williams D (2009) National urban database and access portal tool. Bull Am Meteorol Soc 90(8):1157–1168. https://doi.org/10.1175/2009bams2675.1

Ching J, Mills G, Bechtel B, See L, Feddema J, Wang X, Ren C, Brousse O, Martilli A, Neophytou M, Mouzourides P, Stewart I, Hanna A, Ng E, Foley M, Alexander P, Aliaga D, Niyogi D, Shreevastava A, Bhalachandran P, Masson V, Hidalgo J, Fung J, Andrade M, Baklanov A, Dai W, Milcinski G, Demuzere M, Brunsell N, Pesaresi M, Miao S, Mu Q, Chen F, Theeuwes N (2018) World urban database and access portal tools (WUDAPT): an urban weather, climate, and environmental modeling infrastructure for the Anthropocene. Bull Am Meteorol Soc 99(9):1907–1924. https://doi.org/10.1175/bams-d-16-0236.1

Ching J, Aliaga D, Mills G, Masson V, See L, Neophytou M, Middel A, Baklanov A, Ren C, Ng E, Fung J, Wong M, Huang Y, Martilli A, Brousse O, Stewart I, Zhang X, Shehata A, Miao S, Wang X, Wang W, Yamagata Y, Duarte D, Li Y, Feddema J, Bechtel B, Hidalgo J, Roustan Y, Kim Y, Simon H, Kropp T, Bruse M, Lindberg F, Grimmond S, Demuzure M, Chen F, Li C, Gonzales-Cruz J, Bornstein B, He Q, Tzu P, Hanna A, Erell E, Tapper N, Mall RK, Niyogi D (2019) Pathway using WUDAPT's Digital Synthetic City tool towards generating urban canopy parameters for multi-scale urban atmospheric modeling. Urban Clim 28:100459. https://doi.org/10.1016/j.uclim.2019.100459

Fan H, Zipf A, Fu Q (2014a) Estimation of building types on OpenStreetMap based on urban morphology analysis. In: Huerta J, Schade S, Granell C (eds) Connecting a digital Europe through location and place. Springer International Publishing, Cham, pp 19–35. https://doi.org/10.1007/978-3-319-03611-3_2

Fan H, Zipf A, Fu Q, Neis P (2014b) Quality assessment for building footprints data on OpenStreetMap. Int J Geogr Inf Sci 28(4):700–719. https://doi.org/10.1080/13658816.2013.867495

Garuma GF (2018) Review of urban surface parameterizations for numerical climate models. Urban Clim 24:830–851. https://doi.org/10.1016/j.uclim.2017.10.006

Goetz M, Zipf A (2012) Towards defining a framework for the automatic derivation of 3D CityGML models from volunteered geographic information. Int J 3-D Inform Model (IJ3DIM) 1(2):1–16

Haala N, Brenner C (1999) Extraction of buildings and trees in urban environments. ISPRS J Photogramm Remote Sens 54(2):130–137. https://doi.org/10.1016/S0924-2716(99)00010-6

Hanna SR, Brown MJ, Camelli FE, Chan ST, Coirier WJ, Hansen OR, Huber AH, Kim S, Reynolds RM (2006) Detailed simulations of atmospheric flow and dispersion in downtown manhattan: an application of five computational fluid dynamics models. Bull Am Meteorol Soc 87(12):1713–1726. https://doi.org/10.1175/bams-87-12-1713

Kusaka H, Kondo H, Kikegawa Y, Kimura F (2001) A simple single-layer urban canopy model for atmospheric models: comparison with multi-layer and slab models. Bound-Layer Meteorol 101(3):329–358. https://doi.org/10.1023/A:1019207923078

Lau KK-L, Ren C, Ho J, Ng E (2016) Numerical modelling of mean radiant temperature in high-density sub-tropical urban environment. Energ Buildings 114:80–86. https://doi.org/10.1016/j.enbuild.2015.06.035

Li J, Guan H, Yang X (2011) 3D building reconstruction from airborne Lidar point clouds fused with aerial imagery. In: Urban remote sensing: monitoring, synthesis and modeling in the urban environment

Lindberg F, Holmer B, Thorsson S (2008) SOLWEIG 1.0—modelling spatial variations of 3D radiant fluxes and mean radiant temperature in complex urban settings. Int J Biometeorol 52(7):697–713. https://doi.org/10.1007/s00484-008-0162-7

Liu Z, He C, Zhou Y, Wu J (2014) How much of the world's land has been urbanized, really? A hierarchical framework for avoiding confusion. Landsc Ecol 29(5):763–771. https://doi.org/10.1007/s10980-014-0034-y

Löwner M-O, Benner J, Gröger G, Häfele K-H (2013) New concepts for structuring 3D city models—an extended level of detail concept for CityGML buildings. In: Computational science and its applications—ICCSA 2013. Springer, Berlin Heidelberg, pp 466–480. https://doi.org/10.1007/978-3-642-39646-5_34

Mallet V, Quélo D, Sportisse B, Ahmed De Biasi M, Debry E, Korsakissok I, Wu L, Roustan Y, Sartelet K, Tombette M, Foudhil H (2007) Technical note: the air quality modeling system Polyphemus. Atmos Chem Phys Discuss 7(3):6459–6486

Maronga B, Banzhaf S, Burmeister C, Esch T, Forkel R, Fröhlich D, Fuka V, Gehrke KF, Geletic J, Giersch S, Gronemeier T, Groß G, Heldens W, Hellsten A, Hoffmann F, Inagaki A, Kadasch E, Kanani-Sühring F, Ketelsen K, Khan BA, Knigge C, Knoop H, Krc P, Kurppa M, Maamari H, Matzarakis A, Mauder M, Pallasch M, Pavlik D, Pfafferott J, Resler J, Rissmann S, Russo E, Salim M, Schrempf M, Schwenkel J, Seckmeyer G, Schubert S, Sühring M, von Tils R, Vollmer L, Ward S, Witha B, Wurps H, Zeidler J, Raasch S (2020) Overview of the PALM model system 6.0. Geosci Model Dev 13:1335–1372. https://doi.org/10.5194/gmd-13-1335-2020

Martilli A, Clappier A, Rotach MW (2002) An urban surface exchange parameterisation for mesoscale models. Bound-Layer Meteorol 104(2):261–304. https://doi.org/10.1023/A:1016099921195

Matzarakis A, Rutz F, Mayer H (2010) Modelling radiation fluxes in simple and complex environments: basics of the RayMan model. Int J Biometeorol 54(2):131–139. https://doi.org/10.1007/s00484-009-0261-0

Ng E, Chen L, Wang Y, Yuan C (2012) A study on the cooling effects of greening in a high-density city: an experience from Hong Kong. Build Environ 47:256–271. https://doi.org/10.1016/j.buildenv.2011.07.014

Oke TR, Mills G, Christen A, Voogt JA (2017) Urban climates. Cambridge University Press, Cambridge

Pesaresi M, Ehrlich D, Ferri S, Florczyk A, Freire S, Halkia M, Julea A, Kemper T, Soille P, Syrris V (2016) Operating procedure for the production of the Global Human Settlement

Layer from Landsat data of the epochs 1975, 1990, 2000, and 2014. Publications Office of the European Union

Salamanca F, Krpo A, Martilli A, Clappier A (2009) A new building energy model coupled with an urban canopy parameterization for urban climate simulations—part I. formulation, verification, and sensitivity analysis of the model. Theor Appl Climatol 99(3):331. https://doi.org/10.1007/s00704-009-0142-9

Santiago JL, Martilli A (2010) A dynamic urban canopy parameterization for mesoscale models based on computational fluid dynamics reynolds-averaged Navier–Stokes microscale simulations. Bound-Layer Meteorol 137(3):417–439. https://doi.org/10.1007/s10546-010-9538-4

See L, Mooney P, Foody G, Bastin L, Comber A, Estima J, Fritz S, Kerle N, Jiang B, Laakso M, Liu H-Y, Milčinski G, Nikšič M, Painho M, Pődör A, Olteanu-Raimond A-M, Rutzinger M (2016) Crowdsourcing, citizen science or volunteered geographic information? The current state of crowdsourced geographic information. ISPRS Int J Geo Inf 5(5):55. https://doi.org/10.3390/ijgi5050055

Soulhac L, Salizzoni P, Cierco FX, Perkins R (2011) The model SIRANE for atmospheric urban pollutant dispersion; part I, presentation of the model. Atmos Environ 45(39):7379–7395. https://doi.org/10.1016/j.atmosenv.2011.07.008

Stewart ID, Oke TR (2012) Local climate zones for urban temperature studies. Bull Am Meteorol Soc 93(12):1879–1900. https://doi.org/10.1175/bams-d-11-00019.1

Streets DG, Fu JS, Jang CJ, Hao J, He K, Tang X, Zhang Y, Wang Z, Li Z, Zhang Q, Wang L, Wang B, Yu C (2007) Air quality during the 2008 Beijing Olympic Games. Atmos Environ 41(3):480–492. https://doi.org/10.1016/j.atmosenv.2006.08.046

UN (2016) The World's Cities in 2016, Statistical Papers—United Nations (Ser. A), Population and Vital Statistics Report, United Nations, New York. https://doi.org/10.18356/8519891f-en

Wang Y, Zhu XX, Zeisl B, Pollefeys M (2017) Fusing meter-resolution 4-D InSAR point clouds and optical images for semantic urban infrastructure monitoring. IEEE Trans Geosci Remote Sens 55(1):14–26. https://doi.org/10.1109/TGRS.2016.2554563

Xu Y, Ma P, Ng E, Lin H (2015) Fusion of WorldView-2 Stereo and multitemporal TerraSAR-X images for building height extraction in urban areas. IEEE Geosci Remote Sens Lett 12(8):1795–1799. https://doi.org/10.1109/LGRS.2015.2427738

Chapter 16
The Synergistic Impacts of Urban Air Pollution Compounding Our Climate Emergency

Melissa Anne Hart, Nathan Cooper, Donna Green, and Mathew Lipson

Abstract Urban air pollution is already a massive health problem, the extent of which has only recently been identified. Cities produce the majority of air pollution that affects our health. With new research identifying that there is no safe level of air pollution, trends in rapid urbanisation worldwide suggest that the overwhelming majority of the world's population is breathing air that does not meet international guidelines. However, city dwellers' health is not just affected by air pollution. The problem is compounded by climate extremes exacerbated by both urban heat and global climate change. Combined, these factors have significant detrimental and interconnected synergistic impacts on the health of a city and its inhabitants. This chapter examines how urbanisation can modify the meteorological conditions in a city, and as a consequence, the long-term local climate. In addition, this chapter examines how these regional and global impacts on climate can impact, and be impacted by, air pollution. Although cities make a significant contribution to emissions of both ambient and greenhouse gas air pollutants, they are also pivotal in providing solutions. This interaction between ambient air pollution, urban climate and climate change in our cities is discussed with regard to the seminal role that our cities could take to help, rather than harm, inhabitants' health and well-being through supportive and forward thinking integrated urban design that will contribute to healthier cities.

M. A. Hart (✉)
ARC Centre of Excellence for Climate Extremes & Climate Change Research Centre, University of New South Wales, Sydney, NSW, Australia
e-mail: melissa.hart@unsw.edu.au

N. Cooper · D. Green
Climate Change Research Centre & Digital Grid Futures Institute, University of New South Wales, Sydney, NSW, Australia
e-mail: n.cooper@unsw.edu.au; donna.green@unsw.edu.au

M. Lipson
ARC Centre of Excellence for Climate System Science, University of New South Wales, Sydney, NSW, Australia
e-mail: m.lipson@unsw.edu.au

C. Ren, G. McGregor (eds.), *Urban Climate Science for Planning Healthy Cities*, Biometeorology 5, https://doi.org/10.1007/978-3-030-87598-5_16

Keywords Urban heat island · Air pollution · Climate extremes · Health impacts · Climate change · Energy transitions

Key Messages

- Integrated city transport policy should include no-car policies and congestion zones to mitigate air pollution from transport.
- Promote electric vehicles infrastructure alongside supportive tax policies to increase their use. Introduce mandatory efficiency standards for existing non-electric cars with clear phase out dates to transition to electric vehicles.
- Promote active transport, public transport and bicycle use transport through thoughtful city and landscape design.
- Design urban green spaces to promote health and well-being without trapping air pollution. Increase soft, porous ground covers and select tree and vegetation types to reduce pollen production.
- Lighten road and building surfaces, either with the initial choice of building or surfacing materials, or through retrofit. Require new buildings to have green roofs or solar panels and high energy efficiency.
- Be aware and plan for locally specific concerns including topography, urban density and cultural factors.

16.1 Introduction

Our air is growing increasingly deadly, with estimates of seven million people dying each year from air pollution (WHO 2016). Air pollution is hard to escape, and it is often concentrated in cities. Air pollution has been associated with acute health effects such as allergic respiratory diseases and angina, and chronic health problems such as lung cancer, bronchitis and cardiovascular disease. While the effects on cardiovascular and respiratory systems, as well as mortality from exposure, have been extensively studied, recent studies show that exposure to air pollution affects most, if not all, organs in the body (Schraufnagel et al. 2019a). It is estimated that over 91% of the world's population live in areas where the World Health Organization (WHO) air quality guidelines for air pollution are not met (WHO 2018a); this figure includes more than 90% of children (Watts et al. 2019). Significantly, this figure rises to 96% for people living in large cities (Krzyzanowski et al. 2014). Negative health effects have been documented at levels below the current WHO air quality guidelines suggesting there is no "safe" level of air pollution (Schraufnagel et al. 2019a). Fossil fuel combustion is a major contributor to air pollution and climate change, which means that multiple health co-benefits can arise from a reduction in air pollution from efforts to mitigate greenhouse gas emissions.

One of the most important factors that affect the concentration of air pollutants in the atmosphere is the meteorological conditions influencing the dispersion of pollutants. Urbanisation modifies the meteorological conditions in a city, and as a consequence, the long-term local climate. At the regional scale, urban areas can alter the formation of secondary air pollutants (Voogt 2002) and influence the deposition and dispersion of primary and secondary pollutants (Liu et al. 2015). Climatic conditions and air quality vary greatly within a city due to spatial variability in land-use, surface and geographical characteristics, pollutant emissions and transport infrastructure (Hart and Sailor 2009; Di Virgilio et al. 2018).

Cities themselves play a major role in emitting the greenhouse gases that cause climate change. The United Nations Human Settlement Programme determined a conservative estimate of 40–70% of all greenhouse gas emissions, based on source of emissions, were emitted from cities (UN-HABITAT 2011). That figure increases to 60–70% when indirect emissions are included, indicating the paramount role of cities in reducing climate change and air pollution.

Climate change will worsen this century, causing an increase in extreme weather events such as heatwaves, and an increase in fire weather (IPCC 2014). Through population growth and relocation, the world's urban population is increasing at an unprecedent rate. Much of this growth is likely to be concentrated in just a few countries, with China, India and Nigeria alone predicted to add 900 million urban inhabitants between 2018 and 2050 (UN 2018). This means that in the future, greater numbers of people will be exposed to poor air quality, including vulnerable groups, such as the elderly, and those with chronic respiratory illness. Combined, these factors will have significant detrimental and interconnected synergistic health impacts.

Those living in urban areas bear the brunt of not only poor air quality, but also global (climate change) and regional (urban heat island) heat impacts (Thorpe and Hart 2013; Krayenhoff et al. 2018). This additional burden placed on city dwellers is complicated not only by the combined influence from air pollution, urban climate and global climate change, but by the interaction between these factors. This ominous reality is two-sided, because due to the concentration of people and resources, cities also provide us with a suite of potential solutions through integrated urban development practices. This chapter discusses the interaction between ambient air pollution, urban climate and climate change in our cities, and outlines the seminal role that our cities could take to help, rather than harm, inhabitants' health and well-being through supportive and forward thinking integrated urban design towards developing sustainable, healthy cities.

16.2 The Interaction of Air Quality with Climate

Changes to global (via increased greenhouse gas emissions) and regional (via urbanisation on climate) climate can alter meteorological variables that influence the development, chemical transformation, transport, dispersion and deposition of air pollutants. Climate change will affect air quality by increasing the frequency and severity of bushfires and dust storms, in addition to the need to increase planned hazard reduction burns in some locations to reduce fire risk, all of which can significantly impact air pollution in cities located near and far from the original source of the air pollution (Fig. 16.1).

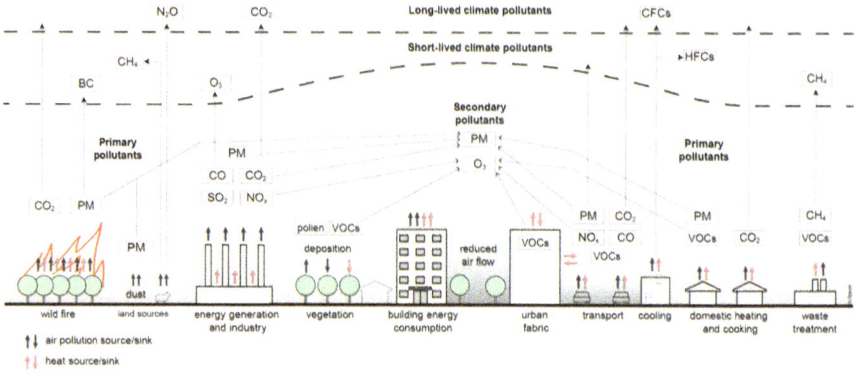

Fig. 16.1 The interaction of urban air pollution with local urban climate and global climate change

16.2.1 Urban Climate

Urban development has a significant influence on surrounding climate and ecosystems. It can influence regional circulation patterns, wind speed, temperature, cloud formation, precipitation, carbon and hydrological cycles. As a result, urban areas have their own characteristic climate (Cleugh and Grimmond 2012). Land-use modifications due to urbanisation modify the energy balance in cities; this in turn affects the urban thermal environment, resulting in urban areas often experiencing higher temperatures than surrounding rural areas. Urban areas have been found to be as much as 10 °C degrees hotter than surrounding rural areas—although this can change depending on how the urban heat island is measured (Baklanov et al. 2016). The influence of urbanisation on surrounding climate varies significantly between cities and is affected by various factors including population density, building materials and morphology, vegetation, season and latitude (Bai et al. 2017).

This impact of urbanisation on the climate within a city can impact a city's air quality. Urbanisation impacts thermal and dynamical properties within the urban and planetary boundary layer and can influence the dispersion of pollutants. Urban street canyons can lead to a decrease in mixing, and subsequent decrease in pollutant dispersion, in the urban boundary layer (Fallmann et al. 2016). High rise buildings can create narrow street canyons that trap pollutants created by street level traffic. This is significant for not only surface concentrations, but for the vertical behaviour of the pollutants within the street canyon as they will impact the concentration of pollutants infiltrating buildings (Wong et al. 2019). Further, higher temperatures in cities can enhance the formation of secondary pollutants, such as tropospheric ozone (Wang and Kwok 2003).

16.2.2 Global Climate Change

Sources of greenhouse gas emissions from urban areas include electricity generation, transportation, building energy use, industry and waste (Fiore et al. 2015). Building and construction accounts for 39% of CO_2 emissions and these emissions are increasing at 1% per year (UN Environment and International Energy Agency 2017).

Heatwaves, that is prolonged periods of excessive heat, have been increasing in duration, intensity, and frequency globally since as far back as the 1950s (Perkins-Kirkpatrick and Gibson 2017). A substantial fraction of these record-breaking heat extremes have been attributed to anthropogenic climate change (King et al. 2016). The most confident impacts of climate model projections are an increase in the spatial scale, frequency, duration and intensity of heatwaves (Cowan et al. 2014). However, these projections only capture increases in heat events due to increased greenhouse gas emissions, they do not include the additional burden placed on inhabitants of our cities by the urban heat island effect. Heatwaves are highly conducive to ozone formation, and the weather conditions associated with heatwaves are similar to those which inhibit dispersion of pollutants. The record-breaking European heatwave of 2003 resulted in more than 70,000 excess deaths (Robine et al. 2008). In England and Wales up to 38% of their excess deaths were attributed to elevated ozone and particulate matter rather than the heat associated with the heatwave (Stedman 2004). Despite the disjointed research in this field, especially in relation to climate impacts on air pollution at varying spatial scales, understanding these interactions between regional and global climate in cities is vital to ensure we are able to quantify future health risks from changes in air pollution and associated potential co-benefits from mitigation.

Future climate projections for specific regions of the world highlight local vulnerabilities. For example, projections for the east coast of Australia, where over 80% of the population live, have shown an increase in the strength of near surface temperature inversions (Ji et al. 2019). Temperature inversions, when the air at the surface is cooler than the air above, play a significant role in inhibiting the dispersion of pollutants. Any increase in the occurrence, duration, or strength of inversions has the capacity to have a substantial influence on the concentration of air pollutants at the surface.

16.3 Air Pollutants

The main cause of poor air quality is the emission of primary, and subsequent formation of secondary, air pollutants. Primary air pollution can come from natural and anthropogenic sources. Natural sources include dust, salt spray from the oceans, volcanic eruptions and bushfires. Anthropogenic sources include emissions from power stations, factories, motor vehicles and hazard reduction burns. While little can be done to reduce natural sources of air pollution, it is possible to control and reduce anthropogenic sources. Secondary pollutants are formed through reactions between primary air pollutants in specific conditions, with increases in temperature generally increasing the formation of pollutants such as tropospheric ozone and secondary particles. The rate of dispersal of primary and secondary air pollution is affected by a range of factors including the source of pollution, local topography and meteorological conditions.

The most common air pollutants that are damaging to human health are: fine and coarse particulate matter (PM); gases such as nitrogen dioxide (NO_2), sulphur dioxide (SO_2), ozone (O_3), and carbon monoxide (CO); along with airborne lead and a class of pollutants called hazardous air pollutants, or air toxics, which exist in very low concentrations in the air but are quite hazardous to plant, animal and human health (Keywood et al. 2016). Some ambient air pollutants also act as short-lived climate pollutants (SLCP) which are atmospheric constituents that have a warming influence on the climate system within 10–30 years of their emission. SLCPs include methane, black carbon and tropospheric ozone. The lifetime of these pollutants varies dramatically, with the estimated lifetime of black carbon and tropospheric ozone varying from days to months, while the lifespan of methane is estimated to be approximately 10 years, however, all these pollutants have short lifespans relative to CO_2, which has a lifespan of approximately 100 years (Fiore et al. 2015).

Two pollutants particularly influenced by both regional and global changes to climate are particulate matter and tropospheric ozone.

16.3.1 Particulate Matter

The main components of PM are sulphate aerosols, nitrate aerosols, carbonaceous aerosols (organic carbon and elemental/black carbon), sea salt and soil dust. Some of these components are released directly from a source with others formed through chemical reactions in the air. The former components are called primary particulates or aerosols and the latter are called secondary aerosols or particulates.

Particulate matter, especially particulate matter below 2.5 μm in diameter ($PM_{2.5}$), is ubiquitously associated with a range of negative health effects including cardiovascular and respiratory outcomes, stroke and premature mortality. These effects have been comprehensively documented from both acute and chronic exposure (Rückerl et al. 2011; Hoek et al. 2013; Atkinson et al. 2014). Current research has found negative health effects at concentrations below WHO and national standards. This suggests there may be no low threshold for a safe concentration of PM (Burnett et al. 2018). Research has found that the health effects of exposure to $PM_{2.5}$, as well as O_3, are modulated by temperature, with the effects of exposure to $PM_{2.5}$ generally increasing in higher temperatures, although the results of studies—which have typically examined urban areas—are not consistent (Lou et al. 2019). For example, elderly people have been found to be the most vulnerable to impacts from air pollution enhanced by high temperatures due to climate change (Lou et al. 2019).

In many locations, the relationship between climate change and PM is more complex than the relationship between climate change and O_3, as the various components of PM are influenced by changes in meteorological variables in different ways. The main meteorological variables likely to impact PM are precipitation frequency and mixing depth. Particulate matter may decrease in regions projected to experience more frequent precipitation, as precipitation is the main PM scavenger, with frequency of precipitation being more important than rate (Jacob and Winner 2009). Climate change will indirectly influence PM levels by increasing the frequency and severity of wildfires and dust storms in regions that are projected to become hotter and drier (Kinney 2008; Takaro et al. 2013).

Black carbon forms from the incomplete combustion of biomass including wood, as well as fossil fuels, and comes from a range of anthropogenic sources, including industrial coal, diesel engines and residential cooking and heating with biomass such as coal and wood (UNEP WMO 2011; Fiore et al. 2015). Natural biomass burning such as from wildfires, savannah and grasslands are major sources of black carbon. Unlike most other components of particulate matter/aerosols—including sulphates, nitrates and organic carbon (OC), black carbon absorbs solar and infrared radiation, thereby having a positive radiative forcing effect which contributes to increased warming. Black carbon has a lifetime in the range of weeks, and so is considered a SLCP (Fiore et al. 2015). Beyond its direct absorption of radiation while in the atmosphere, black carbon can affect radiative forcing in other ways. For example, when deposited on high-albedo surfaces such as ice and snow, black carbon particles reduce the albedo of those surfaces, thereby decreasing their scattering properties and increasing radiative forces.

The largest sources of black carbon are from residential energy sources, such as burning coal and wood for heating and cooking. This is particularly prominent in heavily populated, low- and middle-income countries, especially India and China. Because of the large emissions from these sources and likely increases in populations in these regions, residential sources of black carbon are predicted to significantly increase in their contribution to overall black carbon emissions. Transport is also a major source of black carbon, especially from diesel engines. Technological advances have resulted in decreases in black carbon emissions from on and off-road diesel vehicles; however, these have been offset by an increase in the number of vehicles worldwide. Nonetheless, black carbon emissions from transport are anticipated to decrease in the future as further technological advances and legislation are implemented (UNEP WMO 2011).

16.3.2 Ozone and Its Precursors

Tropospheric ozone is a short-lived climate pollutant that has a residency time of weeks in the atmosphere (Haines et al. 2017). Upper tropospheric and stratospheric ozone are the most significant for climatic forcing, while near-surface ozone is important for public health (UNEP WMO 2011). Tropospheric ozone is a secondary pollutant, produced through a photochemical reaction between sunlight and precursor pollutants such as oxides of nitrogen (NO_x), methane, volatile organic compounds and carbon monoxide. Nitrogen oxides are formed from the organic decomposition of nitrogen compounds in the soil and in oceans. Anthropogenic release of NO_x occurs from the combustion of fossil fuels under pressure and heat, principally from motor vehicle exhausts, resulting in the fixation of nitrogen and oxygen to form NO. In the atmosphere NO readily reacts with hydrocarbons and sunlight to produce nitrogen dioxide (NO_2) which is a yellow-brown coloured gas and is an irritant to humans. Nitrogen dioxide reacts with a host of species to produce ozone. The maximum ground level ozone concentration that can be reached depends on concentrations of volatile organic compounds (VOCs), NO_x, and the ratio between them. Summer conditions favour the formation of O_3 because of increased ultraviolet radiation and temperatures and because of increased availability of precursor emissions including nitrogen oxides from soils and VOCs from vegetation.

Emissions for each precursor pollutant vary regionally and come from a wide range of sources. The transportation sector is one of the largest sources of O_3 precursors including NO_x, along with the NO_x produced by coal-fired power stations. Carbon monoxide is emitted from residential biofuel burning, and is a major health hazard for low-to-middle income countries due to its high toxicity. VOCs are widely prevalent in high-income countries due to the use of solvents in the domestic sector, with VOCs predominantly created by domestic biofuel burning and fossil-fuel use in low-income countries (UNEP WMO 2011).

Tropospheric ozone is directly influenced by changes in climate. In most regions around the world where temperatures are projected to increase, tropospheric ozone will increase in step, especially in locations with high levels of existing precursor pollutants. Studies reviewed by Jacob and Winner (2009) find that climate change projections will cause summertime O_3 levels to increase by between 1 ppb and 10 ppb over the twenty-first century across North America, Europe and Asia, with effects being strongest in more highly polluted regions. These increases are greater in urban areas, which already experience high levels of pollution.

Ozone is detrimental to human health (Ebi and McGregor 2008). Even at low levels, exposure to O_3 increases the risk of mortality, with each 10 ppb increase being associated with an increase in mortality of 0.3% (Bell et al. 2006). A threshold concentration of O_3 below which there are no associated negative health effects has not been found (WHO Regional Office for Europe 2013; COMEAP 2015). Although there is less evidence for associations between long-term exposure to O_3 and mortality (Atkinson et al. 2016), recent research has found evidence of associations between long-term exposure to O_3 and respiratory, cardiovascular and all-cause mortality (Wang et al. 2019). Some quantitative modelling of how future increases in ozone mortality could be avoided has been undertaken. For example, implementing maximum feasible greenhouse gas emission reductions, using currently available technology, compared to a future scenario where little is done to curb emissions, could avoid 5600 premature ozone exposure deaths annually from 2030 onwards in Australia and Japan (West et al. 2007).

Stratospheric ozone is considered beneficial to health, as stratospheric ozone absorbs UV radiation which protects humans from harmful solar radiation (UNEP WMO 2011). However, O_3 from the stratosphere can be transported to the troposphere, and studies have shown it is more likely to be transported to the troposphere under a warming climate (Doherty et al. 2017; Fu and Tian 2019).

16.4 Air Pollution from Cities

Air pollution impacting cities comes from emissions originating from within the city, e.g. transport, industry, and the building sector, and from outside of the city from either natural sources such as dust storms or wildfires, or anthropogenic sources such as hazard reduction burns.

16.4.1 Transport

Motor vehicles release a wide range of air pollutants including more than half of globally emitted NO_x, and significant amounts of PM and black carbon, CO and VOCs (Gouldson et al. 2018; Glazener and Khreis 2019; Watts et al. 2019). Some of these pollutants such as CO and NO_x are emissions from tail-pipe exhausts, while other pollutants arise from of brake and tyre wear, and the re-suspension of particles, such as $PM_{2.5}$ and PM_{10} (Karagulian et al. 2015). It has been estimated that air pollution from motor vehicles is responsible for 184,000 deaths, about 1% of all air pollution-related deaths globally (Glazener and Khreis 2019). As well as being a significant source of air pollutants, road transport is also a major source of greenhouse gas emissions. Road transportation is estimated to be responsible for 30% of NO_x emissions, 10% of PM, 54% of CO and 14% of CO_2 emissions worldwide (Glazener and Khreis 2019).

Relatively recent EU legislation (and legislation in some states in the United States) for more stringent controls on the performance of engines and reductions of emissions has resulted in a decrease in NO_x, black carbon and CO emissions from transport in Europe and North America (UNEP WMO 2011). Nonetheless, the implementation of these standards has not been straightforward. Corporate malfeasance has been found to have been carried out by at least one major international company in recent years. Volkswagen was found guilty of manufacturing vehicles that did not adhere to EU emission standards by using devices to control emissions that were deactivated outside of laboratory testing conditions. A subsequent investigation found nine other European vehicle manufacturers producing cars that did not comply with EU emissions standards (Chossière et al. 2018). Notwithstanding these incidents, improvements in vehicle technology and emission standards are considered to be major factors that will drive down emissions from traffic in the future, and are considered to be key in reducing emissions from low-to-middle income countries as the share of the market slowly increases for electric vehicles as their prices fall (Fiore et al. 2015; Glazener and Khreis 2019).

As a result of increased wealth and population growth, transport emissions from some low-to-middle income countries, particularly China and India, are projected to increase with increasing private car ownership (UNEP WMO 2011; Glazener and Khreis 2019). This scenario is worrying from a health perspective because vehicles in megacities in these countries are already contributing substantially to air pollution. For example, in Beijing, 46–58% of VOCs emissions, 6–22% of PM_{10} emissions and 31–35% of NOx emissions, including 74% of ground-level NOx emissions, come from vehicles. Similarly, in four Indian megacities, vehicles have been estimated to contribute between 20 and 50% of $PM_{2.5}$ emissions (Gouldson et al. 2018).

16.4.2 Indoor Air Quality

Air pollution from domestic fuel sources, particularly biomass burning for heating and cooking, is an especially significant source of air pollutants in low-to-middle income countries. Estimates of the contribution of domestic fuel sources to $PM_{2.5}$ emissions are ~22% for low-to-middle income countries, with some of the largest contributions of domestic fuel in Africa (34%), Central and Eastern Europe (32%) and South America excluding Brazil (25%), although there are large uncertainties tied to these values due to the lack of local studies and inconsistencies in source apportionment (Karagulian et al. 2015). With much of this biomass burning associated with cooking and heating, the indoor air quality issue becomes a gendered one, with women, and young children in their care, having disproportionate regular exposure to high levels of indoor air pollution from cooking stoves.

Residential energy use is likely to be one of the largest contributors to mortality related to air pollution exposure. A study of the causes of ambient air pollution-related mortality found that PM from residential energy use was estimated to cause 32% of air pollution-related mortalities in China, 50% in India, 51% in Vietnam, 55% in Bangladesh and 60% in Indonesia. These estimates increased when the potentially higher toxicity of black carbon found in some regions was accounted for (Lelieveld et al. 2015). In India, while there has been an uptake of LPG-stoves to reduce emissions from household solid or liquid fuel burning, it was reported that 34% of the population in the city of Kolkata (population ~14 million) did not have an LPG-stove, and were cooking using coal, biomass or kerosene (Guttikunda et al. 2014).

In high-income countries, exposure to VOCs is more of a concern, with a wide range of sources from commonly used substances including cleaning products, wood products, air fresheners, paint and paint removers (Shrubsole et al. 2019).

16.4.3 Power Generation

Coal-fired power stations are one of the biggest sources of anthropogenic CO_2, as well as being a substantial source of $PM_{2.5}$, SO_2 and NO_x (UNEP WMO 2011). In the US, emissions from the stationary energy sector are estimated to be the largest contributor to all air pollution-related premature mortality, causing 31% of these deaths (Lelieveld et al. 2015). Over the last few decades, higher quality fuel and increased abatement measures has led to an overall decrease in global SO_2 emissions, with reductions in Europe and North America offsetting increases in East Asia (Aas et al. 2019). On the other hand, black carbon, NO_x, CO emissions are expected to continue to increase (UNEP WMO 2011).

16.5 Air Pollution Impacting Cities

Natural sources of air pollutants from outside of cities can impact air quality within a city under specific meteorological conditions. Human influences on land-use and fire activity can also impact emissions from these sources. The occurrence of both wildfires and dust storms have been shown to be influenced by climate change.

16.5.1 Fire

Air pollution events associated with wildfires have been associated with direct and indirect health impacts, including increased hospital admissions, morbidity and mortality across the world (Black et al. 2017). Weather conditions conducive to wildfires have increased under our changing climate due to increases in concurrent occurrences of atmospheric stability, dryness and near surface fire weather (Dowdy 2017; Di Virgilio et al. 2019). This rise in wildfire occurrence increases dangerous air quality during, and for several days after, these events. In the US this has resulted in an increase in PM in regions impacted by wildfires (McClure and Jaffe 2018). Population exposure to wildfires is increasing, with 77% of countries impacted (Watts et al. 2019). In some regions of Europe, PM pollution caused by an increase in wildfires due to climate change are projected to breach current WHO standards (Knorr et al. 2016).

In many locations, hazard reduction burns, or controlled low intensity fires, are vital to reduce the severity of uncontrolled wildfires. However, if undertaken during unfavourable meteorological conditions, they too have the capacity to trigger extreme air pollution events. In Sydney, Australia, air quality standards are frequently breached due to fire hazard reduction burns. An increase in hospital admissions have been attributed to these burns (Johnston 2017) with one event in 2016 (Fig. 16.2) estimated to have caused 14 premature deaths (Broome et al. 2016).

Fig. 16.2 A view of the city of Sydney on a typical day (left) and a day where the air quality is significantly impacted by smoke from hazard reduction burns (right)

An assessment of the meteorological controls on air pollution in the Sydney region found that planetary boundary layer height, total cloud cover, temperature, and wind speed were the most consistent predictors of PM2.5 during hazard reduction burns. Further, the diurnal variation in these conditions was vital when influencing the concentration of pollutants, with cooler, more stable morning and evening conditions combined with night-time westerly drainage flows resulting in the greatest impact on air quality (Di Virgilio et al. 2018). The findings indicated that altering the timing of the hazard reduction burns by a matter of hours, so that they occur after the break-up of the night-time boundary layer, may significantly reduce the impacts on air quality. An improvement to the planning of hazard reduction burn times could reduce extreme air pollution risk, while still allowing land managers and fire agencies to carry out vital hazard reduction burn activity.

Air pollution is a transboundary issue, affecting city dwellers far away from the initial pollution source. In South East Asia, examples of transboundary pollution include: agricultural burn off on Sumatra and Borneo regularly affecting the health of millions of people living in major cities on other islands in the country such as Jakarta, and other cities in the region, including Kuala Lumpur, Malaysia. During winter months, Taiwan's city dwellers are regularly impacted by air pollution and dust originating from China because of the high atmospheric pressure systems that prevent air pollutants from dispersing, and the winter monsoonal flows that direct this pollution towards Taiwan. Major cities in New Zealand were recently affected by bush fire-related pollution that had travelled from Australia due to north-westerly winds associated with the position of a high pressure system in the North Tasman Sea.

In many regions of the world, agricultural burning has a major impact on air quality. In north-western India the burning of crop residue after harvesting leads to significantly reduced air quality in a very densely populated region. Increased agricultural activity in this region has resulted in 60% more agricultural fire activity over the period 2002 to 2016 (Jethva et al. 2019). In Delhi, the annual PM2.5 concentration is often more than ten times greater than the WHO standard. During some parts of the year, up to 78% of the city's PM2.5 air pollution can be attributed to agricultural burning (Cusworth et al. 2018).

16.5.2 Dust

Around half of all atmospheric dust in the world comes from the Sahara, followed by sources in China and Central Asia (20%), then the Middle East and Australia (Goudie 2014; Middleton 2017). Because of the potential severity of dust events and the large distances they can travel, dust events can potentially cause decreases in overall air quality in urban areas thousands of kilometres away. For example, major dust events in rural China have been documented to affect air quality not only in Beijing and Shanghai, but in Tokyo and Seoul. Saharan desert dust has travelled as far as the Americas, Europe and the Arctic (Goudie 2014). During some of these dust events, levels of PM have exceeded local and WHO standards for air quality by orders of magnitude (Goudie 2014; Middleton 2017). At these elevated levels, concentrations of PM are known to cause a range of acute respiratory and cardiovascular effects including mortality. Exacerbating this problem, other toxic components can be picked up by dust particles and become toxic transboundary air pollution (McGregor 2012). For example, lead in dust particles in China were transported to Japan and Korea, and epidemics of meningococcal meningitis in West Africa have been associated with dust events in the Sahara (Middleton 2017).

Increases in extreme wind events, drought frequency and aridity in some regions due to climate change are likely to increase the frequency and occurrence of dust events (Middleton 2017). There is some evidence of increases in dust events already occurring, such as in the Sahel zone of Africa, the Gobi of Mongolia and the western region of the US. However, other regions have seen a decrease in dust events, such as in China, Central Asia, Pakistan and other parts of the US (Goudie 2014). Goudie (2014) notes that the influence of climate change is likely to affect the occurrence of dust events regionally, with some evidence that increases in predicted wind velocity have led to increases in dust events in the Bodélé depression in Central Sahara (Washington et al. 2009), similarly, in the Colorado plateau, increased temperatures and decreased precipitation are increasing dust events (Munson et al. 2011). Significant uncertainty exists around the location and impact of climate change on dust activity in some regions. For example, decreases in the number of dust events in Northern China over the twentieth century are thought to occur due to changes in atmospheric circulation causing decreases in wind velocity (Goudie 2014).

Anthropogenic impacts on the increase in dust events include increases in: desertification, fires, removal of vegetation for wood, overuse of water from lake beds and bare fields from crops and grazing (Goudie 2014; Middleton 2017). Fiore et al. (2012) note that while the occurrence and magnitude of changes in natural dust is highly uncertain, concentrations of dust may increase as a result of the increasing contribution of anthropogenic dust from construction and disturbed soils. The latest estimates of the fraction of total dust that is anthropogenic is 25%, but estimates vary widely, from 6 to 60% (Fiore et al. 2012).

16.6 Impacts of Urban Air Pollution

Poor air quality is detrimental to human, animal and plant health and has a significant economic cost.

16.6.1 Health

The greatest number of deaths from air pollution exposure occur in low- or middle-income countries with an estimated 91% of premature deaths from exposure to ambient PM (WHO 2018a). Exposure to ambient air pollution causes average life expectancy to decrease by an estimated 1.03 years, with the largest decreases in heavily populated low- and middle-income countries (Apte et al. 2018). Children and the elderly are especially vulnerable to the effects of air pollution exposure due to their physiology (Sacks et al. 2011; Bell et al. 2014; Schraufnagel et al. 2019b). People over 50 years old represent 88% of worldwide premature deaths from exposure to ambient $PM_{2.5}$ (WHO 2018a). In comparison to ambient air pollution, exposure to household and indoor air pollution, especially from cooking with solid fuels, is estimated to cause 3.8 million deaths per year (WHO 2018b).

The number of excess deaths from ambient air pollution has increased overtime, with estimates of global deaths from ambient PM and O_3 increasing 68% and 20%, respectively, between 1990 and 2010 (State of Global Air 2019). These trends are predicted to continue, with an analysis of a business-as-usual emissions scenario, whereby only currently agreed legislation at the time of publication was assumed to be implemented, estimating that premature deaths from ambient air pollution could double by 2050 (Lelieveld et al. 2015). Increases in mortality are predicted to be highest in China and India, and due to increasing urbanisation in these countries, a significant number of these deaths would occur in cities (Lelieveld et al. 2015; WHO Regional Office for Europe 2016).

Current and future increases in global mortality from air pollution are due to the combined increases, and interactions between, global population, aging population and urbanisation (Lelieveld et al. 2015; WHO Regional Office for Europe 2016; State of Global Air 2019). The effects of air pollution are currently most significant in urban areas, with per capita mortality attributable to air pollution estimated to be approximately 50% higher in urban areas compared to rural environments (Lelieveld et al. 2015). With the global urban population estimated to increase to 5.2 billion by 2050, the air pollution exposure is likely to become an even greater public health concern.

16.6.2 Economic

Exposure to air pollution leads to considerable economic impacts, some of which are due to increased healthcare costs caused by hospitalisation, treatment or mortality of exposed people (Marlier et al. 2016). Other costs are due to decreases in productivity, including decreases in the number of working hours, as well as reduction of productivity while at work (Schmidt 2019).

The World Bank estimates the economic welfare impacts of exposure to air pollution by quantifying the trade-offs individuals are willing to make to reduce their risk of mortality. It is estimated that the global welfare losses from ambient and household air pollution was US$5.11 trillion in 2013, which was mostly due to the contribution of ambient air pollution, with ambient $PM_{2.5}$ responsible for over US$3.5 trillion in welfare losses (World Bank 2016). Welfare losses from total air pollution were highest in South Asia, East Asia and the Pacific, which lost the equivalent of 7.4 and 7.5% of GDP, respectively. These figures compare to a loss of 5.1% in Europe, 3% in North America and 2.2% in the Middle East and North Africa (World Bank 2016). In most regions, the welfare losses from total air pollution were mostly from ambient air pollution compared to indoor air pollution, with the only exceptions being South Asia and Sub-Saharan Africa (World Bank 2016).

Labour income losses from premature mortality were estimated to be $225 billion in 2013, with $66 billion from losses in South Asia alone. Measures of labour income loss tend to be highest in younger populations, as the potential income generated over the rest of their lives is much higher compared to older populations. Consequently, labour income losses relative to the rest of the world are higher in Sub-Saharan Africa compared to welfare losses, as this region has a younger population compared to other regions (World Bank 2016).

The cost of mortality places significant burden on national economies, as increasing premature death impacts labour productivity. There are direct and indirect market impacts of air pollution, with the former impacts including losses in labour productivity, health expenditures and the influence of air pollutants, especially O_3, on agricultural crop yields. The indirect impacts include the cost of readjusting economies to account for these losses. The market impacts from air pollution are expected to increase from 0.3% of global GDP in 2015 to 1% in 2060 (WHO Regional Office for Europe 2016). The majority of these losses are from indirect impacts that will significantly increase overtime as economies adjust to the direct impacts of air pollution (WHO Regional Office for Europe 2016).

The economic benefits of mitigating air pollution have also been modelled. A C40 report examining 96 cities across the world estimated that between $198–583 billion could be saved by eliminating particulate matter emissions from buildings and industry, and by reducing traffic emissions by half (C40 2018). Similar studies have been conducted in the US (USGCRP 2016) and globally (Scovronick et al. 2019). All these studies clearly show that the costs of delaying action are much more significant than responding early.

16.7 Air Pollution Mitigation and Recommendations for Achieving Healthy Cities

Cities dwellers are likely to shoulder the majority of the burden of the synergistic impacts of poor air quality and multi-scale climate change. However, city authorities are in a critical position to take the lead when it comes to mitigating these issues, with many mitigation efforts at the city-scale having co-benefits across multiple environmental issues, and with subsequent economic rewards.

16.7.1 Transport

Many cities have implemented no-car policies or congestion zones to mitigate air pollution from transport. As population and urbanisation increased rapidly in the 1980s in Mexico City, so did its air pollution, compounded by its location in a constrained basin (Baklanov et al. 2016). To address these increases, the city implemented a variety of mitigation measures including: the removal of lead in petrol, decreased sulphur content in diesel vehicle fuel, increased use of natural gas in industry and a "no driving day program". As a result, Mexico city's air quality improved significantly, even as its population increased (Baklanov et al. 2016). The London low emissions zone, which has been in place since 2008, and has progressively introduced stricter standards, imposes emissions standards on vehicles in most of Greater London. Vehicles that do not meet the emissions standards are charged. An assessment of school children living in central London found that 6 years after the introduction of the low emissions zone the percentage of children living at addresses exceeding the European Union for annual NO_2 had fallen from 99% to 34% (Mudway et al. 2019). Other cities have also introduced various "car-free policies" with Copenhagen and Brussels introducing the largest car-free areas in Europe. Many cities have introduced car-free days for specific events, with NO_2 concentrations decreasing 40% and 20% in Paris and Leeds on a car-free days introduced for the Tour de France (Glazener and Khreis 2019).

The introduction of Electric Vehicles (EVs) can significantly reduce air pollution and greenhouse gas emissions in urban areas. Various studies have projected that levels of CO_2 in China will decrease by 20%, CO and NO_x by 85% and 27%, respectively, in Taiwan with the introduction and uptake of EVs (Glazener and Khreis 2019). The ability of EV to reduce air pollution depends on the transition to renewable sources of energy generation so that pollutants are not just emitted elsewhere indirectly (Glazener and Khreis 2019). West et al. (2013) found significant co-benefits from greenhouse gases mitigation policies, specifically that there could be 0.5 million avoided deaths by 2030 due to associated reductions in $PM_{2.5}$ and O_3.

Many cities are improving their development of active transport and multi-modal transport (Slovic et al. 2016). Increasing public transport may increase active transport, since any trips by public transport will usually require some walking to reach the intended destination. There are various public health benefits to active transport—as it decreases the risk of developing various poor health outcomes including Type 2 diabetes, obesity, respiratory and cardiovascular disease, cancer and depression. This may reduce some of the health impacts from physical inactivity caused by the use of motor vehicles, which is estimated to cause 3.2 million deaths a year (Glazener and Khreis 2019).

There is a risk of health impacts from increased air pollution exposure as active transport increases in a city; however less than 1% of global cities have ambient air pollution levels high enough to offset the benefits of cycling for 30 min a day. Even in a heavily polluted city such as Delhi, with an annual background $PM_{2.5}$ mean of 153 μg/m^3, an individual could cycle for up to 45 min before the health impacts of air pollution outweighed the physical activity benefits (Glazener and Khreis 2019).

To encourage active transport, interventions are required at multiple levels. This can be a complicated policy challenge because they need to consider changes at the societal level (for example restrictions on car use and decreased speed limits), city-level (changes in population density and planning), route-level (including infrastructure to allow active transport) and individual-level (education). Policies that have used multiple measures to increase active transport, so called integrated policy packages, rather than single measure focused ones, often have the most success with increasing active transport and decreasing greenhouse gas and air pollution from traffic emissions in urban areas (Glazener and Khreis 2019; Watts et al. 2019).

The use of an integrated policy package in Beijing for the 2008 Olympics, which included updating vehicle emissions standards, closing petrol stations, improving public transport and banning diesel and heavy-duty trucks from the city, resulted in significant reductions in air pollution, with SO_2 decreasing 60%, CO by 48%, $PM_{2.5}$ by 27% and NO_2 by 43% (Glazener and Khreis 2019).

16.7.2 Greenspace

Urban vegetation can have a positive or negative impact on urban air quality, depending on the type of vegetation and local context, and should therefore be planned carefully (Abhijith et al. 2017). Vegetation can enhance deposition of air pollutants by capturing particles and gas molecules on the large area of leaf surfaces available. However, vegetation can also inhibit dispersion of pollutants. This process is particularly common in street canyons with tall trees where pollutants may become trapped within the street canyon, or within the tree canopy due to its ability to inhibit inflow of fresh air (Janhäll 2015). Generally, low vegetation planted close to pollutant sources can increase deposition, and considered use of hedges can reduce air pollution exposure for vulnerable groups, for instance, the planting of hedges alongside school playground in urban areas (Vanos 2015). Complexities arise in the variability of different plant species in removing, or trapping, pollutants (Samson et al. 2017), and there may be a seasonal difference in concentrations in and around tree canopies (Salmond et al. 2013). The selection of street tree needs consideration, as some will produce pollen, or biogenic volatile organic compounds (VOCs).

Increasing green cover can reducing urban heat. The mitigation of urban temperatures can help minimise the formation of secondary pollutants like tropospheric ozone. The type and choice of green cover needs to be fit for purpose and location, however, increased canopy cover can reduce urban ambient air temperature, though in some cities it may also increase humidity and subsequently human thermal comfort (Ma et al. 2018).

16.7.3 Integrated Urban Planning

Urban design interventions can serve to significantly hinder, or assist, in active transport use. In particular, dense urban design is frequently the most conducive to active transport, especially where there are short travel distances between work and home, frequent easily accessible parks and recreational space, and bicycle lanes physically separated from other road traffic. The legacy of poorly planned, or car focused cities are, in some places such as the US, hindering faster uptake of integrated planning. The city of London is an example of where a clear strategy to promote integrated urban design and active transport has led to increased local amenity and reduced reliance on private car use (Hickman and Bannister 2014).

The decision to lighten road and building surfaces, either with the initial choice of building or surfacing materials, or through retrofit, can lead to cooler urban temperatures. Some road surfaces in Los Angles have been painted white to reduce local urban heat islands. Other state or national governments are requiring new buildings to have green roofs or solar panels. For example, new commercial buildings in France will need to have green roofs or solar panels to reduce albedo, air pollution and greenhouse gases; and new regulations to come into force in 2020 will require newly built houses in California to have solar panels. Generally, these activities are of great benefit, however, some unintended outcomes of similar activities have recently been identified. A climate modelling study for Stuttgart, Germany, found that painting roof surfaces white served to increase ozone concentrations due to the increase in reflected short wave radiation from the higher albedo surfaces intensifying photochemical reactions.

Locally specific concerns need to also be factored into integrated urban planning. A case in point can be seen in the city of Hong Kong, where lines of high-rise buildings are built along the coastline in order to maximise sea views, known as the "wall effect". The combination of the complex topography and density of buildings in the city results in these walls of buildings blocking sea breezes and overnight drainage flows from the surrounding hillsides, and subsequently inhibiting both pollutant dispersion from the city and ventilation of urban heat (Wong et al. 2011).

16.8 Conclusion

Urban air pollution is already a massive health problem, the extent of which has only recently been identified. Climate change has already exacerbated these problems, and will increasingly do so unless decisive action is taken to implement strategies to mitigate the wicked problem of the synergistic impacts of urban air pollution. Fortunately, one level of government that has been swift to act in comparison with many state or nation governments is the city level, and we have the opportunity to learn from examples across the world of best practice to simultaneously reduce air pollution and greenhouse gas emissions. We know the co-benefits of taking these actions is to our advantage in economic terms, and we cannot afford to delay our response. City-level networks are now well established enabling strong climate action to provide healthier urban environments and more sustainable city futures. With an ever urbanising global population, and to ensure that we continue to develop healthy cities, such initiatives will increasingly need to be at the leading edge of effective collaboration to promote effective and measurable action to reduce the looming health crisis caused by the urban heat island effect, climate change and air pollution.

References

Aas W, Mortier A, Bowersox V, Cherian R, Faluvegi G, Fagerli H, Hand J, Klimont Z, Galy-Lacaux C, Lehmann CMB, Myhre CL, Myhre G, Olivié D, Sato K, Quaas J, Rao PSP, Schulz M, Shindell D, Skeie RB, Stein A, Takemura T, Tsyro S, Vet R, Xu X (2019) Global and regional trends of atmospheric sulfur. Sci Rep 9(1):953. https://doi.org/10.1038/s41598-018-37304-0

Abhijith KV, Kumar P, Gallagher J, McNabola A, Baldauf R, Pilla F, Broderick B, Di Sabatino S, Pulvirenti B (2017) Air pollution abatement performances of green infrastructure in open road and built-up street canyon environments—a review. Atmos Environ 162:71–86. https://doi.org/10.1016/j.atmosenv.2017.05.014

Apte JS, Brauer M, Cohen AJ, Ezzati M, Pope CA (2018) Ambient PM2.5 reduces global and regional life expectancy. Environ Sci Technol Lett 5(9):546–551. https://doi.org/10.1021/acs.estlett.8b00360

Atkinson RW, Kang S, Anderson HR, Mills IC, Walton HA (2014) Epidemiological time series studies of PM2.5 and daily mortality and hospital admissions: a systematic review and meta-analysis. Thorax 69(7):660–665. https://doi.org/10.1136/thoraxjnl-2013-204492

Atkinson RW, Butland BK, Dimitroulopoulou C, Heal MR, Stedman JR, Carslaw N, Jarvis D, Heaviside C, Vardoulakis S, Walton H, Anderson HR (2016) Long-term exposure to ambient ozone and mortality: a quantitative systematic review and meta-analysis of evidence from cohort studies. BMJ Open 6(2):e009493. https://doi.org/10.1136/bmjopen-2015-009493

Bai X, McPhearson T, Cleugh H, Nagendra H, Tong X, Zhu T, Zhu Y-G (2017) Linking urbanization and the environment: conceptual and empirical advances. Annu Rev Environ Resour 42(1):215–240. https://doi.org/10.1146/annurev-environ-102016-061128

Baklanov A, Molina LT, Gauss M (2016) Megacities, air quality and climate. Atmos Environ 126:235–249. https://doi.org/10.1016/j.atmosenv.2015.11.059

Bell ML, Peng RD, Dominici F (2006) The exposure-response curve for ozone and risk of mortality and the adequacy of current ozone regulations. Environ Health Perspect 114(4):532–536. https://doi.org/10.1289/ehp.8816

Bell ML, Zanobetti A, Dominici F (2014) Who is more affected by ozone pollution? A systematic review and meta-analysis. Am J Epidemiol 180(1):15–28. https://doi.org/10.1093/aje/kwu115

Black C, Tesfaigzi Y, Bassein JA, Miller LA (2017) Wildfire smoke exposure and human health: significant gaps in research for a growing public health issue. Environ Toxicol Pharmacol 55:186–195. https://doi.org/10.1016/j.etap.2017.08.022

Broome RA, Johnston FH, Horsley J, Morgan GG (2016) A rapid assessment of the impact of hazard reduction burning around Sydney, May 2016. Med J Aust 205(9):407–408. https://doi.org/10.5694/mja16.00895

Burnett R, Chen H, Szyszkowicz M, Fann N, Hubbell B, Pope CA, Apte JS, Brauer M, Cohen A, Weichenthal S, Coggins J, Di Q, Brunekreef B, Frostad J, Lim SS, Kan H, Walker KD, Thurston GD, Hayes RB, Lim CC, Turner MC, Jerrett M, Krewski D, Gapstur SM, Diver WR, Ostro B, Goldberg D, Crouse DL, Martin RV, Peters P, Pinault L, Tjepkema M, van Donkelaar A, Villeneuve PJ, Miller AB, Yin P, Zhou M, Wang L, Janssen NAH, Marra M, Atkinson RW, Tsang H, Quoc Thach T, Cannon JB, Allen RT, Hart JE, Laden F, Cesaroni G, Forastiere F, Weinmayr G, Jaensch A, Nagel G, Concin H, Spadaro JV (2018) Global estimates of mortality associated with long-term exposure to outdoor fine particulate matter. Proc Natl Acad Sci USA 115(38):9592–9597. https://doi.org/10.1073/pnas.1803222115

C40 (2018) 13 benefits of climate action. Piloting a global approach to measurement. Final report. C40 Cities. https://www.c40.org/benefits

Chossière GP, Malina R, Allroggen F, Eastham SD, Speth RL, Barrett SRH (2018) Country- and manufacturer-level attribution of air quality impacts due to excess NOx emissions from diesel passenger vehicles in Europe. Atmos Environ 189:89–97. https://doi.org/10.1016/j.atmosenv.2018.06.047

Cleugh H, Grimmond S (2012) Chapter 3—Urban climates and global climate change. In: Henderson-Sellers A, McGuffie K (eds) The future of the world's climate, 2nd edn. Elsevier, Boston, pp 47–76

COMEAP (2015) COMEAP: quantification of mortality and hospital admissions associated with ground-level ozone. Committee on the Medical Effects of Air Pollutants, London

Cowan T, Purich A, Perkins S, Pezza A, Boschat G, Sadler K (2014) More frequent, longer, and hotter heat waves for Australia in the twenty-first century. J Clim 27(15):5851–5871. https://doi.org/10.1175/jcli-d-14-00092.1

Cusworth DH, Mickley LJ, Sulprizio MP, Liu T, Marlier ME, DeFries RS, Guttikunda SK, Gupta P (2018) Quantifying the influence of agricultural fires in northwest India on urban air pollution in Delhi, India. Environ Res Lett 13(4):044018. https://doi.org/10.1088/1748-9326/aab303

Di Virgilio G, Hart MA, Jiang N (2018) Meteorological controls on atmospheric particulate pollution during hazard reduction burns. Atmos Chem Phys 18(9):6585–6599. https://doi.org/10.5194/acp-18-6585-2018

Di Virgilio G, Evans JP, Blake SAP, Armstrong M, Dowdy AJ, Sharples J, McRae R (2019) Climate change increases the potential for extreme wildfires. Geophys Res Lett 46(14):8517–8526. https://doi.org/10.1029/2019gl083699

Doherty RM, Heal MR, O'Connor FM (2017) Climate change impacts on human health over Europe through its effect on air quality. Environ Health 16(1):118. https://doi.org/10.1186/s12940-017-0325-2

Dowdy AJ (2017) Climatological variability of fire weather in Australia. J Appl Meteorol Climatol 57(2):221–234. https://doi.org/10.1175/jamc-d-17-0167.1

Ebi KL, McGregor G (2008) Climate change, tropospheric ozone and particulate matter, and health impacts. Environ Health Persp 116(11):1449–1455. https://doi.org/10.1289/ehp.11463

Fallmann J, Forkel R, Emeis S (2016) Secondary effects of urban heat island mitigation measures on air quality. Atmos Environ 125:199–211. https://doi.org/10.1016/j.atmosenv.2015.10.094

Fiore AM, Naik V, Spracklen DV, Steiner A, Unger N, Prather M, Bergmann D, Cameron-Smith PJ, Cionni I, Collins WJ (2012) Global air quality and climate. Chem Soc Rev 41(19):6663–6683. https://doi.org/10.1039/C2CS35095E

Fiore AM, Naik V, Leibensperger EM (2015) Air quality and climate connections. J Air Waste Manage Assoc 65(6):645–685. https://doi.org/10.1080/10962247.2015.1040526

Fu T-M, Tian H (2019) Climate change penalty to ozone air quality: review of current understandings and knowledge gaps. Curr Pollution Rep 5(3):159–171. https://doi.org/10.1007/s40726-019-00115-6

Glazener A, Khreis H (2019) Transforming our cities: best practices towards clean air and active transportation. Curr Environ Health Rep 6(1):22–37. https://doi.org/10.1007/s40572-019-0228-1

Goudie AS (2014) Desert dust and human health disorders. Environ Int 63:101–113. https://doi.org/10.1016/j.envint.2013.10.011

Gouldson A, Sudmant A, Khreis H, Papargyropoulou E (2018) The economic and social benefits of low-carbon cities: a systematic review of the evidence. Coalition for Urban Transitions, Washington, DC

Guttikunda SK, Goel R, Pant P (2014) Nature of air pollution, emission sources, and management in the Indian cities. Atmos Environ 95:501–510. https://doi.org/10.1016/j.atmosenv.2014.07.006

Haines A, Amann M, Borgford-Parnell N, Leonard S, Kuylenstierna J, Shindell D (2017) Short-lived climate pollutant mitigation and the sustainable development goals. Nat Clim Chang 7(12):863–869. https://doi.org/10.1038/s41558-017-0012-x

Hart MA, Sailor DJ (2009) Quantifying the influence of land-use and surface characteristics on spatial variability in the urban heat island. Theor Appl Climatol 95(3):397–406. https://doi.org/10.1007/s00704-008-0017-5

Hickman R, Bannister D (2014) Transport, climate change and the city. Routledge, London

Hoek G, Krishnan RM, Beelen R, Peters A, Ostro B, Brunekreef B, Kaufman JD (2013) Long-term air pollution exposure and cardio- respiratory mortality: a review. Environ Health 12(1):43. https://doi.org/10.1186/1476-069X-12-43

IPCC (2014) Climate change 2014: synthesis report. In: Contribution of working groups I, II and III to the fifth assessment report of the intergovernmental panel on climate change. Intergovernmental Panel on Climate Change, Geneva, Switzerland

Jacob DJ, Winner DA (2009) Effect of climate change on air quality. Atmos Environ 43(1):51–63. https://doi.org/10.1016/j.atmosenv.2008.09.051

Janhäll S (2015) Review on urban vegetation and particle air pollution—deposition and dispersion. Atmos Environ 105:130–137. https://doi.org/10.1016/j.atmosenv.2015.01.052

Jethva H, Torres O, Field RD, Lyapustin A, Gautam R, Kayetha V (2019) Connecting crop productivity, residue fires, and air quality over northern India. Sci Rep 9(1):16594. https://doi.org/10.1038/s41598-019-52799-x

Ji F, Evans JP, Di Luca A, Jiang N, Olson R, Fita L, Argüeso D, Chang LTC, Scorgie Y, Riley M (2019) Projected change in characteristics of near surface temperature inversions for southeast Australia. Clim Dynam 52(3):1487–1503. https://doi.org/10.1007/s00382-018-4214-3

Johnston FH (2017) Understanding and managing the health impacts of poor air quality from landscape fires. Med J Aust 207(6):229–230. https://doi.org/10.5694/mja17.00072

Karagulian F, Belis CA, Dora CFC, Prüss-Ustün AM, Bonjour S, Adair-Rohani H, Amann M (2015) Contributions to cities' ambient particulate matter (PM): a systematic review of local source contributions at global level. Atmos Environ 120:475–483. https://doi.org/10.1016/j.atmosenv.2015.08.087

Keywood M, Hibberd M, Emmerson K (2016) Australia state of the environment: atmosphere. Department of Environment and Energy, Canberra, Australia

King AD, Black MT, Min S-K, Fischer EM, Mitchell DM, Harrington LJ, Perkins-Kirkpatrick SE (2016) Emergence of heat extremes attributable to anthropogenic influences. Geophys Res Lett 43(7):3438–3443. https://doi.org/10.1002/2015gl067448

Kinney PL (2008) Climate change, air quality, and human health. Am J Prev Med 35(5):459–467. https://doi.org/10.1016/j.amepre.2008.08.025

Knorr W, Dentener F, Hantson S, Jiang L, Klimont Z, Arneth A (2016) Air quality impacts of European wildfire emissions in a changing climate. Atmos Chem Phys 16(9):5685–5703. https://doi.org/10.5194/acp-16-5685-2016

Krayenhoff ES, Moustaoui M, Broadbent AM, Gupta V, Georgescu M (2018) Diurnal interaction between urban expansion, climate change and adaptation in US cities. Nat Clim Chang 8(12):1097–1103. https://doi.org/10.1038/s41558-018-0320-9

Krzyzanowski M, Apte JS, Bonjour SP, Brauer M, Cohen AJ, Prüss-Ustun AM (2014) Air pollution in the mega-cities. Curr Environ Health Rep 1(3):185–191. https://doi.org/10.1007/s40572-014-0019-7

Lelieveld J, Evans JS, Fnais M, Giannadaki D, Pozzer A (2015) The contribution of outdoor air pollution sources to premature mortality on a global scale. Nature 525(7569):367–371. https://doi.org/10.1038/nature15371

Liu H, Ma W, Qian J, Cai J, Ye X, Li J, Wang X (2015) Effect of urbanization on the urban meteorology and air pollution in Hangzhou. J Meteorol Res 29(6):950–965. https://doi.org/10.1007/s13351-015-5013-y

Lou J, Wu Y, Liu P, Kota SH, Huang L (2019) Health effects of climate change through temperature and air pollution. Curr Pollut Rep 5(3):144–158. https://doi.org/10.1007/s40726-019-00112-9

Ma S, Pitman A, Yang J, Carouge C, Evans JP, Hart M, Green D (2018) Evaluating the effectiveness of mitigation options on heat stress for Sydney, Australia. J Appl Meteorol Climatol 57(2):209–220. https://doi.org/10.1175/jamc-d-17-0061.1

Marlier ME, Jina AS, Kinney PL, DeFries RS (2016) Extreme air pollution in global megacities. Curr Clim Change Rep 2(1):15–27. https://doi.org/10.1007/s40641-016-0032-z

McClure CD, Jaffe DA (2018) US particulate matter air quality improves except in wildfire-prone areas. PNAS 115(31):7901–7906. https://doi.org/10.1073/pnas.1804353115

McGregor GR (2012) Human biometeorology. Prog Phys Geogr 36(1):93–109. https://doi.org/10.1177/0309133311417942

Middleton NJ (2017) Desert dust hazards: a global review. Aeolian Res 24:53–63. https://doi.org/10.1016/j.aeolia.2016.12.001

Mudway IS, Dundas I, Wood HE, Marlin N, Jamaludin JB, Bremner SA, Cross L, Grieve A, Nanzer A, Barratt BM, Beevers S, Dajnak D, Fuller GW, Font A, Colligan G, Sheikh A, Walton R, Grigg J, Kelly FJ, Lee TH, Griffiths CJ (2019) Impact of London's low emission zone on air quality and children's respiratory health: a sequential annual cross-sectional study. Lancet Public Health 4(1):e28–e40. https://doi.org/10.1016/S2468-2667(18)30202-0

Munson SM, Belnap J, Okin GS (2011) Responses of wind erosion to climate-induced vegetation changes on the Colorado Plateau. PNAS 108(10):3854–3859. https://doi.org/10.1073/pnas.1014947108

Perkins-Kirkpatrick SE, Gibson PB (2017) Changes in regional heatwave characteristics as a function of increasing global temperature. Sci Rep 7(1):12256. https://doi.org/10.1038/s41598-017-12520-2

Robine J-M, Cheung SLK, Le Roy S, Van Oyen H, Griffiths C, Michel J-P, Herrmann FR (2008) Death toll exceeded 70,000 in Europe during the summer of 2003. C R Biol 331(2):171–178. https://doi.org/10.1016/j.crvi.2007.12.001

Rückerl R, Schneider A, Breitner S, Cyrys J, Peters A (2011) Health effects of particulate air pollution: a review of epidemiological evidence. Inhal Toxicol 23(10):555–592. https://doi.org/10.3109/08958378.2011.593587

Sacks JD, Stanek LW, Luben TJ, Johns DO, Buckley BJ, Brown JS, Ross M (2011) Particulate matter-induced health effects: who is susceptible? Environ Health Perspect 119(4):446–454. https://doi.org/10.1289/ehp.1002255

Salmond JA, Williams DE, Laing G, Kingham S, Dirks K, Longley I, Henshaw GS (2013) The influence of vegetation on the horizontal and vertical distribution of pollutants in a street canyon. Sci Total Environ 443:287–298. https://doi.org/10.1016/j.scitotenv.2012.10.101

Samson R, Grote R, Calfapietra C, Cariñanos P, Fares S, Paoletti E, Tiwary A (2017) Urban trees and their relation to air pollution. In: Pearlmutter D, Calfapietra C, Samson R et al (eds) The urban forest: cultivating green infrastructure for people and the environment, Future City, vol 7. Springer International Publishing, Cham, pp 21–30. https://doi.org/10.1007/978-3-319-50280-9_3

Schmidt S (2019) Brain fog: does air pollution make us less productive? Environ Health Perspect 127(5):052001. https://doi.org/10.1289/EHP4869

Schraufnagel DE, Balmes JR, Cowl CT, De Matteis S, Jung S-H, Mortimer K, Perez-Padilla R, Rice MB, Riojas-Rodriguez H, Sood A, Thurston GD, To T, Vanker A, Wuebbles DJ (2019a) Air pollution and noncommunicable diseases: a review by the forum of international respiratory societies' environmental committee, part 2: air pollution and organ systems. Chest 155(2):417–426. https://doi.org/10.1016/j.chest.2018.10.041

Schraufnagel DE, Balmes JR, Cowl CT et al (2019b) Air pollution and noncommunicable diseases: a review by the forum of international respiratory societies' environmental committee, part 1: the damaging effects of air pollution. Chest 155:409–416. https://doi.org/10.1016/j.chest.2018.10.042

Scovronick N, Budolfson M, Dennig F, Errickson F, Fleurbaey M, Peng W, Socolow RH, Spears D, Wagner F (2019) The impact of human health co-benefits on evaluations of global climate policy. Nat Commun 10(1):2095. https://doi.org/10.1038/s41467-019-09499-x

Shrubsole C, Dimitroulopoulou S, Foxall K, Gadeberg B, Doutsi A (2019) IAQ guidelines for selected volatile organic compounds (VOCs) in the UK. Build Environ 165:106382. https://doi.org/10.1016/j.buildenv.2019.106382

Slovic AD, de Oliveira MA, Biehl J, Ribeiro H (2016) How can urban policies improve air quality and help mitigate global climate change: a systematic mapping review. J Urban Health 93(1):73–95. https://doi.org/10.1007/s11524-015-0007-8

State of Global Air (2019) State of global air—2019. Health Effects Institute, Boston, MA

Stedman JR (2004) The predicted number of air pollution related deaths in the UK during the August 2003 heatwave. Atmos Environ 38(8):1087–1090. https://doi.org/10.1016/j.atmosenv.2003.11.011

Takaro TK, Knowlton K, Balmes JR (2013) Climate change and respiratory health: current evidence and knowledge gaps. Expert Rev Respir Med 7(4):349–361. https://doi.org/10.1586/17476348.2013.814367

Thorpe A, Hart M (2013) Changing climates, changing cities? Planning reform and urban sustainability in New South Wales. Australas J Nat Resour Law Policy 16:133–156

UN (2018) 68% of the world population projected to live in urban areas by 2050, says UN. UN DESA, United Nations Department of Economic and Social Affairs. https://www.un.org/development/desa/en/news/population/2018-revision-of-world-urbanization-prospects.html. Accessed 16 Nov 2019

UN Environment and International Energy Agency (2017) Towards a zero-emission, efficient, and resilient buildings and construction sector: Global Status Report 2017. United Nations Environment Programme, Nairobi, Kenya

UNEP WMO (2011) Integrated assessment of black carbon and tropospheric ozone. United Nations Environment Programme and World Meteorological Organization, Nairobi, Kenya

UN-HABITAT (2011) Global reports on human settlements. United Nations Human Settlements Programme, Nairobi, Kenya

USGCRP (2016) The impacts of climate change on human health in the United States: a scientific assessment. U.S. Global Change Research Program, Washington, DC

Vanos JK (2015) Children's health and vulnerability in outdoor microclimates: a comprehensive review. Environ Int 76:1–15. https://doi.org/10.1016/j.envint.2014.11.016

Voogt JA (2002) Urban heat island. In: Douglas I (ed) Causes and consequences of global environmental change. Wiley, Chichester, pp 660–666

Wang T, Kwok JYH (2003) Measurement and analysis of a multiday photochemical smog episode in the pearl river delta of China. J Appl Meteorol 42(3):404–416. https://doi.org/10.1175/1520-0450(2003)042<0404:Maaoam>2.0.Co;2

Wang T, Zhao B, Liou K-N, Gu Y, Jiang Z, Song K, Su H, Jerrett M, Zhu Y (2019) Mortality burdens in California due to air pollution attributable to local and nonlocal emissions. Environ Int 133:105232. https://doi.org/10.1016/j.envint.2019.105232
Washington R, Bouet C, Cautenet G, Mackenzie E, Ashpole I, Engelstaedter S, Lizcano G, Henderson GM, Schepanski K, Tegen I (2009) Dust as a tipping element: the Bodélé Depression, Chad. Proc Natl Acad Sci 106(49):20564–20571. https://doi.org/10.1073/pnas.0711850106
Watts N, Amann M, Arnell N, Ayeb-Karlsson S, Belesova K, Boykoff M, Byass P, Cai W, Campbell-Lendrum D, Capstick S (2019) The 2019 report of The Lancet Countdown on health and climate change: ensuring that the health of a child born today is not defined by a changing climate. Lancet 394(10211):1836–1878. https://doi.org/10.1016/S0140-6736(19)32596-6
West JJ, Szopa S, Hauglustaine DA (2007) Human mortality effects of future concentrations of tropospheric ozone. Compt Rendus Geosci 339(11):775–783. https://doi.org/10.1016/j.crte.2007.08.005
West JJ, Smith SJ, Silva RA, Naik V, Zhang Y, Adelman Z, Fry MM, Anenberg S, Horowitz LW, Lamarque J-F (2013) Co-benefits of mitigating global greenhouse gas emissions for future air quality and human health. Nat Clim Chang 3(10):885–889. https://doi.org/10.1038/nclimate2009
WHO (2016) Ambient air pollution: a global assessment of exposure and burden of disease. World Health Organization, Geneva, Switzerland
WHO (2018a) Air pollution. https://www.who.int/news-room/fact-sheets/detail/ambient-(outdoor)-air-quality-and-health. Accessed 24 Oct 2019
WHO (2018b) Household air pollution and health. https://www.who.int/news-room/fact-sheets/detail/household-air-pollution-and-health. Accessed 24 Oct 2019
WHO Regional Office for Europe (2013) Health and the environment in the WHO European region: creating resilient communities and supportive environments. World Health Organization Regional Office for Europe, Copenhagen, Denmark
WHO Regional Office for Europe (2016) Economic cost of the health impact of air pollution in Europe: clean air, health and wealth. WHO Regional Office for Europe, Copenhagen, Denmark
Wong MS, Nichol J, Ng E (2011) A study of the "wall effect" caused by proliferation of high-rise buildings using GIS techniques. Landsc Urban Plan 102(4):245–253. https://doi.org/10.1016/j.landurbplan.2011.05.003
Wong PPY, Lai P-C, Allen R, Cheng W, Lee M, Tsui A, Tang R, Thach T-Q, Tian L, Brauer M, Barratt B (2019) Vertical monitoring of traffic-related air pollution (TRAP) in urban street canyons of Hong Kong. Sci Total Environ 670:696–703. https://doi.org/10.1016/j.scitotenv.2019.03.224
World Bank (2016) The cost of air pollution: strengthening the economic case for action. The World Bank and Institute for Health Metrics and Evaluation, University of Washington, Seattle

Chapter 17
International Research Collaboration and Programmes for Urban Climate Science for Planning Healthy Cities

Boram Lee, Hunter Jones, and Chao Ren

Abstract Science has shown how climate change fundamentally influences cities—their infrastructure and inhabitants—and how cities may plan for the future. Effective and dedicated support should continue for the international scientific community to improve our knowledge through comprehensive measurements, transdisciplinary science, and quantified attribution, because these have enabled the society to link empirical verification of human health outcomes to corresponding actions and policies.

This chapter informs good practices of the present time for research partnerships, including the needs-driven research, collaborative networks, and effort for co-development of decision-supporting knowledge. Highlighting important findings in urban climate and health sciences, key research priorities, challenges and issues are discussed.

A holistic science-based service for urban health is essential to empower decision-making for sustainable urban life. To this end, there is a critical need for state-of-the-art climate information, tailored for urban areas and addressing their specific needs. It requires to deepen our understanding of, and to predict in all spatial and temporal scales and in resolutions fit to the purpose of decisions, complex climate processes and their impact to health. International effort should be extended for open-access and continuous observational framework for collecting key climate, health and other socio-economic metrics at the city scale.

A core challenge facing decision-makers is identifying and prioritizing climate change interventions in urban health under countervailing influences; therefore

B. Lee (✉)
World Bank, USA, Washington, DC
e-mail: blee10@worldbank.org

H. Jones
Climate Program Office, National Oceanic and Atmospheric Administration, Washington, DC, USA
e-mail: hunter.jones@noaa.gov

C. Ren
Faculty of Architecture, The University of Hong Kong, Pok Fu Lam, Hong Kong
e-mail: renchao@hku.hk

C. Ren, G. McGregor (eds.), *Urban Climate Science for Planning Healthy Cities*, Biometeorology 5, https://doi.org/10.1007/978-3-030-87598-5_17

multi-disciplinary collaboration should extend the understanding and provide actionable knowledge on health co-benefits in the multi-dimensional and cross-sectoral ground. Efforts are to be furthered to improve integrating probabilistic information into operations. Improved interoperability between health and climate information systems will ensure a better understanding of the sensitivity of health to climate factors, as well as improved use of limited resources.

Keywords Climate · Health · Urbanization · Co-benefits · Climate-resilient · Decision-maker · Research · Global agenda · Knowledge · Impacts · Framework · Challenges

Key Messages

Efforts to improve our knowledge and to develop holistic science-based services for urban health continue, through research partnerships and collaboration from global to city scales. Effective and dedicated support for transdisciplinary integrated research and activities to translate research into practice in support of targeted decision-making should be expanded and sustained.

- There is a critical need to improve understanding and prediction of climate processes and their impact on all scales. In particularly expanded research to understand how urban scale interventions to manage climate impacts on health may force or feedback to the urban and global climate systems, creating side effects and/or co-benefits. International effort should be extended for open-access and continuous observational frameworks for collecting key metrics at the city scale.
- For information services to decision-making, future challenges remain in integrating probabilistic information into operations, accounting health co--benefits in the multi-dimensional ground, and assuring interoperability between health and climate information systems.

Our Earth's future is increasingly of our own creation. Humans have expanded their impact on landscapes to a point where only 5% of the land surface of the Earth remains unperturbed while simultaneously increasing the complexity of our impacts through interconnectedness of our economies, cultures, and transportation systems through globalization (Kennedy et al. 2019). Furthermore, though the footprint of our impact on the Earth has expanded, our living spaces have densified through urbanization. The proportion of the world's population living in urban areas is currently at 55% and on track to reach 68% by 2050 (UN DESA 2018).

17.1 Scientific Finding: Health Co-benefits

The combination of a changing climate and urbanization increases the vulnerability of our population, underscoring the criticality of understanding relationships between human health and the environment. The international scientific community strives to improve our knowledge on how cities can cope with a changing climate. Increasingly robust climate science literature and synthesis reports form the basis for understanding, mitigating, and adapting to human-induced climate change. Scientific evidence is mounting that adaptive and mitigative measures support a panoply of co-benefits in the urban environment, and that low-carbon, climate-resilient development leads the society in a more sustainable and equitable manner. For example, health co-benefits include augmented social capital driving decreases in heat illness and other health benefits, while improvements to urban design such as increased walkability contribute to improved physical and mental health, and transitions to clean energy reduce respiratory disease (Bai et al. 2012; Cheng and Berry 2013). It has been proven that preserving access to green spaces in urban areas can have health benefits, not only by increased physical activity and mental health but also by reducing the urban heat island effect during the heatwaves (World Health Organization 2017). The IPCC report indicates that maintaining warming below 1.5 °C could be achieved in tandem with poverty alleviation, improving energy security and health benefits, which, furthermore, could be greater than the costs of mitigation costs (Watts et al. 2018). A recent study found that, in all scenarios, the health benefits of meeting climate goals substantially outweighed the costs of action, particularly in China and India (Markandya et al. 2018).

17.2 Existing Good Practice

There exist established methods for assessing health vulnerability and for defining options for adaptation, integrating policy goals for different societal sectors. As an effort towards such integrated approaches the Urban Health Initiative, launched by the Climate and Clean Air Coalition, promotes assessment tools to building cities in which good health is enabled and encouraged, with a focus on climate change, short-lived climate pollutants and air quality. Decision-makers are to assess the full impact of air pollution and existing urban policies, mapping the health impacts of sectoral emissions (from transport, land use, energy and housing) in different scenarios and calculating the health cost and benefits (World Health Organization 2018).

One of the main challenges in planning healthy cities is found in the gap between health service and climate service providers. The former often lack access to, understanding of and capacity to interpret and apply information on climate, and climate service providers often do not fully recognize the interconnected health impacts of climate change and the role they could play in ensuring more climate-resilient health systems (GFCS 2018). Interoperability between health and climate information systems will ensure better understanding of the sensitivity of health to climate factors, so that limited resources can be better targeted geographically and temporally. The Global Framework for Climate Services (GFCS) of the World Meteorological Organization (WMO; https://gfcs.wmo.int/health) works to improve access to and use of information on climate in climate-vulnerable developing countries, with health as one of the priorities.

Global networks specifically targeted to address urgent impacts are also increasingly important. One such network, the Global Heat Health Information Network (GHHIN 2019), was formed in 2018 to unite the diverse set of practitioners, researchers, and policymakers who are working to protect people from the increasing risk of extreme heat and heat waves across national and organizational boundaries. This network focuses on building capacity through knowledge sharing and expert exchanges, identifying and sharing best practices, identifying and increasing focus on research and operational gaps—such as coverage of Heat Health Action Plans across the globe, and convening bi-annual Forums on Heat and Health to bring network participants together for shared learning and agenda setting. The inaugural Forum was held in Hong Kong in December of 2018, and brought together over 120 participants from 33 countries. This forum concluded with a Call to Action which specifically laid out the actions the network will take in the coming years to address heat health risks in all nations. The forum provided a platform for a broad exchange of research and good practices not only from Hong Kong, but also allowed for the unique expertise and local experiences to be shared globally to form new cross-disciplinary partnerships and collaboration. The biennial Heat Health Forum has been set up to help build community and accelerate the transformation of knowledge to action that can help society better adapt to a warmer world with a joint effort. The second forum will be held in Copenhagen and focus on uncovering issues and shaping high-impact opportunities to protect workers and cities from the health impacts of extreme heat.

The science underlying the complexity of linkages, which consist of core elements of such methods, is yet to be further developed. The aforementioned co-benefits are not without their caveats and health risks. A core challenge facing decision-makers is identifying and prioritizing climate change interventions in specific contexts, such as urban health and well-being, under countervailing influences such as diminished budgets, resistance to change, and unintended side effects. For example, we understand that extreme temperature events are often coupled with high air pollution, and that this can have a compounded effect on respiratory health (Shaposhnikov et al. 2014). However, interventions proposed to mitigate the urban heat island effect, such as planting trees and increasing rooftop albedo, have also been shown to increase allergen counts if species is not carefully considered and to decrease vertical mixing in the urban boundary layer which can reduce air quality (Fallmann et al. 2016; Willis and Petrokofsky 2017).

As such, efforts to deepen our understanding through comprehensive measurements, transdisciplinary science, and quantified attribution in order to link empirical verification of human health outcomes to adaptation measures and corresponding policies should be continued. The international scientific community should also be supported in effecting collaborative efforts to identify key research priorities and to develop science-based integrated urban services supporting safe, healthy and resilient cities.

17.3 Global Frameworks for Research and Actions on Cities and Climate Science

Awareness of the challenges and opportunities posed by increased urbanization has grown across the breadth of societal areas. The international policy communityInternational policy community has renewed the call to build and mobilize science and research in support of the implementation and assessment of global agendas such as the Paris Agreement and the New Urban Agenda (NUA), Aichi Targets, Sendai framework for Disaster Risk Reduction, as well as the Sustainable Development Goals (SDGs). Continuing urbanization makes cities important foci of actions to promote health and protect the climate; city authorities and local governments are critical actors to increase the reach.

A latest synthesis of the research gaps, in view of stimulating the next frontier of research in the IPCC's Seventh Assessment cycle, came out of the dedicated dialogues made at the International Conference on Climate Change and Cities (March 2018, Edmonton, Canada) (UN Habitat 2018). This Global Research and Action Agenda on Cities and Climate Change Science (World Climate Research Programme 2019), discussed at the 48th Plenary of the Intergovernmental Panel on Climate Change (IPCC; October 2018, Sondo, Republic of Korea), describes cross-cutting issues and knowledge gaps; key topical research areas; and suggested approaches for implementation (see Fig. 17.1). The organizing partners[1] and global scientific community emphasized that new knowledge is needed to increase our understanding of the interplay between policies and actions taken at different scales, and how this affects the ability to take effective and coordinated climate action at the city scale (See also Bai et al. 2018). Multi-dimensional urban planning is a crucial tool for addressing climate change adaptation and mitigation in cities, bringing together a breadth of societal contexts, and in doing so further research is needed on how to maximize their potential to improve the health and well-being of urban residents and passively modulate the urban micro-climate.

This figure presents the structure of the Global Research and Action Agenda on Cities and Climate Change Science. The inner circle (orange) presents key crosscutting issues and knowledge gaps for a step-change of knowledge generation on cities and climate change. The middle circle (multi-coloured) presents six topical research

[1] Cities Alliance, City of Edmonton, C40 Cities Leadership Group, Future Earth, ICLEI-Local Governments for Sustainability, IPCC, Sustainable Development Solutions Network (SDSN), United Cities and Local Governments (UCLG), United Nations Environment Programme, United Nations Human Settlements Programme (UN-Habitat), and the World Climate Research Programme (WCRP).

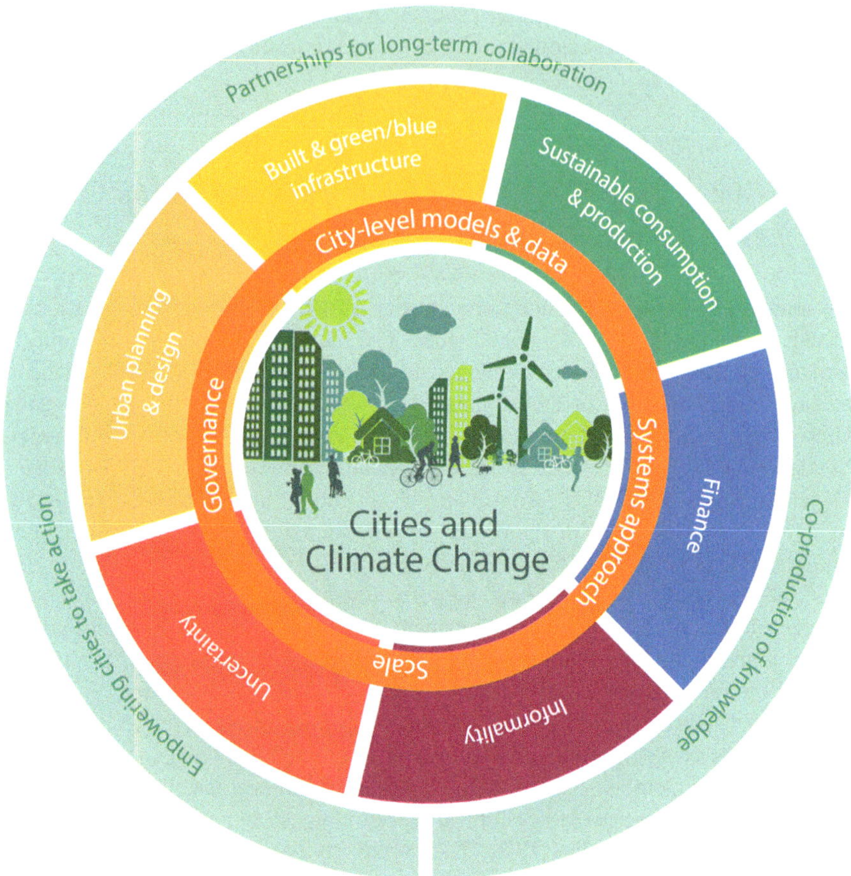

Fig. 17.1 Pathways for climate adaptation and mitigation in cities (WCRP 2019)

areas where more evidence is needed to inform action. The external circle (green) presents three suggested approaches that may facilitate implementation of this Research and Action Agenda.

17.4 International Coordination on Climate Science: WCRP/CMIP

The foundation of science-based services and solutions for urban health impacts is to understanding complex climate processes as well as the complexities of the decisions of users of climate and health information. Often, to respond to this task, strong coordination and cooperation in a global scale is required. The World Climate Research Programme (WCRP 2016) coordinates and facilitates international climate research related to the coupled physical climate system that are too large and too complex to be tackled by a single nation, agency or scientific discipline, and promotes sharing and application of the climate knowledge that contributes to societal well-being. WCRP has been fostering global coordination of downscaling efforts and helping to clarify uncertainties of urban climate, in support of credible research leading towards reliable climate and environmental services for the urban community.

Climate science has achieved remarkable improvements in global- and regional-scale models in their ability to simulate and predict climate variables such as temperature and precipitation at different time scales, though challenges remain. The models have begun to be used to issue predictions operationally, for example, from seasonal to subseasonal scales (Jones 2018), while the decadal models which may be useful for urban planning have yet to be operationalized. Longer-term climate model information is available from WCRP's Coupled Model Intercomparison Project (CMIP) and many derivative products have been developed to be easier to access and use, enabling broad projections on how climate change will affect human health. These are essential to shape policy priorities as well as to cast shorter-term decisions in the context of underlying trends to ensure resilience. However, policy- and decision-makers continue to experience the problem of integrating this probabilistic information into budgets and other aspects of operations. The main sources of uncertainty include: the global response to greenhouse gas emissions driving climate change, understanding impacts on society—including public health, economics, and migration—stemming from climate change and our actions to mitigate and adapt to it, and a thorough understanding of the context of the decisions made by practitioners working to address these issues—where they have leverage to implement novel climate & health insights from improved information.

To comprehensively achieve science-based solutions, urban and health policies require not only global to regional climate projections focused on the mid- or long-term, but also urban-scale information on the effects of climate change at long and short timescales. The foundation of such urban-scale climate information is our current high-resolution urban downscaling expertise, combined with local-impact models, and ensemble dynamical and statistical downscaling. This combination enables the propagation of climate change and uncertainty from the global to the city scale. The WCRP Coordinated Regional Climate Downscaling Experiment (WCRP Coordinated Regional Climate 2020) provides a framework for evaluating and comparing various regional climate downscaling techniques, enhancing regional capabilities for climate prediction. Collaborative and integrative efforts are made also to bridge the temporal gap in prediction; WCRP and the World Weather Research Programme (WWRP) set a collaborative framework for the Subseasonal-to-Seasonal (S2S) Prediction Project (http://s2sprediction.net/) to improve forecast skill and understanding on the subseasonal to seasonal timescale with special emphasis on high-impact weather events.

The precise impacts in specific places are difficult to predict accurately. Further research is required to provide better information to policy- and decision-makers so that they can design effective policies.

Challenges arise with the improvement in integrated very high-resolution modelling, considering the spatial and temporal scales and relevant physical processes, tailored to urban environments. Furthermore, the all-scale output should be made available to urban stakeholders to extract prediction products and information for decision-making processes in cities. In recognition of both the breadth and merit of city's needs that arise from the combination of multiple impacts caused by various climate factors, a critical question is whether climate change projections opted for specific operational decisions would correspond to the purpose and scales of cause/impact. In this context, an ongoing challenge for WCRP is to strengthen its direct contribution as a helpful information resource for urban issues. Efforts are to be furthered for collaboration among climate research communities, city networks and all other relevant urban stakeholders including the frameworks for urban-focused activities in the UN system.

Also, actions and effectiveness of adaptation and mitigation actions at the local (city/neighbourhood) scale are influenced by decisions made at other scales (e.g. provincial/state, national, global) and this is yet to be fully understood.

17.5 Challenges

There is a critical need for international, open-access and continuous observational framework for collecting key climate, health and other socio-economic metrics at the city scale (Ching et al. 2018). The Global Climate Observing System Terrestrial Observation Panel for Climate (GCOS-TOPC) is responsible for identifying and facilitating the monitoring of Essential Climate Variables (ECVs) that feed into IPCC/UNFCCC climate assessments. A recent effort is to review and develop Essential Climate Variables (ECVs) that can represent human adaptation to climate change in cities. The provision of not only physical characteristics of the urban environment but also biological and socio-economic ECV/indicators will feed directly into local and global climate change policy that covers health sectors, for example, through monitoring urban environmental change/adaptation progress through time and possibly against targets. For this, a causative link between chosen ECVs (indicators) and the adaptation action will be required.

Additionally, state-of-the-art climate information, tailored for urban areas and addressing their specific needs, should be developed through high resolution climate models across all relevant temporal and spatial scales.

Realizing a holistic science-based solution for urban health, integration and cross-application are keys to empower decision-making and associated planning. A highly interdisciplinary approach is required—cross-cutting coordination, integration and collaboration—and it must be applied in all stages of research, planning and implementation. The global scientific community reaches out to policy, business, and civil society to convene an engaged research base, focused on the co-design and co-production of knowledge and tools to provide solutions to the grand challenges of the society such as cities and human health. The Knowledge Action

Networks (KANs: http://futureearth.org/knowledge-action-networks)—coordinated by the Future Earth in collaboration with many other global and regional research programmes such as WCRP, IRDR and Urban Health and Wellbeing (UHW)—takes a complex systems approach to integrate diverse knowledge; to encourage research for solutions; and to engage effectively with multiple stakeholders to increase the likelihood of sustainable implementation of new initiatives. The Urban KAN endeavours to work closely with other KANs on Health, extremes and risk management, etc., to reflect the complexity and needs for cross-coordination to address the breadth of urban problems between variables like climate and environmental changes, health, food and transportation.

References

Bai X, Nath I, Capon A, Hasan N, Jaron D (2012) Health and wellbeing in the changing urban environment: complex challenges, scientific responses, and the way forward. Curr Opin Environ Sustain 4(4):465–472. https://doi.org/10.1016/j.cosust.2012.09.009

Bai X, Dawson RJ, Ürge-Vorsatz D, Delgado GC, Barau AS, Dhakal S, Dodman D, Leonardsen L, Masson-Delmotte V, Roberts DC (2018) Six research priorities for cities and climate change. Nature 555:23–25. https://doi.org/10.1038/d41586-018-02409-z

Cheng JJ, Berry P (2013) Health co-benefits and risks of public health adaptation strategies to climate change: a review of current literature. Int J Public Health 58(2):305–311. https://doi.org/10.1007/s00038-012-0422-5

Ching J, Mills G, Bechtel B, See L, Feddema J, Wang X, Ren C, Brousse O, Martilli A, Neophytou M, Mouzourides P, Stewart I, Hanna A, Ng E, Foley M, Alexander P, Aliaga D, Niyogi D, Shreevastava A, Bhalachandran P, Masson V, Hidalgo J, Fung J, Andrade M, Baklanov A, Dai W, Milcinski G, Demuzere M, Brunsell N, Pesaresi M, Miao S, Mu Q, Chen F, Theeuwes N (2018) WUDAPT: an urban weather, climate, and environmental modeling infrastructure for the Anthropocene. Bull Am Meteorol Soc 99(9):1907–1924. https://doi.org/10.1175/bams-d-16-0236.1

Fallmann J, Forkel R, Emeis S (2016) Secondary effects of urban heat island mitigation measures on air quality. Atmos Environ 125:199–211. https://doi.org/10.1016/j.atmosenv.2015.10.094

GFCS (2018) Global framework for climate services, World Meteorological Organization https://www.wmo.int/gfcs. Accessed Nov 2019

GHHIN (2019) Global heat health information network. https://www.ghhin.org. Accessed March 2020

Gutowski WJ, Giorgi F, Timbal B, Frigon A, Jacob D, Kang HS, Raghavan K, Lee B, Lennard C, Nikulin G, O'Rourke E, Rixen M, Solman S, Stephenson T, Tangang F (2016) WCRP coordinated regional downscaling experiment (CORDEX): a diagnostic MIP for CMIP6. Geoscientific Model Development, 9(11), 4087–4095.

Jones HM (2018) Climate change and increasing risk of extreme heat. In: Hosokawa Y (ed) Human health and physical activity during heat exposure. Springer International Publishing, Cham, pp 1–13. https://doi.org/10.1007/978-3-319-75889-3_1

Kennedy CM, Oakleaf JR, Theobald DM, Baruch-Mordo S, Kiesecker J (2019) Managing the middle: a shift in conservation priorities based on the global human modification gradient. Glob Chang Biol 25(3):811–826. https://doi.org/10.1111/gcb.14549

Markandya A, Sampedro J, Smith SJ, Van Dingenen R, Pizarro-Irizar C, Arto I, González-Eguino M (2018) Health co-benefits from air pollution and mitigation costs of the Paris Agreement: a modelling study. Lancet Planet Health 2(3):e126–e133. https://doi.org/10.1016/S2542-5196(18)30029-9

Shaposhnikov D, Revich B, Bellander T, Bedada GB, Bottai M, Kharkova T, Kvasha E, Lezina E, Lind T, Semutnikova E, Pershagen G (2014) Mortality related to air pollution with the Moscow heat wave and wildfire of 2010. Epidemiology 25(3):359–364. https://doi.org/10.1097/EDE.0000000000000090

UN Habitat (2018) International conference on climate change and cities 2018, Edmonton, Canada, 5–7 March 2018

United Nations Department of Economic and Social Affairs (2018) 2018 revision of world urbanization prospects, 16 May 2018. https://www.un.org/development/desa/publications/2018-revision-of-world-urbanization-prospects.html

Watts N, Amann M, Ayeb-Karlsson S, Belesova K, Bouley T, Boykoff M, Byass P, Cai W, Campbell-Lendrum D, Chambers J, Cox PM, Daly M, Dasandi N, Davies M, Depledge M, Depoux A, Dominguez-Salas P, Drummond P, Ekins P, Flahault A, Frumkin H, Georgeson L, Ghanei M, Grace D, Graham H, Grojsman R, Haines A, Hamilton I, Hartinger S, Johnson A, Kelman I, Kiesewetter G, Kniveton D, Liang L, Lott M, Lowe R, Mace G, Odhiambo Sewe M, Maslin M, Mikhaylov S, Milner J, Latifi AM, Moradi-Lakeh M, Morrissey K, Murray K, Neville T, Nilsson M, Oreszczyn T, Owfi F, Pencheon D, Pye S, Rabbaniha M, Robinson E, Rocklöv J, Schütte S, Shumake-Guillemot J, Steinbach R, Tabatabaei M, Wheeler N, Wilkinson P, Gong P, Montgomery H, Costello A (2018) The Lancet Countdown on health and climate change: from 25 years of inaction to a global transformation for public health. Lancet 391(10120):581–630. https://doi.org/10.1016/S0140-6736(17)32464-9

Willis KJ, Petrokofsky G (2017) The natural capital of city trees. Science 356(6336):374–376. https://doi.org/10.1126/science.aam9724

World Climate Research Programme (WCRP) (2016) World climate research programme, World Meteorological Organization. http://www.wcrp-climate.org/. Accessed Nov 2019

World Climate Research Programme (WCRP) (2019) Global research and action agenda on cities and climate change science—full version. Prieur-Richard AH, Walsh B, Craig M, Melamed ML, Colbert M, Pathak M, Connors S, Bai X, Barau A, Bulkeley H, Cleugh H, Cohen M, Colenbrander S, Dodman D, Dhakal S, Dawson R, Espey J, Greenwalt J, Kurian P, Lee B, Leonardsen L, Masson-Delmotte V, Munshi D, Okem A, Delgado Ramos GC, Sanchez Rodriguez R, Roberts D, Rosenzweig C, Schultz S, Seto K, Solecki W, van Staden M, Ürge-Vorsatz D (eds) WCRP Publication No. 13/2019. https://www.wcrp-climate.org/WCRP-publications/2019/GRAA-Cities-and-Climate-Change-Science-Full.pdf

World Health Organization (2017) Urban green space interventions and health: a review of impacts and effectiveness. WHO Regional Office for Europe, Copenhagen

World Health Organization (2018) Urban health initiative. WHO, Geneva

Chapter 18
Conclusions and Moving Forward

Glenn McGregor and Chao Ren

Abstract *Urban Climate Science for Planning Healthy Cities* argues for the recognition of climate as a key element of healthy cities. In doing so, it adds to a legacy of encouragement by urban climatologists and enlightened planners for the inclusion of climate information in decision-making processes related to the planning of cities. Although the call for 'urban climate science into urban planning action' has historical precedents, the case advanced in this book is more than timely, especially in the face of a changing climate (*the climate crisis*), the impacts of which will be felt acutely in the mega-cities of now and the future and, more generally, in a rapidly urbanising world (Wang et al., Urban climate science for planning healthy cities, Springer, London, 2021). In essence, the key idea advanced in this book is, if the attainment of healthy cities from a climate perspective is to be more than a policy aspiration, action towards climate-related health risk management in cities, as can be achieved via climate-sensitive urban planning and design, is imperative. This final chapter draws out some of the common threads running through the material comprising the book's 16 substantive chapters. Further, it takes the opportunity to look to the future and contemplate some of the issues that might form the basis of an evolving urban climate–urban planning research and action agenda.

Keywords Healthy cities · Climate-related health risk management · Climate-sensitive planning and design

G. McGregor (✉)
Department of Geography, Durham University, Durham, UK
e-mail: glenn.mcgregor@durham.ac.uk

C. Ren
Faculty of Architecture, The University of Hong Kong, Pok Fu Lam, Hong Kong
e-mail: renchao@hku.hk

C. Ren, G. McGregor (eds.), *Urban Climate Science for Planning Healthy Cities*, Biometeorology 5, https://doi.org/10.1007/978-3-030-87598-5_18

18.1 Urban Climate Science and Urban Planning and Design: The Relationship and Cross-Disciplinary Collaboration

The pursuit of a thermally comfortable and healthy environment through architecture and urban planning dates back to ancient times. The call for the application of urban climate knowledge and thus the promotion of climate-sensitive urban design and planning, to mitigate the climate vagaries of urban development, is not new. Ever since the discovery of the urban climate effect by Luke Howard in the early 1800s, the fact that cities unintentionally alter the climate of the region in which they lie has begged the question: 'Through intentional alteration of urban configuration and fabric, is it possible to manage urban climate so that a range of benefits arise?' In other words, through active use of climate information in the development of cities, can an apparent urban climate 'disadvantage' be turned to an urban 'advantage', through applying urban planning and design to improve the physical climate environment? By outlining the strategies some cities are piloting to mitigate the unintended climate consequences of urbanisation and describing associated plans for monitoring, modelling and improving urban climate, the material presented in this volume has demonstrated that the potential answer to this question is an unequivocal yes, recognising scale dependency and possible governance constraints moderated by political will. That said, it is important to be realistic. Sadly, urban climate science still has a low impact on the urban planning process in practice. Accordingly, cross-disciplinary/cross-agency collaboration in the development and implementation of climate science-informed urban planning and design needs to be encouraged at all levels of governance (Wang et al. 2021). As revealed by past extreme climate events and the 2020–2021 COVID-19 pandemic, a large chasm exists between high-income and low- to middle-income countries in relation to their ability to respond to pressing environmental health problems. The same holds true concerning progress towards mainstreaming urban climate science into city planning. While the situation is far from where it could be in high-income countries, in many low- to middle-income countries the 'urban climate science into urban planning action' conversation may not have even taken place. This may be because of a range of impediments, not least those associated with the fundamental planning challenges (e.g. provision of services) arising from unprecedented levels of unplanned/uncontrolled urbanisation. In such situations, consideration of the generation of urban climate information and its eventual inclusion in urban development decision-making may well be beyond a luxury, notwithstanding other constrains such as basic human capacity, financial resources and governance issues.

18.2 The Urban Planning: Urban Climate Imperative

There are a number of 'knowns' about the two-way interaction between 'the urban' and climate. Cities as powerhouses of production and consumption generate greenhouse gases. They are therefore important contributors to the observed changes in the global atmospheric composition of gases that increase radiative forcing and drive an enhanced greenhouse effect, the outcome of which are likely changes in the statistical moments of climate. Urban-related emissions also alter air quality, mainly at the urban scale, although 'downwind' effects may be important.

As demonstrated in a number of places in this book, the physical form and materials comprising the built environment alter the exchange of energy, mass and momentum between the land surface occupied by cities and the overlying atmosphere, leading to climate environments that often differ from their rural surrounds. Further, the physical, social, economic and cultural characteristics of urban spaces, often described in combined form as vulnerability, or represented as climate-related risk factors, do and will moderate, the impact of climate variability and change on cities and their inhabitants.

These urban climate 'knowns' bear implications for the achievement of healthy cities and form the basis of what is referred to here as the urban planning–urban climate imperative. This is because a changing set of climate circumstances in cities, driven by human-related local and global scale climate processes, will make it necessary for the inclusion of climate risk mitigation and adaptation strategies as core elements in urban planning and urban design. If climate-related healthy cities are to be more than an ideal, this is an imperative for governments at all scales, with perhaps local governments taking the lead supported by local stakeholder networks. As outlined in a number of places throughout this book, integrating climate risk-related mitigation and adaptation into urban planning and design can help avoid locking a city into counterproductive infrastructure and policies, and the exacerbation of built environment-related climate risks. Added to this is the observation that cities, at diverse stages of urbanisation, face different needs and constraints caused by their own urban planning system and practice. Thus, climate design intervention or implementation must fit into the local context if governments want to mitigate climate-related health issues via creating liveable and comfortable built environments. Critical also is the timing of policy development and implementation as it is often found that the very considerations that would make a positive difference to urban physical environments have been brought to the table too late to be actioned. Key then is the early consideration and integration of heat, air pollution and other hazard mitigation measures into the urban planning and design process (HKGBC 2018).

18.3 Generation of Urban Climate-Related: Health Risk Information

To achieve healthy cities from a climate perspective, knowledge of the extent to which the urban climate condition affects health, at the individual to population scale, along with the level cities alter climate is required. Without such knowledge, the management of climate-related health risk in cities, via the application of urban planning and design, would happen in an information vacuum. Since the turn of the new millennium, a range of approaches have emerged for either generating the data or developing knowledge about the link between urban form and climate and by extension climate-related health risk. Traditionally broadly divided into empirical and numerical modelling approaches, with the former often supporting the latter in relation to model validation, these are now increasingly complemented by 'less formal' citizen science and community-based co-generation (co-production) approaches.

Much of the material presented in this book corroborates the empirical understanding that heat extremes and poor air quality, as defining characteristics of large cities, have a discernible impact on the health of city inhabitants, and there is a link between the scale of urbanisation and the degree of urban climate effect. Notwithstanding the importance of empirical data, its very nature often limits drawing explicit links between particular aspects of urban form and a climate response. This is especially so in relation to how future urban development may affect climate at a range of scales and thus alter climate risk. As demonstrated throughout this book the response to this challenge has been the growing collaboration between urban climate scientists and urban planners in applying increasingly sophisticated urban scale climate models to the problem of projecting future urban climates for a range of urban development scenarios at the local to city scale. When reading across the various city case study chapters, apparent is the global diversity of urban structures. Consequently, to increase the likelihood that modelling studies are situation relevant, we make a call to develop city specific information on current and future urban canopy parameters and urban surface morphology to support model applications to individual cities (Mills et al. 2021).

While some chapters might connote the idea that modelling may become the default for generating climate risk information, material contained in this volume points clearly to emerging complementary non-technocratic approaches. These are visible in the city chapters for New York, London and Manchester where the authors describe activities focused on the co-generation/co-creation of urban climate action plans and solutions and associated enabling information involving a range of stakeholders and disciplines. Such new/non-traditional data streams in combination with traditional measurement and modelling, and developments in big-data science, will facilitate the attainment of a holistic understanding of the interface between urban climate and the lived climate experience and thus create information to better manage urban climate-related risk.

18.4 Urban Climate Interventions to Achieve Healthy Cities

This book asserts that by managing urban climate risk arising from climate extremes, mainly in the form of heat and poor air quality, it is possible to reduce heat and air pollution-related health risk. Of course, such an assertion exists in the knowledge that climate interventions accompany many possible 'non-physical' interventions for achieving a healthy city. The material contained in this book points to a range of theoretical and proven physical climate interventions. In the case of heat as a hazard, most of these are achievable through the alteration of the nature of the urban surface and thus the 'reinstatement' of a more 'natural' coupling of the urban surface and the overlying atmosphere. By reinstatement and natural, we mean nudging the flux of energy, mass and momentum towards values more akin to non-urban or rural surfaces. In radiative flux terms this means altering the absorption of solar radiation by urban surfaces and in energy flux terms, the shift of Bowen ratio values (ratio of sensible to latent heat flux) from those typical of 'hot' desert surfaces to those more aligned with 'cool' vegetated surfaces. It is in the latter context the incorporation of green and blue infrastructure into urban development is highly relevant as demonstrated in a number of the city case studies presented in this book. For example, by considering the climate benefits of domestic gardens in Manchester, Cavan et al. (2021) clearly demonstrate the potential for this unheralded dimension of city green infrastructure to exact positive climate impacts. For Melbourne, Tapper (2021) draws attention to a programme of systematic urban climate research that reveals the importance of single or combined green/blue infrastructure interventions in tempering temperature extremes. Similarly, Ashie (2021) describes a range of options for managing heat at the local scale in Tokyo. These include tree planting, water sprinkling, surface albedo alterations and conductive cooling using groundwater. For Hong Kong, Chan et al. (2021) emphasise the importance of maintaining ventilation corridors for the dissipation of heat and air pollution, a strategy that has been in place for Stuttgart for over 50 years as described in Chap. 12 (Reuter and Kapp 2021). That the efficiency of urban climate intervention levers is conditioned on urban typology, is highlighted by Lemonsu et al. (2021) in the case of Paris. They note that urban greening and, reflective materials and surface watering, are most effective in medium density areas and central Paris respectively. In general, Raven et al. (2021) encourage strategies that target the reduction of waste heat and greenhouse gas emissions via energy efficiency, management of intra-urban climate extremes by modifying building form and layout, the inclusion of heat resistant construction materials and reflective coatings in all aspects of built environment development, and increasing vegetative cover.

Complementing physical interventions focused on realigning surface–atmosphere exchanges of radiation, heat and moisture is the implementation of systems designed to forewarn urban communities and sectors about impending severe weather events such as heatwaves. As an example of this type of intervention Chap. 13 (Golechha et al. 2021) outlines the development, implementation and assessment of a heatwave warning system as part of a wider heat action plan for Ahmedabad,

India. Beyond the specifics of this particular example, Joe et al. (2021), and Lee et al. (2021) in Chaps. 14 and 17, draw attention to warning systems as one of a range of climate services for achieving the health dimensions of the Sustainable Development Goals and by association healthy cities.

18.5 Putting Urban Climate Science into Practice to Attain Healthy Cities

The nature of the material presented in this book provides considerable justification for stating there is evidence of a post-normal turn in the multi-disciplinary urban climate and urban planning communities. That is, unlike much of the normative science typical of the urban climate science literature for some time, outputs from urban climate research are now actively informing the development of urban planning principles and policy, all of which bear implications for achieving healthy cities in a climate sense. Having said that, the experience described for Stuttgart in this book (Reuter and Kapp 2021) has origins that stretch back beyond what is mainly a twenty-first century post-normal turn in urban climate science. In many ways, Stuttgart represents a pre-eminent example of the application of urban climate principles, observations and modelling to urban planning, a position enjoyed by this city because of strong regional to national government support, a key message for all.

Like Stuttgart, the cities of New York, Manchester, Tokyo, London and Paris possess legacies of an interaction between climate science and urban planning. However, in contrast to Stuttgart, their legacies are more ones of a waxing and waning enthusiasm for systematic incorporation of urban climate knowledge into urban planning, as largely determined by issues of the day and political cycles. That of course has now changed with the spectre of climate change and a projected increase in the harshness of urban climate environments, placing the close association between 'the urban' and climate on the agenda, not only in these cities but others such as Singapore, Melbourne and Hong Kong, as is apparent in the associated chapters in this book. Encouragingly, city-based initiatives receive support in principle via international intergovernmental agencies such as the World Meteorological Organisation and international programmes of research collaboration and associated networks, as described by Joe et al. (2021) and Lee et al. (2021) in Chaps. 14 and 17. This all appears quite rosy. However, to be honest, while there is a detectable shift in some quarters in the political willingness to mainstream urban climate science into urban policymaking to achieve healthy cities, there is some distance to go before a climate science–urban planning *Shangri La* is reached at the global level. This gap between normative and positivist worlds is apparent in some of the material presented in this book. For example, Raven et al. (2021) in Chap. 3 call for a clear articulation of the connections between climate change, infrastructure, and public health to city planners, policy-makers and the public. Similarly, Mills et al. (2021) raise the issue of fit for purpose climate information for urban planning.

Clearly, challenges remain in putting urban climate science into urban planning practice. If the future is to be different to the present, these challenges will need addressing.

18.6 The Future

That society adopts a trajectory towards building healthy cities is unequivocal. In the context of the conceptual framework adopted in this book, this means putting in place urban planning and design strategies that lead to the attainment of what is essentially a benign urban climate, or at least a set of event specific actions that minimise health risk arising from climate extremes. Such an aspiration resonates with the wider call to eliminate climate injustices. In the case of the notion of healthy cities, this translates as managing those climate threats that may prejudice the health status of communities who may otherwise have little control over their lived climate experience and consequently their climate-related health futures. Ethically and normatively speaking, we need at all costs, or at least with minimum regret, to avoid the social and physical construction of debilitating climatological worlds in cities.

To pave the way for firmly embedding urban climate science knowledge in urban planning and design, as one of a range of broad interventions to achieve healthy cities, will require a sustained programme of policy relevant research and action and critical reflection upon whether unintended consequences may emerge from policy interactions. In this regard, an urban climate conundrum is the tension between the international level call for the development of compact, high density and therefore energy efficient cities and the goal of a benign urban climate. This is because associated with high density and compact urban structures are high values of urban heat generated by alterations of the partitioning of surface net radiation and/or contributions from anthropogenic heat (e.g. from air conditioning systems) and poor air quality as an outcome of reduced intra-city ventilation. Clearly, a city-by-city challenge for the future will be identifying solutions that strike an acceptable balance between city energy efficiency and a physical climate that does not compromise the attainment of a healthy city.

A requirement for the future is a strong knowledge base and an acute understanding of the nature of the health effects of urban climates. The development and maintenance of urban climate and health observation networks and the betterment of truly integrated urban climate, urban planning and health models is therefore critical. Admittedly, to quench the thirst for data and modelling requires a considerable resource in terms of the technology and the human capacity to collect, analyse and action urban climate, planning and health science. In crude terms to sell the idea to those agencies or indeed individuals who possess the power to make a difference, an economic evaluation of climate-sensitive planning is required. In other words, to garner political uptake of climate informed urban planning and design, the economic quantification of its health benefits will become necessary. This is something clearly beyond the often easily made calls for 'more education' of urban policy developers about the beauties of urban climate science. Of course, the 'catch 22' of all of this is without the requisite evidence base, the argument for mainstreaming climate science into healthy city policymaking may remain more conceptual than economic. Given this, stakeholders in the urban climate–healthy cities space may need to consider alternative approaches to the collection and communication of evidence to support their case for science informed policy. Consequently, initiatives such as citizen science based projects, informal and formal knowledge networks, community-based institutions, non-traditional fund raising, co-creation/production of knowledge and an easily understood lexicon may be increasingly important in the future for providing and communicating the evidence base for climate action towards developing healthy cities. In essence, a re-contemplation of the traditional (current) political and techno-scientific spaces/networks via which knowledge is mediated and thus created and communicated, is needed. This will compel new dialogues between otherwise disparate players in the climate-related healthy city space. Exploring how such alternative spaces and new dialogues come about constitutes a fertile area of future research. Establishing the nature of governance that can best draw on what seems to be a plethora of urban climate research activities of relevance to the achievement of healthy cities is also a worthy endeavour—in resource constrained times replication of effort and reinventing the science wheel should be avoided.

Implicit in the stance adopted in this book is the idea, in simple terms, that 'urban climate is bad', because phenomena such as the urban heat island and associated excess heat during extreme temperature events create an urban disadvantage. Notwithstanding overwhelming evidence of a triple jeopardy shaped by the synergistic effects of vulnerability to heat and extreme outdoor and indoor temperatures and the production of different 'climatological worlds' at the intra-urban scale, as heretical as it may seem, there is a need to also consider what the urban climate advantages might be. For example, outside of the warm season might the urban heat island generate energy savings as well as temper any cold-related health effects? In short, are there healthy city positives associated with urban climate and if so, what are the possible geographical (location and place), scale (local to city) and seasonal dependencies of these? Of course, implicit in this encouragement to consider the two sides of the urban climate coin is how best to monitor the 'climatic performance' of cities in a healthy city context. A question for now and the future is: what might comprise a meaningful set of key performance indicators for healthy cities in a climate sense with such indicators by default not being just physical variables but rather variables that represent health-related climate sensitivity?

Two atmospheric hazards, namely heat and air pollution, and associated interventions to manage these dominate the material presented in this book. Perhaps conspicuous by its absence is urban flooding. While a long tradition of urban hydrology has shed light on the factors that generate a unique set of urban flood characteristics, there remains much scope to investigate the degree to which co-benefits for managing urban heat, air pollution and floods arise from climate-sensitive urban planning and design. Indeed, urban flooding and health-related outcomes might well be hitherto unrecognised diagnostics of the climatic performance of cities.

To date the majority of the published literature on mainstreaming urban climate science into practice emanates from countries with strong economies and research bases. This manifests in this book as a bias towards high-income country experience, something we acknowledge is a weakness. Excuses aside, an immediate imperative is to further fundamental research on urban climate processes and urban built environment (climate)—societal links in low- to middle-income countries, particularly in their mega-cities where unprecedented levels of urbanisation are on a collision course with climate change, the potential climate-related health consequences of which are deeply worrying. Given this, close attention is required regarding what courses of mitigation and adaptation action are feasible in terms of reducing urban climate-related health risk via urban planning and design, along with the identification of other priorities and known and unknown constraints that may work against the achievement of healthy cities. Of course, these limiting dimensions apply equally to high-income countries.

The material presented in this book has made it clear that past polemics associated with environmental determinism do not apply in the case of the link between the urban built environment and people's health. There is now wide acceptance of the concept of the environmental determinants of health and indeed that climate, as one aspect of the biophysical environment, has a discernible impact on health in its widest sense. To repeat, the central argument in this book has been that through climate-sensitive urban planning and design it is possible to manage health-related climate risk arising from the altered relationship between the urban surface and the atmosphere. Furthermore, as demonstrated in this book the urban climate science know-how required to achieve climate-related healthy cities exists. Encouragingly some city to national level governments have mainstreamed or are beginning to incorporate urban climate science principles into urban planning and designs as a response to the challenge of a changing climate. That said, to ensure the future does not turn into the present for large parts of the world, where currently there is a clear disconnect between science and policy, the call to 'urban climate science action', in relation to achieving healthy cities, needs to be made loud and clear in all political spaces. Accordingly, all those involved in the urban climate science and urban planning endeavour and committed to realising the climate ideals associated with a healthy city need to ensure that whatever they do is highly policy relevant. That is not to say that applied urban climate science should eclipse its pure form. Rather, both should work closely together with the urban planning and design communities and cognate disciplines, such as geography, sociology, psychology and public and environmental health, to settle on sustainable climate risk-related mitigation and adaptation strategies. Such common goals and collaboration will enhance our chances of achieving healthy cities or, as imagined by Helmut Landsberg almost 50 years ago, a *Metutopia*, a climate imagining of a healthy city.

References

Ashie Y (2021) Tokyo heat island adaptation measures. In: Ren C, McGregor G (eds) Urban climate science for planning healthy cities. Springer, London

Cavan G, Baker F, Tzoulas K, Smith CL (2021) Manchester—the role of urban domestic gardens in climate adaptation and resilience. In: Ren C, McGregor G (eds) Urban climate science for planning healthy cities. Springer, London

Chan EYY, Googins W, Huang Z, Wong CS (2021) Hong Kong: climatic application in urban planning and design at multiple scales for creating a healthy living environment. In: Ren C, McGregor G (eds) Urban climate science for planning healthy cities. Springer, London

Golechha M, Mavalankar D, Bhan SC (2021) India: heatwave and action plan implementation in Indian cities. In: Ren C, McGregor G (eds) Urban climate science for planning healthy cities. Springer, London

HKGBC (2018) Guidebook on urban microclimate study. Hong Kong Green Building Council, Hong Kong. https://www.hkgbc.org.hk/eng/engagement/file/UMC_Guidebook_amended_reduced.pdf. Accessed 30 Dec 2020

Joe P, Baklanov A, Grimmond S, Bouchet V, Molina LT, Heinke Schluenzen K, Mills G, Tan J, Golding B, Masson V, Ren C, Voogt J, Miao S, Lean H, Heusinkveld B, Hovespyan A, Terrugi G, Parrish P, Christensen JH, Fallmann J, Sokhi R, Spengler R, Yoshitani J (2021) Guidance on integrated urban hydro-meteorological, climate and environmental services: challenges and the way forward. In: Ren C, McGregor G (eds) Urban climate science for planning healthy cities. Springer, London

Lee B, Jones H, Ren C (2021) International Research Collaboration and Programmes for urban climate science for planning healthy cities. In: Ren C, McGregor G (eds) Urban climate science for planning healthy cities. Springer, London

Lemonsu A, de Munck C, Kounkou-Arnaud R, Masson V, Viguié V (2021) What alternatives does Paris have to adapt to future heat waves? In: Ren C, McGregor G (eds) Urban climate science for planning healthy cities. Springer, London

Mills G, Ching J, Bechtel B (2021) Characterising urban morphology for urban climate modelling. In: Ren C, McGregor G (eds) Urban climate science for planning healthy cities. Springer, London

Raven J, Braneon C, Rosenzweig C (2021) Embedding climate change in urban planning and urban design in New York City. In: Ren C, McGregor G (eds) Urban climate science for planning healthy cities. Springer, London

Reuter U, Kapp R (2021) Urban climate in urban planning—the experience from Stuttgart. In: Ren C, McGregor G (eds) Urban climate science for planning healthy cities. Springer, London

Tapper N (2021) Creating cooler, healthier and more liveable Australian cities using irrigated green infrastructure. In: Ren C, McGregor G (eds) Urban climate science for planning healthy cities. Springer, London

Wang L, Jiang X, Li X (2021) Healthy city and its planning. In: Ren C, McGregor G (eds) Urban climate science for planning healthy cities. Springer, London

Correction to: Manchester: The Role of Urban Domestic Gardens in Climate Adaptation and Resilience

Gina Cavan, Fraser Baker, Konstantinos Tzoulas, and Claire L. Smith

Correction to:
Chapter 5 in: C. Ren, G. McGregor (eds.), *Urban Climate Science for Planning Healthy Cities*, Biometeorology 5, https://doi.org/10.1007/978-3-030-87598-5_5

Chapter 5, "**Manchester: The Role of Urban Domestic Gardens in Climate Adaptation and Resilience**" was previously published non-open access. It has now been changed to open access under a CC BY 4.0 license and the copyright holder updated to 'The Author(s)'. The book has also been updated with this change.

The updated original version of this chapter can be found at
https://doi.org/10.1007/978-3-030-87598-5_5

C. Ren, G. McGregor (eds.), *Urban Climate Science for Planning Healthy Cities*, Biometeorology 5, https://doi.org/10.1007/978-3-030-87598-5_19

Index

A
Action levers, 250
Action Plan, 169
Adaptation strategies, 241, 244, 254, 255
Ahmedabad
  extreme heat events and all-cause mortality, 291
  HAP development process, 292–294
  Heat Action Plan implementation, 294
  severe heat wave, 291
Ahmedabad Municipal Corporation (AMC), 291
Air conditioning (AC), 192
  on microclimate, 252, 253
  uses, 253
Air pollutants
  ozone and precursors, 361, 362
  particulate matter, 360, 361
Air Pollution Prevention and Control Action Plan, 168
Air temperature profiles, 228
Andhra Pradesh, 298
Anthropogenic heat, 80
Apparent Temperature (AT), 222
Atmospheric boundary layer, 252
Australian cities, 220
Australian Cooperative Research Centre for Water Sensitive Cites (CRCWSC), 12
Average diurnal cooling, 228
AWESOME project, 83

B
Beijing
  air pollution governance, 169, 171–173
  air quality, 173, 175, 176
  geographical terrain, 176
  industrial transformation and upgrading, 169
  local pollution sources, 176
  ozone levels in, 175
  planning strategy, 183, 184
  regional coordinated development, 182, 183
  regional transport of, 177
Beijing Urban Master Plan, 12, 178–180
Blue corridor, 243
Boundary layer theory, 341
Building Construction Authority (BCA), 192
Building effect parameterization (BEP), 194
Building energy model (BEM), 194, 345
Bus Stop Waiting Room, 131

C
Carbon emissions, 75
Chandrapur city, 296
City Geography Markup Language (CityGML), 346
City of Port Phillip (CoPP), 233
City of Stuttgart
  Climate Atlas Region Stuttgart, 262–265
  climate change, 261
  climate-optimised inner city development, 276, 277
  development areas, 275
  green space planning, 267–270
  orographic situation, 260
  public spaces, construction of, 278
  ventilation, 271, 272, 274
City-urban model (CUM), 200, 204, 206, 207

C. Ren, G. McGregor (eds.), *Urban Climate Science for Planning Healthy Cities*, Biometeorology 5, https://doi.org/10.1007/978-3-030-87598-5

CLIM2 (project), 246
Climate adaptation, 232, 384
Climate Analysis Maps, 264, 265
Climate and Clean Air Coalition, 381
Climate Atlas, 262
Climate challenges, 193, 240
Climate change, 193, 195, 197, 198, 205, 206, 211–213, 240, 243–246, 248, 254
  air quality
    global climate change, 359
    urban climate, 358
  environmental changes, 19
  and health, 20, 21
  human health and well-being, 72
  meteorological parameters, 19
  urban greenhouse gas emissions, 19
  urban planning and construction, 19
Climate change adaptation, 76, 230, 383
Climate Change Risk Assessment (CCRA), 88
Climate projections, 108, 240, 242
Climate-related risk, 76
Climate-resilient health systems, 381
Climate risk, 240
Climate risk mitigation, 391
Climate-sensitive decision-making, 14
Climate-sensitive health and well-being, 9
Climate-sensitive health outcomes, 8, 9
Climate-sensitive planning, 395
Climatic trends, 243
CLIMESPACE Company, 246
Communication methods, 315
Complementing urban planning and design, 13
Complex urban systems
  climate and heat vulnerability, 91
  green infrastructure, 91
  urban sustainability and health policies, 91
Computation Fluid Dynamics (CFD), 345
Contextual matter, 194, 195
Cooling load simulation, 202, 210
Cooperative Research Centre (CRC) program, 232
Cooperative Research Centre for Water Sensitive Cites (CRCWSC), 221, 225–230
Cost-benefit/social impact, 333
Coupled GRSM-MSM, 199–201
COVID-19 pandemic, 390
Cramps, 193

D
Delphi technique, 113
Digital surface model (DSM), 347
Digital Synthetic City (DSC) tool, 350
Digital terrain model (DTM), 347
Direct Model Output (DMO), 197
Disaster risk reduction (DRR), 314
Dynamical-statistical approach, 200, 204, 206, 207

E
Ecosystem services, 252
Energy assessment, 192, 202
Energy behaviour, 76
Energy consumption, 76
EnergyPlus, 205, 210, 211
Energy simulation, 205, 211
ENSEMBLES, 241
Environmental Audit Committee, 74
Environmental determinants of health framework, 6
Environmental determinism, 397
Environmental modelling, 116
Eosystem services (ES), 101
EPICEA (project), 246
Extreme heat events (EHE), 220–223, 226–229, 234
Extreme weather events, 73, 100

F
Federal German Building Code, 261
Federal Nature Conservation Act, 281
Federal, State and Local Government Policy/ Role, 230–234
Financial resources, 333
Food-borne disease (FBD), 10
Fossil fuel combustion, 356

G
Geographic Information Systems (GIS), 77
German Federal Building Code, 281
Global Atmospheric Watch (GAW) programme, 314
Global climate model (GCM), 197
Global Climate Observing System Terrestrial Observation Panel for Climate (GCOS-TOPC), 386
Global Data-Processing and Forecasting System (GDPFS), 319
Global Forecast System (GFS), 199
Global Framework for Climate Services (GFCS), 382
Global Human Settlement project, 349

Global-regional spectral model (GRSM), 197–199
Global warming, 193, 195
Greenbelt, 243
Green-field development, 229
Greenhouse effect, 195
Greenhouse gas emission, 240, 241
Greening, 248, 249
Ground-based vegetation, 248
GRSM-MSM-CUM-CFD, 209

H

Healthy city, concept of
  physical and social environments, 22
  research evolution, 25
  research hotspots, 24
Healthy city planning (HCP)
  atmospheric environment, 32, 33
  built environment, 30
  elderly group, 34
  physical activity and buildings, 34
  research evolution, 28, 29, 35, 36
  research hotspots, 29–31
  urban planning and public health, 26, 27
  urban sustainability, 33
Heat exhaustion, 193
Heat Health Action Plans, 382
Heat-health relationships, 222, 225, 231
Heating/cooling equipment, 194
Heat island mitigation strategies, 12
Heat islands, 74
Heat mitigation, 223–224
Heat-related health risk, 74
Heat-related illnesses (HRI), 193, 205, 206, 211, 212
Heat risk heterogeneity, 76
Heat stroke, 193
Heat vulnerability, 78, 79, 83, 85
  Australian Cities, 221–223
  socio-spatial heat vulnerability indicators, 89
Heat waves, 193
  adaptation strategies, 241
  climate projections, 242
  climatic trends, 243
  daily temperature time series, 241
  Paris's adaptation, 241, 243–246
  severity index, 242
  time periods, 242
  2003 heat wave, 242, 243
Heterogeneous land surfaces, 108
High-density tropical city, 192
High-density urban cities, 4, 5, 12
Hillside Development Plan, 272
Hong Kong
  adaptation strategies, 161, 162
  communicable diseases, 156, 157
  health effect measurements, 153, 154
  medical help-seeking, self-care and mild symptoms, 160, 161
  mitigation strategies, 162, 163
  mortality
    cardiovascular and respiratory infection-related deaths, 154
    urban planning and design, 154
  non-communicable diseases, 158, 159
  urban planning and design, 153
Housing and Development Board (HDB), 192
Human-environment modelling, 200, 208
Hurricane Sandy, 50

I

IMEUM-CFD modelling, 207
India
  Heat Action Plan (HAP)
    in Andhra Pradesh, 298
    Department of Science and Technology, 302
    in Maharashtra, 298
    Ministry of Health and Family Welfare, 301
    by NDMA, 300, 301
    in Odisha, 299
    in Rajasthan, 297
    sample city activities, 295, 296
    state level efforts, 297
    state-level efforts, 297
    in Telangana, 300
  heat waves in, 286, 288
  heat-related mortality in, 288, 290
Individual practices
  AC, 252, 253
  water consumption, 254
Inertial Sub-Layer (ISL), 341
Inner Melbourne Action Plan (IMAP), 233
Institute of Planning and Urban Development of the Parisian region (IAU), 246
Integrated 3D geospatial platform, 209
Integrated multi-scale environmental urban model (IMEUM)
  CFD modelling, 207
  CUM, 200, 204, 206, 207
  downscaling process, 197
  dynamical-statistical approach, 200, 204, 206, 207

Integrated multi-scale environmental urban model (IMEUM) (*cont.*)
evaluation, 199–201
GCM, 197
GRSM, 197–199
human-environment modelling, 200, 208
methodology workflow, 197
MSM, 197–199
OTCI, 200, 208
Singapore's IMEUM, 12
surface air temperature, 197
sustainable planning purpose, 197
Integrated Urban Hydrometeorological, Climate and Environmental Services (IUS) initiative, 13
Integrated Urban Services (IUS)
applications, 325
capacity development, 329
city requirements/local tailoring, 320
concept and components of, 335
databases and data sharing, 323
decision support, 326
evaluation and methodologies, 328
implementation, 329–331
resources for, 331–333
modelling and prediction, 323–325
observation sparse, 321, 322
research and development, 328
service delivery, communications and outreach, 327
Integrating passive cooling strategies, 78
Integration, 319
Intergovernmental Panel on Climate Change (IPCC), 4, 73, 212
IPCC Fourth Assessment Report, 221
IPCC Sixth Assessment Report, 221
International community, 4
International Research Collaboration and Programmes
countervailing influences, 382
health co-benefits, 381
WCRP/CMIP, 384, 385
Irrigated green infrastructure
adoption, 221
EHE (*see* Extreme heat events (EHE))
heat mitigation, 223–224
heat-related mortality and morbidity, 220
heat vulnerabilities of Australian Cities, 221–223
LID, 221
SuDS, 221
urban cooling, 225–230
urban heat mitigation, 230–234
WSUD (*see* Water sensitive urban design (WSUD))
Island-state Singapore, 193

J
Japanese Alps, 132

K
Knowledge Action Networks (KANs), 387

L
Land use modification, 252
Landscape planning, 252
Large-scale territorial planning policies
land use modification, 252
landscape planning, 252
soil-sealing, 250
urban expansion trends, 250, 251
Light Detection and Ranging (LIDAR) systems, 347
Local Climate Zone (LCZ) scheme, 349
Localized weather data, 203, 204, 210, 211
Local-scale environmental analysis, 208
Local urban planning, 241
London buildings
epidemiological research, 81
HPRU project, 82
outdoor pollutants and heat, 82
risk of overheating, 83
London care homes, 88, 89
London local authorities, 89, 90
London's local urban climate
embedding passive cooling strategies, 81
heat island intensity, 80
heat-related mortality risk, 81
residential space heating loads, 81
spatio-temporal variation, 79
London Unified Model (LondUM), 79
Low Impact Development (LID), 221
Low- to middle-income countries, 5

M
Maharashtra, 298
Manchester Green and Blue Infrastructure Strategy (MGBIS), 102, 114
Manchester's gardens, 106
Mapping interactions, 78

Mast and temperature sensor, 227
Mega cities, 4, 5
Melbourne Water, 232
Mesoscale spectral model (MSM), 197–199
Meteorological Service Singapore (MSS), 192
Microclimate analysis
  energy simulation, 205, 211
  EnergyPlus, 205, 210, 211
  multiple measurement points, 202, 210
  QUEST Output, 203, 204, 210, 211
  QUEST workflow, 202, 210
  weather sensors, 210
Microclimate condition, 194
Micro-mesoscale coupling, 194
Microscale, 227
Millennium Drought, 232
Model-driven techniques, 347
Model nesting, 345
Model Output Statistics (MOS) technique, 204, 205
Monin-Obukhov Similarity Theory (MOST), 342
Multi-Hazard Early Warning Systems (MHEWS) projects, 333
Multi-scale and multi-criteria analyses, 243, 244
Multi-scale UHI, 193, 194
MUSCADE project, 244, 248

N

Nagpur, 296
National Disaster Management Act, 297
National Disaster Management Authority (NDMA), 300
National Environment Agency (NEA), 192
National Hydrological and Meteorological Services (NHMS), 314, 334–335
National Parks Board (NParks), 192
National Urban Database and Access Portal Tool, 349
National Urban Policy (2011), 232
National Water Initiative (2004), 230
Natural soils, 241
New York City climate
  annual average precipitation ranges, 51
  daily maximum summer temperature, 51
  extreme events, 53
  extreme precipitation timescales, 53
  historical and future extreme heat in, 56
  New York metropolitan region, 53
  resilience and infrastructure plans, 55
  sea breeze effects, 52
  stakeholder engagement, 56, 58
Non-physical determinants of urban health, 11
Normalised digital surface model (nDSM), 347

O

Odisha, 299
Open Geospatial Consortium (OGC), 346
OpenStreetMap (OSM), 348
Outdoor air temperature, 244
Outdoor thermal comfort, 209
Outdoor thermal comfort index (OTCI), 200, 208

P

Paris Agreement, 8, 124
Paris's adaptation
  multi-scale and multi-criteria analyses, 243, 244
  research initiatives, 244, 245
  urban design strategies, 246–250
Pavement watering, 248–250
Physical environment, 6, 8
Planning recommendations, 265, 266
Policy development, 230
Pollution, 6, 12, 14
Probabilistic Quantitative Precipitation forecasts (PQPFs), 199
Probability of detection (POD), 199–201
Programme resources, 331
Proportion correct (PC), 199–201
Public health/emergency management, 232
Public health initiatives, 75

Q

Quantitative Precipitation Forecasts (QPFs), 199
Quantitative urban environment simulation tool (QUEST)
  climate change, 205, 206, 211, 212
  components, 195, 196
  design process, 195
  development, 195
  global temperatures, 195
  HRI, 205, 206, 211, 212
  IMEUM (*see* Integrated multi-scale environmental urban model (IMEUM))
  microclimate analysis, 202, 210
  urban planning, 195
  workflow, 208–210

QUEST workflow
ArcGIS, 208
baseline climate selection interface, 202
development, 208
feature, 208
flowchart, 201, 208
GRSM-MSM-CUM-CFD, 209
integrated 3D geospatial platform, 209
outdoor thermal comfort, 209
simulation, 201
urban planning and design, 209

R
Rajasthan, 297
Rapid laser pulses, 347
Reflective materials, 247, 248
Regional warming, 74
Relative humidity (RH), 211
Remote sensing, 77
Representative Concentration Pathways (RCPs), 334
Resilient Cities, 316

S
Sendai Framework for Disaster Risk Reduction 2015–2030, 8, 313
Service resources, 333
Severity index, 242
Shared Socio-Economic Pathways (SSP), 334
Singapore, 191, 192
Singapore Land Authority (SLA) standard, 209
Smart city, 317
Soil-sealing, 250
Solar radiation, 211
South Asian Association for Regional Cooperation (SAARC) Region, 302, 303
South Asian Climate Outlook Forum (SASCOF), 302
STAR tools, 108
State Disaster Management Authority (APSDMA), 298
State of Baden-Württemberg, 262
STEVE tool, 205
Street-level thermal comfort, 191
Strong heat stress (SHS), 251
Subseasonal-to-Seasonal (S2S) Prediction Project, 385
Sunburn, 193
Sunnyside Yard, NYC
testing strategic locations, 63, 64
UDCW planning process, 58, 61, 64, 65
urban climate, 62
urban microclimate comparisons, 62
Surat city, 296
Surface air temperature, 197
Surface layer (SL), 342
Surface temperature patterns, 192
SURFEX climate model, 227
Sustainable Development Goals of Agenda 2030, 8
Sustainable Urban Drainage Systems (SuDS), 221
Systems-based approach, 78

T
Technical expertise, 332
Telangana, 300
Thermal comfort, 193, 195, 196, 207–209, 212, 213
Thermal maps, 263
Tokyo
annual average air temperature, 120, 121
climograph, 127
environmental impact, 122, 124
extremely hot days and tropical nights, 121
Marathon course heat countermeasures, 128, 129
mitigation and adaptation measures, 124, 125
radiant temperature, 126, 127
sensible temperature in urban spaces, 125, 126
wind paths (*see* Wind paths)
Tokyo Bay, 132
"Triple Heat Jeopardy" mapping framework, 85, 86
2003 European heatwave, 13
2003 Heat wave, 242, 243

U
UCCRN urban design climate workshops (UDCW), 58, 61, 64, 65
UK Climate Projections, 73, 108
United Nations (UN) HABITAT-III conference, 315
Universal thermal climate index (UTCI), 249
Urban adaptation, 240
Urban air pollution
cardiovascular and respiratory systems, 356

climatic conditions and air quality, 357
fossil fuel combustion, 356
impact
dust, 367
economic impacts, 369
fire, 365, 366
health, 368
indoor air quality, 364
mitigation and recommendations, 370–372
population growth and relocation, 357
transport, 363
Urban boundary layer (UBL), 342
Urban canopy layer (UCL), 343
Urban canopy model TEB, 244
Urban Canopy Parameters (UCPs), 344
Urban centres, 220
Urban climate, 392
interventions, 393, 394, 396
Urban Climate Change Research Network (UCCRN), 11
Urban climate conundrum, 395
Urban climate factors
heat-resistant construction materials, 50
layout of buildings and urban districts, 49
urban systems, 48
vegetative cover, 50
Urban climate intervention, 393
Urban climate models, 79
and health, 350
heterogeneous topography, 343
surface-air exchange processes, 344
Urban climate research, 394
Urban climate resilience, 92
Urban climate science, 6, 7, 394
conceptual framework, 8, 9
factors linking regional climate to urban health outcomes, 10
for healthy living cities, planning policy, 5
non-physical determinants of urban health, 11
Urban climatology, 7
Urban cooling, 221, 224–230, 232, 234
Urban design strategies
action levers, 250
greening, 248, 249
pavement watering, 248–250
reflective materials, 247, 248
urban development and sensitive-climate, 246
Urban development, 392
Urban energy exchange model, 107
Urban energy performance, 194, 195
Urban expansion trends, 250, 251
Urban form
defined, 340
Urban green infrastructure
average garden land surface, 105
climate regulation and biodiversity, 112
cooling functions, 104
CSS tool, 105
domestic gardens, 101
ecosystem services (ES), 101
garden land surface, 106
garden use and maintenance, 104
health and well-being benefits, 101
image analysis, 105
lack of regulation, 101
Manchester's Gardens, 113, 114
Manchester Strategy, 102, 103
spatial variation, 106
surface temperatures, 112
vegetation type and structure, 104
Urban greening, 224
Urban Health Initiative, 381
Urban heat island (UHI), 220, 240, 241, 243, 244, 246, 247, 250, 251, 254, 350
contextual matter, 194, 195
develop multi-scale urban microclimate, 195
diversification, 194
irrigated green infrastructure (*see* Irrigated green infrastructure)
mitigation, 244
multi-scale, 193, 194
in Singapore, 191, 192
urban energy performance, 194, 195
Urban heat mitigation
Federal, State and Local Government Policy/Role, 230–234
Urban impact, 243
Urban infrastructure, 315
Urban livability, 230
Urban Meteorology Science Advisory Group (GURME SAG), 314
Urban microclimate, 47
Urban microscale models, 345
Urban morphological data
characteristics, 347
spatial and temporal variations, 346
Urban morphology, 49, 194
Urban planning, 195, 390, 391, 394
contract, 282
and design, 209
and urban design, 243, 246, 250, 254
Urban population, 74, 250
Urban-related emissions, 391

Urban service requirements, 317
Urban sprawl, 250
Urban surface, 7
Urban warming, 75
Urban warmth, 220

V
Vector-borne disease (VBD), 10
Victorian policy, 232
Volunteered Geographic Information (VGI), 348
VURCA project, 244, 250

W
Water-borne disease (WBD), 10
Water consumption
  watering gardens, 254
Watering gardens, 254
Waterproof surfaces, 223
Water security, 232
Water sensitive urban design (WSUD), 221, 224–226, 229–231, 233, 234
Weather research forecasting (WRF) system, 194, 228, 345
Weather sensors, 210
Web of Science (WOS), 23
Wind paths
  and evaluation level, 134, 136
  and heat countermeasures, 143–148
  heat island countermeasure map, 136, 137
  Nihonbashi/Daimaruyu District, 137, 139
  Osaki Station, 140
  Shinagawa Station and Tamachi Station, 141, 142
World Meteorological Organization (WMO), 13, 199
World Urban Database and Portal Tools (WUDAPT) project, 349
World Weather Research Programme (WWRP), 199, 385

Z
Zagreb Declaration for Healthy Cities, 72

The manufacturer's authorised representative in the EU is Springer Nature Customer Service Centre GmbH, Europaplatz 3, 69115 Heidelberg, Germany. If you have any concerns regarding our products, please contact ProductSafety@springernature.com

Printed and bound by CPI Group (UK) Ltd, Croydon, CR0 4YY

30/09/2026

02225491-0016